AF356165

Editors

Manoj Kumar Nallapaneni
School of Energy and
Environment
City University of Hong Kong
Kowloon
Hong Kong

Subrata Hait
Department of Civil and
Environmental Engineering
Indian Institute of Technology
Patna
India

Anshu Priya
School of Energy and
Environment
City University of Hong Kong
Kowloon
Hong Kong

Varsha Bohra
Department of Applied Science
Hong Kong Metropolitan
University
Ho Man Tin
Hong Kong

Editorial Office
MDPI
St. Alban-Anlage 66
4052 Basel, Switzerland

This is a reprint of articles from the Special Issue published online in the open access journal *Sustainability* (ISSN 2071-1050) (available at: www.mdpi.com/journal/sustainability/special_issues/ Resources_Conservation_Recycling_Waste_Management).

For citation purposes, cite each article independently as indicated on the article page online and as indicated below:

Lastname, A.A.; Lastname, B.B. Article Title. *Journal Name* **Year**, *Volume Number*, Page Range.

ISBN 978-3-7258-0904-2 (Hbk)
ISBN 978-3-7258-0903-5 (PDF)
doi.org/10.3390/books978-3-7258-0903-5

Contents

About the Editors

Manoj Kumar Nallapaneni

Dr. Nallapaneni is a transdisciplinary energy and sustainability engineer with a PhD in Digital Circular Economy and Circular Power System from the School of Energy and Environment, City University of Hong Kong. He has obtained two Masters degrees, one in Renewable Energy Technologies from Karunya University, India, and the other in Environmental Economics from Annamalai University, India. He holds a Bachelor's degree in Electrical and Electronics Engineering from GITAM University. Before joining the CityU, he worked as a Research Fellow at Universiti Malaysia Pahang, Malaysia, on a project which focused on using solar photovoltaics as urban and rural infrastructure. Prior to this, he worked as an Assistant Professor in the Department

of Electrical and Electronics Engineering at the Bharat Institute of Engineering and Technology, Hyderabad, India, and Energy Engineer at Atiode Solar Systems Limited, Benin City, Nigeria.

Dr. Nallapaneni's research works focus on the topics of simulation-, experimental-, real-time empirical and location-, or climate-specific studies mainly focused on building sustainable and resilient systems (decentralized, networked, and centralized) across critical infrastructure sectors by adopting nexus thinking and systems innovation with an emphasis on circular business models and digitalization. He worked on key sustainability challenges that include design, performance modeling, and analysis of a wide range of clean energy and environmental systems, food–energy–water–waste nexus, industrial symbiosis, waste valorization and material passports, carbon accounting, and pricing. He possess an interdisciplinary skill set that includes performance analytics, techno-economics, life cycle assessment, resilience assessment, leveraging digital innovation (blockchain, IoT, smart contracts, and AI), business model innovation, and nexus systems design with better conceptualization skills.

Subrata Hait

Dr. Subrata Hait is an Associate Professor in the Department of Civil and Environmental Engineering at Indian Institute of Technology (IIT) Patna, Bihar. He obtained his Ph.D. Degree in Civil Engineering from IIT Kanpur in the year 2011. Prior to joining as an Assistant Professor at IIT Patna in June 2012, he worked as a Senior Project Engineer at IIT Kanpur within a consortia of seven IITs towards the Preparation of Ganga River Basin Environment Management Plan (GRB EMP), a project sponsored by the Ministry of Environment, Forest and Climate Change (MoEFCC), and Govt. of India (GoI) for the abatement of pollution and restoration of continuous and unpolluted flow in the Ganga River. Presently, he is serving as Associate Dean Academics (PG) at IIT Patna. He has also served as the Head of the Department (HoD) of Civil and Environmental Engineering at IIT Patna for three years (2016–2019). His current research interests in the broad area of Pollution Prevention and Resource Recovery (P2R2) include e-waste treatment and metal recovery, organic waste management, the removal of micro-plastics and emerging contaminants from aqueous matrices. He is a Fellow of the Environmental Engineering division of the Institution of Engineers, India (IEI). Also, he is affiliated to the organizations like the American Society of Civil Engineers (ASCE), American Chemical Society (ACS), International Solid Waste Association (ISWA), and International Water Association (IWA). Apart from editing two books, he has authored many papers at his credit in various international journals. Apart from serving as a Reviewer for different international journals published by the leading publishers including Elsevier, Dr. Hait is serving as an Academic Editor of *PLOS One* and *PLOS Water*, and an Editorial Board Member of *SN Applied Sciences*, Springer Nature.

Anshu Priya

Anshu Priya, PhD, is an environmental and microbial biotechnologist working towards waste management and the establishment of a green and circular economy through biotechnological interventions. She has experience in leading, supervising, and undertaking research in the broader areas of Waste and Biomass Valorization with a focus on Biohydrometallurgy, Hazardous Waste Management, and Biorefinery. Dr. Priya earned her PhD from the Indian Institute of Technology Patna and worked as researcher at City University of Hong Kong. She has experience in both teaching and research and is the recipient of various scientific awards, grants, and fellowships. Dr. Priya is also an editor and reviewer of various Journals of international repute.

Varsha Bohra

Dr. Varsha Bohra is a Postdoctoral fellow in the Department of Applied Science, Hong Kong Metropolitan University, Hong Kong, SAR China. She holds a PhD in Microbiology from CSIR-National Environmental Engineering Research Institute, Nagpur, India. Her research interests include the valorisation of agricultural waste into biobased useful products employing molecular and omics approach. Currently, she is motivated to apply her bioinformatics knowledge in analyzing various omics datasets from mangrove ecosystems. The research aims at understanding the ever-evolving community structure and ecosystem functions of the nutrient-deficient yet robust environment. She is parallelly associated with research inspecting the role of microbial assemblage on harmful algal bloom and toxicity.

Preface

Welcome to "Resources Conservation, Recycling, and Waste Management". As you hold this reprint in your hands, you are embarking on a transformative journey exploring the profound significance of responsible resource management and its pivotal role in shaping a sustainable future.

Throughout history, humanity has fueled its progress and development by harnessing the Earth's resources. However, this relentless pursuit of growth has exacted a toll on our planet. Excessive consumption, pollution, and the depletion of finite resources have left us grappling with the urgent need for change. The time has come for a paradigm shift—a resounding call to action to safeguard our planet for generations to come.

Within the pages of this reprint, we delve into the intricate dimensions of resource conservation, recycling, and waste management. Drawing upon the expertise and insights of dedicated experts, researchers, and practitioners, we illuminate the path towards sustainability. Each chapter offers a holistic view of the subject, covering a diverse array of topics that span various sectors.

As you delve into these pages, you will explore the fundamental principles of resource conservation. You will discover the significance of reducing our ecological footprint and optimizing resource allocation, paving the way for a more balanced and efficient use of our planet's resources. This reprint also delves deeply into the principles and practices of recycling, revealing the transformative potential of innovative technologies and processes that can turn waste into valuable resources. Furthermore, we examine waste management systems, from collection and sorting to treatment and disposal, emphasizing the importance of responsible waste management practices and the role of digitalization in mitigating environmental harm.

The topics explored in this reprint are far-reaching, encompassing a wide range of sectors. Our aim is to empower individuals, communities, and policymakers with the knowledge and strategies needed to shift from a linear model of resource use to a circular economy. By upholding the principles of sustainability, efficiency, and equity, we can forge a future where resources are treasured, waste is minimized, and the delicate balance of our planet is restored.

As you embark on this enlightening journey, I encourage you to approach these topics with an open mind and a willingness to embrace change. Together, we hold the power to create a world where responsible resource management is at the forefront of our collective consciousness. Let us embark on this transformative journey, armed with knowledge, passion, and a deep commitment to preserving the beauty of our Earth.

Manoj Kumar Nallapaneni, Subrata Hait, Anshu Priya, and Varsha Bohra
Editors

sustainability

MDPI

Editorial

From Trash to Treasure: Unlocking the Power of Resource Conservation, Recycling, and Waste Management Practices

Manoj Kumar Nallapaneni [1,2,3,*], Subrata Hait [4], Anshu Priya [1,5] and Varsha Bohra [6,7,8]

1   School of Energy and Environment, City University of Hong Kong, Kowloon, Hong Kong; anshupriya26@gmail.com
2   Center for Circular Supplies, HICCER—Hariterde International Council of Circular Economy Research, Palakkad 678631, Kerala, India
3   Swiss School of Business and Management Geneva, Avenue des Morgines 12, 1213 Genève, Switzerland
4   Department of Civil and Environmental Engineering, Indian Institute of Technology Patna, Bihta 801106, Bihar, India; shait@iitp.ac.in
5   Centre for Plant and Environmental Biotech, Amity Institute of Biotechnology, Amity University, Noida 201313, Uttar Pradesh, India
6   Department of Biology, Hong Kong Baptist University, Kowloon Tong, Hong Kong; bohravarsha11@gmail.com
7   Council of Scientific and Industrial Research (CSIR)–National Environmental Engineering Research Institute, Nagpur 440020, Maharashtra, India
8   School of Science and Technology, Hong Kong Metropolitan University, Ho Man Tin, Hong Kong
*   Correspondence: mnallapan2-c@my.cityu.edu.hk

Citation: Nallapaneni, M.K.; Hait, S.; Priya, A.; Bohra, V. From Trash to Treasure: Unlocking the Power of Resource Conservation, Recycling, and Waste Management Practices. *Sustainability* **2023**, *15*, 13863. https://doi.org/10.3390/su151813863

Received: 14 August 2023
Accepted: 21 August 2023
Published: 18 September 2023

1. Introduction

"Trash to Treasure" refers to transforming discarded or unwanted items, often considered trash or waste, into valuable or desirable products. It involves repurposing, upcycling, or creatively reusing materials typically disposed of in landfills. For instance, repurposing or upcycling trash can create distinctive and valuable products while reducing the demand for new raw materials. Various materials can be involved in the Trash to Treasure initiatives; for example, discarded wood can be transformed into stylish furniture pieces [1], old fabrics can be repurposed into trendy fashion accessories (Figure 1a) [2], pamphlets in the daily newspaper into decorative place mat (Figure 1b) [3] or glass bottles can be turned into decorative vases or lamps [4].

(a)

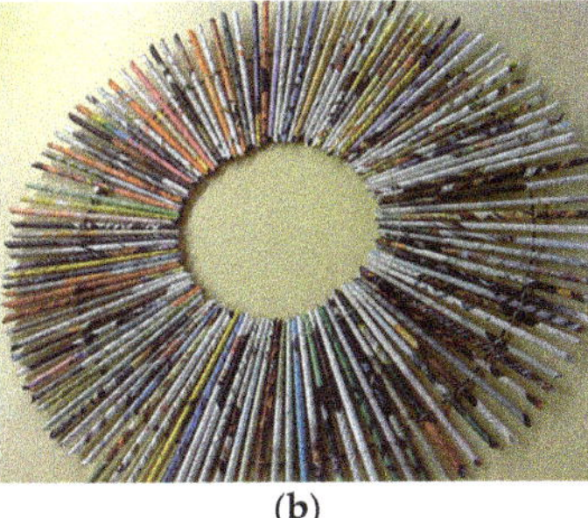
(b)

Figure 1. (**a**). Jean into a trendy tote bag (Picture Credits: iHanna) [2]; (**b**). Pamphlets in the daily newspaper into decorative placemats (Picture Credits: Dollar Store Crafts) [3].

Like the ones said above and the ones seen in Figure 1, there are many possible options, but all these fall under general examples; nevertheless, when we talk from an advanced scientific angle, we can create much more valuable products, such as food waste to value-added products like biosurfactants [5], plastic waste to activated carbons [6], spent batteries into buffer storage units for emergency purposes or stationary energy storage in variety of application [7,8] and others like this. This way, *Trash to Treasure* promotes sustainability, resourcefulness, and environmental consciousness. Additionally, when we look at the Trash to Treasure projects with more profound thought, they promote environmental sustainability and encourage creativity, craftsmanship, and entrepreneurship [2–5]. Overall, Trash to Treasure represents a shift toward embracing sustainability, encouraging resourcefulness, and fostering a culture of various activities that broadly fall under a circular economy. However, for such transformation to become a new normal, we must unlock the power of *resource conservation, recycling, and waste management* practices.

"Resource Conservation, Recycling, and Waste Management" are interconnected practices that aim to reduce the consumption of resources by promoting the reuse and recycling of materials and effectively managing waste to minimize environmental impact, enable sustainable development, conserve energy, offer economic benefits, promote climate change adaptation and mitigation techniques, and promote public health and safety. As a result, the importance of *resource conservation, recycling, and waste management* cannot be ignored in today's world. Hence, to understand how the solutions related to *Trash to Treasure* are being proposed and practiced, we opened a Special Issue titled *"Resource conservation, Recycling, and Waste Management"*, calling for contributions *(www.mdpi.com/journal/sustainability/ special_issues/Resources_Conservation_Recycling_Waste_Management*, accessed 11 August 2023*)*. The response was positive with a wide range of contributions; based on which we (the editors) carried out a discussion (Section 2), paving the way for some actions, that we listed (in Section 3) as a call-to-action for individuals, call-to-action for industries/corporates, and call-to-action for governments and policymakers for better realization and implementation of Trash to Treasure initiatives by unlocking the power of resource conservation, recycling, and waste management.

2. Resource conservation, Recycling and Waste Management

The Special Issue *"Resource conservation, Recycling, and Waste Management"* includes fifteen insightful contributions covering different aspects of sustainable development achieved through efficient approaches, and the gist of all the contributions is summarized categorically in Sections 2.1–2.3 below.

2.1. Resource conservation

Resource conservation refers to the awareness and application of practices to preserve and manage natural resources sustainably. It involves recognizing the finite nature of resources and the need to protect them for the benefit of current and future generations [9]. This Special Issue attracted few studies that fall under these conditions; for instance, the article *"Combined Effects of Biochar and Inhibitors..."* (https://www. mdpi.com/2071-1050/15/7/6100, accessed on 10 August 2023) highlights the potential of combined effects of biochar and inhibitors like nitrification inhibitors (methyl 3-(4-hydroxyphenyl) propionate), and urease inhibitors (n-butyl phosphorothioate triamine) on greenhouse gas (GHG) emissions, global warming potential and nitrogen use efficiency in roasted tobacco cropping systems as adequate soil GHG mitigation strategies in agroecosystems. Further, there was a significant increase in the crop yield and nitrogen uptake potential. The study could be a valuable and practical option for improving crop yield and mitigating climate change, thereby boosting sustainable agriculture. There is another article *"Developing a Sustainable Omnichannel Strategic Framework toward Circular Revolution..."* (https://www.mdpi.com/2071-1050/14/18/11578, accessed on 10 August 2023) that empirically examined the relationship between the quality of integration (INQ) and brand loyalty (BL), perceived quality (PQ), Brand awareness (BAW), brand association and brand equity (BE) in the context of Omnichannel Marketing (OM). This study brings a conceptual extension to the literature on omnichannel strategies, INQ and OM. In contrast, they presented the necessary reasons for managers to provide INQ in an omnichannel environment to increase brand equity with an empirical application. This study could benefit brand owners, managers, and marketers by showing how to set up the omnichannel system toward a circular revolution. Another article, *"A Critical Assessment on Functional Attributes and Degradation Mechanism of Membrane Electrode Assembly…"* (https://www.mdpi.com/2071-1050/13/24/13938, accessed on 10 August 2023), discussed direct methanol fuel cells (DMFC), a subset of polymer electrolyte membrane fuel cells (PEMFC) that possess benefits such as fuel flexibility, reduction in plant balance, and benign operation. The DMFCs have the potential to play an essential role in the future, specifically in replacing lithium-ion batteries (LiBs) for able and military applications. Inadequate reliability can potentially impede the commercialization of DMFCs. Therefore, the present article aims to assess the general degradation mechanism of MEA components of DMFCs with the basic structured procedure while excluding the system modeling and quantitative/qualitative analysis. Another article, *"Increased Digital Resource Consumption in Higher Educational Institutions…"* (https://www.mdpi.com/2071-1050/14/4/2377, accessed on 10 August 2023), is a novel study in a sense that it assesses the role of artificial intelligence (AI) as a booming technology, for understanding student behavior and evaluating their performance. The article discusses an AI-based analytics tool, the Random-Forest-based classification model, which can predict student performance early in their courses by allowing for early intervention. The study has very strong practical implications in forecasting the behavioral elements of teaching and e-learning for students in virtual education systems.

2.2. Recycling

Recycling is collecting, sorting, processing, and transforming used or discarded materials into new products or raw materials. It involves converting waste materials into reusable resources, thereby reducing the need for extracting and manufacturing new materials from virgin sources [10]. It's important to note that the recyclability of materials can vary depending on factors such as local recycling infrastructure, market demand, and the specific composition of the materials [11]. Following local recycling guidelines and practices ensures effective and efficient recycling processes [10,11]. This Special Issue attracted a few studies that fall under these conditions; for instance, the article *"… Clean Energy Vehicles in Japan Considering Copper Recycling"* (https://www.mdpi.com/2071-1050/15/3/2113, accessed on 10 August 2023) deals with the scope of introduction of clean energy vehicles (CEV) in the transportation sector to achieve carbon neutrality. The research highlights the need for strategies and policies that consider metal resources recycling and supply constraints in addition to factors such as carbon dioxide (CO_2) emissions during CEV promotion. The optimization model used in this study provides valuable insights into the sustainable consumption of copper resources through recycling and reducing CO_2 emissions. Another article, *"Development, Critical Evaluation, and Proposed Framework: End-of-Life Vehicle Recycling…"* (https://www.mdpi.com/2071-1050/14/22/15441, accessed on 10 August 2023), deals with end-of-life vehicle (ELV) recycling for sustainable development. This study has been performed through a mixed-method approach: a literature and policy review accompanied by detailed structured interviews with major stakeholders and industrial visits. This investigation reveals that India's ELV recycling system is embryonic and struggling against numerous inherent impediments. This research could be beneficial in assisting the government in implementing regulatory and legal frameworks. Another article, *"Development of a Reverse Logistics Modeling for End-of-Life Lithium-Ion Batteries…"* (https://www.mdpi.com/2071-1050/14/22/15321, accessed on 10 August 2023), discusses spatial modeling framework to quantify the environmental and economic effects of the expansion of the supporting infrastructure network for electric vehicle (EV) end-of-life LiBs management and sustainable recycling in Canada. The reverse logistics study presented in the manuscript integrates the geographic information system, material flow analysis for estimating the availability of spent LiBs stocks, and the life cycle assessment approach to assess the environmental impact. Along similar lines, an insightful article, *"An Assessment of Drivers and Barriers to Implementation of Circular Economy in the ELV Recycling…"* (https://www.mdpi.com/2071-1050/14/20/13084, accessed on 10 August 2023), highlighted the impediments and drivers regarding implementing circular economy in India's ELV recycling sector in India. According to the research, economic viability, environmental degradation, and global agenda are the three leading primary drivers. In contrast, limited technology, financial constraints, and lack of knowledge and expertise are significant barriers that thwart circular economy implementation in India's ELV recycling sector. This study could be constructive in assisting the Indian authorities in devising appropriate policies and strategies for developing a regulatory and legal framework conducive to both circular economy and sustainability. In another article, *"Recent studies and technologies in the separation of polyvinyl chloride for resources recycling…"* (https://www.mdpi.com/2071-1050/15/18/13842, accessed on 13 August 2023), researchers try to manage the plastic waste containing polyvinyl chloride (PVC), which is often destined for landfills as it poses particular difficulties for thermal treatment because of its additives, such as chloride (Cl–), which can negatively impact the refractory materials used in boilers. However, recognizing the value of PVC in PVC-bearing mixed plastics as a valuable resource, the authors focused on understanding the technologies for separating the PVC. Their systematic review stressed various technologies such as selective comminution, gravity separation, magnetic separation, electrical separation, flotation, and other advanced technologies such as sorting and density-surface-based separation. They also mentioned that, out of all these, flotation seems to be a widely used method for PVC separation from mixed plastic, thus promoting mixed plastic waste recycling.

2.3. Waste Management

Waste management strategies refer to the approaches and practices used to handle and manage waste in an environmentally responsible and sustainable manner. These strategies aim to minimize waste generation, maximize resource recovery, and reduce the negative impacts of waste on human health and the environment [12]. This Special Issue attracted few studies that fall under these conditions, for instance the article *"Adsorption of Fatty Acid on Beta-Cyclodextrin…"* (https://www.mdpi.com/2071-1050/15/2/1559, accessed on 10 August 2023), focuses on the role of β-cyclodextrin (β-CD) functionalized cellulose nanofiber (CNF) to adsorb the long chain fatty acids (LCFA), palmitic acid. The adsorption kinetics and isotherms were also elucidated to describe the adsorption behavior precisely. The study has practical implications in wastewater treatment for removing LCFAs from the wastewater. Another article on *"COVID-19 Biomedical Plastics Wastes…"* (https://www.mdpi.com/2071-1050/14/11/6466, accessed 10 August 2023) discussed the steep rise in plastic waste and management of biomedical plastic waste generated because of COVID-19 outbreak. The article elaborates on the issues of safe biomedical plastic waste disposal strategies. The article explicitly highlights the measurement of environmental issues in terms of plastic waste footprint and the strategy for safe disposal. The article also discusses sustainable techniques to reduce plastic waste and the need for incorporating Personal protective equipment (PPE) management policies into fiscal policies, to encourage green technology and find and implement safer practices. Another article, *"Heavy Metal, Waste, COVID-19, and Rapid Industrialization in This Modern Era…"* (https://www.mdpi.com/2071-1050/14/8/4746, accessed on 10 August 2023), presents a comparative valuation of the COVID-19 pandemic, heavy metal, waste and their effect on the atmosphere, humans, and economy. The review article highlights poor waste management practices and environmental and health disasters due to COVID-19-generated waste. Further, the article emphasizes the need for future studies on new policies for waste management, pollution monitoring, and waste recycling. Another article, *"Exploring Industry-Specific Research Themes on E-Waste…"* (https://www.mdpi.com/2071-1050/15/16/12244, accessed on 10 August 2023), did a thorough literature search on e-waste literature contributions in MDPI *Sustainability* journal to identify the prominent research themes, publication trends, research evolution, research clusters, and industries related to e-waste through descriptive analysis. The study gave four major research themes and clusters: closed-loops supply chains, e-waste, sustainable development, and waste electrical and electronic equipment. Overall, this review can be a foundation for subsequent scholarly pursuits toward e-waste management and fresh lines of inquiry for the journal to focus further scientific collection in this field. In another article, *"A Bibliometric Analysis of Sustainable Product Design Methods from 1999 to 2022..."* (accessed 10 August 2023), researchers explored the importance of effective product design strategies in promoting sustainable production, consumption, and disposal practices. They mainly highlighted the challenges of determining the most effective design approaches from a sustainability point of view to identify the current research trends, progress, and disparities between China and the rest of the world. They observed that the Chinese studies emphasized digital-driven development, rural revitalization, and system design. On the other hand, research from other countries highlighted a circular economy, distribution, additive manufacturing, and artificial intelligence. Notably, Chinese and international studies lacked quantitative research methods concerning socio-cultural sustainability. Another article, *"Leveraging Blockchain and Smart Contract Technologies to Overcome Circular Economy Implementation Challenges"* (https://www.mdpi.com/2071-1050/14/15/9492, accessed on 10 August 2023), is something different than other articles published in this Special Issue, which provided an appropriate digital solution for circular economy (CE) implementation. The article presents a thorough investigation of challenges under five barrier categories, Technological, Financial, Infrastructural, Institutional, and Societal, to address CE challenges and solutions to challenges raised in the earlier information and communication technologies-based solutions for CE. This perspective further explores the role of blockchain smart contract technology in overcoming CE challenges and presents a circular economy blockchain (CEB) architecture development.

3. Call-to-Action

Based on the Special Issue contributions discussion in Section 2, we can argue that research related to resource conservation, recycling, and waste management practices is extensively happening on various waste products or resources and the systems or processes surrounded by them across different sectors of society. However, when we see things from deeper insights, many of these *Trash to Treasure* activities are still in the early stages, at least in the global south. In contrast, there are many instances of greenwashing in the name of waste management and sustainable transition in the global north. So, as a best practice, with this editorial, we came up with call-to-action (CtA) statements, which are directives that encourage or prompt individuals or groups to take a specific course of action. In our context, CtA serves as an invitation, urging people to participate, engage, or support initiatives or movements in Trash to Treasure activities by unlocking the power of resource conservation, recycling, and waste management. A well-crafted CtA should be action-oriented; that is what we did here, see Boxes 1–4.

Box 1. Call-to-action for Researchers and Academicians.

Call-to-action initiatives for researchers and academicians to realize resource conservation, recycling, and waste management are crucial for generating knowledge, innovation, and evidence-based solutions. Here are some key initiatives that researchers and academicians can undertake:

Conduct Research on Sustainable Practices: Research to explore and analyze sustainable practices in resource conservation, recycling, and waste management. Investigate waste reduction strategies, recycling technologies, circular economy models, and the environmental impacts of different waste management approaches. Generate knowledge that can inform policies, practices, and technological advancements.
Develop Innovative Technologies: Focus on developing innovative technologies and processes that improve resource conservation, recycling, and waste management. This can include advancements in recycling techniques, waste-to-valuable materials, waste-to-energy conversion, waste sorting and separation methods, and sustainable materials design. Collaborate with industry partners, gov-ernment agencies, and other stakeholders to translate research into practical applications.

Box 1. *Cont.*

Collaborate in Interdisciplinary and Transdisciplinary Projects: Foster collaboration among researchers and academicians from various disciplines to address the complex challenges of resource conservation, recycling, and waste management. Encourage interdisciplinary and transdisciplinary projects integrating engineering, environmental science, social sciences, economics, and policy studies ex-pertise. This can lead to holistic and comprehensive solutions.
Share Knowledge and Best Practices: Publish research findings in academic journals and share knowledge through conferences, seminars, news articles, and workshops. Disseminate information on best practices, case studies, and successful waste management initiatives.
Engage with Stakeholders: Collaborate with stakeholders, including government agencies, NGOs, industry representatives, and communities, to understand their needs, challenges, and perspectives. Engage in participatory research approaches that involve stakeholders in problem-solving, decision-making, and implementing sustainable waste management practices.
Educate and Mentor Students: Integrate resource conservation, recycling, and waste management topics into academic curricula across various disciplines. Educate students about the importance of sustainability and equip them with the knowledge and skills needed to contribute to sustainable waste management practices. Mentor students in research projects focused on waste management and encourage them to pursue careers in the field.
Policy and Advocacy: Engage in policy discussions and advocacy efforts related to resource conservation, recycling, and waste management. Provide expert input to policymakers, contribute to developing evidence-based policies, and advocate for sustainable practices at local, national, and international levels.
Foster Industry-Academia Collaboration: Collaborate with industry partners to bridge the gap between research and practice. Work together to develop and implement sustainable waste management solutions, test new technologies, and conduct pilot projects. Facilitate knowledge transfer and technology transfer to ensure research outcomes have real-world impact.
Continuous Learning and Improvement: Stay updated with the latest advancements, technologies, and best practices in resource conservation, recycling, and waste management. Foster a culture of con-tinuous learning and improvement within academic institutions. Encourage collaboration among researchers through conferences, seminars, and research networks to share experiences and learn from each other.
Influence Research Funding Priorities: Advocate for research funding priorities that support resource conservation, recycling, and waste management. Engage with funding agencies, policymakers, and research councils to emphasize the importance of research in these areas and secure funding for innovative projects.

Box 2. Call-to-action for Individuals.

Call-to-action initiatives for individuals in realizing resource conservation, recycling, and waste management are essential for promoting sustainable practices at the individual level. Here are some key initiatives individuals can take:

Practice Reduce, Reuse, Recycle: Practice the three R's of waste management. Reduce waste by avoiding unnecessary packaging, opting for reusable products, and buying only what is needed. Reuse items whenever possible, such as using refillable water bottles, bringing your shopping bags, and donating or selling items instead of throwing them away. Finally, recycle materials accepted in your local recycling program to ensure they are properly processed and turned into new products.
Separate and Sort Waste: Properly segregate waste at home by separating recyclable materials, such as paper, plastic, glass, and metal, from non-recyclable items. This makes it easier for recycling facilities to process and recover valuable resources.
Composting: Establish a composting system for organic waste, such as fruit and vegetable scraps, yard trimmings, and coffee grounds. Composting helps reduce the amount of waste sent to landfills while producing nutrient-rich compost that can be used in gardens or landscaping.
Responsible Consumption: Make conscious choices when purchasing products. Consider the environmental impact of the items you buy, such as their packaging, durability, and recyclability. Opt for products made from recycled materials or those with minimal packaging.
Energy Conservation: Conserve energy by practicing energy-efficient habits. Turn off lights when not in use, unplug electronics when they are not being used, use energy-efficient appliances, and adjust thermostats for optimal energy use. Energy conservation reduces the demand for fossil fuels and helps mitigate climate change.

Box 2. *Cont.*

> *Educate Yourself*: Stay informed about your community's waste management practices, recycling guidelines, and resource conservation initiatives. Familiarize yourself with local recycling programs, collection schedules, and guidelines to ensure proper waste disposal.
> *Spread Awareness*: Share your knowledge and enthusiasm for resource conservation and waste management with others. Encourage friends, family, and colleagues to adopt sustainable practices and participate in recycling programs. Organize community events, workshops, or educational campaigns to raise awareness about the importance of recycling and waste reduction.
> *Support Recycling Infrastructure*: Advocate for and support the development of recycling infrastructure in your community. Engage with local authorities, businesses, and organizations to promote the expansion of recycling programs, collection centers, and facilities.
> *Participate in Clean-up and Recycling Initiatives*: Get involved in local clean-up events and recycling drives. Join community efforts to clean up parks, beaches, and other public spaces while promoting recycling and proper waste disposal.
> *Engage in Policy Advocacy*: Support policies and regulations that promote resource conservation, recycling, and waste management. Stay informed about relevant legislation and advocate for initiatives prioritizing local, regional, and national sustainable practices.

Box 3. Call-to-action for Industries/Corporates.

> Call-to-action initiatives for industries/corporates in realizing resource conservation, recycling, and waste management are crucial for promoting sustainable practices at a larger scale. Here are some key initiatives that industries and corporations can undertake:
>
> *Implement Sustainable Supply Chains*: Evaluate and optimize supply chains to minimize waste generation and resource consumption. Incorporate sustainable practices such as sourcing materials from responsible suppliers, reducing packaging waste, and promoting the use of recycled or renewable materials.
> *Adopt Circular Economy Principles*: Embrace the principles of the circular economy by designing products for durability, repairability, and recyclability. Implement strategies such as product take-back programs, remanufacturing, and incorporating recycled content into new products.
> *Waste Reduction and Recycling Programs*: Establish comprehensive waste reduction and recycling programs within the organization. Set ambitious waste reduction targets, promote waste segregation and sorting, and invest in efficient recycling infrastructure. Encourage employees to actively participate in recycling and waste management initiatives.
> *Resource Efficiency and Conservation*: Implement measures to optimize resource efficiency, such as energy-efficient technologies, water conservation practices, and waste minimization strategies. Conduct regular audits to identify areas for improvement and implement energy-saving initiatives throughout operations.
> *Adopt Life Cycle Assessments*: Conduct life cycle assessments (LCAs) to analyze the environmental impacts of products and processes. Use the results to identify opportunities for waste reduction, resource conservation, and environmental improvements throughout the supply chain.
> *Collaboration and Partnerships*: Collaborate with suppliers, customers, and other stakeholders to promote sustainability and waste management initiatives. Engage in partnerships to develop innovative recycling, waste reduction, and resource conservation solutions.
> *Employee Education and Engagement*: Educate employees about the importance of resource conservation, recycling, and waste management. Provide training on waste reduction practices, recycling guidelines, and the organization's sustainability goals. Encourage employee participation through recognition programs, incentives, and internal communication channels.
> *Transparent Reporting*: Provide transparent and accurate reporting on waste generation, recycling rates, and resource conservation efforts. Publicly disclose sustainability performance to stakeholders, shareholders, and customers. Use sustainability reports to demonstrate progress and set future targets.
> *Extended Producer Responsibility (EPR)*: Take responsibility for the entire life cycle of products by implementing EPR programs. Develop strategies to collect, recycle, or safely dispose of products at the end of their life, ensuring their proper management and reducing the burden on waste management systems.
> *Innovation and Research*: Invest in research and development to explore innovative technologies and processes that promote resource conservation and waste management. Support initiatives that enhance recycling capabilities, develop new recycling methods, and improve the recyclability of materials.
> *Advocacy and Policy Engagement*: Engage in advocacy efforts to support policies and regulations that promote sustainable practices, resource conservation, and waste management. Collaborate with industry associations and organizations to influence and shape policies that drive sustainability.

Box 4. Call-to-action for Governments and Policy Makers.

> Call-to-action initiatives for governments and policymakers in realizing resource conservation, recycling, and waste management are crucial for creating an enabling environment and driving systemic change. Here are some key initiatives that governments and policymakers can undertake:
>
> *Develop Comprehensive Waste Management Policies*: Establish comprehensive waste management policies prioritizing waste reduction, recycling, and resource conservation. Set clear targets and timelines for waste diversion, recycling rates, and landfill reduction. Ensure alignment with international best practices and commitments.
>
> Strengthen Legislative Frameworks: Enact and enforce legislation that supports sustainable waste management practices. This includes regulations on waste segregation, recycling requirements, EPR, and landfill restrictions. Strengthen penalties for illegal dumping and improper waste disposal practices.
>
> *Promote and Implement EPR Programs*: Implement EPR programs that hold producers responsible for the entire life cycle of their products, including post-consumer waste management. Encourage producers to design products for recyclability, establish take-back systems, and support recycling infrastructure development.
>
> *Invest in Recycling Infrastructure*: Allocate resources and funding to develop and enhance recycling infrastructure, including collection systems, sorting facilities, and recycling plants. Support the es-tablishment of recycling hubs and material recovery facilities to enable efficient and effective recy-cling processes.
>
> *Provide Incentives and Support Mechanisms*: Offer financial incentives, tax breaks, grants, and subsidies to businesses and industries that adopt sustainable waste management practices, invest in recycling technologies, and reduce waste generation. Support research and development efforts focused on waste management innovation.
>
> *Foster Public-Private Partnerships*: Facilitate partnerships between the government, private sector, and civil society organizations to promote collaboration and knowledge sharing. Encourage joint initiatives to develop sustainable waste management solutions, share best practices, and leverage resources.
>
> *Education and Awareness Campaigns*: Implement public education and awareness campaigns to inform citizens about the importance of resource conservation, recycling, and waste management. Guide proper waste segregation, recycling practices, and the benefits of sustainable waste management. Promote behavior change through targeted messaging and community engagement.
>
> *Support Research and Innovation*: Invest in research and development to explore new technologies, processes, and materials that enhance recycling capabilities, improve waste management practices, and promote resource efficiency. Support pilot projects and innovation centers focused on waste management and recycling.
>
> *International Cooperation and Knowledge Exchange*: Foster international cooperation and knowledge exchange on best practices, policies, and technologies for waste management. Partner with other countries and international organizations to share experiences, collaborate on research, and promote global solutions.
>
> *Monitoring and Reporting*: Establish monitoring and reporting systems to track progress toward waste management targets and evaluate the effectiveness of policies and initiatives. Regularly publish reports on waste generation, recycling rates, and resource conservation efforts to promote transparency and accountability

4. Conclusions

Overall, this editorial argues that we will all see a new normal in the Trash to Treasure initiatives in the near future as the knowledge around this subject is increasing and widely accepted, of course, with some uncertainties. However, it can also be understood that realizing this new normal is not easy without unlocking the power of *resource conservation, recycling, and waste management* practices, in which the actions that stakeholders of society should follow play an important role. For instance, researchers and academicians under-taking CtAs can contribute significantly to advancing resource conservation, recycling, and waste management. Their work can drive innovation, inform policies, and provide evidence-based solutions to address the environmental challenges associated with waste and resource management. Individuals undertaking CtAs can be significant in realizing resource conservation, recycling, and waste management. Collectively, these actions can substantially impact conserving resources, reducing waste, and protecting the environment for future generations. Similarly, industries/corporations undertaking CtAs can make significant strides in realizing resource conservation, recycling, and waste management. By integrating sustainability into their operations, they can reduce environmental impacts, enhance their reputation, and contribute to a more circular and sustainable economy. Lastly, when Governments and policymakers undertaking the CtAs can create an enabling environment for resource conservation, recycling, and waste management. They can drive systemic change, set the direction for sustainable practices, and ensure the long-term well-being of communities and the environment.

Author Contributions: Conceptualization, M.K.N.; writing—original draft preparation, M.K.N. and A.P.; writing—review and editing, M.K.N., A.P., S.H., and V.B.; visualization, M.K.N.; supervision, M.K.N. All authors have read and agreed to the published version of the manuscript.

Funding: This research received no external funding.

Acknowledgments: The Special Issue Editors would like to take the opportunity to thank the authors who responded to the call and the MDPI Sustainability Journal for approving the topic and allowing us to be in the guest editor roles. We are also deeply indebted to the reviewers whose input was indispensable in selecting the published papers.

Conflicts of Interest: The authors declare no conflict of interest.

References

1. Gleason, C. *Wood Pallet Projects: Cool and Easy-to-Make Projects for the Home and Garden*; Fox Chapel Publishing: Mount Joy, PA, USA, 2013.
2. IHanna. Handbag Made Out of Old Jeans. 2008. Available online: https://www.ihanna.nu/blog/2008/01/handbag-from-old-jeans/ (accessed on 10 August 2023).
3. Pamphlets in the Daily Newspaper into Decorative Placemat. Dollar Store Crafts. Available online: https://dollarstorecrafts.com/ (accessed on 10 August 2023).
4. Muneneh, M. An Exploration Into Barriers To Use Of Waste Glass in Interior Spaces in Nairobi. Doctoral Dissertation, University of Nairobi, Nairobi, Kenya, 2013.
5. V-Surf Limited. Available online: https://vsurf.co/ (accessed on 10 August 2023).
6. Yuan, X.; Kumar, N.M.; Brigljević, B.; Li, S.; Deng, S.; Byun, M.; Lee, B.; Lin, C.S.K.; Tsang, D.C.W.; Lee, K.B.; et al. Sustainability-inspired upcycling of waste polyethylene terephthalate plastic into porous carbon for CO_2 capture. *Green Chem.* **2022**, *24*, 1494–1504. [CrossRef]
7. Kumar, N.M.; Chopra, S.S. Blockchain-assisted spent electric vehicle battery participation for load frequency control problems in interconnected power systems is resilient, low-carbon, and offers revenues to the operators. *Sustain. Energy Technol. Assessments* **2023**, *57*, 103209. [CrossRef]
8. Nallapaneni, M.K. Electric vehicle revolution among Hong Kong residents: Ways to handle the inevitable surge in battery waste in line with circular economy principles. In Proceedings of the Research Assistant Professors Selection Panel Committee Meeting, Department of Building Environment and Energy Engineering, The Hong Kong Polytechnic University (PolyU), Hong Kong, China, 21 April 2023. [CrossRef]
9. Foo, D. *Process Integration for Resource Conservation*; CRC Press: Boca Raton, FL, USA, 2016.
10. Kumar, N.M.; Chopra, S.S. Insights into material recovery, revenue, and global warming potential of India's end-of-life photovoltaic installations reveals the urgent need for blockchain-based solar passports. *Sustain. Energy Technol. Assess.* **2023**, *58*, 103326.
11. Carlson, A.E. Recycling norms. *Calif. L. Rev.* **2001**, *89*, 1231. [CrossRef]
12. Hamer, G. Solid waste treatment and disposal: Effects on public health and environmental safety. *Biotechnol. Adv.* **2003**, *22*, 71–79. [CrossRef] [PubMed]

 *sustainability*

Article

Combined Effects of Biochar and Inhibitors on Greenhouse Gas Emissions, Global Warming Potential, and Nitrogen Use Efficiency in the Tobacco Field

Tongkun Zhang [1,2], Yuan Tang [3], Weichang Gao [4], Xinqing Lee [1], Huan Li [1,2], Wei Hu [3] and Jianzhong Cheng [1,*]

[1] State Key Laboratory of Environmental Geochemistry, Institute of Geochemistry, Chinese Academy of Sciences, Guiyang 550081, China
[2] University of Chinese Academy of Sciences, Beijing 100049, China
[3] School of Public Health, The Key Laboratory of Environmental Pollution Monitoring and Disease Control, Ministry of Education, Guizhou Medical University, Guiyang 550025, China
[4] Guizhou Academy of Tobacco Science, Guiyang 550081, China
* Correspondence: chengjianzhong@vip.gyig.ac.cn

Abstract: Biochar (BC), nitrification inhibitors (methyl 3-(4-hydroxyphenyl) propionate, MHPP), and urease inhibitors (n-butyl phosphorothioate triamine, NBPT) have emerged as effective soil greenhouse gas (GHG) mitigation strategies in agroecosystems. However, the combined use of BC and inhibitors in karst areas has no available data. Therefore, the combined effects of BC, MHPP, and NBPT on GHG emissions, global warming potential (GWP) and nitrogen use efficiency (NUE) in roasted tobacco cropping systems were studied to improve the understanding in climate mitigation. CO_2, CH_4, and N_2O emissions from soils were measured using static chamber-gas chromatography. Results showed that the combined use of BC and inhibitors significantly increased soil total nitrogen, available potassium, electric conductivity, pH, and soil organic matter compared to the control. The combined use of BC and MHPP or NBPT significantly increased cumulative soil CO_2 emissions by 33.95% and 34.25%, respectively. The exponential–exponential function of soil CO_2 fluxes with soil moisture and temperature demonstrated good fit (R^2: 0.506–0.836). The combination of BC and NBPT increased the cumulative soil CH_4 emissions by 14.28% but not significantly compared to the fertiliser treatment. However, the combination of BC and MHPP resulted in a significant reduction in cumulative soil CH_4 emissions by 80.26%. In addition, the combined use of BC and MHPP or NBPT significantly reduced the cumulative soil N_2O emissions by 26.55% and 40.67%, respectively. The inhibition effect of NBPT was better than MHPP. Overall, the combined use of BC and inhibitors significantly reduced the yield-scaled GWP, markedly increased crop yield and NUE, and mitigated climate change in the southwest karst region.

Keywords: biochar; nitrification inhibitor; urease inhibitor; greenhouse gas; global warming potential; nitrogen use efficiency

Citation: Zhang, T.; Tang, Y.; Gao, W.; Lee, X.; Li, H.; Hu, W.; Cheng, J. Combined Effects of Biochar and Inhibitors on Greenhouse Gas Emissions, Global Warming Potential, and Nitrogen Use Efficiency in the Tobacco Field. *Sustainability* **2023**, *15*, 6100. https://doi.org/10.3390/su15076100

Academic Editors: Nallapaneni Manoj Kumar, Subrata Hait, Anshu Priya and Varsha Bohra

Received: 8 March 2023
Revised: 28 March 2023
Accepted: 29 March 2023
Published: 31 March 2023

1. Introduction

Terrestrial ecosystem emissions have dramatically increased atmospheric CO_2, CH_4, and N_2O, which contribute to global warming [1]. Since 2010, atmospheric greenhouse gas (GHG) concentrations have continued to increase; in 2017, the annual average concentrations of CO_2, CH_4, and N_2O reached 410.53, 1853, and 328.9 mm^3/m^3, respectively [2]. Consequently, effective measures to change agricultural management programmes to mitigate GHG emissions are urgently needed.

The following effective field measures are currently recommended to reduce GHG emissions without decreasing crop yields: biochar (BC) amendments [3] and dual-inhibitor applications [4]. BC is a highly aromatic carbon sequestration material produced by pyrolytic carbonisation of biomass under anoxic or oxygen-limited conditions [5]. BC is rich in N, P, and K and has high pH, high porosity, large specific surface area, high carbon content, high cation exchange capacity, and high thermal stability [6]. Some studies have shown that BC has a relatively low turnover rate and may survive in the soil for more than 100 years [7]. Therefore, if BC interacts with the soil for a long time, then it can be used as a means to mitigate climate change by enhancing soil carbon sequestration [8]. The carbon sequestration pathways of BC include the conversion of readily decomposable plant carbon to stable BC, increased capability of plants to capture atmospheric CO_2 due to increased biomass, and inhibition of decomposition of readily decomposable soil organic carbon [9]. Simultaneously, the application of BC may impact soil CO_2 emissions by changing soil structure, cation exchange, enzyme activity, and microbial communities [10]. Numerous studies have shown that soil GHG emissions are reduced with BC addition. Agegnehu et al. (2016) found that the BC treatment group had significantly lower GHG emissions than the control group [11]. Xie et al. (2013) found that replacing straw amendments with BC reduced CH_4 emissions and increased soil organic carbon storage [12]. However, several studies have shown that BC amendments are ineffective and may even promote GHG emissions. For example, Zhang et al. (2010) found that BC significantly increased CH_4 emissions [13]. In addition, the application of BC at high rates increased CO_2 and N_2O emissions from forest soils [14]. The effect of BC on soil GHG emissions depends on many factors, such as feedstock, pyrolysis temperature, and BC application rate, according to a series of meta-analyses [15,16]. Furthermore, environmental conditions, such as soil texture, fertiliser application, and climate change, can affect GHG emissions from BC-applied soils [17].

Moreover, urease and nitrification inhibitors are frequently used in soil amendments. Studies have shown that nitrification inhibitors decrease N_2O generation by directly lowering nitrification and indirectly reducing denitrification of NO_3^- [18,19]. Urease inhibitors diminish the concentration of NH_4^+ in the soil and the likelihood of N_2O volatilisation by retarding the conversion of NH_4^+ [20]. Zaman et al. (2008) observed that the application of inhibitors significantly reduced N_2O emissions and NO_3^- leaching [21]. A meta-analysis showed that nitrification inhibitors reduce direct N_2O emissions by 44% [22]. However, inconsistent results have also been discovered with urease and nitrification inhibitors in the reduction of N losses. Lam et al. (2018) found that the application of urease inhibitors in acid soils increased N_2O emissions by 17% [23]. In addition, the effects of urease and nitrification inhibitors on soil CH_4 emissions are variable because they may increase CH_4 emissions, enhance CH_4 oxidation, or have no impact at all [24,25]. These different results may be attributable to soil type, inhibitor type, inhibitor application rate, and application method [26,27]. Meanwhile, in crop nitrogen (N) transformation and utilisation, the use of nitrification and urease inhibitors significantly increased crop yields and promoted N uptake [28].

Studies recently found that the combined application of BC and inhibitors is more effective than single applications in mitigating GHG emissions and improving crop yields [29,30]. Most of these studies were conducted on agricultural land and focused on food crops, such as maize, rice, and wheat, with minimal research on cash crops, such as tobacco [31,32]. In southwest China, namely, the karst area, tobacco is a significant cash crop [33]. Karst areas are known to be highly susceptible to soil degradation caused by human activities [34], whilst the unique geological environment and climatic circumstances of karst areas constitute an important production prerequisite for high-quality tobacco. The application of BC and inhibitors to the soil can increase soil carbon sink, promote soil fertility, and mitigate climate change. Therefore, investigating the combined effects of BC, urease, and nitrification inhibitors on soil GHG fluxes in roasted tobacco cropping systems in karst areas is crucial. Given the uncertainty of the combined effects of BC, methyl 3-(4-hydroxyphenyl) propionate (MHPP), and n-butyl phosphorothioate triamine (NBPT) on GHG emissions from the soil, a full accounting of the global warming potential (GWP) of soil GHG emissions is necessary during assessment of their climate impact in karst areas. A field experiment was set up using static chamber-gas chromatography (GC) techniques in the roasted tobacco cropping system in karst areas to simultaneously measure soil CO_2, CH_4, and N_2O fluxes. This study aimed to understand the combined effects of BC and MHPP or NBPT on soil GHG emissions from roasted tobacco cropping systems in karst areas and those on crop production and N conversion.

2. Material and Methods

2.1. Experimental Site

The field trial was conducted at the Pingba Tobacco Experimental Base of the Guizhou Academy of Tobacco Science ($26°26'193''$ N, $106°14'166''$ E; 1391 m above sea level). The region has a typical subtropical humid monsoon climate with an annual average temperature of 14.5 °C and an annual average precipitation of 122 mm. Referring to the Genetic Soil Classification of China, the soil in this study was yellow soil. The soil texture of the site was clay loam. Table 1 lists the basic properties of the different treatments.

Table 1. Effect of BC, MHPP, and NBPT on soil properties.

	T1	T2	T3	T4
pH	6.58 ± 0.15 [b]	6.47 ± 0.02 [b]	7.04 ± 0.09 [a]	6.94 ± 0.17 [a]
EC (μS/cm)	151.73 ± 33.42 [c]	388.67 ± 74.57 [b]	881.67 ± 52.37 [a]	895.00 ± 86.26 [a]
TC (g/kg)	19.19 ± 0.89 [a]	19.19 ± 1.77 [a]	21.49 ± 2.60 [a]	20.18 ± 2.34 [a]
TN (g/kg)	1.76 ± 0.12 [b]	1.83 ± 0.04 [b]	2.26 ± 0.16 [a]	2.07 ± 0.10 [a]
TP (g/kg)	0.92 ± 0.06 [a]	1.03 ± 0.05 [a]	1.07 ± 0.02 [a]	0.96 ± 0.03 [a]
TK (g/kg)	13.83 ± 0.83 [a]	14.37 ± 0.76 [a]	12.17 ± 0.78 [b]	11.67 ± 0.71 [b]
AN (mg/kg)	123.08 ± 6.94 [b]	146.43 ± 5.91 [a]	143.89 ± 19.65 [ab]	133.97 ± 11.61 [ab]
AP (mg/kg)	13.14 ± 3.89 [b]	23.94 ± 3.40 [ab]	27.43 ± 9.72 [a]	20.51 ± 5.44 [ab]
AK (mg/kg)	289.29 ± 122.78 [b]	467.88 ± 72.14 [b]	1382.41 ± 328.31 [a]	1042.48 ± 138.50 [a]
SOM (g/kg)	31.20 ± 4.50 [b]	29.66 ± 5.91 [b]	45.23 ± 2.07 [a]	41.48 ± 6.17 [a]

T1, T2, T3, and T4 represent CK-, F-, F + BC + MHPP-, and F + BC + NBPT-amended soils, respectively. EC: electric conductivity; TC: total carbon content; TN: total nitrogen content; TP: total phosphorus content; TK: total potassium content; AN: rapid-acting nitrogen content; AP: effective phosphorus content; AK: rapid-acting potassium content; SOM: soil organic matter content. Different letters within the line indicate significant differences between treatments at $p < 0.05$.

2.2. Field Experiments

The experiment was conducted in a completely randomised block design with three replicated four-treatment field trials. In April 2022, roasted tobacco-specific fertiliser, BC, nitrification inhibitor (MHPP), and urease inhibitor (NBPT) were uniformly mixed into the 0–20 cm soil layer and applied at 675 kg/ha, 20 t/ha, 13.5 kg/ha, and 337.5 g/ha, respectively. The treatment settings were as follows:

(1) T1: No fertiliser application (CK);
(2) T2: Single application of special fertiliser for roasted tobacco (F);
(3) T3: Special fertiliser for roasted tobacco (F) + biochar (BC) + nitrification inhibitor (MHPP);
(4) T4: Special fertiliser for roasted tobacco (F) + biochar (BC) + urease inhibitor (NBPT).

Special fertiliser for roasted tobacco (N: P_2O_5: K_2O = 10:10:25) was applied in all plots before the high row monopoly (30 cm). After sowing, uniformly sized tobacco seedlings (Yunyan87) were transplanted from the seedbed to the row monopoly. For tobacco transplantation, 1.1 m row spacing and 0.6 m row distance were adapted. A total of 60 plants were planted in each plot, which had an area of 39.6 m^2. Tobacco yields in the field trial were remarkably dependent on natural precipitation. The additional fertiliser (272 kg/ha) was applied 1 month after the basic fertiliser (675 kg/ha). Consistent management measures were applied during the growing season. Soil temperature and moisture sensors were placed at soil depths of 5, 10, and 20 cm, and the data were recorded with a logger (TR-6, Shuncoda, China). The entire growth cycle of roasted tobacco is approximately 120 days, which is divided into three stages: root extension period (REP, 30 days), vigorous period (VP, 30 days), and mature period (MP, 60 days).

2.3. Gas Sampling and Measurements

Soil GHG fluxes were measured using static chamber-GC techniques from May to August 2020. A chamber base (10 cm high, 30 cm diameter) with deep circular depressions was placed inside each area. These bases were kept in place for the duration of the experiment. When the opaque removable polyvinyl chloride (PVC) chambers (50 cm high) were put on the bases, the depressions were filled with water to prevent air leakage. The chamber has a thermometer that records the air temperature during sampling time. An electric fan was also installed in the chamber to ensure that the gases were well mixed during the sampling process. A 20 mL sample of gas was taken from each chamber at 0, 6, 12, and 18 min after each closure using a syringe, and then filled into evacuated 12 mL glass bottles. The samples were then analysed within 24 h after collection. Soil GHG emissions were assessed every 7 days. Measurements between 8:00 and 11:00 a.m. were selected to limit the daily volatility in flow patterns.

A modified GC (Agilent 7890), including a flame ionisation detector (FID) and an electron capture detector (ECD), was used to detect CO_2, CH_4, and N_2O in soil [35]. N_2 was used as the carrier gas. Nitrous oxide was separated using two stainless-steel columns (column 1 with 1 m length and 2.2 mm i.d.; column 2 with 3 m length and 2.2 mm i.d.) fitted with 80–100 mesh Porapack Q. CH_4 and N_2O were identified using FID and ECD, respectively. The oven, ECD, and FID were run at 55 °C, 330 °C, and 200 °C, respectively [36].

The GC was calibrated using three recognised reference gases to assure the precision and dependability of the continuous readings. If linear regression values of $R^2 > 0.90$ were not provided, then they were not included in the gas sample data set. Soil GHG emission fluxes were determined using the following equation [37]:

$$F_{GHG} = \frac{60}{100} \times \rho_0 \times H \times \frac{273}{273 + T} \times \left(\frac{dc}{dt}\right),$$

where F_{GHG} is the flux of CO_2, CH_4, or N_2O ($mg \cdot m^{-2} \cdot h^{-1}$); 60 and 100 are unit conversion factors for calculating GHG fluxes; ρ_0 is the density of CO_2, CH_4, or N_2O in standard conditions (g/L); H is the height of the sampling chamber (cm); T is the temperature of the chamber (°C); dc/dt is the rate of change ($\mu L/L$) of the CO_2, CH_4, or N_2O concentration in the sampling chamber over time t (min), with positive and negative values indicating emissions and absorption, respectively.

The emissions of CO_2, CH_4, and N_2O are accumulated over two adjacent measurement periods during the growth cycle of roasted tobacco. These emissions are determined using the following equation:

$$E_{GHG} = \sum_{i=1}^{n} (F_i + F_{i+1})/2 \times (t_{i+1} - t_i) \times 24,$$

where E_{GHG} is the cumulative emission of soil GHGs; F_i and F_{i+1} represent the soil GHG fluxes at sampling periods i and i + 1, respectively; $t_{i+1} - t_i$ is the interval (d) between the i-th and (i + 1)-th measurement times; n is the total number of gas samples collected.

The relationship between soil CO_2 emission fluxes and soil temperature and moisture can be fitted with the help of the following exponential–exponential function [38]:

$$F_{CO_2} = ae^{bW}e^{cT},$$

where a, b, and c are fitting constants; W is the soil volumetric water content (SVWC) (%); T is the soil temperature (°C).

The N_2O emission factor (EF_{N2O}) (kg/ha) from the application of N fertiliser was estimated by applying the following equation:

$$EF_{N2O} = (EF - E_0)/N_{ap} \times 100\%,$$

where EF (kg/ha) is the cumulative N_2O emission from the N fertiliser treatment; E_0 (kg/ha) is the cumulative N_2O emission from the unfertilised treatment (T1); N_{ap} is the amount of N fertiliser (kg N/ha) applied to N treatment plots.

2.4. Determination of Crop Yield, Nitrogen Use Efficiency, and GWP

The yield of tobacco is the mass of dry matter collected during the growing season. Tobacco yield was assessed on three tobacco plants from each plot. Tobacco samples were divided into three parts: root, stem, and leaf. The samples were then washed with tap water, followed by deionised water. These samples were placed in an oven at 105 °C for 30 min and finally baked at 65 °C for 48 h to maintain a uniform weight. The yield was calculated by weighing the dried tobacco samples, and the total nitrogen (TN) content was assessed using an elemental analyser (Vario MACRO Analyser, Elementar Analysensysteme GmbH, Hanau, Germany). The amount of N uptake and the nitrogen use efficiency (NUE) in the crop were obtained by calculation.

The NUE is calculated as follows:

$$NUE = (N_{DMF} - N_{DMU})/N_{ap},$$

where N_{DMF} is the N content of dry matter obtained from the fertilised treatment; N_{DMU} is the N content of dry matter obtained from the unfertilised control treatment; N_{ap} is the amount of N applied.

Cumulative emissions of CO_2, CH_4, and N_2O were analysed to balance the net greenhouse effect. Cumulative GHG emissions were converted into CO_2 equivalents (CO_2-eq) using their specific GWP values. These emissions are an index defined as the cumulative radiative forcing caused by a unit quantity of current gas release and at a selected time frame in the future. The GWP (based on a 100 year time horizon) is 25, 298, and 1 for CH_4, N_2O, and CO_2, respectively. The following equation was used to estimate the GWP of different treatments [39]:

$$GWP = CO_2 \text{ emission} + CH_4 \text{ emission} \times 25 + N_2O \text{ emission} \times 298,$$

where GWP is expressed in kg CO_2-eq/ha; CO_2, CH_4, and N_2O fluxes are expressed in kg/ha.

The yield-scaled GWP (t CO_2-eq/t) is calculated in accordance with the following method [40]:

$$Yield - scaled\ GWP = GWP/yield.$$

2.5. Soil Collection and Analyses

Soil samples (0–20 cm) were collected 1 week before the construction of the plots and at the end of the field experiment. Five soil samples were collected from each plot and then mixed thoroughly to form a composite sample. Soil pH was determined with a pH meter in a soil–water suspension at a ratio of 1/2.5. Soil electric conductivity (EC) was assessed with a conductivity meter (Orion, CM-180) at a soil/water ratio of 1/2.5. Soil total carbon (TC) and TN were determined with an elemental analyser [41]. Soil total phosphorus (TP) and available phosphorus (AP) were measured using the H_2SO_4–$HClO_4$ ablation and $NaHCO_3$ extraction techniques, respectively. The NaOH fusion method and the CH_3COONH_4 extraction technique were used to estimate total potassium (TK) and available potassium (AK), respectively [42]. Soil available nitrogen (AN) was analysed by the alkaline solution diffusion method [43]. Soil NH_4^+ and NO_3^- were extracted with 2 M KCl solution (1:5, *w/v*). NH_4^+ and NO_3^- were determined by the flow analyser system [29]. Soil inorganic nitrogen (SIN) was determined by 1 M KCl extraction followed by Kjeldahl distillation in the presence of MgO + Devarda's alloy, followed by titration for TN [44]. Soil organic matter (SOM) was determined by oxidation with potassium dichromate.

2.6. Data Analyses

All data are provided as the means $\pm$ standard error. One-way analysis of variance was used to evaluate the effects of N fertiliser, BC, MHPP, and NBPT on soil CO_2, CH_4, and N_2O emissions and their related GWP, yield-scaled GWP, crop yield, EF_{N2O}, N uptake, and NUE. At the 0.05 significance level, the LSD test was used to evaluate the significance of the observed differences. The relationship between GHG fluxes and environmental factors was explored using correlation analysis. Multiple linear regression analysis was used to determine the significance of the fit of soil CO_2 emissions concerning soil temperature and water content at different depths. All statistical analyses were performed using SPSS 16.0 (SPSS Inc., Chicago, IL, USA).

3. Results and Discussion

3.1. Effects of BC, MHPP, and NBPT on Soil Physicochemical Properties

Results showed that the combined use of BC and MHPP or NBPT significantly increased TN by 13.11% and 23.50% ($p < 0.05$), respectively, whilst AK levels significantly increased by 1.23-fold to 1.95-fold compared to T2 (Table 1). Regarding TC, the mean values of T3 and T4 were 11.99% and 5.16% higher than T2 but insignificant, respectively. In addition, fertilisation treatments (T2, T3, and T4) all increased the mean values of AP concentrations in the soil. T2, T3, and T4 significantly increased the soil EC values by 1.56, 4.81, and 4.90 times compared to T1. The combined effect of BC and inhibitors significantly increased the mean soil pH by 7.26% to 8.81% due to the alkalinity of BC and the neutrality of the inhibitor. By contrast, no significant changes were observed in AN, TP, and TK levels in unamended and amended soils after BC and inhibitor addition. SOM levels were mainly associated with the combination of BC and inhibitors. No significant change was observed in T2 compared to the control, whilst SOM levels increased significantly by 44.97% and 32.95% in T3 and T4 treatments, respectively. The meta-analysis found that the application of BC and inhibitors had different effects on soil physicochemical properties, which were related to factors, such as feedstock, pyrolysis temperature, application rate, and soil texture [45]. Overall, the combination of BC and inhibitor significantly increased soil pH, EC, and SOM and improved soil fertility, which is consistent with the results of Singh et al. (2022) [46]. This combination also significantly increased TN, markedly reduced NO_3^- leaching, and improved the effectiveness of N fertiliser [47].

SVWC increased significantly with rising soil depth during the roasted tobacco growth cycle ($p < 0.05$, Table 2). Compared to the control, T2, T3, and T4 treatments increased the SVWC at a depth of 5 cm by an average of 4.39%, 2.12%, and 4.09%, respectively, during the root extending period. T2, T3, and T4 treatments increased the SVWC at 10 cm depth by an average of 11.54%, 6.90%, and 5.16%, respectively. Moreover, soil water content at 5, 10, and 20 cm was significantly different in either treatment. At the vigorous period, the combined effect of the BC and inhibitors on SVWC was consistent with the root extending period but varied relatively minimally (from 0.38% to 9.06%) compared to the control. T3 and T4 treatments increased the SVWC at 5 cm depth by an average of 6.47% and 2.64% compared to the control. Meanwhile, T3 and T4 treatments increased the SVWC at 20 cm depth by an average of 1.83% and 0.38%, respectively. Notably, the combination of BC and inhibitors had a larger effect on soil water content at 5 and 10 cm than at 20 cm. At the mature period, T3 and T4 treatments reduced SVWC at 5, 10, and 20 cm depths compared to the control (with T3 at -3.22%, -1.70%, and -3.24% and T4 at -5.24%, -0.16%, and -5.94%, respectively).

Throughout the roasted tobacco growth cycle, SVWC showed remarkably similar trends with transplanting time for the different treatments (Figure 1). The mean SVWC at 5 cm was 16.37% $\pm$ 0.93%, 16.92% $\pm$ 0.87%, 16.61% $\pm$ 0.94%, and 16.35% $\pm$ 0.74% for T1, T2, T3 and T4, respectively. Moreover, although not dramatically, soil temperature declined with soil depth and was negatively linked with SVWC (R: -0.39 to -0.50, $p < 0.05$). The mean 5 cm soil temperature for T1, T2, T3, and T4 was 25.50 $\pm$ 0.68 °C, 24.42 $\pm$ 0.78 °C, 24.15 $\pm$ 0.75 °C, and 24.24 $\pm$ 0.55 °C, respectively, whilst the mean 20 cm soil temperature was 23.42 $\pm$ 0.52 °C,

22.60 ± 0.57 °C, 22.37 ± 0.51 °C, and 22.64 ± 0.44 °C, respectively. At the mature period, soil temperatures for T3 and T4 treatments at different depths were significantly lower than for T1, which may have been due to the combination of BC and inhibitors [48,49]. However, as the application rate of BC was low, the difference in soil temperature between treatments during the root extension and vigorous periods was insignificant.

Table 2. Effect of BC, MHPP, and NBPT on soil moisture and temperature at various depths during the roasted tobacco growth cycle.

Growth Periods	Treatment	SVWC5	SVWC10	SVWC20	ST5	ST10	ST20
REP	T1	13.20 ± 0.99 [aC]	17.24 ± 1.03 [aB]	23.79 ± 1.42 [aA]	25.54 ± 0.89 [aA]	24.06 ± 0.72 [aAB]	22.52 ± 0.64 [aB]
	T2	13.78 ± 0.91 [aC]	19.23 ± 0.90 [aB]	25.75 ± 1.16 [aA]	24.85 ± 1.14 [aA]	22.69 ± 0.88 [aAB]	21.53 ± 0.73 [aB]
	T3	13.48 ± 1.03 [aC]	18.43 ± 1.15 [aB]	24.26 ± 1.08 [aA]	24.12 ± 0.96 [aA]	22.62 ± 0.76 [aA]	21.17 ± 0.67 [aA]
	T4	13.74 ± 0.88 [aC]	18.13 ± 0.84 [aB]	23.32 ± 0.94 [aA]	24.91 ± 0.70 [aA]	23.24 ± 0.64 [aA]	22.02 ± 0.66 [aA]
VP	T1	20.85 ± 1.14 [aC]	25.85 ± 1.17 [aB]	31.11 ± 2.67 [aA]	24.81 ± 0.41 [aA]	23.61 ± 0.37 [aA]	22.58 ± 0.22 [aA]
	T2	22.74 ± 1.24 [aC]	26.73 ± 0.99 [aB]	33.13 ± 2.09 [aA]	24.28 ± 0.50 [aA]	22.90 ± 0.35 [aA]	22.04 ± 0.29 [aA]
	T3	22.20 ± 1.15 [aC]	27.58 ± 1.33 [aB]	31.68 ± 2.37 [aA]	23.94 ± 0.57 [aA]	22.63 ± 0.43 [aAB]	21.68 ± 0.32 [aB]
	T4	21.40 ± 1.03 [aC]	26.91 ± 1.06 [aB]	31.23 ± 2.29 [aA]	23.83 + 0.54 [aA]	22.90 ± 0.36 [aA]	22.38 ± 0.28 [aA]
MP	T1	15.84 ± 0.75 [aC]	18.85 ± 0.73 [aB]	22.22 ± 0.78 [aA]	25.93 ⊥ 0.69 [aA]	25.24 ± 0.61 [aA]	24.68 ± 0.64 [aA]
	T2	15.48 ± 0.60 [aC]	17.97 ± 0.57 [aB]	20.84 ± 0.81 [aA]	24.18 ± 0.69 [bA]	23.93 ± 0.62 [bA]	23.80 ± 0.63 [bA]
	T3	15.33 ± 0.71 [aC]	18.53 ± 0.86 [aB]	21.50 ± 1.01 [aA]	24.32 ± 0.71 [bA]	24.14 ± 0.57 [bA]	23.76 ± 0.51 [bA]
	T4	15.01 ± 0.43 [aC]	18.82 ± 0.65 [aB]	20.90 ± 0.98 [aA]	23.99 ± 0.43 [bA]	23.61 ± 0.28 [bA]	23.28 ± 0.37 [bA]

SVWC5, SVWC10, and SVWC20 represent the soil volumetric water content at 5, 10, and 20 cm depths, respectively; ST5, ST10 and ST20 represent the soil temperature at 5, 10, and 20 cm depths, respectively. Different lowercase letters within a column indicate significant differences between treatments at $p < 0.05$. Different capital letters within a row indicate significant differences between soil depths at $p < 0.05$.

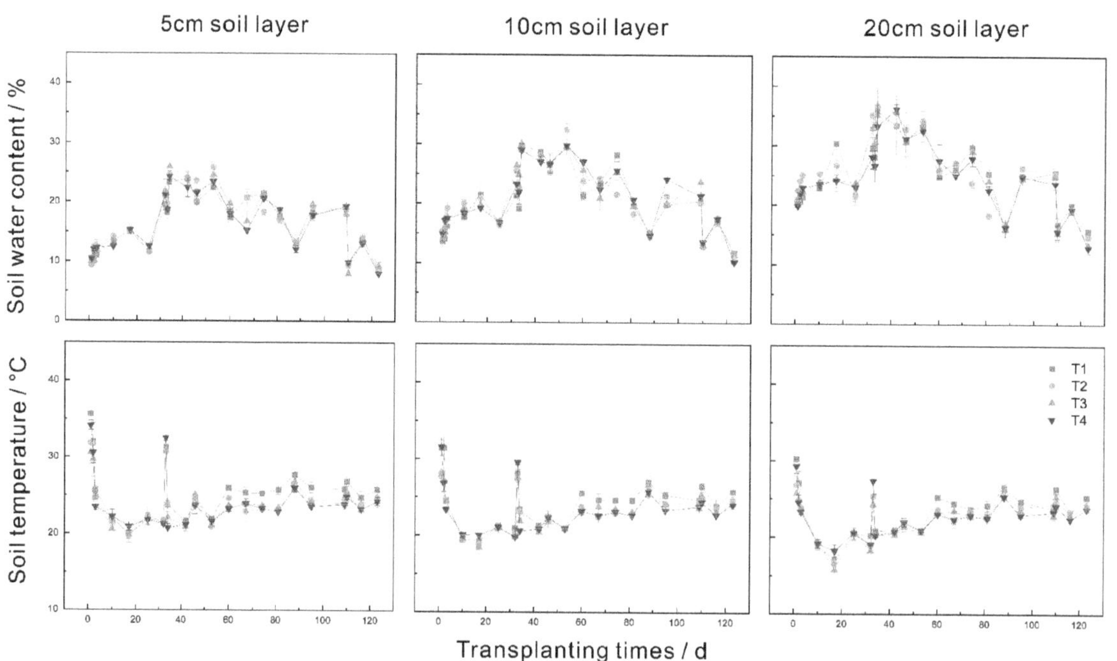

Figure 1. Changes in SVWC and soil temperature at three depths (5, 10, and 20 cm) for different treatments during the roasted tobacco growth cycle.

3.2. Effects of BC, MHPP, and NBPT on Soil CO_2 Emission

In roasted tobacco cropping systems, soil CO_2 fluxes showed similar seasonal dynamics between treatments after the application of BC, MHPP, and NBPT (Figure 2a). Fertiliser application significantly increased soil CO_2 emissions compared to the control. Soil CO_2 emissions were significantly increased in T3 and T4 treatments compared to T2, but no significant difference was observed between the two treatments. At present, studies conducted on the combined use of BC and inhibitors have mainly focused on N_2O, with few studies on CO_2. Suggestions indicate that BC, as an exotic carbon source, has a portion of its active carbon pool [50]. In addition, the soil has a portion of the active carbon pool, and the combination of BC and inhibitors may lead to the decomposition of the respective active carbon pools of BC and soil. BC promotes soil CO_2 emissions [51], and MHPP or NBPT may promote or inhibit these emissions. Notably, the combined use of BC and inhibitors contributed significantly to high soil CO_2 fluxes.

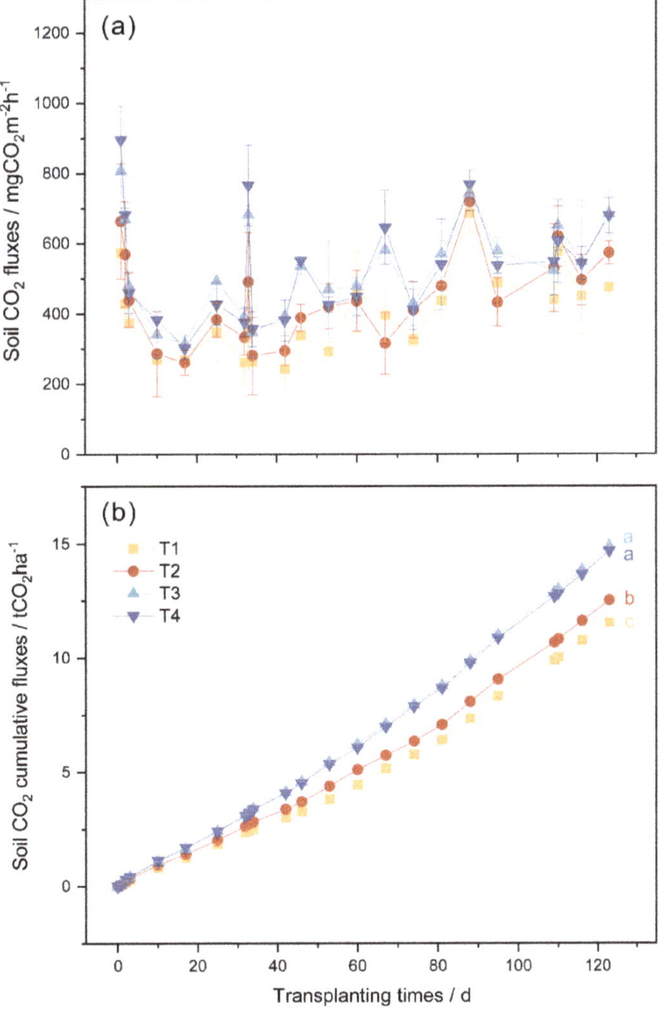

Figure 2. Changes in soil CO_2 flux (**a**) and cumulative CO_2 flux (**b**) for different treatments during the roasted tobacco growth cycle. Different letters indicate significant differences at $p < 0.05$.

Results showed that soil CO_2 fluxes varied significantly with transplanting time for the treatments from T1 to T4, with average daily CO_2 emission fluxes of 397.03, 445.82, 531.83, and 533.03 mg $CO_2 \cdot m^{-2} \cdot h^{-1}$ at 120 days after transplanting. Meanwhile, the average daily CO_2 emission fluxes at 30 days after transplanting were 360.21, 418.76, 499.29, and 503.70 mg $CO_2 \cdot m^{-2} \cdot h^{-1}$ for treatments from T1 to T4, respectively (Figure 2b). For the majority of the roasted tobacco growth cycle, CO_2 emissions from soils with the application of BC and inhibitors were significantly higher than the control whilst those from soils with BC and inhibitors were occasionally lower than those from the control. No significant difference was observed in average daily CO_2 emissions between T3 and T4 treatments. Compared to the control, the average daily soil CO_2 fluxes were substantially larger in T2, T3, and T4 treatments, increasing by 12.29%, 33.95%, and 34.25%, respectively ($p < 0.05$). Cumulative CO_2 emissions from the control (11.54 ± 0.00 t $C \cdot ha^{-1}$) were significantly lower than those from T2 (12.51 ± 0.00 t $C \cdot ha^{-1}$), T3 (14.89 ± 0.00 t $C \cdot ha^{-1}$), and T4 (14.68 ± 0.00 t $C \cdot ha^{-1}$) treatments. However, no significant difference in cumulative CO_2 emissions was observed between T3 and T4 treatments, suggesting that the combined effect of BC and MHPP was not significantly different from that of BC and NBPT.

Correlation analysis results revealed a strong and positive correlation between soil CO_2 emission fluxes and soil pH but not soil EC. Soil pH is considered to be the important factor influencing CO_2 emissions, and the combined application of BC and inhibitors significantly enhanced soil pH and increased soil CO_2 emission fluxes [52]. This finding is also consistent with that of Li et al. (2021), wherein the combined use of BC and inhibitors raised soil pH, and soil CO_2 emission fluxes were associated with changes in soil properties [53]. Sheng et al. (2016) also found that the carbon sequestration potential of BC in soils decreased with soil pH, especially in the short term [54]. Notably, soil CO_2 emission fluxes are also related to soil temperature and moisture. Significantly negative correlations existed between soil CO_2 emission fluxes and SVWC (Figure 3a). The linear function described a variation in CO_2 fluxes from 25.0% to 36.0%, with larger correlation coefficients at 20 cm depth than at 5 and 10 cm depth. In addition, soil CO_2 fluxes were significantly positively correlated with soil temperature at different soil depths, with the linear function explaining 39.7% to 51.8% of the variation in CO_2 fluxes. Moreover, the correlation coefficients for 20 cm depth were larger than those for 5 and 10 cm depths. The exponential–exponential function accounted for 50.6% to 70.1%, 61.1% to 78.8%, and 78.6% to 83.3% of the variance in soil respiration at the 5, 10, and 20 cm soil depths, respectively, when soil CO_2 emission fluxes were evaluated by soil temperature and moisture (Table 3). R^2 values were larger in the 5 and 10 cm soil layers for T3 and T4 treatments than for T1 and T2. This condition indicates that an exponential model effectively explained soil respiration by soil temperature and moisture after the application of BC and inhibitors. Furthermore, the association of soil respiration with soil temperature and moisture in the 20 cm soil layer did not differ significantly between the treatments. Overall, soil moisture and temperature conditions may be the main variables influencing soil CO_2 emissions from agroecosystems in karst areas [38,55].

Table 3. Exponential–exponential functions of soil respiration with soil moisture and temperature for different treatments at various soil depths after the application of BC, MHPP, and NBPT.

Treatments	Soil Depths		
	5 cm	10 cm	20 cm
T1	$F = 199.14e^{-0.023W}e^{0.040T}$ ($R^2 = 0.506$)	$F = 196.96e^{-0.022W}e^{0.045T}$ ($R^2 = 0.611$)	$F = 197.55e^{-0.024W}e^{0.054T}$ ($R^2 = 0.826$)
T2	$F = 180.91e^{-0.020W}e^{0.049T}$ ($R^2 = 0.517$)	$F = 108.09e^{-0.019W}e^{0.076T}$ ($R^2 = 0.727$)	$F = 83.60e^{-0.013W}e^{0.087T}$ ($R^2 = 0.836$)
T3	$F = 136.46e^{-0.012W}e^{0.063T}$ ($R^2 = 0.701$)	$F = 106.27e^{-0.010W}e^{0.077T}$ ($R^2 = 0.788$)	$F = 88.77e^{-0.005W}e^{0.084T}$ ($R^2 = 0.786$)
T4	$F = 147.53e^{-0.012W}e^{0.059T}$ ($R^2 = 0.689$)	$F = 105.32e^{-0.010W}e^{0.076T}$ ($R^2 = 0.763$)	$F = 90.74e^{-0.011W}e^{0.088T}$ ($R^2 = 0.835$)

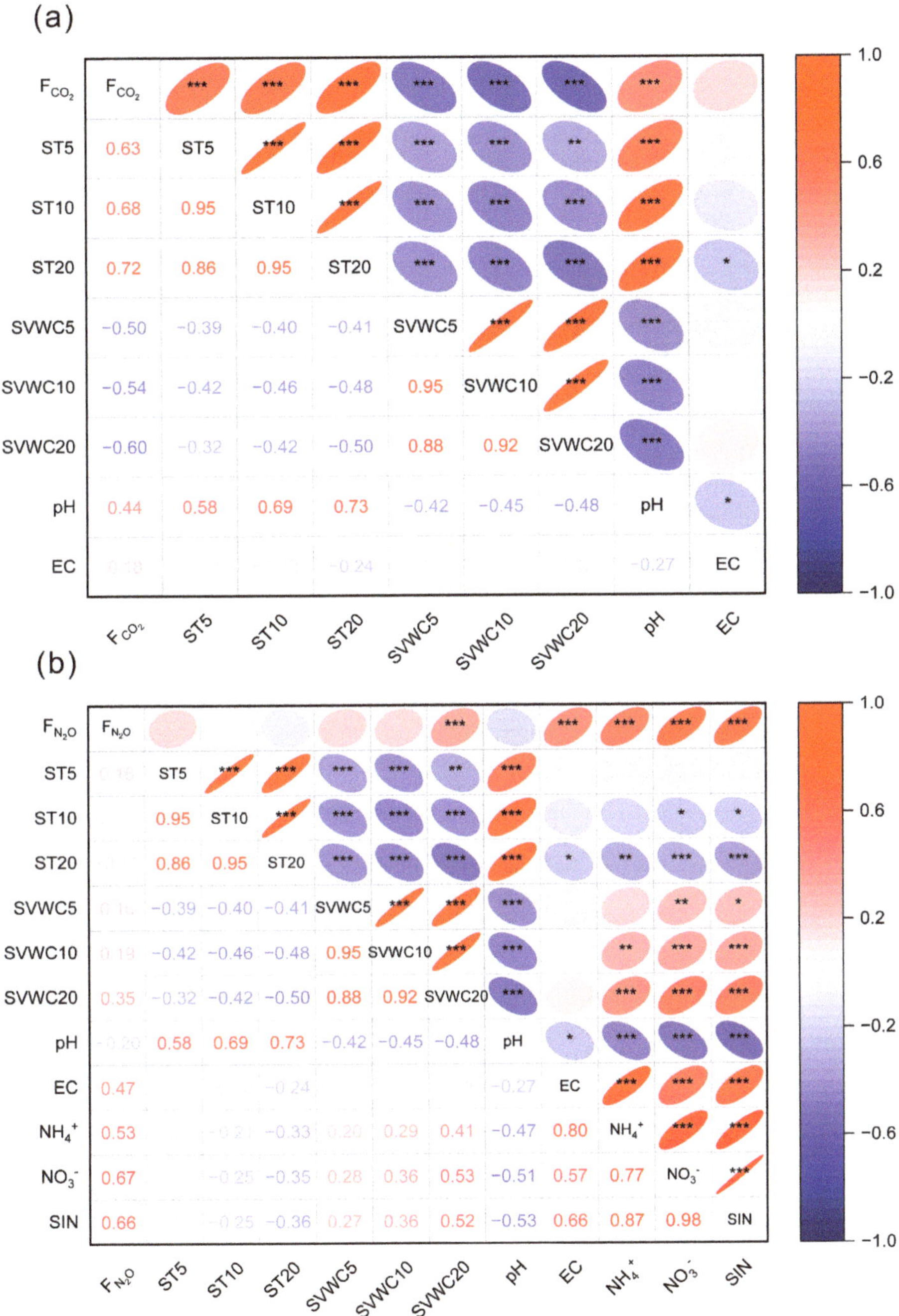

Figure 3. Correlation analysis of soil CO_2 (**a**) and N_2O (**b**) fluxes with environmental factors. FCO_2 and FN_2O represent the fluxes of CO_2 and N_2O, respectively. SIN: soil inorganic nitrogen. Asterisks indicate statistical significance with significance levels of * $p \leq 0.05$, ** $p \leq 0.01$, and *** $p \leq 0.001$ for *p* values.

3.3. Effects of BC, MHPP, and NBPT on Soil CH$_4$ Emission

Throughout the roasted tobacco growth cycle, CH$_4$ fluxes in all field treatments showed similar seasonal variations, demonstrating the occurrence of positive fluxes (showing release) and negative fluxes (showing uptake) (Figure 4a). Linear regression analysis revealed that soil CH$_4$ fluxes were positively correlated with SVWC, particularly at 5 cm (Figure 5). Soil was converted from a CH$_4$ sink to a CH$_4$ source by increasing soil moisture, whilst the combination of BC and inhibitors only enhanced or diminished the CH$_4$ source sink effect under different soil moisture conditions. Under high soil moisture conditions, additional anaerobic zones are formed in the soil, and methanogenic bacteria are active. Simultaneously, the organic carbon added by BC provides a rich and effective substrate for methanogenic bacteria; thus, CH$_4$ emissions are high [56]. On the contrary, under low soil moisture conditions, the anaerobic zone is small, and the oxidative zone is large. Methane-oxidising bacteria are remarkably active, and most of the CH$_4$ produced is oxidised by methane-oxidising bacteria, even absorbing CH$_4$, creating a negative flux [57]. In this study, the mean daily CH$_4$ emission fluxes at 120 days after transplanting for the treatments from T1, T2, T3, and T4 were -0.88, 0.04, -2.02 and 2.48 µg CH$_4 \cdot$m$^{-2} \cdot$h^{-1}, respectively. Regarding cumulative soil CH$_4$ emissions, compared to the control with $66.16 + 20.05$ g CH$_4 \cdot$ha^{-1}, T4 and T2 treatments were the highest at 120.38 ± 21.20 and 105.34 ± 17.40 g CH$_4 \cdot$ha^{-1}, respectively, followed by T3 at 20.79 ± 19.27 g CH$_4 \cdot$ha^{-1}, indicating that fertiliser application significantly increased cumulative soil CH$_4$ emissions ($p < 0.05$, Figure 4b). In addition, a significant difference in cumulative soil CH$_4$ emissions was observed between the T3 and T4 treatments. The plots with BC and NBPT had a 14.28% increase in mean cumulative soil CH$_4$ emissions; however, the plots with BC and MHPP had a significant 80.26% reduction. Thus, the combined effect of BC and inhibitors had differences in influencing CH$_4$ emissions.

On the one hand, BC application was found to increase soil CH$_4$ uptake significantly [58]. However, Yu et al. (2013) discovered that the effect of BC application on CH$_4$ emissions depended on soil moisture, reducing CH$_4$ emissions at low and medium soil moisture levels but stimulating CH$_4$ production at the highest soil moisture levels [56]. On the other hand, inhibitors had contradictory effects on CH$_4$ emissions, including inhibition [59], promotion [60], and no impact [61]. In addition, for T2, the high ammonium content during fertiliser hydrolysis may have increased the abundance of methanogenic bacteria, which may have indirectly promoted soil CH$_4$ production when C was taken up by microorganisms as a substrate [28]. Similar to the findings of Li et al. (2009), the combined application of BC and MHPP reduced CH$_4$ emissions by 80.26% compared to T2 [59]. An increase in root biomass was observed for T3 treatment, which is consistent with the results of Xu et al. (2002) [62]. An increase in crop root biomass can markedly raise oxygen availability at the inter-root level [63], which, in turn, inhibits methanogenic activity [64,65] and increases methanotroph activity [66,67]. Therefore, the combined application of BC and MHPP in this research likely inhibited CH$_4$ generation and increased CH$_4$ oxidation by raising crop biomass and oxygen availability, resulting in a decrease in CH$_4$ emissions. By contrast, the effects of BC and NBPT may have been different. NBPT had a considerable inhibitory effect on CH$_4$ oxidation in T4 due to the high ammonium retention after fertiliser application, resulting in an increase in nitrification relative to methane-oxidising bacteria and an overall decrease in CH$_4$ oxidation. At this time, nitrification is less efficient than CH$_4$ oxidation by methane-oxidising bacteria [68]. Consequently, the combined use of BC and NBPT may enhance CH$_4$ emissions, mostly due to their positive effects on crop production, increasing C input to the soil through BC and root secretions [69]. The net impact of BC in combination with MHPP or NBPT on CH$_4$ emissions depends mainly on the final outcome of their synthesis, oxidation, and transport processes.

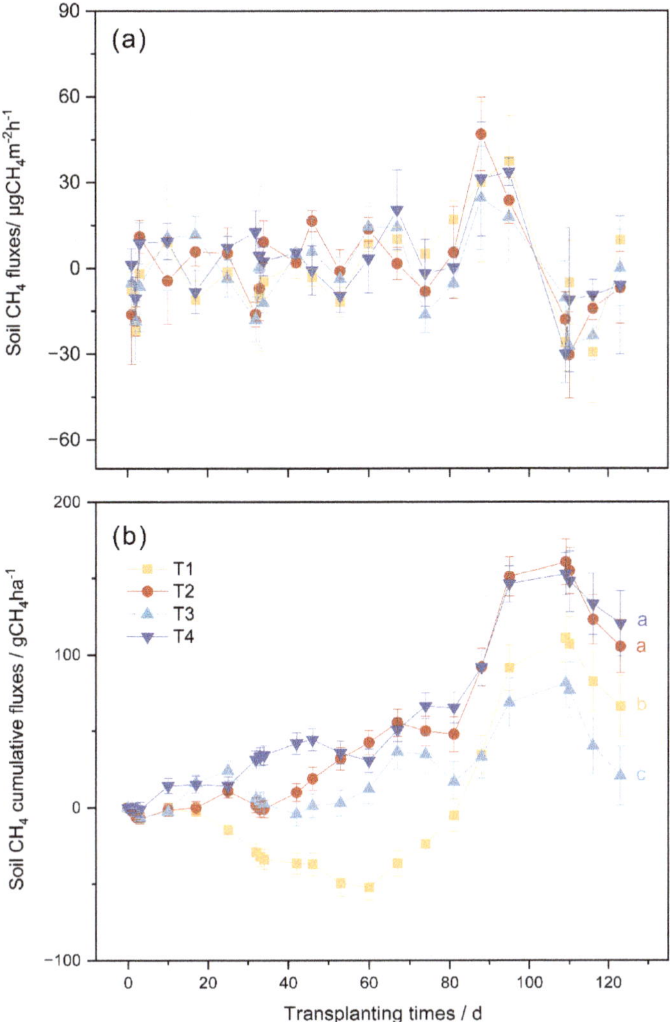

Figure 4. Changes in soil CH$_4$ flux (**a**) and cumulative CH$_4$ flux (**b**) for different treatments during the roasted tobacco growth cycle. Different letters indicate significant differences at $p < 0.05$.

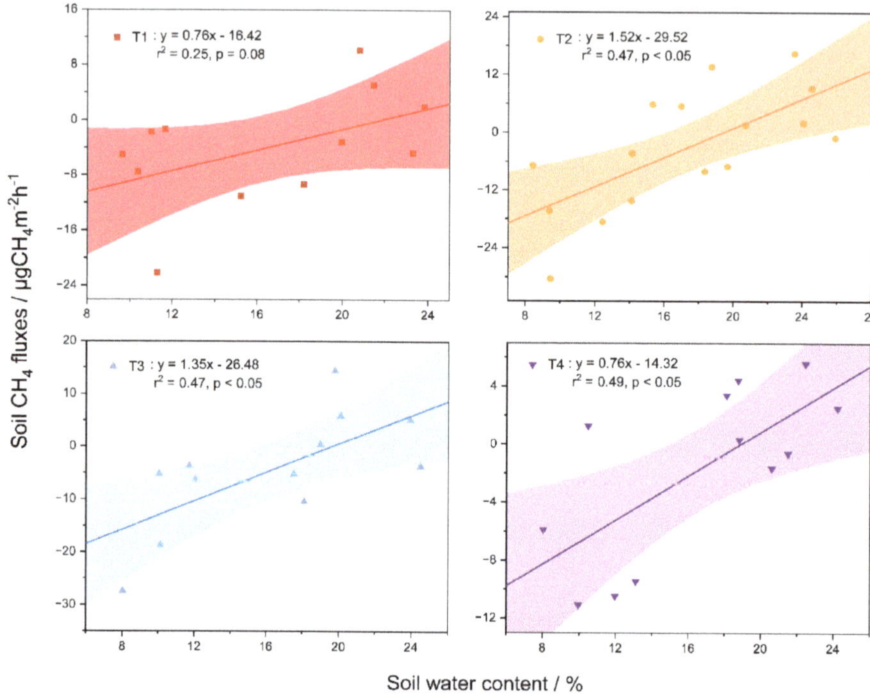

Figure 5. Linear regression of soil CH_4 emission fluxes against SVWC for each field treatment during the roasted tobacco growing cycle. The coloured areas indicate the 95% confidence interval of the regression line.

3.4. Effects of BC, MHPP, and NBPT on Soil N_2O Emission

Notably, a trade-off connection was discovered between soil N_2O emissions and the existence of BC, MHPP, and NBPT. The combination of BC and MHPP revealed that NBPT reduced soil N_2O emissions compared to T2, which is consistent with past field observations in non-karst areas [70,71]. Seasonal fluctuations in soil N_2O fluxes throughout the roasted tobacco growth cycle depended mainly on N fertiliser application and were also influenced by SVWC (Figure 6a). The mean daily N_2O emission fluxes for treatments from T1, T2, T3, and T4 were 54.98, 350.26, 274.77, and 232.45 µg $N_2O \cdot m^{-2} \cdot h^{-1}$, respectively, at 120 days after transplanting. Cumulative soil N_2O emissions for T2, T3, and T4 treatments were 7.72 ± 0.00, 5.67 ± 0.00, and 4.58 ± 0.00 kg $N_2O \cdot ha^{-1}$, which were 6.49, 4.50, and 3.45 times higher than the control, respectively (Figure 6b). Moreover, cumulative soil N_2O emission fluxes were significantly reduced by 26.55% and 40.67% for the combination of BC and MHPP or NBPT, respectively, compared to the plots with fertiliser application. The combined effect of BC and NBPT was more effective than the combination of BC and MHPP in suppressing soil N_2O emissions. In addition, correlation analysis revealed a positive connection between soil N_2O fluxes and 20 cm SVWC but not significantly with soil temperature. Notably, soil N_2O fluxes had significant positive correlations with EC, NH_4^+, NO_3^-, and SIN (Figure 3b). In all field treatments, soil N_2O fluxes increased with rising SIN content, and the linear function depicted the change in soil N_2O fluxes from 53.0% to 76.0% (Figure 7). Combining the effects of BC and inhibitor combination on soil physicochemical properties, BC had high N content and NBPT contained amino groups, but MHPP did not contain amino groups. Moreover, for the sample experimental site, the amount of BC applied was substantially higher than the amount of both inhibitors. Therefore, the combination of BC and inhibitors significantly increased TN. The BC and

NBPT combination increased soil TN to a larger extent than BC and MHPP. However, NH_4^+, NO_3^-, and SIN were more effective than TN in indicating N_2O emissions. SIN contains two main components, namely, NH_4^+ and NO_3^-, which are important substrates for nitrification and denitrification, respectively. A significant correlation was also observed between the dynamics of N_2O emission fluxes and those of NH_4^+ and NO_3^- concentrations. In particular, NO_3^- and SIN had a considerable impact on controlling soil N_2O emissions in karst areas [72,73].

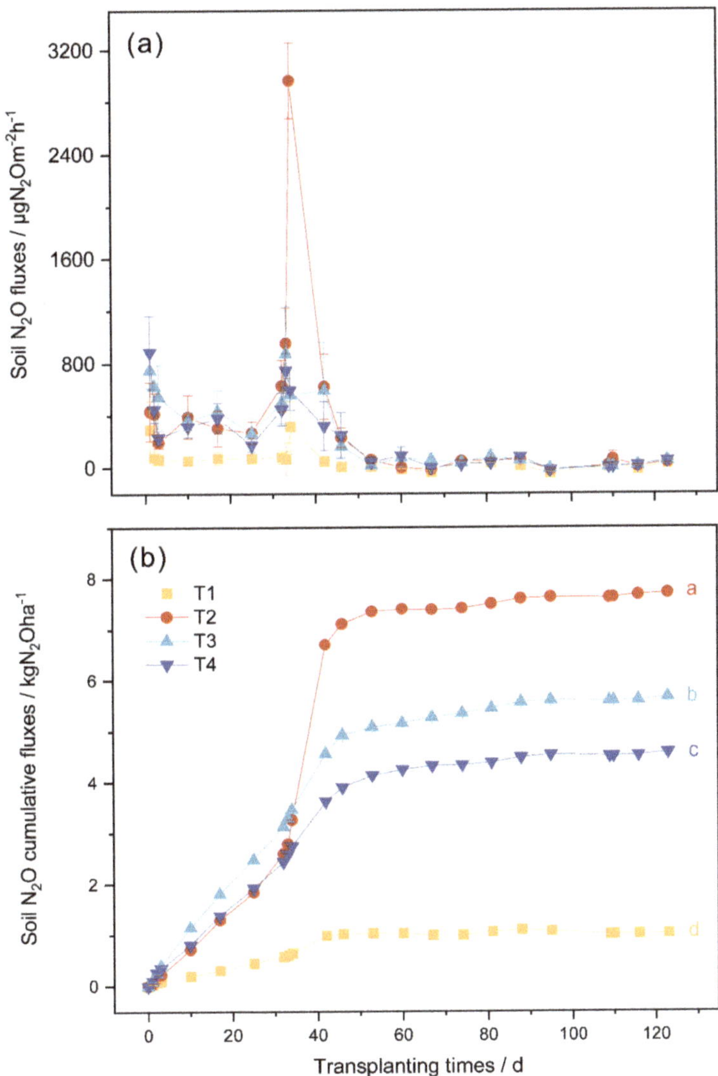

Figure 6. Changes in soil N_2O flux (**a**) and cumulative N_2O flux (**b**) for different treatments during the roasted tobacco growth cycle. Different letters indicate significant differences at $p < 0.05$.

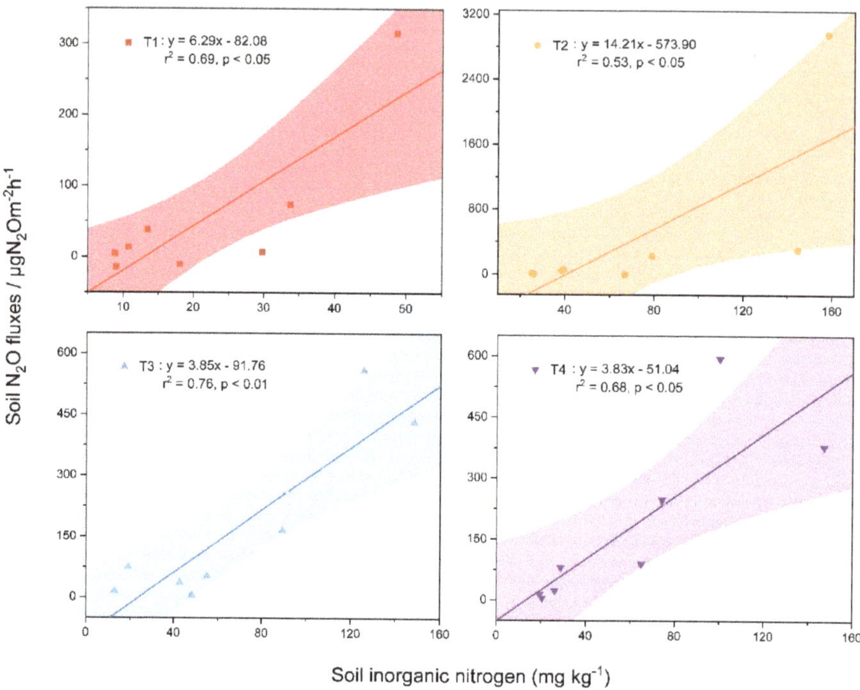

Figure 7. Linear regression of soil N_2O emission fluxes with soil inorganic N for each field treatment during the roasted tobacco growing cycle. The coloured areas indicate the 95% confidence interval of the regression line.

Urease inhibitors reduce the effectiveness of substrates for nitrification, which is dominant in agricultural dry-crop soils, by inhibiting the active site of the urease molecule and, thus, delaying urea hydrolysis [26,74]. Nitrification inhibitors directly inhibit the nitrification process in soils by limiting the oxidation of NH_4^+ to NO_2^- by ammonia-oxidising bacteria, thereby slowing the formation of NO_3^- in soils [75]. This phenomenon lowers the availability of NO_3^- to denitrifying bacteria [76]. The inhibition of soil N_2O emissions by the combined use of BC and MHPP or NBPT is evident in the unique geological background conditions of karst (Table 4). Recent studies revealed numerous reasons for the reduction in N_2O emissions due to the application of BC, MHPP, and NBPT. Shen et al. (2014) observed that BC addition increased N_2O emissions by 13–82% [77]. According to Troy et al. (2013), the stimulatory impact of BC on N_2O emissions was related to an increase in denitrification caused by BC-derived unstable organic carbon in significant concentrations [78]. Liu et al. (2019) discovered that the application of BC improved the abundance of NH_3-oxidising bacteria and accelerated nitrification, providing denitrifying bacteria with the capability to produce N_2O and NO_3^- [79]. Thus, BC addition promoted N_2O emissions, which may be due to increased nitrification and denitrification processes. Furthermore, according to the meta-analysis, the application of inhibitors reduced N_2O emissions by 33–58% [22]. Xia et al. (2017) found that N_2O emissions were reduced by 31% after the application of urease inhibitors [80], which suggested that the reduction in nitrite effectiveness with the addition of nitrification inhibitors inhibited the soil nitrification process and reduced N_2O emissions. In addition, the combined effect of BC and inhibitors may markedly reduce the supply of NO_3^-, thereby decreasing the activity of denitrifying bacteria and inhibiting N_2O formation [81]. Chen et al. (2019) discovered that the decrease in N_2O emissions under the combination of BC and inhibitors is caused by the increased abundance of nosZI, which accelerates the conversion of N_2O to N_2 [82]. Results revealed that the combined use of BC and inhibitors was highly effective in reducing N_2O emissions and boosting crop yield and NUE [31]. Li et al. (2022) found that the combined application of BC and inhibitors could further reduce soil N_2O emissions, but the addition of BC had the potential to reduce the effectiveness of inhibitors. Thus, the mitigating effect of the combination of BC and inhibitors needs further study [29].

Table 4. Cumulative N_2O flux (kg/ha) and N_2O emission factor (EF_{N2O}) (%) for the different treatments during the roasted tobacco growth cycle.

Growth Periods	T1	T2	T3	T4
	Cumulative N_2O flux (kg/ha)			
REP	0.58 ± 0.00 [c]	2.60 ± 0.00 [b]	3.13 ± 0.00 [a]	2.44 ± 0.00 [b]
VP	0.42 ± 0.00 [d]	4.79 ± 0.00 [a]	2.15 ± 0.00 [b]	1.88 ± 0.00 [c]
MP	0.03 ± 0.00 [b]	0.32 ± 0.00 [a]	0.39 ± 0.00 [a]	0.26 ± 0.00 [a]
Total	1.03 ± 0.00 [d]	7.72 ± 0.00 [a]	5.67 ± 0.00 [b]	4.58 ± 0.00 [c]
	EF_{N2O} (%)			
REP	/	1.35 ± 0.00 [b]	1.70 ± 0.00 [a]	1.24 ± 0.00 [b]
VP	/	2.08 ± 0.00 [a]	0.82 ± 0.00 [b]	0.69 ± 0.00 [b]
MP	/	0.14 ± 0.00 [a]	0.17 ± 0.00 [a]	0.11 ± 0.00 [a]
Total	/	3.17 ± 0.00 [a]	2.20 ± 0.00 [b]	1.69 ± 0.00 [c]

Different letters within a row indicate significant differences between treatments at $p < 0.05$.

EF_{N2O} varied considerably between treatments throughout the growth cycle of roasted tobacco (Table 4). EF_{N2O} varied from 0.11% to 2.08% at different growth stages, with T2 treatment having the largest EF_{N2O} in the vigorous period and T3 and T4 treatments with the largest EF_{N2O} in the root extension period. Except for T1, all three treatments applied basal fertiliser before the root extension period, and the cumulative N_2O emissions of treatments from T2 to T4 did not change significantly during the root extension period. This finding may be attributed to the partially developed root system and the insufficient root exudates, resulting in the absence of significant differences under the combined effect with microorganisms. Thus, the combined effect of BC and inhibitors was insignificant [83]. EF_{N2O} generally decreased over time. However, similar to EF_{N2O}, cumulative N_2O emissions from T2 treatment were largest at this time, which was due to secondary fertilisation during the vigorous period. Compared to T2, cumulative N_2O emissions were substantially low in T3 and T4 treatments, suggesting that the combination of BC, MHPP, and NBPT played a significant suppressive role. No significant difference in EF_{N2O} was observed between treatments during the mature period. Throughout the roasted tobacco growth cycle, treatments with the addition of BC and MHPP or NBPT had substantially lower mean EF_{N2O} with significant reductions in EF_{N2O} of 30.60% and 46.69%, respectively, compared to treatments with only N fertiliser applied. This difference was mainly observed during the vigorous period, wherein no significant difference was found during the root extension and mature periods. Furthermore, EF_{N2O} was larger in T3 treatment than in T4, suggesting that the combined effect of BC and NBPT was superior to the BC and MHPP combination considering N_2O emission inhibition [84,85].

3.5. GWP and Crop Yield Response to BC, MHPP, and NBPT

The combined effects of BC, MHPP, and NBPT on crop yields in roasted tobacco cropping systems are of concern as the demand for crop production in karst areas increases, and the accompanying changes in GWP could provide an important indication of climate change mitigation. Significant differences ($p < 0.05$) in GWP were found between treatments throughout the roasted tobacco growing cycle. The mean GWP values were 11.85, 14.81, 16.58, and 16.05 t CO_2-eq/ha for T1, T2, T3, and T4, respectively. T4 treatments had substantially larger mean GWP than T1 and T2 treatments. However, no significant difference was observed between T3 and T4 ($p < 0.05$), which was related to the contribution of CO_2, N_2O, and CH_4 to the mean GWP of the roasted tobacco growing cycle. The four treatments were dominated by the contribution of CO_2, which ranged from 84.46% to 97.39%. The contribution of N_2O was considerably reduced in T3 and T4 treatments using inhibitors compared to T2, with 10.19% and 8.51%, respectively. By contrast, the contribution of CH_4 to the mean GWP was negligible compared to CO_2.

Similar to earlier studies [86–88], the combined use of BC and inhibitors played an important role in increasing crop yield. The mean crop yield was 1.96, 3.62, 4.83, and 5.03 t for T1, T2, T3, and T4, respectively. Compared to T1, the addition of BC and inhibitors significantly increased tobacco biomass by 146.43% and 156.63% in T3 and T4 treatments, respectively. This result is comparable to the average increase ($\leq$10% to $\geq$200%) observed by meta-analysis [89,90]. No significant difference in crop yield was found between T3 and T4 treatments. By contrast, T2 treatment with fertiliser application only increased tobacco biomass by 84.69%. These results may be attributed to the high nutrient content of BC and its rapid dissolution in the soil solution, thus increasing soil fertility, especially AN, AP, AK, and SOM contents. Consequently, the inhibitor increased the effectiveness of SIN, plant N uptake, and N fertiliser. Therefore, the combination of BC and inhibitors can promote tobacco growth and increase biomass.

Notably, the yield-scaled GWP value is a key indicator [40]. Over the roasted tobacco growth cycle, mean yield-scaled GWP values were 6.04, 4.10, 3.43, and 3.19 t CO_2-eq/t for T1, T2, T3, and T4, respectively (Table 5). Compared to the fertilised treatment, the mean yield-scaled GWP values were significantly reduced by 16.34% and 22.20% for the T3 and T4 treatments, respectively, but were not significantly different. The combined use of BC and inhibitors increased mean GWP throughout the roasted tobacco growth cycle but reduced the mean yield-scaled GWP. By contrast, the latter is more important than the former [91]. The NBPT outperformed the MHPP considering mean yield-scaled GWP in roasted tobacco cropping systems. Therefore, the combined application of BC and inhibitors should be actively promoted to increase crop yield whilst reducing the yield-scaled GWP and mitigating global warming [26].

Table 5. The contribution of the GHGs, crop yield, and yield-scaled GWP for different treatments during the roasted tobacco growth cycle.

	T1	T2	T3	T4
CO_2 (kg/ha)	11,536.58 $\pm$ 0.23 [c]	12,507.42 $\pm$ 0.33 [b]	14,886.81 $\pm$ 0.64 [a]	14,683.24 $\pm$ 0.50 [a]
CH_4 (g/ha)	66.16 $\pm$ 20.05 [b]	105.34 $\pm$ 17.40 [a]	20.79 $\pm$ 19.27 [c]	120.38 $\pm$ 21.20 [a]
N_2O (g/ha)	1030.54 $\pm$ 0.02 [d]	7716.30 $\pm$ 0.07 [a]	5666.31 $\pm$ 0.04 [b]	4583.23 $\pm$ 0.03 [c]
GWP (tCO_2-eq/ha)	11.85 $\pm$ 0.00 [c]	14.81 $\pm$ 0.00 [b]	16.58 $\pm$ 0.00 [a]	16.05 $\pm$ 0.00 [a]
CO_2 contribution (%)	97.39 $\pm$ 0.00 [a]	84.46 $\pm$ 0.00 [c]	89.81 $\pm$ 0.00 [b]	91.47 $\pm$ 0.00 [b]
CH_4 contribution (%)	0.01 $\pm$ 0.00 [a]	0.02 $\pm$ 0.00 [a]	0.00 $\pm$ 0.00 [a]	0.02 $\pm$ 0.00 [a]
N_2O contribution (%)	2.59 $\pm$ 0.00 [d]	15.53 $\pm$ 0.00 [a]	10.19 $\pm$ 0.00 [b]	8.51 $\pm$ 0.00 [c]
Crop yield (t/ha)	1.96 $\pm$ 0.15 [c]	3.62 $\pm$ 0.14 [b]	4.83 $\pm$ 0.35 [a]	5.03 $\pm$ 0.24 [a]
Yield-Scaled GWP (tCO_2-eq/t)	6.04 $\pm$ 0.00 [a]	4.10 $\pm$ 0.01 [b]	3.43 $\pm$ 0.00 [c]	3.19 $\pm$ 0.00 [c]

The contribution of CO_2 refers to the contribution of GWP_{CO2} to GWP_{GHG}; that of CH_4 and N_2O is the same as above. Different letters within a row indicate significant differences between treatments at $p < 0.05$.

3.6. Effects of BC, MHPP, and NBPT on N Uptake and NUE

The net utilisation of N fertiliser currently used in the field is remarkably low and is accompanied by serious environmental problems, such as GHG emissions and N leaching. The ideal way to improve net utilisation is to use N boosters in conjunction with controlled N fertiliser use [92]. The main N boosters are MHPP and NBPT. MHPP prolongs the retention of ammonium N in the soil whilst NBPT allows urea to remain in the soil as NH_4^+ [93], thereby increasing the NUE of fertiliser and reducing the environmental impacts of N fertiliser runoff [94].

Throughout the roasted tobacco growth cycle, compared to T2 (Table 6), the combination of BC and inhibitors significantly increased N uptake by 33.52% and 39.11% in T3 and

T4 treatments, respectively. Meanwhile, the combination of BC and inhibitors significantly increased NUE by 19.85% and 23.16% in T3 and T4 treatments, respectively. By contrast, no significant difference in N uptake and NUE was observed between T3 and T4 treatments. No significant changes in NUE and N uptake were found between treatments during the root extension period. On the contrary, NUE and N uptake were substantially larger in T3 and T4 treatments compared to T2 during the vigorous and mature periods. This phenomenon may be attributed to the partially grown root system during the root extension period. Upon its entrance into the vigorous and mature period, the root system is fully developed and can produce sufficient root secretions, which can have an inhibitory effect on nitrification and benefit N uptake by the crop, thereby increasing NUE [95]. Therefore, the combined effect of BC and inhibitors can significantly increase plant N uptake and NUE.

Table 6. N uptake and NUE in different treatments during the roasted tobacco growth cycle.

Growth Periods	Treatment	Dry Weight Accumulation (kg/ha)	Amount of N Uptake (kg/ha)	NUE (%)
REP	T1	198.38 ± 29.15 [b]	6.84 ± 1.01 [b]	/
	T2	503.91 ± 27.74 [a]	17.38 ± 0.96 [a]	7.03 ± 0.03 [a]
	T3	527.21 ± 84.35 [a]	18.19 ± 2.91 [a]	7.56 ± 0.58 [a]
	T4	428.48 ± 54.22 [a]	14.78 ± 1.87 [a]	5.29 ± 0.29 [b]
VP	T1	353.03 ± 8.23 [c]	12.18 ± 0.32 [b]	/
	T2	453.03 ± 13.02 [b]	15.63 ± 0.42 [b]	1.64 ± 0.00 [b]
	T3	842.47 ± 15.12 [a]	29.07 ± 0.48 [a]	8.02 ± 0.02 [a]
	T4	847.47 ± 11.12 [a]	29.24 ± 0.40 [a]	8.10 ± 0.00 [a]
MP	T1	1408.18 ± 107.87 [c]	48.58 ± 3.72 [c]	/
	T2	2659.23 ± 100.99 [b]	91.74 ± 3.48 [b]	20.49 ± 0.03 [b]
	T3	3458.60 ± 255.84 [a]	119.32 ± 8.83 [a]	33.59 ± 0.54 [a]
	T4	3754.34 ± 176.94 [a]	129.52 ± 6.10 [a]	38.43 ± 0.25 [a]
Total	T1	1959.60 ± 149.14 [a]	67.61 ± 5.15 [c]	/
	T2	3616.16 ± 140.84 [a]	124.76 ± 4.86 [b]	27.14 ± 0.03 [b]
	T3	4828.28 ± 354.16 [a]	166.58 ± 12.22 [a]	46.99 ± 0.75 [a]
	T4	5030.30 ± 243.29 [a]	173.55 ± 8.39 [a]	50.30 ± 0.34 [a]

Different letters within a column indicate significant differences between treatments at $p < 0.05$.

A study showed that concentrations of the NH_4^+ form of mineral N were higher than NO_3^- a few days after the application of BC, resulting in increased nutrient effectiveness, N uptake, and crop yield. The retention of ammonium N in the soil due to BC addition not only provides environmental benefits through reduced N_2O emissions and NO_3^- leaching [96] but also agronomic benefits through increased NUE, especially in soils with low N [97]. N retention by BC is associated with its pore structure and microporous electrostatic interaction [98]. Furthermore, a high pyrolysis temperature leads to the large NO_3^- adsorption capacity of BC [99]; functional groups on BC may improve its NO_3^- adsorption capacity by extending its soil contact time [49,100]. N taken by BC may be gradually released for crop absorption over time [101]. Several studies have shown that BC amendments may indirectly influence plant N absorption by modifying the N conversion efficiency of the soil and positively impacting crop yields [102,103]. Concerning inhibitors, Drulis et al. (2022) found that the use of urease inhibitors reduced nutrient leaching, improved NUE, and significantly increased crop yields [104]. Moreover, MHPP may limit the activity of ammonia-oxidising archaea (AOA) and ammonia-oxidising bacteria (AOB), as well as their associated enzymes, thereby delaying the oxidation of NH_4^+ to NO_3^- and lowering NO_3^- accumulation and leaching losses whilst maintaining high soil NH_4^+ concentrations [105,106]. Thus, nitrification inhibitors can similarly significantly increase NUE and crop yield [107]. In addition, studies have shown that the combined use of BC and inhibitors performed well considering yield enhancement, efficient N use, and reduction in N losses [31,108]. He et al. (2022) discovered that the combined application of

BC and inhibitors significantly improved plant N absorption and crop yield by decreasing residual inorganic N concentrations and N_2O losses [32], which is consistent with the obtained results. He et al. (2018) found that the combined application of BC and inhibitors had no effect on NUE in the first year but increased significantly in the second year [109]. Therefore, the combined application of BC and MHPP or NBPT may be an effective practice to increase crop NUE, and this combination may lead to high fertiliser utilisation and reduce the amount of applied N fertiliser [110].

4. Conclusions

A comprehensive analysis of GHG emissions, GWP, and NUE of roasted tobacco cropping systems in the southwest karst area was conducted in this study to verify the combined effects of BC, MHPP, and NBPT. Considering soil physicochemical properties, the results showed that the combined use of BC and MHPP or NBPT significantly increased TN by 23.50% and 13.11%, respectively, whilst AK levels significantly increased by 1.95-fold and 1.23-fold compared to T2, respectively. Moreover, soil EC increased significantly by 1.56, 4.81, and 4.90 times in T2, T3, and T4 compared to T1. The combination of BC and inhibitors significantly increased the mean soil pH by 7.26% to 8.81%, whilst SOM levels were significantly increased by 44.97% and 32.95%, respectively. Considering soil GHG, the average daily soil CO_2 fluxes were substantially larger in T2, T3, and T4 treatments compared to the control, increasing by 12.29%, 33.95%, and 34.25%, respectively. The cumulative CO_2 emissions of the treatments with the addition of BC and inhibitors were significantly higher than the control by 27.21% to 29.03%. The exponential–exponential function of soil CO_2 emission fluxes with soil moisture and temperature was well fitted, and R^2 reached 0.506 to 0.836. The combination of BC and NBPT increased the cumulative soil CH_4 emissions by 14.28% compared to the fertiliser treatment. However, the combination of BC and MHPP significantly reduced the cumulative soil CH_4 emissions by 80.26%. Thus, the combined effect of BC and inhibitors on CH_4 emissions depends mainly on the results of their synthesis, oxidation, and transport processes. Furthermore, the combination of BC and MHPP or NBPT resulted in a significant reduction in cumulative soil N_2O emissions by 26.55% and 40.67%, respectively. Notably, NO_3^- and SIN are of considerable importance for limiting soil N_2O emissions in karst areas. Throughout the roasted tobacco growth cycle, the combination of BC and MHPP or NBPT significantly reduced EF_{N2O} by 30.60% and 46.69%, respectively, compared to the fertiliser treatment. With the combination of BC and inhibitors, the GWP increased significantly by 8.37% to 11.95%, with the contribution of CO_2 dominating the GWP ranging from 84.46% to 97.39%. Meanwhile, the crop yield increased significantly by 146.43% and 156.63% compared to the control. However, the mean yield-scaled GWP values were significantly reduced by 16.34% and 22.20% for T3 and T4 treatments, respectively, compared to the fertiliser treatment. In addition, N uptake was markedly enhanced by BC and inhibitor application, resulting in a significant increase in NUE of 19.85% to 23.16%, and reducing the residual inorganic N concentration and N_2O loss in the soil. Therefore, the combined use of BC and inhibitors in tobacco fields in the southwest karst region is an effective option for improving crop yield and mitigating climate change.

Author Contributions: Study conceptualization and design, T.Z., J.C. and Y.T.; acquisition of data, T.Z. and Y.T.; analysis and interpretation of data, T.Z. and J.C.; drafting of manuscript, T.Z.; critical revision, T.Z., Y.T., W.G., X.L., H.L., W.H. and J.C. All authors have read and agreed to the published version of the manuscript.

Funding: This study was funded by the Strategic Priority Research Program of the Chinese Academy of Sciences (Grant no. XDB40020201), the National Natural Science Foundation of China (42263013), the Science and Technology Program of Guizhou Province (Grant nos. [2021]187 and ZK[2022]047), the Science and Technology Project of the Guizhou Company of the China Tobacco Corporation (Grant no. 2020XM08), the Key Research and Development Program of the China Tobacco Corporation (Grant no. 110202102038), and the Science and Technology Project of the Bijie Company of Guizhou

Tobacco Corporation (Grant no. 2022520500240192). Jianzhong Cheng was supported by the "Light of West China" Program of Chinese Academy of Sciences.

Institutional Review Board Statement: Not applicable.

Informed Consent Statement: Not applicable.

Data Availability Statement: Not applicable.

Conflicts of Interest: The authors declare no conflict of interest.

References

1. Melillo, J.M.; Steudler, P.A.; Aber, J.D.; Newkirk, K.; Lux, H.; Bowles, F.P.; Catricala, C.; Magill, A.; Ahrens, T.; Morrisseau, S. Soil Warming and Carbon-Cycle Feedbacks to the Climate System. *Science* **2002**, *298*, 2173–2176. [CrossRef] [PubMed]
2. Rhodes, C.J. World Meteorological Organisation (WMO) Report: Global Greenhouse Gas Concentrations Highest in 800,000 Years. *Sci. Prog.* **2017**, *100*, 428–429. [CrossRef] [PubMed]
3. Saarnio, S.; Heimonen, K.; Kettunen, R. Biochar addition indirectly affects N2O emissions via soil moisture and plant N uptake. *Soil Biol. Biochem.* **2013**, *58*, 99–106. [CrossRef]
4. Ni, K.; Pacholski, A.; Kage, H. Ammonia volatilization after application of urea to winter wheat over 3 years affected by novel urease and nitrification inhibitors. *Agric. Ecosyst. Environ.* **2014**, *197*, 184–194. [CrossRef]
5. Lehmann, J. A handful of carbon. *Nature* **2007**, *447*, 143–144. [CrossRef]
6. Gul, S.; Whalen, J.K.; Thomas, B.W.; Sachdeva, V.; Deng, H. Physico-chemical properties and microbial responses in biochar-amended soils: Mechanisms and future directions. *Agric. Ecosyst. Environ.* **2015**, *206*, 46–59. [CrossRef]
7. Wang, J.; Xiong, Z.; Kuzyakov, Y. Biochar stability in soil: Meta-analysis of decomposition and priming effects. *GCB Bioenergy* **2016**, *8*, 512–523. [CrossRef]
8. Lorenz, K.; Lal, R. Biochar application to soil for climate change mitigation by soil organic carbon sequestration. *J. Plant Nutr. Soil Sci.* **2014**, *177*, 651–670. [CrossRef]
9. Oliveira, F.R.; Patel, A.K.; Jaisi, D.P.; Adhikari, S.; Lu, H.; Khanal, S.K. Environmental application of biochar: Current status and perspectives. *Bioresour. Technol.* **2017**, *246*, 110–122. [CrossRef]
10. Lehmann, J.; Rillig, M.C.; Thies, J.; Masiello, C.A.; Hockaday, W.C.; Crowley, D. Biochar effects on soil biota—A review. *Soil Biol. Biochem.* **2011**, *43*, 1812–1836. [CrossRef]
11. Agegnehu, G.; Bass, A.M.; Nelson, P.N.; Bird, M.I. Benefits of biochar, compost and biochar–compost for soil quality, maize yield and greenhouse gas emissions in a tropical agricultural soil. *Sci. Total. Environ.* **2016**, *543*, 295–306. [CrossRef]
12. Xie, Z.; Xu, Y.; Liu, G.; Liu, Q.; Zhu, J.; Tu, C.; Amonette, J.E.; Cadisch, G.; Yong, J.W.H.; Hu, S. Impact of biochar application on nitrogen nutrition of rice, greenhouse-gas emissions and soil organic carbon dynamics in two paddy soils of China. *Plant Soil* **2013**, *370*, 527–540. [CrossRef]
13. Zhang, A.; Cui, L.; Pan, G.; Li, L.; Hussain, Q.; Zhang, X.; Zheng, J.; Crowley, D. Effect of biochar amendment on yield and methane and nitrous oxide emissions from a rice paddy from Tai Lake plain, China. *Agric. Ecosyst. Environ.* **2010**, *139*, 469–475. [CrossRef]
14. Hawthorne, I.; Johnson, M.S.; Jassal, R.S.; Black, T.A.; Grant, N.J.; Smukler, S.M. Application of biochar and nitrogen influences fluxes of CO2, CH4 and N2O in a forest soil. *J. Environ. Manag.* **2017**, *192*, 203–214. [CrossRef]
15. He, Y.; Zhou, X.; Jiang, L.; Li, M.; Du, Z.; Zhou, G.; Shao, J.; Wang, X.; Xu, Z.; Bai, S.H.; et al. Effects of biochar application on soil greenhouse gas fluxes: A meta-analysis. *GCB Bioenergy* **2017**, *9*, 743–755. [CrossRef]
16. Wu, Z.; Zhang, X.; Dong, Y.; Li, B.; Xiong, Z. Biochar amendment reduced greenhouse gas intensities in the rice-wheat rotation system: Six-year field observation and meta-analysis. *Agric. For. Meteorol.* **2019**, *278*, 107625. [CrossRef]
17. Zhou, G.; Zhou, X.; Zhang, T.; Du, Z.; He, Y.; Wang, X.; Shao, J.; Cao, Y.; Xue, S.; Wang, H.; et al. Biochar increased soil respiration in temperate forests but had no effects in subtropical forests. *For. Ecol. Manag.* **2017**, *405*, 339–349. [CrossRef]
18. Majumdar, D.; Kumar, S.; Pathak, H.; Jain, M.; Kumar, U. Reducing nitrous oxide emission from an irrigated rice field of North India with nitrification inhibitors. *Agric. Ecosyst. Environ.* **2000**, *81*, 163–169. [CrossRef]
19. Sun, H.; Zhang, H.; Powlson, D.; Min, J.; Shi, W. Rice production, nitrous oxide emission and ammonia volatilization as impacted by the nitrification inhibitor 2-chloro-6-(trichloromethyl)-pyridine. *Field Crops Res.* **2015**, *173*, 1–7. [CrossRef]
20. Dawar, K.; Zaman, M.; Rowarth, J.; Blennerhassett, J.; Turnbull, M. Urease inhibitor reduces N losses and improves plant-bioavailability of urea applied in fine particle and granular forms under field conditions. *Agric. Ecosyst. Environ.* **2011**, *144*, 41–50. [CrossRef]
21. Zaman, M.; Nguyen, M.L.; Blennerhassett, J.D.; Quin, B.F. Reducing NH3, N2O and NO3—N losses from a pasture soil with urease or nitrification inhibitors and elemental S-amended nitrogenous fertilizers. *Biol. Fertil. Soils* **2007**, *44*, 693–705. [CrossRef]
22. Qiao, C.; Liu, L.; Hu, S.; Compton, J.E.; Greaver, T.L.; Li, Q. How inhibiting nitrification affects nitrogen cycle and reduces environmental impacts of anthropogenic nitrogen input. *Glob. Chang. Biol.* **2015**, *21*, 1249–1257. [CrossRef] [PubMed]
23. Lam, S.K.; Suter, H.; Bai, M.; Walker, C.; Davies, R.; Mosier, A.R.; Chen, D. Using urease and nitrification inhibitors to decrease ammonia and nitrous oxide emissions and improve productivity in a subtropical pasture. *Sci. Total. Environ.* **2018**, *644*, 1531–1535. [CrossRef] [PubMed]

24. Ghosh, S.; Majumdar, D.; Jain, M. Methane and nitrous oxide emissions from an irrigated rice of North India. *Chemosphere* **2003**, *51*, 181–195. [CrossRef]
25. Malla, G.; Bhatia, A.; Pathak, H.; Prasad, S.; Jain, N.; Singh, J. Mitigating nitrous oxide and methane emissions from soil in rice–wheat system of the Indo-Gangetic plain with nitrification and urease inhibitors. *Chemosphere* **2005**, *58*, 141–147. [CrossRef]
26. Bharati, K.; Mohanty, S.R.; Padmavathi, P.; Rao, V.R.; Adhya, T.K. Influence of Six Nitrification Inhibitors on Methane Production in a Flooded Alluvial Soil. *Nutr. Cycl. Agroecosyst.* **2000**, *58*, 389–394. [CrossRef]
27. Li, X.; Zhang, X.; Xu, H.; Cai, Z.; Yagi, K. Methane and nitrous oxide emissions from rice paddy soil as influenced by timing of application of hydroquinone and dicyandiamide. *Nutr. Cycl. Agroecosyst.* **2009**, *85*, 31–40. [CrossRef]
28. Folina, A.; Tataridas, A.; Mavroeidis, A.; Kousta, A.; Katsenios, N.; Efthimiadou, A.; Travlos, I.; Roussis, I.; Darawsheh, M.; Papastylianou, P.; et al. Evaluation of Various Nitrogen Indices in N-Fertilizers with Inhibitors in Field Crops: A Review. *Agronomy* **2021**, *11*, 418. [CrossRef]
29. Li, Z.; Xu, P.; Han, Z.; Wu, J.; Bo, X.; Wang, J.; Zou, J. Effect of biochar and DMPP application alone or in combination on nitrous oxide emissions differed by soil types. *Biol. Fertil. Soils* **2023**, *59*, 123–138. [CrossRef]
30. He, T.; Yuan, J.; Luo, J.; Lindsey, S.; Xiang, J.; Lin, Y.; Liu, D.; Chen, Z.; Ding, W. Combined application of biochar with urease and nitrification inhibitors have synergistic effects on mitigating CH4 emissions in rice field: A three-year study. *Sci. Total. Environ.* **2020**, *743*, 140500. [CrossRef]
31. He, T.; Liu, D.; Yuan, J.; Luo, J.; Lindsey, S.; Bolan, N.; Ding, W. Effects of application of inhibitors and biochar to fertilizer on gaseous nitrogen emissions from an intensively managed wheat field. *Sci. Total. Environ.* **2018**, *628–629*, 121–130. [CrossRef]
32. He, T.; Yuan, J.; Xiang, J.; Lin, Y.; Luo, J.; Lindsey, S.; Liao, X.; Liu, D.; Ding, W. Combined biochar and double inhibitor application offsets NH$_3$ and N$_2$O emissions and mitigates N leaching in paddy fields. *Environ. Pollut.* **2022**, *292*, 118344. [CrossRef]
33. Wu, H.; Liu, Q.; Ma, J.; Liu, L.; Qu, Y.; Gong, Y.; Yang, S.; Luo, T. Heavy Metal(loids) in typical Chinese tobacco-growing soils: Concentrations, influence factors and potential health risks. *Chemosphere* **2020**, *245*, 125591. [CrossRef]
34. Li, D.; Wen, L.; Yang, L.; Luo, P.; Xiao, K.; Chen, H.; Zhang, W.; He, X.; Chen, H.; Wang, K. Dynamics of soil organic carbon and nitrogen following agricultural abandonment in a karst region. *J. Geophys. Res. Biogeosci.* **2017**, *122*, 230–242. [CrossRef]
35. Wang, Y.S.; Wang, Y.H. Quick measurement of CH$_4$, CO$_2$ and N$_2$O emissions from a short-plant ecosystem. *Adv. Atmos. Sci.* **2003**, *20*, 842–844.
36. Zou, J.; Huang, Y.; Jiang, J.; Zheng, X.; Sass, R.L. A 3-year field measurement of methane and nitrous oxide emissions from rice paddies in China: Effects of water regime, crop residue, and fertilizer application. *Glob. Biogeochem. Cycles* **2005**, *19*, 1–9. [CrossRef]
37. Kroon, P.S.; Hensen, A.; van den Bulk, W.C.M.; Jongejan, P.A.C.; Vermeulen, A. The importance of reducing the systematic error due to non-linearity in N$_2$O flux measurements by static chambers. *Nutr. Cycl. Agroecosyst.* **2008**, *82*, 175–186. [CrossRef]
38. Tang, Y.; Gao, W.; Cai, K.; Chen, Y.; Li, C.; Lee, X.; Cheng, H.; Zhang, Q.; Cheng, J. Effects of biochar amendment on soil carbon dioxide emission and carbon budget in the karst region of southwest China. *Geoderma* **2021**, *385*, 114895. [CrossRef]
39. Piva, J.T.; Dieckow, J.; Bayer, C.; Zanatta, J.A.; de Moraes, A.; Pauletti, V.; Tomazi, M.; Pergher, M. No-till reduces global warming potential in a subtropical Ferralsol. *Plant Soil* **2012**, *361*, 359–373. [CrossRef]
40. Bah, H.; Ren, X.; Wang, Y.; Tang, J.; Zhu, B. Characterizing Greenhouse Gas Emissions and Global Warming Potential of Wheat-Maize Cropping Systems in Response to Organic Amendments in Eutric Regosols, China. *Atmosphere* **2020**, *11*, 614. [CrossRef]
41. Cheng, J.; Lee, X.; Gao, W.; Chen, Y.; Pan, W.; Tang, Y. Effect of biochar on the bioavailability of difenoconazole and microbial community composition in a pesticide-contaminated soil. *Appl. Soil Ecol.* **2017**, *121*, 185–192. [CrossRef]
42. Wang, S.; Chen, H.Y.H.; Tan, Y.; Fan, H.; Ruan, H. Fertilizer regime impacts on abundance and diversity of soil fauna across a poplar plantation chronosequence in coastal Eastern China. *Sci. Rep.* **2016**, *6*, 20816. [CrossRef] [PubMed]
43. Chen, J.; Chen, Z.; Ai, Y.; Xiao, J.; Pan, D.; Li, W.; Huang, Z.; Wang, Y. Impact of Soil Composition and Electrochemistry on Corrosion of Rock-cut Slope Nets along Railway Lines in China. *Sci. Rep.* **2015**, *5*, 14939. [CrossRef] [PubMed]
44. Dawar, K.; Rahman, S.U.; Fahad, S.; Alam, S.S.; Alam Khan, S.; Dawar, A.; Younis, U.; Danish, S.; Datta, R.; Dick, R.P. Influence of variable biochar concentration on yield-scaled nitrous oxide emissions, Wheat yield and nitrogen use efficiency. *Sci. Rep.* **2021**, *11*, 16774. [CrossRef] [PubMed]
45. Cai, Y.; Akiyama, H. Effects of inhibitors and biochar on nitrous oxide emissions, nitrate leaching, and plant nitrogen uptake from urine patches of grazing animals on grasslands: A meta-analysis. *Soil Sci. Plant Nutr.* **2017**, *63*, 405–414. [CrossRef]
46. Singh, H.; Northup, B.K.; Rice, C.W.; Prasad, P.V.V. Biochar applications influence soil physical and chemical properties, microbial diversity, and crop productivity: A meta-analysis. *Biochar* **2022**, *4*, 8. [CrossRef]
47. Deng, B.-L.; Wang, S.-L.; Xu, X.-T.; Wang, H.; Hu, D.-N.; Guo, X.-M.; Shi, Q.-H.; Siemann, E.; Zhang, L. Effects of biochar and dicyandiamide combination on nitrous oxide emissions from Camellia oleifera field soil. *Environ. Sci. Pollut. Res.* **2019**, *26*, 4070–4077. [CrossRef]
48. Kannan, P.; Paramasivan, M.; Marimuthu, S.; Swaminathan, C.; Bose, J. Applying both biochar and phosphobacteria enhances Vigna mungo L. growth and yield in acid soils by increasing soil pH, moisture content, microbial growth and P availability. *Agric. Ecosyst. Environ.* **2021**, *308*, 107258. [CrossRef]
49. Haider, G.; Steffens, D.; Moser, G.; Müller, C.; Kammann, C.I. Biochar reduced nitrate leaching and improved soil moisture content without yield improvements in a four-year field study. *Agric. Ecosyst. Environ.* **2017**, *237*, 80–94. [CrossRef]

50. Sagrilo, E.; Jeffery, S.; Hoffland, E.; Kuyper, T.W. Emission of CO_2 from biochar-amended soils and implications for soil organic carbon. *GCB Bioenergy* **2015**, *7*, 1294–1304. [CrossRef]
51. Fidel, R.B.; Laird, D.A.; Parkin, T.B. Impact of Biochar Organic and Inorganic Carbon on Soil CO_2 and N_2O Emissions. *J. Environ. Qual.* **2017**, *46*, 505–513. [CrossRef]
52. Sheng, Y.; Zhu, L. Biochar alters microbial community and carbon sequestration potential across different soil pH. *Sci. Total. Environ.* **2018**, *622–623*, 1391–1399. [CrossRef]
53. Li, J.; Kwak, J.-H.; Chang, S.; Gong, X.; An, Z.; Chen, J. Greenhouse Gas Emissions from Forest Soils Reduced by Straw Biochar and Nitrapyrin Applications. *Land* **2021**, *10*, 189. [CrossRef]
54. Sheng, Y.; Zhan, Y.; Zhu, L. Reduced carbon sequestration potential of biochar in acidic soil. *Sci. Total. Environ.* **2016**, *572*, 129–137. [CrossRef]
55. Yerli, C.; Cakmakci, T.; Sahin, U. CO_2 emissions and their changes with H_2O emissions, soil moisture, and temperature during the wetting–drying process of the soil mixed with different biochar materials. *J. Water Clim. Chang.* **2022**, *13*, 4273–4282. [CrossRef]
56. Yu, L.; Tang, J.; Zhang, R.; Wu, Q.; Gong, M. Effects of biochar application on soil methane emission at different soil moisture levels. *Biol. Fertil. Soils* **2013**, *49*, 119–128. [CrossRef]
57. Walkiewicz, A.; Kalinichenko, K.; Kubaczyński, A.; Brzezińska, M.; Bieganowski, A. Usage of biochar for mitigation of CO_2 emission and enhancement of CH_4 consumption in forest and orchard Haplic Luvisol (Siltic) soils. *Appl. Soil Ecol.* **2020**, *156*, 103711. [CrossRef]
58. Wang, E.; Yuan, N.; Lv, S.; Tang, X.; Wang, G.; Wu, L.; Zhou, Y.; Zhou, G.; Shi, Y.; Xu, L. Biochar-Based Fertilizer Decreased Soil N_2O Emission and Increased Soil CH_4 Uptake in a Subtropical Typical Bamboo Plantation. *Forests* **2022**, *13*, 2181. [CrossRef]
59. Li, X.; Zhang, G.; Xu, H.; Cai, Z.; Yagi, K. Effect of timing of joint application of hydroquinone and dicyandiamide on nitrous oxide emission from irrigated lowland rice paddy field. *Chemosphere* **2009**, *75*, 1417–1422. [CrossRef]
60. Datta, A.; Adhya, T. Effects of organic nitrification inhibitors on methane and nitrous oxide emission from tropical rice paddy. *Atmos. Environ.* **2014**, *92*, 533–545. [CrossRef]
61. Jumadi, O.; Hartono, H.; Masniawati, A.; Iriany, R.N.; Makkulawu, A.T.; Inubushi, K. Emissions of nitrous oxide and methane from rice field after granulated urea application with nitrification inhibitors and zeolite under different water managements. *Paddy Water Environ.* **2019**, *17*, 715–724. [CrossRef]
62. Xu, X.; Boeckx, P.; Van Cleemput, O.; Zhou, L. Urease and nitrification inhibitors to reduce emissions of CH_4 and N_2O in rice production. *Nutr. Cycl. Agroecosyst.* **2002**, *64*, 203–211. [CrossRef]
63. Jiang, Y.; Qian, H.; Huang, S.; Zhang, X.; Wang, L.; Zhang, L.; Shen, M.; Xiao, X.; Chen, F.; Zhang, H.; et al. Acclimation of methane emissions from rice paddy fields to straw addition. *Sci. Adv.* **2019**, *5*, eaau9038. [CrossRef] [PubMed]
64. Jiang, Y.; van Groenigen, K.J.; Huang, S.; Hungate, B.A.; van Kessel, C.; Hu, S.; Zhang, J.; Wu, L.; Yan, X.; Wang, L.; et al. Higher yields and lower methane emissions with new rice cultivars. *Glob. Chang. Biol.* **2017**, *23*, 4728–4738. [CrossRef] [PubMed]
65. Zhu, Z.; Ge, T.; Liu, S.; Hu, Y.; Ye, R.; Xiao, M.; Tong, C.; Kuzyakov, Y.; Wu, J. Rice rhizodeposits affect organic matter priming in paddy soil: The role of N fertilization and plant growth for enzyme activities, CO_2 and CH_4 emissions. *Soil Biol. Biochem.* **2018**, *116*, 369–377. [CrossRef]
66. Watanabe, I.; Hashimoto, T.; Shimoyama, A. Methane-oxidizing activities and methanotrophic populations associated with wetland rice plants. *Biol. Fertil. Soils* **1997**, *24*, 261–265. [CrossRef]
67. Nie, T.; Huang, J.; Zhang, Z.; Chen, P.; Li, T.; Dai, C. The inhibitory effect of a water-saving irrigation regime on CH_4 emission in Mollisols under straw incorporation for 5 consecutive years. *Agric. Water Manag.* **2023**, *278*, 108163. [CrossRef]
68. Crill, P.; Martikainen, P.; Nykanen, H.; Silvola, J. Temperature and N fertilization effects on methane oxidation in a drained peatland soil. *Soil Biol. Biochem.* **1994**, *26*, 1331–1339. [CrossRef]
69. Wang, X.; Zhang, L.; Zou, J.; Liu, S. Optimizing net greenhouse gas balance of a bioenergy cropping system in southeast China with urease and nitrification inhibitors. *Ecol. Eng.* **2015**, *83*, 191–198. [CrossRef]
70. Xu, X.; Boeckx, P.; Wang, Y.; Huang, Y.; Zheng, X.; Hu, F.; Van Cleemput, O. Nitrous oxide and methane emissions during rice growth and through rice plants: Effect of dicyandiamide and hydroquinone. *Biol. Fertil. Soils* **2002**, *36*, 53–58. [CrossRef]
71. Jiang, J.; Hu, Z.; Sun, W.; Huang, Y. Nitrous oxide emissions from Chinese cropland fertilized with a range of slow-release nitrogen compounds. *Agric. Ecosyst. Environ.* **2010**, *135*, 216–225. [CrossRef]
72. Thangarajan, R.; Bolan, N.S.; Kunhikrishnan, A.; Wijesekara, H.; Xu, Y.; Tsang, D.C.; Song, H.; Ok, Y.S.; Hou, D. The potential value of biochar in the mitigation of gaseous emission of nitrogen. *Sci. Total. Environ.* **2018**, *612*, 257–268. [CrossRef] [PubMed]
73. Case, S.D.; McNamara, N.P.; Reay, D.S.; Stott, A.W.; Grant, H.K.; Whitaker, J. Biochar suppresses N2O emissions while maintaining N availability in a sandy loam soil. *Soil Biol. Biochem.* **2015**, *81*, 178–185. [CrossRef]
74. Manunza, B.; Deiana, S.; Pintore, M.; Gessa, C. The binding mechanism of urea, hydroxamic acid and N-(N-butyl)-phosphoric triamide to the urease active site. A comparative molecular dynamics study. *Soil Biol. Biochem.* **1999**, *31*, 789–796. [CrossRef]
75. Zerulla, W.; Barth, T.; Dressel, J.; Erhardt, K.; von Locquenghien, K.H.; Pasda, G.; Rädle, M.; Wissemeier, A. 3,4-Dimethylpyrazole phosphate (DMPP)—A new nitrification inhibitor for agriculture and horticulture—An introduction. *Biol. Fertil. Soils* **2001**, *34*, 79–84. [CrossRef]
76. Subbarao, G.V.; Ito, O.; Sahrawat, K.L.; Berry, W.L.; Nakahara, K.; Ishikawa, T.; Watanabe, T.; Suenaga, K.; Rondon, M.; Rao, I.M. Scope and Strategies for Regulation of Nitrification in Agricultural Systems—Challenges and Opportunities. *Crit. Rev. Plant Sci.* **2006**, *25*, 303–335. [CrossRef]

77. Shen, J.; Tang, H.; Liu, J.; Wang, C.; Li, Y.; Ge, T.; Jones, D.L.; Wu, J. Contrasting effects of straw and straw-derived biochar amendments on greenhouse gas emissions within double rice cropping systems. *Agric. Ecosyst. Environ.* **2014**, *188*, 264–274. [CrossRef]

78. Troy, S.M.; Lawlor, P.G.; Flynn, C.J.O.; Healy, M.G. Impact of biochar addition to soil on greenhouse gas emissions following pig manure application. *Soil Biol. Biochem.* **2013**, *60*, 173–181. [CrossRef]

79. Liu, X.; Ren, J.; Zhang, Q.; Liu, C. Long-term effects of biochar addition and straw return on N_2O fluxes and the related functional gene abundances under wheat-maize rotation system in the North China Plain. *Appl. Soil Ecol.* **2018**, *135*, 44–55. [CrossRef]

80. Xia, L.; Lam, S.K.; Chen, D.; Wang, J.; Tang, Q.; Yan, X. Can knowledge-based N management produce more staple grain with lower greenhouse gas emission and reactive nitrogen pollution? A meta-analysis. *Glob. Chang. Biol.* **2016**, *23*, 1917–1925. [CrossRef]

81. Cavigelli, M.; Robertson, G. Role of denitrifier diversity in rates of nitrous oxide consumption in a terrestrial ecosystem. *Soil Biol. Biochem.* **2001**, *33*, 297–310. [CrossRef]

82. Chen, H.; Yin, C.; Fan, X.; Ye, M.; Peng, H.; Li, T.; Zhao, Y.; Wakelin, S.A.; Chu, G.; Liang, Y. Reduction of N_2O emission by biochar and/or 3,4-dimethylpyrazole phosphate (DMPP) is closely linked to soil ammonia oxidizing bacteria and nosZI-N_2O reducer populations. *Sci. Total. Environ.* **2019**, *694*, 133658. [CrossRef]

83. Wang, X.; Bai, J.; Xie, T.; Wang, W.; Zhang, G.; Yin, S.; Wang, D. Effects of biological nitrification inhibitors on nitrogen use efficiency and greenhouse gas emissions in agricultural soils: A review. *Ecotoxicol. Environ. Saf.* **2021**, *220*, 112338. [CrossRef]

84. Park, S.H.; Lee, B.R.; Kim, T.H. Urease and nitrification inhibitors with pig slurry effects on ammonia and nitrous oxide emissions, nitrate leaching, and nitrogen use efficiency in perennial ryegrass sward. *Anim. Biosci.* **2021**, *34*, 2023–2033. [CrossRef]

85. Saud, S.; Wang, D.; Fahad, S. Improved Nitrogen Use Efficiency and Greenhouse Gas Emissions in Agricultural Soils as Producers of Biological Nitrification Inhibitors. *Front. Plant Sci.* **2022**, *13*, 854195. [CrossRef]

86. Liu, C.; Wang, K.; Zheng, X. Effects of nitrification inhibitors (DCD and DMPP) on nitrous oxide emission, crop yield and nitrogen uptake in a wheat–maize cropping system. *Biogeosciences* **2013**, *10*, 2427–2437. [CrossRef]

87. Farhangi-Abriz, S.; Torabian, S.; Qin, R.; Noulas, C.; Lu, Y.; Gao, S. Biochar effects on yield of cereal and legume crops using meta-analysis. *Sci. Total. Environ.* **2021**, *775*, 145869. [CrossRef]

88. Vitkova, J.; Kondrlova, E.; Rodny, M.; Surda, P.; Horak, J. Analysis of soil water content and crop yield after biochar application in field conditions. *Plant Soil Environ.* **2017**, *63*, 569–573. [CrossRef]

89. Jeffery, S.; Verheijen, F.G.A.; van der Velde, M.; Bastos, A.C. A quantitative review of the effects of biochar application to soils on crop productivity using meta-analysis. *Agric. Ecosyst. Environ.* **2011**, *144*, 175–187. [CrossRef]

90. Bai, S.H.; Omidvar, N.; Gallart, M.; Kämper, W.; Tahmasbian, I.; Farrar, M.B.; Singh, K.; Zhou, G.; Muqadass, B.; Xu, C.-Y.; et al. Combined effects of biochar and fertilizer applications on yield: A review and meta-analysis. *Sci. Total. Environ.* **2022**, *808*, 152073. [CrossRef]

91. Ali, M.A.; Kim, P.; Inubushi, K. Mitigating yield-scaled greenhouse gas emissions through combined application of soil amendments: A comparative study between temperate and subtropical rice paddy soils. *Sci. Total. Environ.* **2015**, *529*, 140–148. [CrossRef] [PubMed]

92. Galloway, J.N.; Townsend, A.R.; Erisman, J.W.; Bekunda, M.; Cai, Z.; Freney, J.R.; Martinelli, L.A.; Seitzinger, S.P.; Sutton, M.A. Transformation of the Nitrogen Cycle: Recent Trends, Questions, and Potential Solutions. *Science* **2008**, *320*, 889–892. [CrossRef] [PubMed]

93. Bremner, J.M. Recent research on problems in the use of urea as a nitrogen fertilizer. *Fertil. Res.* **1995**, *42*, 321–329. [CrossRef]

94. Wang, D.; Guo, L.; Zheng, L.; Zhang, Y.; Yang, R.; Li, M.; Ma, F.; Zhang, X.; Li, Y. Effects of nitrogen fertilizer and water management practices on nitrogen leaching from a typical open field used for vegetable planting in northern China. *Agric. Water Manag.* **2019**, *213*, 913–921. [CrossRef]

95. Tanaka, J.P.; Nardi, P.; Wissuwa, M. Nitrification inhibition activity, a novel trait in root exudates of rice. *AoB PLANTS* **2010**, *2010*, plq014. [CrossRef]

96. Qin, S.; Wang, Y.; Hu, C.; Oenema, O.; Li, X.; Zhang, Y.; Dong, W. Yield-scaled N2O emissions in a winter wheat–summer corn double-cropping system. *Atmos. Environ.* **2012**, *55*, 240–244. [CrossRef]

97. Dawar, K.; Khan, H.; Zaman, M.; Muller, C.; Alam, S.S.; Fahad, S.; Alwahibi, M.S.; Alkahtani, J.; Saeed, B.; Saud, S.; et al. The Effect of Biochar and Nitrogen Inhibitor on Ammonia and Nitrous Oxide Emissions and Wheat Productivity. *J. Plant Growth Regul.* **2021**, *40*, 2465–2475. [CrossRef]

98. Ye, Z.; Liu, L.; Tan, Z.; Zhang, L.; Huang, Q. Effects of pyrolysis conditions on migration and distribution of biochar nitrogen in the soil-plant-atmosphere system. *Sci. Total. Environ.* **2020**, *723*, 138006. [CrossRef]

99. Alsewaileh, A.S.; Usman, A.R.; Al-Wabel, M.I. Effects of pyrolysis temperature on nitrate-nitrogen ($NO_3{}^-$-N) and bromate ($BrO_3{}^-$) adsorption onto date palm biochar. *J. Environ. Manag.* **2019**, *237*, 289–296. [CrossRef]

100. He, L.; Shan, J.; Zhao, X.; Wang, S.; Yan, X. Variable responses of nitrification and denitrification in a paddy soil to long-term biochar amendment and short-term biochar addition. *Chemosphere* **2019**, *234*, 558–567. [CrossRef]

101. Taghizadeh-Toosi, A.; Clough, T.J.; Sherlock, R.R.; Condron, L.M. Biochar adsorbed ammonia is bioavailable. *Plant Soil* **2012**, *350*, 57–69. [CrossRef]

102. Rajkovich, S.; Enders, A.; Hanley, K.; Hyland, C.; Zimmerman, A.R.; Lehmann, J. Corn growth and nitrogen nutrition after additions of biochars with varying properties to a temperate soil. *Biol. Fertil. Soils* **2012**, *48*, 271–284. [CrossRef]

103. Güereña, D.; Lehmann, J.; Hanley, K.; Enders, A.; Hyland, C.; Riha, S. Nitrogen dynamics following field application of biochar in a temperate North American maize-based production system. *Plant Soil* **2013**, *365*, 239–254. [CrossRef]

104. Drulis, P.; Kriaučiūnienė, Z.; Liakas, V. The Effect of Combining N-Fertilization with Urease Inhibitors and Biological Preparations on Maize Biological Productivity. *Agronomy* **2022**, *12*, 2264. [CrossRef]

105. Chen, Z.; Luo, X.; Hu, R.; Wu, M.; Wu, J.; Wei, W. Impact of Long-Term Fertilization on the Composition of Denitrifier Communities Based on Nitrite Reductase Analyses in a Paddy Soil. *Microb. Ecol.* **2010**, *60*, 850–861. [CrossRef]

106. Guo, Y.J.; Di, H.J.; Cameron, K.; Li, B.; Podolyan, A.; Moir, J.; Monaghan, R.M.; Smith, L.C.; O'Callaghan, M.; Bowatte, S.; et al. Effect of 7-year application of a nitrification inhibitor, dicyandiamide (DCD), on soil microbial biomass, protease and deaminase activities, and the abundance of bacteria and archaea in pasture soils. *J. Soils Sediments* **2013**, *13*, 753–759. [CrossRef]

107. Shoji, S.; Delgado, J.; Mosier, A.; Miura, Y. Use of Controlled Release Fertilizers and Nitrification Inhibitors to Increase Nitrogen Use Efficiency and to Conserve Air and Water Quality. *Commun. Soil Sci. Plant Anal.* **2001**, *32*, 1051–1070. [CrossRef]

108. Sun, L.; Li, J.; Fan, C.; Deng, J.; Zhou, W.; Aihemaiti, A.; Yalkun, U.J. The effects of biochar and nitrification inhibitors on reactive nitrogen gas (N_2O, NO and NH_3) emissions in intensive vegetable fields in southeastern China. *Arch. Agron. Soil Sci.* **2021**, *67*, 836–848. [CrossRef]

109. He, T.; Liu, D.; Yuan, J.; Ni, K.; Zaman, M.; Luo, J.; Lindsey, S.; Ding, W. A two years study on the combined effects of biochar and inhibitors on ammonia volatilization in an intensively managed rice field. *Agric. Ecosyst. Environ.* **2018**, *264*, 44–53. [CrossRef]

110. Hailegnaw, N.S.; Mercl, F.; Kulhánek, M.; Száková, J.; Tlustoš, P. Co-application of high temperature biochar with 3,4-dimethylpyrazole-phosphate treated ammonium sulphate improves nitrogen use efficiency in maize. *Sci. Rep.* **2021**, *11*, 5711. [CrossRef]

 *sustainability*

Article

Developing a Sustainable Omnichannel Strategic Framework toward Circular Revolution: An Integrated Approach

Tuğba Yeğin [1] and **Muhammad Ikram** [2,*]

1   Faculty of Economics and Administrative Sciences, Demir Celik Campus, Karabuk University, 78050 Karabuk, Turkey
2   School of Business Administration, Al Akhawayn University in Ifrane, Avenue Hassan II, P.O. Box 104, Ifrane 53000, Morocco
*   Correspondence: i.muhammad@aui.ma

Abstract: One of the contributions of digitalization to cyclical change is the adoption of Omnichannel Marketing (OM) as a new marketing strategy for brands. In this research, we examined whether the quality of integration (INQ) in omnichannel environments has an effect on brand equity (BE) and its dimensions (brand loyalty (BL), brand association and brand awareness (BAS), and perceived quality (PQ)) within the framework of a structural model. We aim to expand the limited number of INQ research areas. In this context, in the first stage of our research, we conducted an online survey consisting of three parts with the consumers of the Nike luxury sportswear brand, which is in 11th place in the global brand value ranking, residing in Turkey from the developing countries. In the second stage of the analysis, we performed CFA for scale reliability and validity. Crobach's alpha, AVE and CR values for all factors of the scale exceeded the threshold values in the literature. In addition, the goodness-of-fit values of the scale, which were checked for compliance with the research, exceeded the threshold values. In the third stage of the analysis, we performed SEM analysis to test the model of the study and the assumptions of the study. The SEM results of our research confirmed the assumptions established between INQ and BE and its components in the context of OM. SEM results revealed that INQ had the highest effect (0.93) on BAS and PQ and the least effect (0.86) on BL, and INQ affected BE with 0.90. The results of this research, which examines the predictors of brand equity and its components, offer implications for OM, INQ, BE subject areas that have not been empirically analyzed despite increasing knowledge and still having limitations in theoretical information. Our research is unique, as it is the first study to empirically examine the relationship between INQ and BE and its components in the context of OM. The research on omnichannel applications is quite limited. This study brings a conceptual extension to the literature on omnichannel strategies, INQ and OM, whereas they presented the necessary reasons for managers to provide INQ in an omnichannel environment in order to increase brand equity, with an empirical application. In addition, the most important benefit of this research is that it shows brand owners and managers and brand marketers a way to set up the omnichannel system toward circular revolution.

Keywords: sustainable omnichannel; strategic framework; circular economy; consumer behavior; sustainable development; integrated channel

Citation: Yeğin, T.; Ikram, M. Developing a Sustainable Omnichannel Strategic Framework toward Circular Revolution: An Integrated Approach. *Sustainability* **2022**, *14*, 11578. https://doi.org/10.3390/su141811578

Academic Editors: Nallapaneni Manoj Kumar, Subrata Hait, Anshu Priya and Varsha Bohra

Received: 31 August 2022
Accepted: 14 September 2022
Published: 15 September 2022

Publisher's Note: MDPI stays neutral with regard to jurisdictional claims in published maps and institutional affiliations.

1. Introduction

Brand equity, competitive advantage, global recognition, and high net worth are important because they bring global recognition as a competitive advantage and increase financial value. The number of studies investigating the antecedents of brand equity as an abstract indicator is a marketing concept that needs to be reexamined according to each new marketing strategy (Oh et al., 2020; Aaker, 1991). According to [1], while Apple is the highest technology brand in 2020, Amazon (USD 200.667 billion) is in second place, followed by Microsoft (USD 166.001 billion) with Nike (USD 42.538 billion) in 11th place. Existing research on multichannel marketing suggests that brand equity can be strengthened by providing cross-channel service quality [2–4]. Although there are many multichannel studies investigating the effects of channel integration quality on brand equity (e.g., [5–15]), since the omnichannel field is very new, there are limitations, literature gaps on this subject and within the provision of INQ in omnichannel marketing. It is extremely important to investigate the changes that may occur in the brand value. Brand equity, as an intangible asset created by marketing activities, is the sum of values that increase or decrease the value of products and services offered by the company to consumers, depending on the brand's distinguishing characteristics such as name or symbol [16–18]. Brand loyalty (BL) is the repeated purchase of products from the same brand continuously and consistently in the future [19]. Perceived quality (PQ) is the perception formed in the minds of consumers regarding the quality of the brand's products [20]. The Brand Association (BA) states that it is the unique characteristics (in the minds of consumers) that distinguish a brand from other brands [21]. Brand awareness (BAW) is the consumer's capability to realize, separate, identify and call to mind the product category to which the brand belongs [16,22].

The pressure of the sustainable development plan applied to prevent global warming at a global level, protect the environment, purify the air from CO_2 emissions to zero waste on brands, businesses and marketers force them to change their marketing and management strategies in harmony with technology [23–32]. This change is expected to enable the production and marketing of products that are produced, distributed, priced, and promoted with less cost, less waste, and more recyclable raw materials and green energy [23–30,33]. In this context, it is important for cyclical change that brands provide maximum service quality with minimum environmental pollution, energy loss, and time-saving in the channels where they deliver their products to their customers [23,25,29,30,33]. In this context, the solution is now omnichannel strategies and the quality of inter-channel integration that is expected to be provided in omnichannel channels.

Omnichannel marketing (OM), which provides brands with sustainable growth opportunities across many channels, is increasingly preferred by more brands, including channel harmony beyond multichannel marketing [34–37]. During 2021, 0.14% of all multi-channel marketing sales were single-channel marketing, and 83% of them were omnichannel marketing orders [38]. These comparative figures show that it is significant for marketers, brand owners, managers, retailers and to understand and apply OM well [2].

Omnichannel marketing, which offers the advantage of offering brands products to their customers through many channels, differs from multichannel marketing in that all channels are evaluated as a whole, allows customers to like a product through the brand's online channels and receive it from the physical store, or to discover the product in the physical store and buy it online [39]. However, due to the reasons arising from the lack of coordination between these channels, customers can give up and return the products they purchased. At the same time, these returns cause an increase in carbon emissions from waste products at the local and global level [39,40]. As a way to avoid the potential harms of OM, which is increasingly being adopted by consumers, the researchers recommend that brands implement marketing strategies to maximize the level of inter-channel coordination [41–43]. There is a lack of research in the literature on whether brands' INQ dimensions (Channel–Service Configuration (CSC), Content Consistency (CC), Process Consistency (PC), Assurance Quality (AQ)) provision affects the brands' intangible assets (BE, BL, BAW, BA, PQ) in omnichannel environments [39,43]. However, further investigation of these intangible assets, which the brand is trying to acquire, protect and increase with tangible and moral sacrifices in the long term, for omnichannel environments will provide both theoretical and practical contributions. In this context, despite the increasing use of OM, we aim to narrow this gap in the literature with the empirical results of this study in the context of developing countries.

As a result, the studies investigating the impact of INQ on BE and its dimensions for omnichannel are insufficient in the literature, and most of them are conceptual [43–45].

Brand equity, as an intangible asset created by marketing activities, is the sum of values that increase or decrease the value of products and services offered by the company to consumers, depending on the brand's distinguishing characteristics such as name or symbol [16,17]. BL is the repeated purchase of products from the same brand continuously and consistently in the future [19]; PQ is the perception formed in the minds of consumers regarding the quality of the brand's products [20]; BA states that it is the unique characteristics (in the minds of consumers) that distinguish a brand from other brands [17]. BAW is the consumer's capability to realize, separate, identify and call to mind the product category to which the brand belongs [16,22].

There are some cases where companies already have the infrastructure in place. Customers are encouraged to register products online by many fast-moving goods brands. Providing consumers with the ability to return products instead of scrapping them, as Apple already does, could increase the culture of recycling. Meanwhile, consumer power—driven by social media—could influence manufacturers to take circular actions [25,29]. Omnichannel service has emerged since consumers turn to brands that offer this service and desire to store across many channels simultaneously. OM studies are therefore based on consumers (consumer behavior, customer loyalty, purchasing behavior, etc.). For the last 12 years, researchers have been carried out in this examining the relations [43,46–49].

Channel synchronization enables products to be delivered across multiple channels of the same quality to consumers [36,50,51]. That is, it is essential to integrate all channels for OM to achieve its purpose [43,49,52]. INQ and factors have been discussed in previous studies as a predictor of Omnichannel strategies, customer, purchase, brand, service quality, perceived value and fluency, brand equity, customer loyalty (Table 1).

Table 1. Summary of omnichannel strategies related to past studies.

Research	Omnichannel	Customer	Purchase	Brand	Service Quality	Perceived Value	Perceived Fluency	INQ	Effect on	Conceptualizing	Moderator	Analysis
[53,54]	✓		✓						✓	Online Perceived Value and Online Purchase Intention			SEM
[54]	✓								✓	Perceived Service Quality and Perceived Risk			Diffusion Theory and Scenario Analysis
[55]	✓								✓		Determinants of Cross-Channel Integration of Retailers		Regression
[56]	✓								✓	Customer Satisfaction			
[43]	✓	✓	✓		✓	✓			✓	Cross buying intentions and perceived value			

Table 1. *Cont.*

Research	Omnichannel	Customer	Purchase	Brand	Service Quality	Perceived Value	Perceived Fluency	INQ	Effect on	Conceptualizing	Moderator	Analysis
[57]	✓	✓	✓	✓				✓	Customer engagement			PLS-SEM
[58]	✓	✓			✓		✓	✓	perceived fluency			SEM
[59]	✓	✓			✓			✓	consumers' satisfaction and loyalty			SEM
[45]	✓	✓	✓						Customer Equity			Literature Review
[42]	✓	✓			✓			✓		Service Quality Measurement for Omnichannel Retail: Scale Development and Validation		SEM
[60]	✓	✓							Customer Loyalty and Mediating Role of Customer Engagement and Relationship Program Receptiveness			PLS SEM
[41]	✓	✓			✓			✓	Customer Loyalty		Omni-Channel Shopping Experience	SEM
[61]	✓	✓			✓				Customers' Attitudinal Loyalty and Positive Affect Experience			PLS SEM

✓indicates the indicator has been used in prior studies.

However, there are aspects of this research that differ from the articles given in Table 1. First, this research is quantitative and tests assumptions with Structural Equation Model (SEM) analysis. Second, this study claims that INQ is a predictor of BE and its components. This claim was previously implicitly claimed in the 2017 qualitative study by Hossain et al., but this is the first time it has been examined with direct and quantitative analysis. As a result, the studies investigating the impact of integration quality (INQ) on brand equity (BE) and its dimensions for omnichannel are insufficient in the literature, and most of them are conceptual [43–45].

Although studies on the definition, dimensions, impacts, and strategies of OM affect the retail and service sectors in many ways, there is limited evidence in the literature on the impact of the quality of channel integration, which should be carefully considered in an omnichannel system, on brands and consumers [43,57,62]. Studies on the effects of the high level of inter-channel integration quality of the brands that perform OM that will bring value to the brand are quite insufficient. Although the compatibility of multiple channels of a brand with each other forms the basis of OM, most of the studies on OM do not focus on INQ and do not observe how the components of brand equity are affected by INQ. Statistical reports are revealing that there are some brands that have lost share value and lost sales turnover as a result of failure to integrate [12,13,43]. These financial impacts on INQ's brand are immediate, but it may take longer to see the change in consumer-based brand equity. Brand equity is more intangible than financial values, in other words, the loyalty of the customers to the brand, the degree of recognition of the brand, the perceived quality of the brand, etc. With the realization of the total change in the components and the sub-components associated with these components, a change in brand value can be seen. However, with a relational model to be established between INQ and brand equity and its components, and testing this model with empirical analysis, the effects that may occur on the value of the brand in the absence of INQ can be presented to marketers and brand managers to take precautions. Moreover, the research of [45] suggested that future studies investigating the relationship between INQ and brand equity, brand image, knowledge, and loyalty by creating a scale and a data-oriented model are recommended, which is the reason for the emergence of this research. Moreover, this research relies on the research of [43], which conceptualizes the dimensions and sub-dimensions of INQ and empirically tests the data using dynamic capability theory to incorporate brand equity dimensions [43,50,51]. In this context, in this research, we establish a model between INQ and brand equity and its components, and within the framework of this model, we make four assumptions and reduce this gap in the literature by performing an empirical analysis.

In this research, we develop a model that provides experimental proof for the conceptional dimensions, presents the relationship between INQ and BE and its sub-dimensions (BL, PQ, BA /BAW) for omnichannel, and seeks answers to the following questions:

a.   What is the relationship between INQ and BE for OM?
b.   What is the relationship between INQ and the dimensions BL, PQ, and BA /BAW for OM?

In this context, in this study, we aim to discuss the effects of INQ on BE, BL, BAS, PQ and the results from this effect. For this, we develop a scale consisting of three parts, which is suitable for the purpose of our study, from the literature. The first part of this scale consists of the items that we aim to obtain for the demographic information of the participants. The second part consists of the scale items of INQ, and the third part consists of the scale items of BE and its components. Sources of [43] for the INQ scale (21 items) and [22] for brand equity (10 items) were used. We did not develop scales but simply adapted the scales from these two widely cited studies to our research. However, we tested the accuracy and reliability of these scales in the methodological part of the research. We used a 5-point Likert scale to measure the items. We had help from the statistics company, which has the second largest consumer portfolio in the country, to apply this survey to Nike sportswear consumers residing in Turkey. The company applied our online survey to 1549 customers, which we selected randomly from 198,000 Nike brand customers within its body. To our survey, 847 consumers responded. Of these, survey data were excluded from participants under 18 years of age, those who had no experience with all Nike brand channels, and those who provided inconsistent responses to the three attention control questions (ACQs). As a result, survey data from 626 participants were valid for the final analysis. We did not detect any response bias in the data of 626 participants. Thus, the number of participants in our study exceeded the lower limit of the literature that it is appropriate to carry out with five times as many samples as the number of scale items for SEM. Thus, we first tested the reliability and validity of the scale of our research with the data obtained. We applied Cronbach's alpha analysis, which is widely used as a basis for the measurement of scale reliability. Cronbach's alpha values of all factors were above the desired value of (0.70) in the literature. Second, we performed (confirmatory factor analysis) CFA to see the validity of our scale and its compatibility with our research. All factor loadings of INQ ranged from 0.66 to 0.97, and for multiple correlation squares (R^2), values ranged from 0.44 to 0.94, and t > 2.58 was significant at the $p < 0.001$ significance level. All factor loadings of BE were significant between 0.79–0.88, and multiple correlation squares (R^2) values were between 0.62–0.77, and t > 2.58, $p < 0.001$ significance level. These values showed a strong relationship with the corresponding structures, in compliance with the literature [63–65]. In addition, AVE and CR values were important for us in terms of the reliability of the constructs. AVE values of 0.80 and CR values of 0.50 were obtained for all structures for the above-predicted values (Appendix A, Table A1) [64,66]. Thus, the CFA results and Cronbach's Alpha results show that the scales are consistent with research and are accurate and reliable. Then, the Pearson correlation analysis results showed that we examined the significance of the relationships between the first-order sub-dimensions of INQ (independent variable) and the first-order sub-dimensions of BE (dependent variable). Thus, the results showed that the correlation coefficients between the relation of INQ and its sub-dimensions and BE and its dimensions were statistically significant ($p < 0.01$) [67]. In addition, we obtained the discriminant validity of the scale structures with the results of discriminant analysis. The discriminant validity of our scale, which provided the necessary literature values for [68], was verified. The SEM results of our research confirmed the assumptions established between INQ and BE and their components in the context of OM. SEM results revealed that INQ had the highest effect (0.93) on BAS and PQ and the least effect (0.86) on BL, and INQ affected BE with 0.90.

The results of this research, which examine the predictors of brand equity and its components, offer implications for OM, INQ, BE subject areas that have not been empirically analyzed despite increasing knowledge and still having limitations in theoretical information. Our research is unique, as it is the first study to empirically examine the relationship between INQ and BE and its components in the context of OM. In addition, it is unique because it is the first study to examine the relationship between omnichannel strategies and brand equity in detail in the context of Turkey, a developing country with very limited studies on the Circular Revolution. Moreover, our research expands the limitations of the studies of [43,45].

In addition, with the results of this study, we present numerical data that provide cross-channel integration in omnichannel marketing that can be effective for marketers and brand managers who are looking for ways to increase the main brand value from the past. In other words, cross-channel integration can be used as a predictor of brand equity and its components. Brand managers who want to increase their brand equity can see the necessity of including INQ increasing strategies in their omnichannel strategies with the SEM results of this research. Thus, with this study, we not only shed light on future research in the field of OM, INQ, and BE, but also provide important advice to brand managers who carry out OM.

In this context, our research continues as follows: Section 2: Literature Review, Section 3: Exploratory conceptual framework and hypothesis, Section 4: Research methodology, Section 5: Results, Section 6: Discussion and implications, Section 7: Conclusions.

2. Literature Review

2.1. Omnichannel Marketing (OM)

In the literature, it is called "omnichannel marketing" when a consumer makes simultaneous purchases through many alternative channels such as physical stores, computers, smart devices (tablets, phones, wearable technology, kiosks, etc.), virtual stores with three-dimensional glasses, social media sites, and e-commerce platforms (Amazon, Alibaba.com, etc.), which includes searching for the product before purchase, researching product features, comparing multiple brands that make the same product, viewing the product in various physical and virtual retail locations, touching the product, purchasing and receiving the product, and returning it. All these channels and this great multichannel service are offered to consumers by marketers [69–73].

International research shows that OM, which is still evolving and reflects consumers' desires to store on multiple channels simultaneously, has been particularly adopted by retailers [37,74–76]. OR provides retailers with an integrated retail experience by combining all the services [76,77].

OR not only enables the simultaneous use of channels but also integrates all existing channels through various digital technologies for a retailer or brand to provide great customer knowledge [43,77–79]. Customer touchpoints, i.e., all channels that are common areas for OM and OR, should be viewed and organized like the harmony of all musicians in an orchestra [80]. The need to expand the concept of INQ for omnichannel, multiply its examples, and adapt it to retailers in all industries is supported by many studies [43,60,81,82].

2.2. Integration Quality (INQ)

Previous research has found that INQ is a brand evaluation tool for consumers that influences them and has valuable and positive outcomes for marketers [60,83–85]. The concept of channel integration refers to the coordination between all customer touchpoints (websites, physical stores, kiosks, smartphones, etc.) of a brand or retailer to provide consumers with a sustainable and flawless experience [51,57,60,86]. INQ is considered in the literature as the performance of channel integration [43,45,57,58,60,87–91].

INQ is the ability to provide seamless service to customers across all channels, and its importance for omnichannel marketing has been repeatedly highlighted in the literature [51]. Brands can profit from a superior competitive advantage in the marketplace by integrating all the channels they use [92].

2.3. Dimensions of INQ

Research on INQ and its dimensions claim that a retailer's channels can have not only physical but also virtual quality, and cross-channel inconsistency can reduce overall quality perceptions [43,50,51,57]. The research in [51], among the phenomenon researchers who make this claim, argues that INQ is an important factor in omnichannel services so that consumers do not have problems buying services through more than one channel. Relevant research has confirmed the importance of INQ [43]. In this study, we used the research of [43] for the INQ dimensions (CSC, CC, PC, AQ). The detail description of previous studies that used different approaches and methods reflects the Omnichannel is presented in Table 2. The definitions of the dimensions are given in Table 3 in the next sections.

2.4. The Outcome of Integration Quality, Brand Equity, and Dimensions

INQ has been involved in some customer-focused studies in recent research in the Omnichannel and multichannel context.

Table 2. Related research methods, approach, analysis, implications and differences of this research belonging to the related research.

Research	Factor	Approach	Analysis	Aim and Results of Research
[90,93]	-	qualitative	-	This study, which examines past research to examine the continuity strategies of OM qualitatively, examines mostly integration quality dimensions, horizontal and vertical strategies.
[94]	Promotion, Product and Price, Information Access, Order Fulfillment Customer Service, Transparency, Freedom, and Synchronization	quantitative	SEM Analysis	The study explores the customer experience in the OM environment. This study, which investigates the effect of Transparency, Freedom, and Synchronization dimensions on customer loyalty, shows that customer experience mismatch has a negative impact on customer retention, but channel transparency, ease and smoothness can effectively mitigate this negative impact.

Table 2. *Cont.*

Research	Factor	Approach	Analysis	Aim and Results of Research
[95]	Content Consistency, Marketing Consistency, Synchronization	quantitative	SEM Analysis	Expanding the knowledge in the field of customer experience and channel integration, this research revealed that the 4Ps are more effective than emotional in improving the cognitive customer experience.
[43]	Breadth, Transparency, Content Consistency, Process Consistency	quantitative	SEM Analysis	Focusing on the integration quality area in the context of OM, the study presents cross-buying intent as a result of INQ.
[96]	-	quantitative	MANOVA	This research, which revealed that the brand image can be increased due to the digital customer data obtained in the omnichannel environment, was built on shopping scenarios.
[97]	-	qualitative	-	The case study was conducted with 15 retail store managers. The results of the analysis revealed that the use of technology in physical stores strengthens brand loyalty and brand image.
[48]	Content Consistency, Process Consistency, Marketing Consistency, Freedom, Synchronization, Working Together	quantitative	SEM Analysis	The study carried out in the retail industry conceptualized the customer experience in the omnichannel environment as a multidimensional construct and found that perceived compliance and risk play a mediating role in the plot of omnichannel shopping intention with the omnichannel experience.
[45]	-	qualitative	-	In this study, which conceptualizes the integration quality dimensions for Omnichannel marketing and presents its impact on customer equality through a literature review, and INQ is seen as a predictor of customer equality.
This Research	Channel-Service Configuration (CSC), Content Consistency (CC), Process Consistency (PC), Assurance Quality (AQ)	quantitative	SEM Analysis	For OM, we explore the impact of INQ on brand equity and its dimensions with an empirical analysis for the Nike brand. We conceptualize INQ as a predictor of BE and its components BL, BAW, BA, PQ.

Previous research has shown a relationship between service integration and customer equity [45,98]. However, there is no quantitative research in the omnichannel integration quality literature that uses data to examine the relationship between INQ and the drivers of customer equity. Customer equity is examined in the qualitative research of [45]. Here, the concept of equality is the feeling of getting what one deserves, which ensures mutual equity in relationships [45]. It is the feeling on which long-term relationships, especially between the brand and the customer, are based [99–101]. Customer equity, however, is the sum of the lifetime value of all customers of a company [86,102,103] and has three driving forces: value equity, relationship equity, and brand equity [104]. Value equity encompasses the customer's actual valuation of the brand and is their perception based on what they gave up when they purchased the product during their buying behavior [104]. Relationship equity deals with what the customer gets from the relationship based on the perceived total inputs in the relationship between a customer and a brand [105]. Brand equity is the sum of the customer's tangible and intangible perceptions of the brand. The brand's marketing strategies influence it. It has been used as part of customer equity since the 2000s. While brand equity can make the brand attractive to new customers, it contains the miraculous power to create repurchase behavior among existing customers and build an emotional attachment to the brand.

The formation of brand equity is related to the following four dimensions [16]: BL, PQ, BA, and BAW. Although brand equity has been studied since the 1980s, it has been a driving force in customer equity studies since 2000 [102,106]. Several multichannel studies have examined the relationship between multichannel service quality and brand equity [107–110]. However, in an omnichannel marketing environment, a brand may be more influenced by brand equity as part of customer equity than in a multichannel environment [45,52,71]. Since omnichannel allows customers to store on all channels simultaneously, there will be differences compared to brands that do not use omnichannel that will affect not only the customer's emotions but also the buying experience, relationship quality, perceived value, and brand equity [111]. In this context, [45] recommended further research on omnichannel marketing efforts and data-based analysis of customer equity drivers (value equity, relationship equity, brand equity). At the same time, [43] suggested that it would be beneficial to research the dimensions and impacts of INQ in different domains, consumer groups, and industries. Due to the lack of research focusing on these recommendations and their relationship with integration quality (INQ) and brand equity, there is an opportunity to measure the dimensions of brand equity (BL, PQ, BA, and BAW) as INQ outcomes.

3. Exploratory Phase: Conceptual Framework and Hypotheses Development

In the first phase, before quantitative data analysis with Confirmatory Factor Analysis (CFA) and Structural Equation Modeling (SEM), the INQ is composed of four main dimensions: Channel Service Configuration, Content Consistency, Process Consistency, and Assurance Quality. We determined them through qualitative data analysis based on articles [43,45,50,51,57,58,93]. In the second phase, before CFA and SEM, brand equity consisted of four main dimensions: BL, PQ, BA, and BAW. The dimensions of BA and BAW were combined under a single factor from these dimensions. Finally, brand equity was influenced by the qualitative data analysis of three main dimensions.

In the third phase, we discussed the accuracy of this qualitative data analysis results in online meetings with two professors, three associate professors, and researchers in omnichannel and brand equity in the institutes of three different countries with high q rankings.

In the last phase, we have established hypotheses that we will analyze with SEM using the extensive research we have conducted to determine the impact of INQ on brand equity and its dimensions, as well as the content of articles with a high number of citations scanned in Scopus and the Web of Science. Being aware that qualitative data analysis is the research's dynamics, we paid special attention to this phase. In this way, the research contributed to expanding information about INQ and omnichannel research areas and reproducing limited knowledge through qualitative data analysis.

This section aims to discuss the findings of the qualitative phase and the model's hypotheses presented in Figure 1, which were developed for the research of quantitative data analysis, with the supporting literature sources in the following sections.

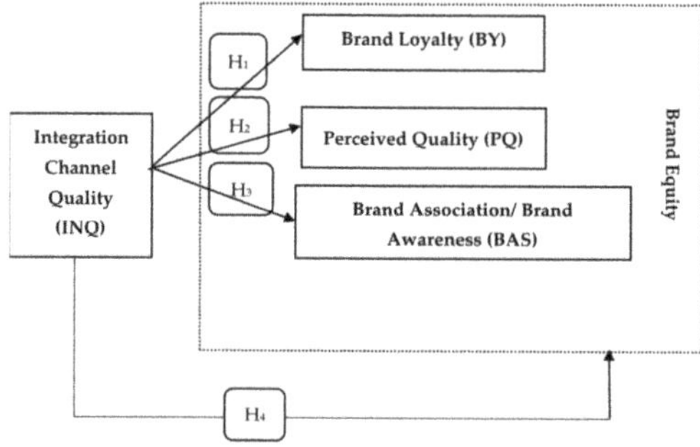

Figure 1. Sustainable omnichannel strategic framework.

3.1. Integration Quality (INQ) and Brand Loyalty (BL)

Consumer-created perceptions of service quality offered by the brand are believed to have a significant impact on actions such as brand satisfaction, repurchase, the recommendation to the surrounding community, and brand dependence [112–115]. Several multichannel studies have examined the relationship between multichannel service quality and BL [62,107,110,116,117]. Although there are many studies on INQ for OM, few studies have examined its effects on building BL [62]. BL exists when a customer repeatedly and consistently buys products from the same brand in the future [19]. With OM, which involves an integrated service approach across all channels, it may be more feasible than multichannel marketing to strengthen existing customers' relationships with the brand, thereby creating dependency among customers and thus ensuring that loyal customers repeatedly purchase the brand's products, i.e., creating BL [62,118–120]. In this context, we make the following assumption:

H₁: *Integration quality (INQ) positively influences brand loyalty (BL).*

3.2. Integration Quality (INQ) and Perceived Quality (PQ)

PQ is a dynamic of brand equity. Zeithaml (1988) defines PQ as the consumer's sense of a product's superiority. PQ can be influenced by factors such as the general sense of the consumer, their experiences with the product, their prejudices, their needs, their environment, the brand's offerings, their relationship with the brand, etc., and it is a competitive advantage for brands as well as more sales [121–127]

Despite a large body of literature on PQ [20,128,129], there are few studies on integration quality or the relationship between service quality and PQ [126]. The quality of services offered through brands' online and offline channels and customer satisfaction when shopping across all channels influence PQ [22,130–132]. The consumer realizes the brand's superiority when he feels high quality through prolonged product use [133]. Therefore, the consumer will prefer the brand's product that he perceives as having higher quality than other brands [20]. In this context, we make the following assumption:

H₂: *Integration quality (INQ) positively influences perceived quality (PQ).*

3.3. Integration Quality (INQ) and Brand Association/Awareness (BAS)

BA, defined as everything that appears in consumers' minds about a brand and whose importance for brand equity has been noted by brand authors [134] and [116], is combined with BAW [22]. BAW is the definition, meaning, and information about the brand in the consumer's mind [21]. BAW is the antecedent component of brand equity, and increasing consumer BAW leads to brand preference and differentiation from competitors [17]. Consumers' positive experiences with the brand result in positive BAW/BA. Furthermore, being able to shop from all brand channels can impact BAW/BA as a positive reflection of the same quality service and integration quality [135,136]. Several studies on multichannel marketing have proven that service quality impacts the BA and BAW [110]. Online and offline service quality perceptions combine with consumers' overall service perceptions to create a positive judgment for BAW/BA [137]. Consumers' positive experiences with shopping options and services offered by traditional and e-commerce sites lead to increased BAW/BA [133]. In this context, we make the following assumption:

H₃: *Integration Quality (INQ) positively influences Brand Association/Awareness (BAS).*

3.4. Integration Quality (INQ) and Brand Equity (BE)

Brand equity is the assets and identities associated with a brand, brand name, or symbols added to or subtracted from the value of products or services of a company or its customers [138]. While the research of [139] and [17] evaluate the concept of brand equity as the added value that the brand name adds to products, Ref. [140] defines it as consumer confidence in brand identity and image. Central to the brand equity literature definitions is that brand equity combines BL, PQ, BAW, and associations [22,138]. In a sense, brand equity is an important structure for the brand that represents customers' evaluation of the brand, attracts customers to the brand, enables them to recognize and associate with the brand, and is influenced by its marketing strategies and 4Ps [17]. Customers' recognition of a brand can have a different effect on BAW by creating a brand's positive image in their memory. Positive, reciprocal, and long-term relationships between customers and the brand can be built due to high-quality and integrated service quality across all online and offline channels. If this relationship can be built, it can lead the customer to respond to the brand's marketing mix's 4P elements (product, price, place, promotion). The brand's feeling and positive perception of the product desired from all customer touchpoints, such as price, ease of payment, speed of delivery, campaign, special discount, customer service, and the purchasing process at the same level can enable BAW, superiority and direct and positive effects on brand equity [20,22,136,137]. This is because studies on omnichannel marketing, consumer attitudes, multichannel marketing, and service quality point in this direction [9,110,135,141–143]. In this context, we make the following assumption:

H₄: *Integration Quality (INQ) positively influences brand equity (BE).*

4. Research Methodology

4.1. Measures

As a result of the literature review, the research assumed that INQ had four first-order sub-dimensions: CSC, CC, PC, and AQ [43]. Furthermore, as a result of INQ, it was assumed that Brand Equity (BE) consisted of three first-order sub-dimensions: BL, PQ, and BAS (BAS abbreviation used instead of BAW/BA abbreviation) [22].

Sources of [43] for the INQ scale (21 items) and [22] for Brand Equity (10 items) were used. We did not develop scales but simply adapted the scales from these two widely cited studies to our research. However, we tested the accuracy and reliability of these scales in the methodological part of the research. We used a 5-point Likert scale to measure the items. Table 3 provides brief conceptual definitions of the structures that make up the scales, their scales, and the resources used in detail.

Table 3. Functionalization of structures.

Structures	Definitions	Sub-Dimension	Definitions
INQ- [43]	It is the performance of a brand or retailer in coordinating all customer touchpoints (websites, physical stores, kiosks, smartphones, etc.) owned by the retailer to provide consumers with a sustainable and flawless experience [43,51,57].	(CSC)	Channel performance in terms of the same quality and consistency across all channels [43,50,57,58].
		(CC)	It refers to the consistency of information going and coming through the brand's different channels [51,57,60].
		(PC)	Consistency of consumer elements (sense of service, wait time, image, level of employee appreciation) across channels [43,50].
		(AQ)	It refers to the security provided to consumers across all channels. ASQ encompasses the totality of privacy, security, and e-service quality and is conceptualized as a dimension of INQ [43,60].

Table 3. *Cont.*

Structures	Definitions	Sub-Dimension	Definitions
BE- [22]	Brand equity is the assets and identities associated with a brand, brand name, or symbols added to or subtracted from the value of products or services of a company or its customers [138].	(BL)	It is the repeated purchase of products of the same brand by a consumer in the future [16,21].
		(PQ)	It is a consumer's sense of a product's superiority [20].
		(BA)	Everything about a brand appears in the consumer's memory [21].
		(BAW)	It is the set of definitions, meanings, and information about the brand in consumers' memory [21].

4.2. Data Collection

Survey data were collected from March–January 2022 with the support of research companies with a database of approximately 198.000 consumers in Turkey. We had help from the statistics company, which has the second largest consumer portfolio of the country, to apply this survey to Nike sportswear consumers residing in Turkey. The company applied our online survey to 1549 customers, which we selected randomly from 4.800 Nike brand customers within its body. To our survey, 847 consumers responded. Of these, survey data were excluded from participants under 18 years of age, those who had no experience with all Nike brand channels, and those who provided inconsistent responses to the three attention control questions (ACQs). As a result, survey data from 626 participants were valid for the final analysis. We did not detect any response bias in the data of 626 participants. Thus, the number of participants in our study exceeded the lower limit of the literature that it is appropriate to carry out with five times as many samples as the number of scale items for SEM [63,144].

5. Results

5.1. Demographic Results

The demographic profile of the participants is presented as a percentage in Table 4.

Table 4. Demographic profile of the participants.

Gender		Marital Status	
Female	%50.2	Single	%47.4
Male	%49.8	Married	%52.6
Education		**Age**	
High School and Below	%39.1	18–24	%26.2
University	%33.5	25–35	%21.7
Graduate	%27.3	36–44	%26.4
		Over 45 years old	%25.7
Job		**Distribution Channels Used**	
Student	%12.6	I buy the products of this brand online (from the website), they are delivered to my address.	%31.6
Private Sector Employee	%37.2	I buy products of this brand online (website), delivered from store.	%33.1
Officer	%54.2	I buy the products of this brand from the store; they are delivered to my address.	%35.3
Academical Personal	%4.6		
Teacher	%7.2		
Small Business	%8.1		
Engineer	%3.2		
Military Officer	%2.2		
Nurse	%5.9		
Financial Advisor	%4.6		
Doctor	%5.0		
Lawyer	%5.1		

According to Table 4, the ratio of male and female participants was almost equal. Likewise, the rates of single and married participants were very close to each other. The majority of the participants were university and master's graduates. It was found that 26.2% of participants were 18–24 years old, 21.7% were 25–35 years old, 26.4% were 36–44, and 25.7% were 45 or older, while the majority of participants were classified as private-sector employees at 37.2% (n = 233). The channel usage alternative rates of the participants were very close to each other.

5.2. Result for Measurement

In this research, the reliability analysis and confirmatory factor analysis methods were first used to test the consistency and accuracy of the multilevel scales (INQ scale and brand equity scale) created using the literature. CFA specifies multifactorial structures and tests the fit of the model to the data [145]. We iteratively tested both scales to determine the goodness-of-fit values and to obtain a final model. In assessing goodness of fit, we followed the recommendations of [65] ($X^2 < 3$; GFI > 0.90; AGFI $\geq$ 0.900; NFI > 0.90; RMSEA < 0.01, 0.05 and values below 0.08 indicate excellent, good, and moderate fit, respectively; 0 < CFI < 1) which are widely accepted in the literature [65]. We also performed the Cronbach's alpha test for scale reliability.

We analyzed the validity and reliability of INQ constructs (CSC, CC, PC, AQ) and BE (BL, PQ, BAS) in two separate path diagrams and reported the significant values of the constructs in the same table (Table 5). All factor loadings of INQ ranged from 0.66 to 0.97, and for multiple correlation squares (R^2), values ranged from 0.44 to 0.94, and t > 2.58 was significant at the $p < 0.001$ significance level. All factor loadings of BE were significant between 0.79–0.88, and for multiple correlation squares (R^2), values were from 0.62–0.77, and t > 2.58, $p < 0.001$ significance level. These values showed a strong relationship with the corresponding structures, in compliance with the literature [63–65].

Table 5. Inter-order relations values.

	X^2/df	*p*	RMSEA	CFI	GFI	AGFI	NNFI	NFI	RMR	SRMR
INQ	1.932	0.000	0.039	0.99	0.97	0.97	0.99	0.99	0.017	0.013
BE	1.374	0.000	0.020	0.99	0.99	0.99	0.99	0.99	0.009	0.008

AVE and CR values were important for us in terms of the reliability of the constructs. AVE values of 0.80 and CR values of 0.50 were obtained for all structures for the above-predicted values (Table 6) [64,66].

Table 6. AVE and CR Values.

Variables	CR	AVE	Cronbach's Alpha
CSC	0.90	0.54	0.937
CC	0.78	0.47	0.858
PC	0.77	0.45	0.850
AQ	0.79	0.49	0.891
BL	0.88	0.72	0.881
BAS	0.90	0.65	0.902
PQ	0.79	0.65	0.787

For reliability analysis, Cronbach's Alpha results show that the reliability values of both the INQ and BE scales are above 0.70. Thus, the CFA results and Cronbach's Alpha results show that the scales are consistent with research, accurate, and reliable.

Then, the Pearson correlation analysis showed that we examined the significance of the relationships between the first-order sub-dimensions of INQ (independent variable) and the first-order sub-dimensions of BE (dependent variable). Thus, the results showed that the correlation coefficients between INQ and its sub-dimensions and BE and its dimensions were statistically significant ($p < 0.01$) (Table 7) [67].

Table 7. Pearson correlation analysis results.

Variables	CSC	CC	PC	AQ	BL	BAS	PQ
CSC	1	0.901 **	0.910 **	0.916 **	0.786 **	0.851 **	0.783 **
CC		1	0.873 **	0.880 **	0.747 **	0.817 **	0.744 **
PC			1	0.887 **	0.755 **	0.820 **	0.766 **
AQ				1	0.746 **	0.817 **	0.760 **
BL					1	0.893 **	0.820 **
BAS						1	0.843 **
PQ							1

** $p < 0.01$, CSC: Channel-Service Configuration, CC: Content Consistency, PC: Process Consistency, AQ: Assurance Quality, BL: Brand Loyalty, BAS: Brand Awareness and Brand Association, PQ: Perceived Quality.

We used the Fornell–Larcker Criterion (FLC) to calculate the discriminant validity of the scale with discriminant validity analysis [68]. The discriminant validity of our constructs was ensured because the square roots of the extracted mean variance (AVE) values of the scale variables of our study were higher than the correlations between the constructs [68]. Thus, the discriminant validity of the scale of the study was ensured (Table 8).

Table 8. Discriminant validity results.

Variables	CSC	CC	PC	AQ	BL	BAS	PQ
CSC	**0.93**						
CC	0.91	**0.92**					
PC	0.92	0.90	**0.93**				
AQ	0.88	0.90	0.91	**0.92**			
BL	0.91	0.89	0.85	0.91	**0.93**		
BAS	0.89	0.91	0.91	0.83	0.89	**0.89**	
PQ	0.89	0.90	0.90	0.90	0.89	0.88	**0.87**

Notes: The square root of AVE is indicated in bold; CSC: Channel-Service Configuration, CC: Content Consistency, PC: Process Consistency, AQ: Assurance Quality, BL: Brand Loyalty, BAS: Brand Awareness and Brand Association, PQ: Perceived Quality.

5.3. Result for Structural Models

For the analysis of our hypotheses, we used the method SEM (structural equation modeling), which is hypothesized due to its compatibility with the research, is recommended for analysis with small or medium sample sizes, and does not require a normal distribution of the data. For SEM, we used the program LISREL 8.7 [44,64,146,147].

SEM analysis values of hypothesis testing were found in two different models. The first path diagram includes each degree of relationship between INQ and the dimensions BL, PQ, and BAS separately (Figure 2). The statistical results showed a standardized beta of 0.86 (INQ-BL), 0.93 (INQ-PQ), and 0.93 (INQ-BAS), respectively (Figure 3). These values were significant at $p < 0.001$ (Table 9). According to [148], the R^2 value was accepted as a significant effect size. Furthermore, the goodness-of-fit values were found to be excellent according to the literature: $X^2/df = 4.492$, RMSEA = 0.077, CFI = 0.99, IFI = 0.99, RMR = 0.041, SRMR = 0.040, GFI = 0.93, AGFI = 0.91, NFI = 0.99, NNFI = 0.99. Thus, hypotheses H_1, H_2, and H_3 were confirmed, which stated that overall integration quality had a significant and positive influence in the sub-dimensions of BL, perception quality, and BAW/BA (BAS), which are combined into one dimension.

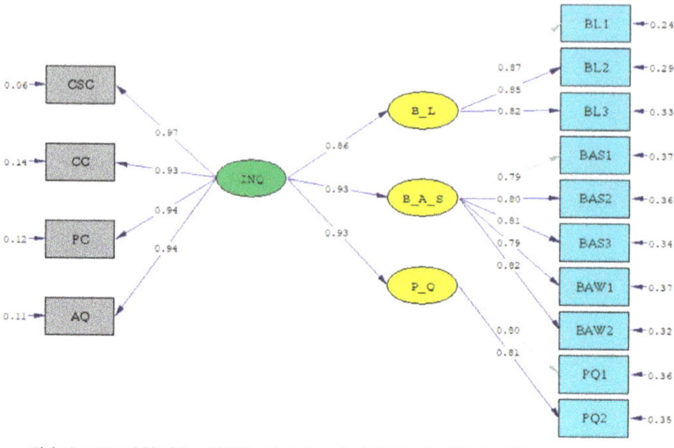

Figure 2. The first structural model.

Table 9. Findings for the first structural model.

Paths	β/Path Coefficients	T Statistics	*p* Value	Result
INQ-BL	0.86	22.39 **	0.000	Accepted
INQ-PQ	0.93	22.11 **	0.000	Accepted
INQ-BAS	0.93	21.66 **	0.000	Accepted

** *p* < 0.01.

The other path diagram shows the relationship between INQ and brand equity within the structural model. (Figure 3). The standardized beta value for INQ-BE was 0.90, and *p* < 0.001 was significant (Table 10). The model explained the total variance by R^2. According to [148], the R^2 value was accepted as a significant effect size. Thus, hypothesis H_4 was confirmed, which states that the overall quality of integration has a significant and positive effect on brand equity.

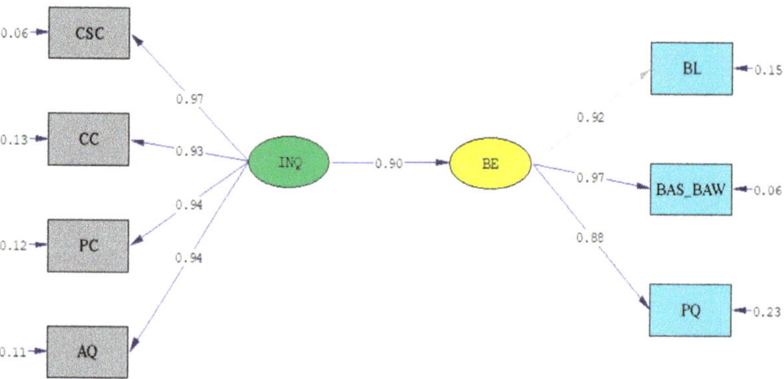

Figure 3. The second structural model.

Table 10. Findings for the second structural model.

	β/Path Coefficients	T Statistics	*p* Value	Result
INQ-BE	0.90	25.56 **	0.000	Accepted

Source: From the authors. ** $p < 0.01$.

6. Discussion and Implications

Based on the dimensions of integration quality proposed by Hossain et al., in 2020, we presented a model of the brand effect of INQ and tested it empirically using SEM. While INQ consists of four sub-dimensions, namely channel-service configuration, content consistency, process consistency, and assurance quality, brand equity consists of four sub-dimensions: BL, PQ, BAW, and Brand Association. We confirmed with CFA that the size of the scales decreased to one, which is consistent with the literature.

The results of the structural models have confirmed that INQ is an important predictor and promoter of brand equity and its sub-dimensions (BL, PQ, BAW, and BA combined in one dimension). In other words, our research proves that brand equity and its dimensions (BL, PQ, BAW and BA-(BAS)) for Nike, a luxury sportswear brand that uses omnichannel, are a result of integration quality, which is a performance indicator of the harmony of all services provided at all customer touchpoints of the brand. For this reason, this research has revealed that brands that use omnichannel marketing care about integration quality and have higher brand equity than other brands competing in the same market that do not care about omnichannel marketing integration quality.

Considering the results of the first hypothesis test from the research model results in Table 10, it was found that integration quality had a positive and significant effect on BL (t = 22.39, $p < 0.01$). When the level of integration quality increases by one unit in the channels of Nike, a luxury sportswear brand that engages in omnichannel marketing, the loyalty level of the brand is positively affected by 0.86. This relationship between INQ and BL has not been previously examined in the omnichannel study. However, in a multichannel environment, the relationship between the factor of service quality used instead of INQ and BL has been established in a few studies [10,42,51,52]. For example, [149] research conducted in Malaysia found that service quality is effective in creating brand loyalty. However, unlike multichannel, this hypothesis finding makes a unique contribution to the literature, as omnichannel means delivering the same service to customers in all channels at the same time [10]. With the provision of cross-channel INQ in an omnichannel environment, customer satisfaction, which is an important antecedent of brand loyalty in the literature, can be gained. The common entity that is the factor in the emergence of OM is consumer desire. In this context, the service that will be provided with the same transparency, harmony and strategies in all channels, which will lead to an increase in customer satisfaction and thus brand loyalty [18,150].

When examining the result of the second hypothesis test, it was found that the integration quality of Nike brand's sales channels had a positive and significant effect on BA and BAW, which are combined into a single factor (t = 2.11, $p < 0.01$). Thereby, there will be a positive effect of 0.93 on BA and BAW when the integration quality increases by one unit, summarized in a single factor. Brand awareness and brand association components are actually influenced by brand image [16,17,106,108,151]. The image of a brand in society can change what consumers think about the brand. Therefore, if a brand has a good image, the consumer's awareness of the brand is positively affected. In this context, the positive effect of INQ, which means the performance measurement of the harmony to be achieved between channels, on brand association and awareness in this study offers an important opportunity for brands to improve their brand image, brand awareness and brand association for omnichannel environments [43,45,149]. In addition, it has been revealed in previous studies that communication has a great contribution on brand awareness. Based on this finding, inter-channel communication and one-to-one communication at contact points with customers will be beneficial both in increasing the level of INQ and in improving brand awareness and attitude toward the brand [33,152,153]. In this context, this SEM result, which reveals the effect of INQ on brand awareness and brand association in the OM environment, opens the way for a fairly new research area and reduces the literature gap.

As a result of the third hypothesis test, it was found that the integration quality of the Nike brand distribution channels had a positive and significant effect on the quality of the brand perceived by the consumer ($t = 21.66$, $p < 0.01$). A one-unit increase in the level of integration quality attributable to the compliance performance of all brand channels has a positive effect of 0.93 on the quality that the brand evokes in the minds of consumers. The impact of service quality on the perceived quality of the brand was previously discussed for multi-channel environments [53,91,125,127,132,154]. However, to the best of the author's knowledge, we are not the first research to directly reveal this relationship, as there are already very limited studies for omnichannel. Some antecedents have been provided in the literature for a brand trying to increase its perceived quality in the minds of its customers, which is related to the excellence of a brand in the most general sense. This study obtained INQ as a predictor of the perceived quality of the brand. The finding of our research, which allows us to redefine an influencer of the concept of perceived quality, which is an abstractly felt quality, for the OM environments brought by digitalization, is a first in its field. For OM, there is no system that concretely measures the services offered through the channels. In other words, the approximation of the inter-channel compatibility performance to the maximum will lead to improvements in the perceived quality of the brand. Past research has presented evidence that perceived quality increases brand loyalty [22,130–132]. In addition to the satisfaction of customers with product quality, product, price, and promotions, purchasing the brand's products from all online and offline channels of the brand with the same quality, price, customer relations, after-sales service, discounts and payment facilities will increase the perceived quality of the brand.

Finally, we were curious to know whether brand equity, which is important as a component of customer equity, is affected by the integration quality of the brand's channels and the integration quality of a brand in the case of omnichannel marketing application. Considering the results of the first hypothesis test from the research model results in Table A1, in the fourth hypothesis, we examined the impact of integration quality on brand equity and found that integration quality had a positive and significant impact on brand equity ($t = 25.57$, $p < 0.01$). If the brand achieves an improvement of one unit in integration quality, which is an indicator of the harmonization performance of the operation in the brand's channels, the equality of this brand will increase by 0.90 units in the positive direction. In other words, we reduce the literature gap with this hypothesis, which was asked to be discussed by [45]. With this study, which explores the relationship between INQ and brand equity for the first time, we provide theoretical contributions and expand the explored research area in customer equity subject areas with INQ. Brand managers and marketers have historically resorted to many innovations to increase brand equity. Time intervals and innovations brought by digitalization have changed the strategies of 4P (Product, Price, Place, Promotion) while also offering new markets for brands to present their products. What did not change with all this was the effort to increase the value of the brand's tangible and intangible assets. We pave the way for future BE research with our research, where we present a very usable result for omnichannel environments to investigate the factors that may be effective in increasing BE.

6.1. Theoretical Implications

The findings of this study have some unique theoretical implications.

First, research on omnichannel applications is quite limited. This study brings a conceptual extension to the literature on omnichannel strategies, INQ, OM.

Second, this research has demonstrated that INQ is a provider of brand equity in an omnichannel environment that has thus far been tested with customer-oriented relationships and purchase intention, and as far as the authors know, its relationship with brand assets has not been quantitatively studied. Therefore, it is the first research to examine these structures and relationships in the field of OM.

Third, the H_1, H_2, H_3 assumptions of this study, in which we set up an impact analysis with each of the INQ and BE dimensions (BL, BAS, PQ) were accepted. In this context, the work to discover the predecessors of BE has been extended with the INQ adopted for OM.

Fourth, we narrowed the literature gap in the context of BE by providing a validation for the antecedents of INQ (CSC, CC, PC, AQ), whose validation is not yet fully clarified in the literature.

6.2. Managerial Implications

This study has valuable contributions to marketers and brand managers. First of all, there are brands that are rapidly transitioning to omnichannel environments that have not yet been fully adopted, reaching customers through all channels, but experiencing losses compared to single-channel marketing in this direction. In this sense, we provided marketers with ways to increase INQ with this study. We also presented the necessary reasons for managers to provide INQ in an omnichannel environment in order to increase brand equity, with an empirical application. In addition, the most important benefit of this research is that it shows brand owners and managers and brand marketers a way to set up the omnichannel system.

7. Conclusions

In this research, we empirically investigated the relationships between INQ and BL, PQ, BAW, brand association and BE. The reliability, divergence and convergent validity results of the scale of our research exceeded the literature values. In other words, the scale we used for the empirical application of this research and the model of the research matched. Path coefficients and goodness-of-fit values obtained by SEM analysis recommended in the literature in impact studies were in agreement with the literature. Based on the problems of our research, four different hypotheses, which we constructed theoretically, were accepted. That is, the SEM results confirmed the assumptions we made between INQ and BE and their components in the context of OM. SEM results revealed that INQ had the highest effect (0.93) on BAS and PQ and the least effect (0.86) on BL, and INQ affected BE with 0.90.

The results of this research, which examine the predictors of brand equity and its components, have not been empirically analyzed despite increasing knowledge, and they offer implications for the subject areas OM, INQ, BE, which still have limitations in theoretical knowledge. Our research is unique, as it is the first study to empirically examine the relationship between INQ and BE and its components in the context of OM. It is also unique in that it is the first study to examine in detail the relationship between omni-channel strategies and brand equity in the context of Turkey, a developing country with very limited work on the Circular Revolution. Furthermore, our research results expand the limitations of some studies in the literature.

The results of INQ, which is one of the dynamics of service quality, and brand equity, which is a dimension of customer equity, will help managers, customers, companies, marketers and brand consultants to establish an effective omnichannel system. It will contribute to all available information. It has been revealed that the efforts of the brand to realize the integration of all services offered in the omnichannel environment have an effect on all information, awareness and associations in the mind of the consumer, the provision of INQ in this framework, the provision of the consumer's loyalty to the brand, the perceived quality of the brand in the mind of the customer, and the positive if it is negative, or already positive. If it is positive, it will have beneficial returns for the brand, such as reinforcement.

One of the broadest contributions of this article is that it offers a solution for omnichannel environments to reduce the cost of textile products returned due to inter-channel incompatibility, as well as the amount of waste generated by brands throwing away returned products. Thus, we expand the field of research for the protection of the environment and reduce carbon emissions, as well as shed light on the research to be conducted in this field.

Our research has several limitations. First, the research data were collected only in Turkey. This limits the research, as the results cannot be generalized to other countries. In the research, the participants' data were collected in a single time frame in 2022, and the time limit is another limitation of the research. Furthermore, the research was only applied to Nike, a famous sportswear brand that operates in the apparel industry. Although generalization for the apparel sector is easy, the inability to extend the research to other sectors that require more services is the final limitation of the research. Expanding this research to overcome these three limitations is recommended for future studies. The impact of cross-channel integration quality on brand equity and dimensions can be studied for omnichannel in different income countries, periods, and industries. Moreover, the brand value was a dynamic of customer equity. Therefore, whether customer equity is an outcome of INQ could be a new research topic for the future.

Author Contributions: Conceptualization, T.Y. and M.I.; Formal analysis, T.Y.; Investigation, T.Y. and M.I.; Methodology T.Y. and M.I.; Project administration and supervision M.I.; Validation, T.Y. and M.I.; Writing—original draft, T.Y.; Writing—review and editing, T.Y. and M.I. All authors have read and agreed to the published version of the manuscript.

Funding: This research received no specific grant from any funding agency in the public, commercial, or not-for-profit sectors.

Institutional Review Board Statement: The study was conducted in accordance with the Social and Human Sciences Research Ethics Committee of Karabuk University, Turkey, and was approved by the Social and Human Sciences Research Ethics Committee of Karabuk University (2022/04 and 11 May 2022).

Informed Consent Statement: Not Applicable.

Data Availability Statement: Not applicable.

Conflicts of Interest: The authors declare no conflict of interest.

Abbreviations

OM	Omnichannel Marketing
INQ	Integration Quality
CSC	Channel-Service Configuration
CC	Content Consistency
PC	Process Consistency
AQ	Assurance Quality
BE	Brand Equity
BL	Brand Loyalty
BAW	Brand Awareness
BA	Brand Association
BAS	Brand Awareness and Association
PQ	Perceived Quality
CFA	Confirmatory Factor Analysis
SEM	Structural Equation Model
AVE	Subtracted Average Variance
CR	Composite Reliability

Appendix A

Table A1. Results of constructs.

Constructs	Items	CR	AVE	Cronbach's Alpha
Channel-service choice [57,58]	This brand offers its products through multiple channels.	0.90	0.54	0.93
	When buying products of this brand, I can choose from more than one channel.			
	If I cannot buy products of this brand from a particular channel, I can use other channels.			
	I am aware of the products offered by this brand's multiple channels (website, physical store and mobile application).			
	I know how to use the features of this brand's multiple channels (website, physical store and mobile app).			
	This brand kept me well informed about the various features of its multiple channels (website, physical branch and mobile app).			
	This brand does not force me to use a particular channel for a particular purpose.			
	Services provided through different channels of this brand are suitable for these channels.			
Content Consistency [53,57,58]	The product prices of this brand are consistent across all channels.	0.78	0.47	0.85
	This brand provides consistent promotional information across different channels.			
	This brand provides consistent product information across different channels.			
	In general, the information on multiple channels of this brand is consistent.			
Process Consistency [57,58]	All channels of this brand (website, physical store and mobile application) are easy to use.	0.77	0.45	0.85
	All channels of this brand (website, physical store and mobile application) have a flexible system to meet my needs.			
	The service experience is consistent across all channels of this brand (website, physical store and mobile app).			
	This brand maintains a consistent brand image across all its channels (website, mobile app and physical store).			
Assurance quality [43]	My personal information is protected in all channels of this brand.	0.79	0.49	0.89
	My personal information in all channels of this brand isn't shared with others.			
	My financial information on all channels of this brand isn't shared with others.			
	All channels of this brand have sufficient security features.			
	All channels of this brand provide the means by which I can express my complaints.			
Brand Loyalty [22]	I plan to stay loyal to this brand.	0.88	0.72	0.88
	When purchasing a product, the products of this brand are always my first choice.			
	If there is a channel where I can reach this brand, I won't turn to other brands.			
Brand Association & Brand Awareness [22]	Some features of this brand come to my mind quickly.	0.90	0.65	0.90
	I can quickly remember the symbol or logo of this brand.			
	I have a hard time imagining this brand in my mind.			
	I easily recognize this brand among other competing brands.			
	I am aware of this brand.			
Perceived Quality [22]	The quality of the products of this brand is extremely high.	0.79	0.65	0.78
	The products of this brand are very likely to be functional.			

References

1. Trevail, C. *Interbrand Report, Best Global Brands 2021*; Interbrand: New York, NY, USA, 2021.
2. Kartajaya, H.; Kotler, P.; Setiawan, I. *Marketing 4.0: Moving from Traditional to Digital*; John Wiley & Sons: Hoboken, NJ, USA, 2016.
3. Kotler, P. Why Broadened Marketing Has Enriched Marketing. *AMS Rev.* **2018**, *8*, 20–22. [CrossRef]
4. Kotler, P. Philip Kotler: Some of My Adventures in Marketing. *J. Hist. Res. Mark.* **2017**, *9*, 203–208. [CrossRef]
5. Rubio, N.; Villaseñor, N.; Yagüe, M.J. Influence of Value Co-Creation on Virtual Community Brand Equity for Unichannel vs. Multichannel Users. *Sustainability* **2021**, *13*, 8403. [CrossRef]
6. Hyun, H.; Park, J.; Hawkins, M.A.; Kim, D. How Luxury Brands Build Customer-Based Brand Equity through Phygital Experience. *J. Strateg. Mark.* **2022**, *392*, 1–25. [CrossRef]

7. Asare, C.; Majeed, M.; Cole, N.A. Omnichannel Integration Quality, Perceived Value, and Brand Loyalty in the Consumer Electronics Market: The Mediating Effect of Consumer Personality. In *Advances in Information Communication Technology and Computing*; Springer: Singapore, 2022; pp. 29–45.
8. Gartner, J.; Fink, M.; Floh, A.; Eggers, F. Service Quality in Social Media Communication of NPOs: The Moderating Effect of Channel Choice. *J. Bus Res.* **2021**, *137*, 579–587. [CrossRef]
9. Becagli, C.; Milanesi, M. Omnichannel Retailing and Brand Equity: A New Balance to Achieve. In *The Art of Digital Marketing for Fashion and Luxury Brands*; Springer International Publishing: Cham, Switzerland, 2021; pp. 31–49.
10. Park, J.; Kim, R.B. Importance of Offline Service Quality in Building Loyalty of OC Service Brand. *J. Retail. Consum. Serv.* **2022**, *65*, 102493. [CrossRef]
11. Badenhop, A.; Frasquet, M. Online Grocery Shopping at Multichannel Supermarkets: The Impact of Retailer Brand Equity. *J. Food Prod. Mark.* **2021**, *27*, 89–104. [CrossRef]
12. O'Grady, M.; Wu, S.; Kumar, S. *UK Multichannel Retailers Must Adjust Their Store Portfolios to Remain Profitable*; Forrester: London, UK, 2018.
13. Sharma, N. H&M Dragged down by Lack of Online-Offline Integration: Analyst. 2017. Available online: https://www.cnbc.com/video/2017/12/15/hm-dragged-down-by-lack-of-online-offline-integration-analyst.html (accessed on 14 July 2022).
14. Ghosh, S.K.; Seikh, M.R.; Chakrabortty, M. Pricing Strategy and Channel Co-Ordination in a Two-Echelon Supply Chain Under Stochastic Demand. *Int. J. Appl. Comput. Math.* **2020**, *6*, 28. [CrossRef]
15. Ghosh, S.K.; Seikh, M.R.; Chakrabortty, M. Analyzing a Stochastic Dual-Channel Supply Chain under Consumers' Low Carbon Preferences and Cap-and-Trade Regulation. *Comput. Ind. Eng.* **2020**, *149*, 106765. [CrossRef]
16. Aaker, D. *Managing Brand Equity*; Free Press: New York, NY, USA, 1991.
17. Keller, K.L. Conceptualizing, Measuring, and Managing Customer-Based Brand Equity. *J. Mark.* **1993**, *57*, 1–22. [CrossRef]
18. Gürbüz, A.; Yeğin, T. Consumer Based on The Impact of Relationship Marketing Brand Equity Components: A Model Proposal. *J. Soc. Sci.* **2018**, *31*, 599–622. [CrossRef]
19. Oliver, R.L. Whence Consumer Loyalty? *J. Mark.* **1999**, *63*, 33–44. [CrossRef]
20. Zeithaml, V.A. Consumer Perceptions of Price, Quality, and Value: A Means-End Model and Synthesis of Evidence. *J. Mark.* **1988**, *52*, 2–22. [CrossRef]
21. Keller, K.L. Understanding Brands, Branding and Brand Equity. *Interact. Mark.* **2003**, *5*, 7–20. [CrossRef]
22. Yoo, B.; Donthu, N. Developing and Validating a Multidimensional Consumer-Based Brand Equity Scale. *J. Bus. Res.* **2001**, *52*, 1–14. [CrossRef]
23. Kumar, N.M.; Chopra, S.S. Leveraging Blockchain and Smart Contract Technologies to Overcome Circular Economy Implementation Challenges. *Sustainability* **2022**, *14*, 9492. [CrossRef]
24. Mohammed, M.A.; Abdulhasan, M.J.; Kumar, N.M.; Abdulkareem, K.H.; Mostafa, S.A.; Maashi, M.S.; Khalid, L.S.; Abdulaali, H.S.; Chopra, S.S. Automated Waste-Sorting and Recycling Classification Using Artificial Neural Network and Features Fusion: A Digital-Enabled Circular Economy Vision for Smart Cities. *Multimed. Tools Appl.* **2022**. [CrossRef]
25. Nallapaneni, M.K.; Chopra, S.S. Sustainability and Resilience of Circular Economy Business Models Based on Digital Ledger Technologies. In Proceedings of the In Waste Management and Valorisation for a Sustainable Future, Seoul, Korea, 26–28 October 2021.
26. Kumar, N.M.; CHOPRA, S.S. Blockchain Technology for Tracing Circularity of Material Flows. In Proceedings of the 10th International Conference of the International Society for Industrial Ecology (ISIE 2019): Industrial Ecology for Eco-civilization, Beijing, China, 7–11 July 2019.
27. Kumar, N.; Chopra, S. Blockchain Based Digital Infrastructure for Circular Economy. In Proceedings of the 8th International World Conference on Applied Science, Engineering and Management (WCSEM 2019), Fukuoka, Japan, 5–6 December 2019.
28. Nallapaneni, M.; Xu, Z.; Chopra, S. Blockchain-Empowered Online Information Sharing Platform for Enhancing Resilience of Circular Energy Supplies. In Proceedings of the International Symposium on Sustainable Systems and Technology (ISSST 2021), Virtual, 21–25 June 2021.
29. Ikram, M. Transition toward Green Economy: Technological Innovation's Role in the Fashion Industry. *Curr. Opin. Green Sustain. Chem.* **2022**, *37*, 100657. [CrossRef]
30. Ikram, M.; Sroufe, R.; Awan, U.; Abid, N. Enabling Progress in Developing Economies: A Novel Hybrid Decision-Making Model for Green Technology Planning. *Sustainability* **2021**, *14*, 258. [CrossRef]
31. Yeğin, T. The Place and Future of Artificial Intelligence in Marketing Strategies. *Ekev Akad. Derg.* **2020**, *81*, 489–506. [CrossRef]
32. Yeğin, T. The Relationship of Brands' Advertising Campaigns and Consumers' Digital Shopping Frenzy Behavior: A Review for Big Sale Days. In *Current Studies in Social Sciences IV*; Balcıoğulları, A., Ed.; Akademisyen Kitabevi: Ankara, Turkey, 2022; pp. 113–131, ISBN 9786258399561.
33. Yeğin, T.; Ikram, M. Performance Evaluation of Green Furniture Brands in the Marketing 4.0 Period: An Integrated MCDM Approach. *Sustainability* **2022**, *14*, 644. [CrossRef]
34. Käuferle, M.; Reinartz, W. Distributing through Multiple Channels in Industrial Wholesaling: How Many and How Much? *J. Acad. Mark. Sci.* **2015**, *43*, 746–767. [CrossRef]
35. Kumar, V.; Venkatesan, R. Who Are the Multichannel Shoppers and How Do They Perform?: Correlates of Multichannel Shopping Behavior. *J. Interact. Mark.* **2005**, *19*, 44–62. [CrossRef]

36. Montoya-Weiss, M.M.; Voss, G.B.; Grewal, D. Determinants of Online Channel Use and Overall Satisfaction with a Relational, Multichannel Service Provider. *J. Acad. Mark. Sci.* **2003**, *31*, 448–458. [CrossRef]
37. Verhoef, P.C.; Kannan, P.K.; Inman, J.J. From Multi-Channel Retailing to Omni-Channel Retailing. *J. Retail.* **2015**, *91*, 174–181. [CrossRef]
38. Meyer, B. What We Can Learn from Omnichannel Statistics for 2022. 2021. Available online: https://www.omnisend.com/blog/omnichannel-statistics/ (accessed on 28 August 2022).
39. Jena, S.K.; Meena, P. Competitive Sustainable Processes and Pricing Decisions in Omnichannel Closed-up Supply Chains under Different Channel Power Structures. *J. Retail. Consum. Serv.* **2022**, *69*, 103114. [CrossRef]
40. Borthakur, A.; Govind, M. Computer and Mobile Phone Waste in Urban India: An Analysis from the Perspectives of Public Perception, Consumption and Disposal Behaviour. *J. Environ. Plan. Manag.* **2019**, *62*, 717–740. [CrossRef]
41. Chen, X.; Su, X.; Li, Z.; Wu, J.; Zheng, M.; Xu, A. The Impact of Omni-Channel Collaborative Marketing on Customer Loyalty to Fresh Retailers: The Mediating Effect of the Omni-Channel Shopping Experience. *Oper. Manag. Res.* **2022**. [CrossRef]
42. Zhang, M.; He, X.; Qin, F.; Fu, W.; He, Z. Service Quality Measurement for Omni-Channel Retail: Scale Development and Validation. *Total Qual. Manag. Bus. Excell.* **2019**, *30*, S210–S226. [CrossRef]
43. Hossain, T.M.T.; Akter, S.; Kattiyapornpong, U.; Dwivedi, Y. Reconceptualizing Integration Quality Dynamics for Omnichannel Marketing. *Ind. Mark. Manag.* **2020**, *87*, 225–241. [CrossRef]
44. Lee, S.-Y.; Wang, S.J. Sensitivity Analysis of Structural Equation Models. *Psychometrika* **1996**, *61*, 93–108. [CrossRef]
45. Taufique Hossain, T.M.; Akter, S.; Kattiyapornpong, U.; Wamba, S.F. The Impact of Integration Quality on Customer Equity in Data Driven Omnichannel Services Marketing. *Procedia Comput. Sci.* **2017**, *121*, 784–790. [CrossRef]
46. Barwitz, N.; Maas, P. Understanding the Omnichannel Customer Journey: Determinants of Interaction Choice. *J. Interact. Mark.* **2018**, *43*, 116–133. [CrossRef]
47. Nica, E.; Sabie, O.-M.; Mascu, S.; Lutan (Petre), A.G. Artificial Intelligence Decision-Making in Shopping Patterns: Consumer Values, Cognition, and Attitudes. *Econ. Manag. Financ. Mark.* **2022**, *17*, 31. [CrossRef]
48. Shi, S.; Wang, Y.; Chen, X.; Zhang, Q. Conceptualization of Omnichannel Customer Experience and Its Impact on Shopping Intention: A Mixed-Method Approach. *Int. J. Inf. Manag.* **2020**, *50*, 325–336. [CrossRef]
49. Yang, S.; Zhou, Y.; Yao, J.; Chen, Y.; Wei, J. Understanding Online Review Helpfulness in Omnichannel Retailing. *Ind. Manag. Data Syst.* **2019**, *119*, 1565–1580. [CrossRef]
50. Banerjee, M. Misalignment and Its Influence on Integration Quality in Multichannel Services. *J. Serv. Res.* **2014**, *17*, 460–474. [CrossRef]
51. Sousa, R.; Voss, C.A. Service Quality in Multichannel Services Employing Virtual Channels. *J. Serv. Res.* **2006**, *8*, 356–371. [CrossRef]
52. Akter, S.; Hossain, M.I.; Lu, S.; Aditya, S.; Hossain, T.M.T.; Kattiyapornpong, U. Does Service Quality Perception in Omnichannel Retailing Matter? A Systematic Review and Agenda for Future Research. In *Exploring Omnichannel Retailing*; Springer International Publishing: Cham, Switzerland, 2019; pp. 71–97.
53. Wu, J.-F.; Chang, Y.P. Multichannel Integration Quality, Online Perceived Value and Online Purchase Intention. *Internet Res.* **2016**, *26*, 1228–1248. [CrossRef]
54. Herhausen, D.; Binder, J.; Schoegel, M.; Herrmann, A. Integrating Bricks with Clicks: Retailer-Level and Channel-Level Outcomes of Online–Offline Channel Integration. *J. Retail.* **2015**, *91*, 309–325. [CrossRef]
55. Cao, L.; Li, L. The Impact of Cross-Channel Integration on Retailers' Sales Growth. *J. Retail.* **2015**, *91*, 198–216. [CrossRef]
56. Hammerschmidt, M.; Falk, T.; Weijters, B. Channels in the Mirror. *J. Serv. Res.* **2016**, *19*, 88–101. [CrossRef]
57. Lee, Z.W.Y.; Chan, T.K.H.; Chong, A.Y.-L.; Thadani, D.R. Customer Engagement through Omnichannel Retailing: The Effects of Channel Integration Quality. *Ind. Mark. Manag.* **2019**, *77*, 90–101. [CrossRef]
58. Shen, X.-L.; Li, Y.-J.; Sun, Y.; Wang, N. Channel Integration Quality, Perceived Fluency and Omnichannel Service Usage: The Moderating Roles of Internal and External Usage Experience. *Decis. Support Syst.* **2018**, *109*, 61–73. [CrossRef]
59. Hamouda, M. Omni-Channel Banking Integration Quality and Perceived Value as Drivers of Consumers' Satisfaction and Loyalty. *J. Enterp. Inf. Manag.* **2019**, *32*, 608–625. [CrossRef]
60. Gao, M.; Huang, L. Quality of Channel Integration and Customer Loyalty in Omnichannel Retailing: The Mediating Role of Customer Engagement and Relationship Program Receptiveness. *J. Retail. Consum. Serv.* **2021**, *63*, 102688. [CrossRef]
61. Mainardes, E.W.; de Moura Rosa, C.A.; Nossa, S.N. Omnichannel Strategy and Customer Loyalty in Banking. *Int. J. Bank Mark.* **2020**, *38*, 799–822. [CrossRef]
62. Kim, S.; Connerton, T.P.; Park, C. Transforming the Automotive Retail: Drivers for Customers' Omnichannel BOPS (Buy Online & Pick up in Store) Behavior. *J. Bus. Res.* **2022**, *139*, 411–425. [CrossRef]
63. Hair, J.F.; Ringle, C.M.; Sarstedt, M. PLS-SEM: Indeed a Silver Bullet. *J. Mark. Theory Pract.* **2011**, *19*, 139–152. [CrossRef]
64. Hair, J.F.; Anderson, R.E.; Babin, B.J.; Black, W.C. *Multivariate Data Analysis: A Global Perspective*; Pearson: Upper Saddle River, NJ, USA, 2010; Volume 7.
65. Kline, R.B. Software Review: Software Programs for Structural Equation Modeling: Amos, EQS, and LISREL. *J. Psychoeduc. Assess.* **1998**, *16*, 343–364. [CrossRef]
66. Fornell, C.; Larcker, D.F. Structural Equation Models with Unobservable Variables and Measurement Error: Algebra and Statistics. *J. Mark. Res.* **1981**, *18*, 382–388. [CrossRef]

67. Benesty, J.; Chen, J.; Huang, Y.; Cohen, I. Pearson Correlation Coefficient. In *Noise Reduction in Speech Processing*; Springer: Berlin/Heidelberg, Germany, 2009; pp. 1–4.
68. Fornell, C.; Larcker, D.F. Evaluating Structural Equation Models with Unobservable Variables and Measurement Error. *J. Mark. Res.* **1981**, *18*, 39–50. [CrossRef]
69. Berman, B.; Thelen, S. A Guide to Developing and Managing a Well-integrated Multi-channel Retail Strategy. *Int. J. Retail. Distrib. Manag.* **2004**, *32*, 147–156. [CrossRef]
70. Cui, L.; Wang, Y.; Chen, W.; Wen, W.; Han, M.S. Predicting Determinants of Consumers' Purchase Motivation for Electric Vehicles: An Application of Maslow's Hierarchy of Needs Model. *Energy Policy* **2021**, *151*, 112167. [CrossRef]
71. Manser Payne, E.; Peltier, J.W.; Barger, V.A. Omni-Channel Marketing, Integrated Marketing Communications and Consumer Engagement. *J. Res. Interact. Mark.* **2017**, *11*, 185–197. [CrossRef]
72. Verhoef, P.C.; Neslin, S.A.; Vroomen, B. Multichannel Customer Management: Understanding the Research-Shopper Phenomenon. *Int. J. Res. Mark.* **2007**, *24*, 129–148. [CrossRef]
73. Vinerean, S.; Budac, C.; Baltador, L.A.; Dabija, D.-C. Assessing the Effects of the COVID-19 Pandemic on M-Commerce Adoption: An Adapted UTAUT2 Approach. *Electronics* **2022**, *11*, 1269. [CrossRef]
74. Brynjolfsson, E.; Hu, Y.J.; Rahman, M.S. Competing in the Age of Omnichannel Retailing. In *Cambridge: MIT Sloan Management Review*; Massachusetts Institute of Technology: Cambridge, MA, USA, 2013; pp. 1–7.
75. Gao, F.; Su, X. Online and Offline Information for Omnichannel Retailing. *Manuf. Serv. Oper. Manag.* **2017**, *19*, 84–98. [CrossRef]
76. Rigby, D. The Future of Shopping. *Harv. Bus. Rev.* **2011**, *89*, 65–76.
77. Popa, I.D.; Dabija, D.-C.; Grant, D.B. Exploring Omnichannel Retailing Differences and Preferences Among Consumer Generations. In *Griffiths School of Management and IT Annual Conference on Business, Entrepreneurship and Ethics*; Springer: Cham, Switzerland, 2019; pp. 129–146.
78. Lazaris, C.; Vrechopoulos, A. From Multichannel to "Omnichannel" Retailing: Review of the Literature and Calls for Research. In Proceedings of the In 2nd International Conference on Contemporary Marketing Issues, (ICCMI), Athens, Greece, 18–20 June 2014; pp. 1–6.
79. Levy, M.; Weitz, B.; Grewal, D. *Retailing Management*, 9th ed.; McGraw-Hill Education: New York, NY, USA, 2013.
80. Ailawadi, K.L.; Farris, P.W. Managing Multi- and Omni-Channel Distribution: Metrics and Research Directions. *J. Retail.* **2017**, *93*, 120–135. [CrossRef]
81. Bahar, V.S.; Nenonen, S.; Starr, R.G. From Channel Integration to Platform Integration: Capabilities Required in Hospitality. *Ind. Mark. Manag.* **2021**, *94*, 19–40. [CrossRef]
82. Li, Z.; Yang, W.; Jin, H.S.; Wang, D. Omnichannel Retailing Operations with Coupon Promotions. *J. Retail. Consum. Serv.* **2021**, *58*, 102324. [CrossRef]
83. Emrich, O.; Paul, M.; Rudolph, T. Shopping Benefits of Multichannel Assortment Integration and the Moderating Role of Retailer Type. *J. Retail.* **2015**, *91*, 326–342. [CrossRef]
84. Seck, A.M.; Philippe, J. Service Encounter in Multi-Channel Distribution Context: Virtual and Face-to-Face Interactions and Consumer Satisfaction. *Serv. Ind. J.* **2013**, *33*, 565–579. [CrossRef]
85. Sombultawee, K.; Wattanatorn, W. The Impact of Trust on Purchase Intention through Omnichannel Retailing. *J. Adv. Manag. Res.* **2022**, *19*, 513–532. [CrossRef]
86. Kliestik, T.; Kovalova, E.; Lăzăroiu, G. Cognitive Decision-Making Algorithms in Data-Driven Retail Intelligence: Consumer Sentiments, Choices, and Shopping Behaviors. *J. Self-Gov. Manag. Econ.* **2022**, *10*, 30–42.
87. Chang, Y.P.; Li, J. Seamless Experience in the Context of Omnichannel Shopping: Scale Development and Empirical Validation. *J. Retail. Consum. Serv.* **2022**, *64*, 102800. [CrossRef]
88. Gasparin, I.; Panina, E.; Becker, L.; Yrjölä, M.; Jaakkola, E.; Pizzutti, C. Challenging the "Integration Imperative": A Customer Perspective on Omnichannel Journeys. *J. Retail. Consum. Serv.* **2022**, *64*, 102829. [CrossRef]
89. Mishra, S.; Malhotra, G.; Chatterjee, R.; Shukla, Y. Consumer Retention through Phygital Experience in Omnichannel Retailing: Role of Consumer Empowerment and Satisfaction. *J. Strateg. Mark.* **2021**, 1–18. [CrossRef]
90. Neslin, S.A. The Omnichannel Continuum: Integrating Online and Offline Channels along the Customer Journey. *J. Retail.* **2022**, *98*, 111–132. [CrossRef]
91. Quach, S.; Barari, M.; Moudrý, D.V.; Quach, K. Service Integration in Omnichannel Retailing and Its Impact on Customer Experience. *J. Retail. Consum. Serv.* **2022**, *65*, 102267. [CrossRef]
92. Wakolbinger, L.M.; Stummer, C. Multi-Channel Management: An Exploratory Study of Current Practices. *Int. J. Serv. Econ. Manag.* **2013**, *5*, 112. [CrossRef]
93. Oh, L.-B.; Teo, H.-H. Consumer Value Co-Creation in a Hybrid Commerce Service-Delivery System. *Int. J. Electron. Commer.* **2010**, *14*, 35–62. [CrossRef]
94. Gao, W.; Li, W.; Fan, H.; Jia, X. How Customer Experience Incongruence Affects Omnichannel Customer Retention: The Moderating Role of Channel Characteristics. *J. Retail. Consum. Serv.* **2021**, *60*, 102487. [CrossRef]
95. Gao, W.; Fan, H.; Li, W.; Wang, H. Crafting the Customer Experience in Omnichannel Contexts: The Role of Channel Integration. *J. Bus. Res.* **2021**, *126*, 12–22. [CrossRef]
96. Blom, A.; Lange, F.; Hess, R.L. Omnichannel-Based Promotions' Effects on Purchase Behavior and Brand Image. *J. Retail. Consum. Serv.* **2017**, *39*, 286–295. [CrossRef]

97. Savastano, M.; Bellini, F.; D'Ascenzo, F.; de Marco, M. Technology Adoption for the Integration of Online–Offline Purchasing. *Int. J. Retail. Distrib. Manag.* **2019**, *47*, 474–492. [CrossRef]
98. Wang, S.; Fan, J.; Zhao, D.; Yang, S.; Fu, Y. Predicting Consumers' Intention to Adopt Hybrid Electric Vehicles: Using an Extended Version of the Theory of Planned Behavior Model. *Transportation* **2016**, *43*, 123–143. [CrossRef]
99. Adams, J.S. Inequity in Social Exchange. In *Advances in Experimental Social Psychology*; Academic Press: Cambridge, MA, USA, 1965; pp. 267–299.
100. Homans, G.C. The Humanities and the Social Sciences. *Am. Behav. Sci.* **1961**, *4*, 3–6. [CrossRef]
101. Low, B.; Johnston, W.J. Relationship Equity and Switching Behavior in the Adoption of New Telecommunication Services. *Ind. Mark. Manag.* **2006**, *35*, 676–689. [CrossRef]
102. Aravindakshan, A.; Rust, R.T.; Lemon, K.N.; Zeithaml, V.A. Customer Equity: Making Marketing Strategy Financially Accountable. *J. Syst. Sci. Syst. Eng.* **2004**, *13*, 405–422. [CrossRef]
103. Rust, R.T.; Lemon, K.N.; Zeithaml, V.A. *Driving Customer Equity: Linking Customer Lifetime Value to Strategic Marketing Decisions*; Marketing Science Institute: Cambridge, MA, USA, 2001; Volume 108.
104. Lemon, K.N.; Rust, R.T.; Zeithaml, V.A. What Drives Customer Equity? *Mark. Manag.* **2001**, *10*, 20–25.
105. Wang, B.; Wang, P.; Tu, Y. Customer Satisfaction Service Match and Service Quality-Based Blockchain Cloud Manufacturing. *Int. J. Prod. Econ.* **2021**, *240*, 108220. [CrossRef]
106. Keller, K.L.; Brexendorf, T.O. Measuring Brand Equity. In *Handbuch Markenführung*; Springer: Berlin/Heidelberg, Germany, 2019; pp. 1409–1439.
107. Esmaeilpour, M.; Mohamadi, Z.; Rajabi, A. Effect of Dimensions of Service Quality on the Brand Equity in the Fast Food Industry. *Stud. Bus. Econ.* **2016**, *11*, 30–46. [CrossRef]
108. Keller, K.L. Leveraging Secondary Associations to Build Brand Equity: Theoretical Perspectives and Practical Applications. *Int. J. Advert.* **2020**, *39*, 448–465. [CrossRef]
109. Rezaei, J.; Nispeling, T.; Sarkis, J.; Tavasszy, L. A Supplier Selection Life Cycle Approach Integrating Traditional and Environmental Criteria Using the Best Worst Method. *J. Clean Prod.* **2016**, *135*, 577–588. [CrossRef]
110. White, R.C.; Joseph-Mathews, S.; Voorhees, C.M. The Effects of Service on Multichannel Retailers' Brand Equity. *J. Serv. Mark.* **2013**, *27*, 259–270. [CrossRef]
111. Hole, Y.; Pawar, M.S.; Khedkar, E.B. Omni Channel Retailing: An Opportunity and Challenges in the Indian Market. *J. Phys. Conf. Ser.* **2019**, *1362*, 12121. [CrossRef]
112. Khan, M.A.; Ashraf, R.; Malik, A. Do Identity-Based Perceptions Lead to Brand Avoidance? A Cross-National Investigation. *Asia Pac. J. Mark. Logist.* **2019**, *31*, 1095–1117. [CrossRef]
113. Ladhari, R.; Leclerc, A. Building Loyalty with Online Financial Services Customers: Is There a Gender Difference? *J. Retail. Consum. Serv.* **2013**, *20*, 560–569. [CrossRef]
114. Nawafleh, S. Factors Affecting the Continued Use of E-Government Websites by Citizens. *Transform. Gov. People Process Policy* **2018**, *12*, 244–264. [CrossRef]
115. Pandey, S.; Chawla, D. Using Qualitative Research for Establishing Content Validity of E-Lifestyle and Website Quality Constructs. *Qual. Mark. Res. Int. J.* **2016**, *19*, 339–356. [CrossRef]
116. Keller, K.L. Brand Equity Management in a Multichannel, Multimedia Retail Environment. *J. Interact. Mark.* **2010**, *24*, 58–70. [CrossRef]
117. Rezaei, S.; Valaei, N. Branding in a Multichannel Retail Environment. *Inf. Technol. People* **2017**, *30*, 853–886. [CrossRef]
118. Algharabat, R.; Abdallah Alalwan, A.; Rana, N.P.; Dwivedi, Y.K. Three Dimensional Product Presentation Quality Antecedents and Their Consequences for Online Retailers: The Moderating Role of Virtual Product Experience. *J. Retail. Consum. Serv.* **2017**, *36*, 203–217. [CrossRef]
119. Kalia, P.; Paul, J. E-Service Quality and e-Retailers: Attribute-Based Multi-Dimensional Scaling. *Comput. Hum. Behav.* **2021**, *115*, 106608. [CrossRef]
120. Xu, X.; Jackson, J.E. Examining Customer Channel Selection Intention in the Omni-Channel Retail Environment. *Int. J. Prod. Econ.* **2019**, *208*, 434–445. [CrossRef]
121. Aaker, D.A.; Jacobson, R. The Financial Information Content of Perceived Quality. *J. Mark. Res.* **1994**, *31*, 191–201. [CrossRef]
122. Gotlieb, J.B.; Grewal, D.; Brown, S.W. Consumer Satisfaction and Perceived Quality: Complementary or Divergent Constructs? *J. Appl. Psychol.* **1994**, *79*, 875–885. [CrossRef]
123. Homer, P.M. Perceived Quality and Image: When All Is Not "Rosy". *J. Bus. Res.* **2008**, *61*, 715–723. [CrossRef]
124. Kirmani, A.; Zeithaml, V. Advertising, Perceived Quality, and Brand Image. In *Brand Equity and Advertising: Advertising's Role in Building Strong Brands*; Fuqua School of Business, Duke University: Durham, NC, USA, 1993; pp. 143–161.
125. Snoj, B.; Pisnik Korda, A.; Mumel, D. The Relationships among Perceived Quality, Perceived Risk and Perceived Product Value. *J. Prod. Brand Manag.* **2004**, *13*, 156–167. [CrossRef]
126. Völckner, F.; Hofmann, J. The Price-Perceived Quality Relationship: A Meta-Analytic Review and Assessment of Its Determinants. *Mark. Lett.* **2007**, *18*, 181–196. [CrossRef]
127. Vranešević', T.; Stančec, R. The Effect of the Brand on Perceived Quality of Food Products. *Br. Food J.* **2003**, *105*, 811–825. [CrossRef]
128. Collins-Dodd, C.; Lindley, T. Store Brands and Retail Differentiation: The Influence of Store Image and Store Brand Attitude on Store Own Brand Perceptions. *J. Retail. Consum. Serv.* **2003**, *10*, 345–352. [CrossRef]

129. Richardson, P.S.; Dick, A.S.; Jain, A.K. Extrinsic and Intrinsic Cue Effects on Perceptions of Store Brand Quality. *J. Mark.* **1994**, *58*, 28–36. [CrossRef]
130. Huang, Y.-T.; Cheng, F.-F. The Effect of Online Sales Promotion Strategies on Consumers' Perceived Quality and Purchase Intention: A Moderating Effect of Brand Awareness. In Proceedings of the 2013 Fifth International Conference on Service Science and Innovation, Kaohsiung, Taiwan, 29–31 May 2013; pp. 91–95.
131. Krishnamurthi, L.; Papatla, P. Accounting for Heterogeneity and Dynamics in the Loyalty–Price Sensitivity Relationship. *J. Retail.* **2003**, *79*, 121–135. [CrossRef]
132. Liu, J.; Ye, Q. How Price Affect Online Purchase Behavior in Online Healthcare Consulting? Perceived Quality as a Mediator. In Proceedings of the International Conference, ICSH 2015, Phoenix, AZ, USA, 17–18 November 2015; Springer: Cham, Switzerland, 2016; pp. 196–203.
133. Yoo, B.; Donthu, N.; Lee, S. An Examination of Selected Marketing Mix Elements and Brand Equity. *J. Acad. Mark. Sci.* **2000**, *28*, 195–211. [CrossRef]
134. Aaker, J. The Negative Attraction Effect? A Study of the Attraction Effect Under Judgment and Choice. In *NA–Advances in Consumer Research Volume*; Rebecca, H., Holman, M., Solomon, R., Eds.; ACR International, LLC: Provo, UT, USA, 1991; Volume 18, pp. 462–469.
135. Bagozzi, R.P. The Self-Regulation of Attitudes, Intentions, and Behavior. *Soc. Psychol. Q.* **1992**, *55*, 178. [CrossRef]
136. Brady, M.K.; Cronin, J.J. Some New Thoughts on Conceptualizing Perceived Service Quality: A Hierarchical Approach. *J. Mark.* **2001**, *65*, 34–49. [CrossRef]
137. Parasuraman, A.; Zeithaml, V.A.; Malhotra, A. E-S-QUAL. *J. Serv. Res.* **2005**, *7*, 213–233. [CrossRef]
138. Aaker, D.A.; Jacobson, R. The Value Relevance of Brand Attitude in High-Technology Markets. *J. Mark. Res.* **2001**, *38*, 485–493. [CrossRef]
139. Farquhar, P.H. Managing Brand Equity. In *Marketing Research*; Kaufman, M., Ed.; Scientific Research: Beijing, China, 1989; Volume 1.
140. Lassar, W.; Mittal, B.; Sharma, A. Measuring Customer-based Brand Equity. *J. Consum. Mark.* **1995**, *12*, 11–19. [CrossRef]
141. Ajzen, I.; Fishbein, M. A Bayesian Analysis of Attribution Processes. *Psychol. Bull.* **1975**, *82*, 261–277. [CrossRef]
142. Helson, H. Current Trends and Issues in Adaptation-Level Theory. *Am. Psychol.* **1964**, *19*, 26–38. [CrossRef]
143. Valos, M.J.; Ewing, M.T.; Powell, I.H. Practitioner Prognostications on the Future of Online Marketing. *J. Mark. Manag.* **2010**, *26*, 361–376. [CrossRef]
144. Hair, J.F.; Black, W.C.; Babin, B.J.; Anderson, R.E.; Tatham, R.L. Multivariate Data Analysis. *Up. River* **1998**, *5*, 207–219.
145. Brown, T.A.; Moore, M.T. *Confirmatory Factor Analysis. Handbook of Structural Equation Modeling, 361, 379*; Boston University: Boston, MA, USA, 2012.
146. Becker, J.-M.; Klein, K.; Wetzels, M. Hierarchical Latent Variable Models in PLS-SEM: Guidelines for Using Reflective-Formative Type Models. *Long Range Plann.* **2012**, *45*, 359–394. [CrossRef]
147. Raykov, T. Equivalent Structural Equation Models and Group Equality Constraints. *Multivar. Behav. Res.* **1997**, *32*, 95–104. [CrossRef]
148. Cohen, J. *Statistical Power Analysis for the Behavior Science*; Academic Press: Hilsdale, MI, USA, 1988.
149. Hanaysha, J. Testing the Effect of Service Quality on Brand Equity of Automotive Industry: Empirical Insights from Malaysia. *Glob. Bus. Rev.* **2016**, *17*, 1060–1072. [CrossRef]
150. Yeğin, T. Brand Loyalty in Creating Relationship Marketing Practices: A Study on Gsm Operators. *Elektron. Sos. Bilim. Derg.* **2020**, *77*, 201–216. [CrossRef]
151. Oh, T.T.; Keller, K.L.; Neslin, S.A.; Reibstein, D.J.; Lehmann, D.R. The Past, Present, and Future of Brand Research. *Mark. Lett.* **2020**, *31*, 151–162. [CrossRef]
152. Chang, Y.; Li, Y.; Yan, J.; Kumar, V. Getting More Likes: The Impact of Narrative Person and Brand Image on Customer–Brand Interactions. *J. Acad. Mark. Sci.* **2019**, *47*, 1027–1045. [CrossRef]
153. Yeğin, T.; Durmaz, Y. The Effect of Trust, Commitment, and Relational Satisfaction among the Relationship Quality Dimensions on Consumer-Based Brand Equity Dimensions: A Model Proposal. *Turk. Stud.-Soc. Sci.* **2020**, *15*, 3207–3234. [CrossRef]
154. Cheung, R.; Lam, A.Y.C.; Lau, M.M. Drivers of Green Product Adoption: The Role of Green Perceived Value, Green Trust and Perceived Quality. *J. Glob. Sch. Mark. Sci.* **2015**, *25*, 232–245. [CrossRef]

 *sustainability*

MDPI

Review

A Critical Assessment on Functional Attributes and Degradation Mechanism of Membrane Electrode Assembly Components in Direct Methanol Fuel Cells

Arunkumar Jayakumar [1,*], Dinesh Kumar Madheswaran [1] and Nallapaneni Manoj Kumar [2,*]

1   Green Vehicle Technology Research Centre, Department of Automobile Engineering, SRM Institute of Science and Technology, Chennai 603203, India; mdineshautomobile@gmail.com
2   School of Energy and Environment, City University of Hong Kong, Kowloon, Hong Kong
*   Correspondence: arunkumj1@srmist.edu.in (A.J.); mnallapan2@cityu.edu.hk (N.M.K.)

Abstract: Direct methanol fuel cells (DMFC) are typically a subset of polymer electrolyte membrane fuel cells (PEMFC) that possess benefits such as fuel flexibility, reduction in plant balance, and benign operation. Due to their benefits, DMFCs could play a substantial role in the future, specifically in replacing Li-ion batteries for portable and military applications. However, the critical concern with DMFCs is the degradation and inadequate reliability that affect the overall value chain and can potentially impede the commercialization of DMFCs. As a consequence, a reliability assessment can provide more insight into a DMFC component's attributes. The membrane electrode assembly (MEA) is the integral component of the DMFC stack. A comprehensive understanding of its functional attributes and degradation mechanism plays a significant role in its commercialization. The methanol crossover through the membrane, carbon monoxide poisoning, high anode polarization by methanol oxidation, and operating parameters such as temperature, humidity, and others are significant contributions to MEA degradation. In addition, inadequate reliability of the MEA impacts the failure mechanism of DMFC, resulting in poor efficiency. Consequently, this paper provides a comprehensive assessment of several factors leading to the MEA degradation mechanism in order to develop a holistic understanding.

Keywords: direct methanol fuel cells (DMFCs); polymer electrolyte membrane fuel cells (PEM-FCs); membrane electrode assembly (MEA); methanol crossover; polarization; methanol oxidation; flooding; Nafion®; platinum (Pt); ruthenium (Ru)

Citation: Jayakumar, A.; Madheswaran, D.K.; Kumar, N.M. A Critical Assessment on Functional Attributes and Degradation Mechanism of Membrane Electrode Assembly Components in Direct Methanol Fuel Cells. *Sustainability* 2021, 13, 13938. https://doi.org/10.3390/su132413938

Academic Editor: Nicu Bizon

Received: 15 November 2021
Accepted: 15 December 2021
Published: 16 December 2021

Publisher's Note: MDPI stays neutral with regard to jurisdictional claims in published maps and institutional affiliations.

1. Introduction

Accelerated climate change and environmental degradation are being suffered globally due to the widespread use of fossil fuels [1]. Fuel cells fall under the category of potential reliable energy systems [2], as most of the renewable energy systems suffer due to their intermittent operational nature [3]. Amongst different types of fuel cells, polymer electrolyte membrane fuel cells (PEMFCs) are arguably the fastest-growing and most likely to be used in the near future because of their unique attributes such as high power density, low operating temperature (60–80 °C), quick start-up, and dynamic response [4]. In particular, when compared to redox flow batteries [5] and Li-ion batteries [6], PEMFC is deemed to be a superior candidate considering numerous factors such as capital cost, electrical efficiency, dynamic response, and power density. Incidentally, the widely used fuel for PEM fuel cells is hydrogen, which poses challenges and concerns in terms of storage and safety that subsequently impede its widespread commercialization [7]. Therefore, significant endeavors have been committed to incorporate direct alcohol fuel cells (DAFC), considering its distinctive benefits such as simple operation, capability, high energy density, safe fuel handling and reasonably low environmental impact. Additionally, DAFC was established to tackle the storage crisis of hydrogen as well as to avoid the necessity of a reformer for the conversion of alcohol to hydrogen [8].

DAFC that uses methanol as a fuel is normally referred to as direct methanol fuel cells (DMFC). Methanol is the extensively utilized fuel for DAFC applications and on a volumetric basis, it has 50% higher specific energy density (6.09 kWh kg^{-1}) than liquid hydrogen (3.08 kWh kg^{-1}), and it is simple alcohol, with only one carbon atom, which oxidizes more effectively than other liquid hydrocarbon fuels. In addition, the methanol infrastructure is well established and safer, unlike hydrogen. [9]. Low boiling point (65 °C), high flammability, and ability to flow through the membrane from the anode to the cathode side (i.e., high crossover) are the limitations of using methanol as a fuel for DAFCs, because the chemical oxidation of methanol and the fuel crossover throughout DMFC operation culminates in a reduced power output, which substantially limits the widespread usage of DMFCs. Additionally, the methanol feed is diluted with CO_2 and probably nitrogen, which may comprise CO traces that act as a catalyst poison [10]. CO could be removed from fuel feed using water gas shift and oxidation reactors; however, the removal adversely affects the system efficiency and leads to an increase in volume, weight, start-up period, and response to variations in energy demand of the system [9,11]. Ethanol, on the other hand, can be used as an alternative fuel because of its advantages such as higher energy density than methanol, and its relatively low toxicity. Ethanol possesses additional benefits such as lower cross-over rates that have less detrimental impacts on DAFCs performance; however, ethanol has two carbon atoms (C-C) bonded, and this bond is quite complex to break at the lower operating temperatures of DAFCs [12]. Ethylene glycol is an alternative alcohol fuel for DAFCs, which is less toxic, safer in terms of handling, and has high energy density and lower volatility because of its high boiling point (about 198 °C) compared to methanol. Moreover, ethylene glycol has a theoretical energy density of 4.8 Ah mL^{-1}, which is 17% greater than that of methanol, i.e., 4 Ah mL^{-1}, which is particularly noteworthy for portable electronic applications [13]. Although various types of alcohol are employed as fuel for DAFC, the Direct Methanol Fuel Cells (DMFC) exhibit higher performance and lower crossover compared to DEFCs. This is because, when using ethanol as a fuel, there is a rapid decline in cell voltage at higher current densities, due to the rapid anode poisoning [14]. Owing to the aforementioned unique advantages, it is evident why DMFCs are the most widely employed alcohol-based fuel cells.

From the structural configuration, the DMFC comprises an anode/cathode gas diffusion layer (GDL), anode/cathode catalyst layers (CL), and an electrolyte (i.e., a proton conducting membrane), and bipolar plates (BPPs) on either side. Frequently, these components are engineered separately and pressed all together as a single unit at high pressure and temperature. The CL and GDL integrated with the membrane, as a single unit is referred to as membrane electrode assembly (MEA). This DMFCs' configuration is similar to that of a PEMFC stack. Methanol with water is fed through the flow channels of BPPs on the anode side of the cell through the flow channel of the bipolar plates and oxygen is fed on the cathode side.

The methanol and water react electrochemically and generate protons, electrons, and CO_2 at the anode, due to the methanol oxidation at the CL, as given in Equation (1). The membrane (Nafion®) conducts protons to the cathode and impede the electrons since the membrane is ionically conductive. The membrane also aids in CO_2 rejection, because insoluble carbonates are produced in alkaline electrolytes. The electrons thus travel through an external circuit, producing electrical energy and recombining with protons on the cathode side with oxygen atoms and producing water, as given in Equation (2). The overall reactant flow and the ion transfer mechanism is depicted in Figure 1 and given in Equation (3) [15–18].

$$\text{Anode reaction}: CH_3OH + H_2O \rightarrow CO_2 + 6H^+ + 6e^- \rightarrow E^{\circ}_{anode} 0.046 \text{ V} \tag{1}$$

$$\text{Cathode reaction}: \frac{3}{2}O_2 + 6H^+ + 6e^- \rightarrow 3H_2O \rightarrow E^{\circ}_{cathode} = 1.23 \text{ V} \tag{2}$$

$$\text{Overall reaction}: CH_3OH + \frac{3}{2}O_2 + H_2O \rightarrow CO_2 + 3H_2O \rightarrow E_{cell} = 1.18\ V \qquad (3)$$

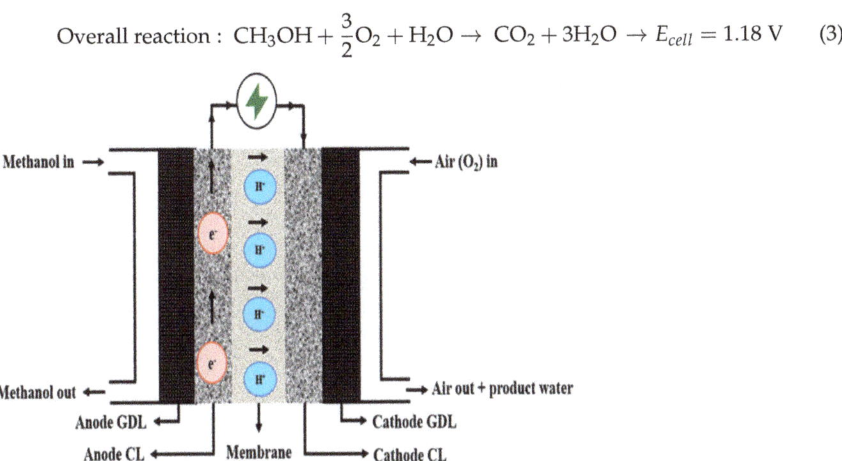

Figure 1. Schematic of DAFC and the Ion transfer Mechanism.

The most broadly employed membrane for DMFCs is Nafion®, as it demonstrates exceptional proton conductivity, as well as outstanding mechanical and chemical stability. However, Nafion® possess high fuel crossovers throughout the membrane and they are costly, have inadequate device lifespans, owing to the chemical and mechanical degradation. Hence, the potential constraint of the DMFC system is that it deteriorates from destitute anode kinetics, high catalyst loadings, low power density, and high cost [19–21]. Thus, DMFC and DEFC belong to the same family with marginally different characteristics.

In general, methanol should oxidize readily when the anode potential exceeds 0.046 V in proportion to the reversible hydrogen electrode (RHE). Correspondingly, when the cathode potential falls under 1.23 V, oxygen should be decreased gradually. In practice, the sluggish electrode kinetics (kinetic losses) drive electrode reactions to vary from their ideal thermodynamic values, resulting in a substantial reduction of the theoretical cell efficiency [22]. For any fuel cells, the performance, a good grade of stability is a vital requirement and subsequently, the degradation mechanisms for DMFC components have a direct influence on the performance of DMFCs [23]. In addition, the operating circumstances and the degradation processes are directly co-related, given that the degradation of active-type DMFC performance during cyclic voltage loading is significantly greater than under continuous voltage or current operation [24]. However, minimal consideration has been committed to studying the degradation mechanism of the membrane electrode assembly of DMFCs. The present paper provides prominence to the general MEA degradation mechanism of the DMFC system. Considering all these aspects, the paper is categorized into the following sections: Section 2 elaborates on the functional attributes of the various DAFC stack components including the membrane, CL, and the GDL. Section 3 expounds on the degradation involved in the MEA. Section 4 exemplifies the degradation mechanism in the DMFC component arranged systematically. Section 5 summarizes the significance of the work along with the critical assessment and limitations.

2. DMFC Functional Components

The critical components of the DMFC are the membrane electrode assembly which encompasses the membrane, CL and GDL. To be empathetic on the root causes in the MEA degradation involves the understanding of its functional attributes which are elaborated in the subsequent section.

2.1. Membrane

The membrane acts as a separating layer between the anode and cathode layer, which conducts protons and impedes the flow of electron. The membrane ought to have high proton conductivity, high chemical stability, and must be minimally permeable to methanol and water while possessing high durability at a reasonable cost [25,26]. The membranes for DMFC are normally based on fluorinated polymer and sulfonic acid groups similar to the PEMFC systems [27]. Nafion® is the widely used membrane, which is a perfluorinated polymer, primarily composed of a polytetrafluoroethylene (PTFE) support and lengthy perfluoro vinyl ether pendant side chains that are wrapped up by sulfonic functional groups. The polytetrafluoroethylene (PTFE) support offers chemical and thermal conductivity, while the sulfonated group ensures proton conductivity. The membrane is normally 175 microns in thickness, and the electrodes are typically 2 mm in thickness [28]. Many commercially available membranes are below 175 microns in thickness, but they are not reported to be efficient in proton conduction in the DMFC [29]. The chemical structure of a Nafion® perfluorinated ionomer is given in Figure 2 [30].

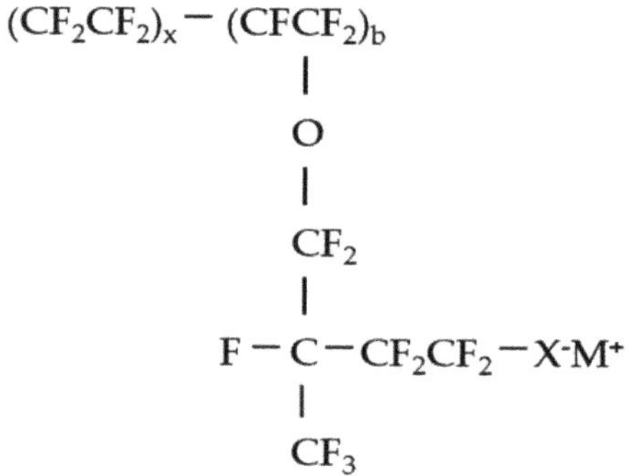

Figure 2. Typical structure of Nafion® perfluorinated ionomer [30].

The X in the structure of Nafion® predominantly refers to the sulfonic ionic functional groups and the M refers to the metal cation in neutral form of a proton (H^+) in the acidic form. Typically, a perfluoro sulfonyl fluoride copolymer-based Nafion® from DuPont, Delaware, USA is observed as the best-fit membrane for DMFCs, due to its high proton conductivity, exceptional mechanical properties, excellent chemical stability, and easy availability. However, it has shortcomings such as high manufacture expense, and at low humidity or high temperatures, these membranes are less proton conductive, have low mechanical property, high alcohol permeability, and are limited to operating temperatures [31]. As a result, such Nafion® membranes are susceptible to rapid dehydration at high temperatures, resulting in loss of proton conductivity, and in certain circumstances, causes irreversible changes in the microstructure of the membrane [32]. As a consequence, the key difficulties in contemporary DMFC research are to produce an alternate membrane capable of operating at higher temperatures or with low humidification of reactants. Choices of Nafion®, on the other hand, are frequently cheaper, i.e., sulfonated polyether ether ketone (sPEEK) membranes [33] and sulfonated poly aryl ethers (SPAEs) [34]. Few commercially available membranes for DMFCs are Aciplex (Asahi Kasei Chemicals, Tokyo, Japan) [35], Flemion (Asahi Glass, Chiba, Japan) [36], Gore-select (W. L. Gore & associates, Newark, DE, USA) [37] and the perfluoro sulfonic acid (PFSA) based Fumapen F-1850 and E-730, (Fumatech Bietigheim-Bissingen, Germany), [38] though Nafion from Dupont is the most prevalent.

2.2. Gas Diffusion Electrode (GDE)

The GDL and the electrode (catalyst) as a single unit is referred to as gas diffusion electrode (GDE) [39]. The GDE for DMFC is typically made of a porous mixture of carbon-backed platinum (Pt) or Ruthenium (Ru) [40]. Pt is predominantly used as the electrocatalyst for DMFC. However, with methanol as a fuel, there could be "CO poisoning." Therefore, Ru is added to promote the electrocatalyst activity by adsorption of OH and stripping off adsorbed CO from nearby Pt sites. Thus, at the anode, the catalyst, namely, Pt/Ru, in appropriate proportion initiates the methanol electro-oxidation to generate protons and electrons [41]. For an effective electrochemical reaction, the catalyst particles should be in close proximity with the protonic and electronic conductors. Additionally, there should be sections for reactants (i.e., porosity) to reach the catalyst zone and for reaction products to leave the cell [42]. The contact point of reactants, catalyst, and the electrolyte is usually indicated as the three-phase interface [43]. To accomplish an adequate rate of reaction, the area of catalyst zones must be several times greater than the geometrical area of the catalyst. Consequently, the catalysts are made porous to form a three-dimensional network, where the three-phase interfaces are established [44]. The catalysts are typically 0.45 mm thick (before hot-pressing), with a catalyst loading that lies within the range of 0.2 to 0.5 mg/cm^2. The catalyst loading is one of the cost-hindering aspects of the DMFC, compared to other MEA components, given that Pt-Ru/C anode catalyst constitutes 36% and Pt/C cathode catalyst constitute 21% of the overall cost of a single DMFC stack in mass production (10,000 units per year), while the membrane and GDL const 12% and 8%, respectively [45]. The DMFC stack cost could not be downsized unless the amount of Pt is reduced by any cost-effective element combined with it, or by entirely replacing Pt with any non-noble elements. Alternative catalysts can be investigated in addition to lessening the CO contamination, thereby reducing the overall cost of the DMFC [46]. Though in certain circumstances, the catalysts are coated onto the membrane, as in the present context, it is taken into the account the CL and GDL as the gas diffusion electrode. The precious catalyst is typically Pt, which has the highest activity towards oxygen reduction reaction (ORR) compared to all catalysts grounded on the Sabatier principle. This systematic configuration is termed as volcano plot and is illustrated in Figure 3 [47].

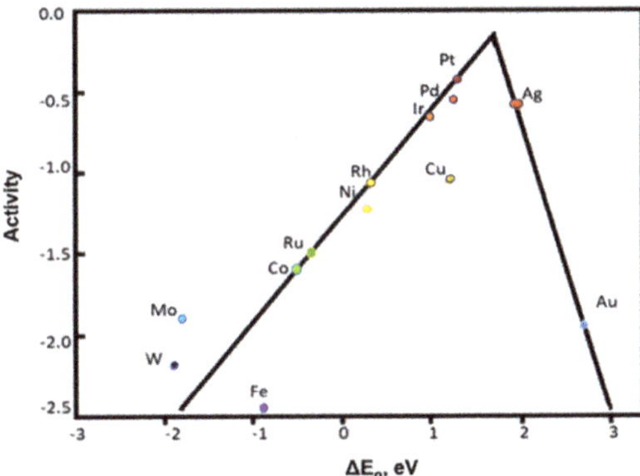

Figure 3. Volcano plots showing the catalytic activity trends as a function of the binding energy [47].

The GDL in general comprises a macroporous backing layer, usually made of porous, conductive carbon cloth or carbon cloth, and a microporous layer (MPL) [48]. Good diffusion characteristics (facilitating reactants to come in contact with the catalyst site); good electrical conductivity; stability in the DMFC environment; high permeability/pore size distribution for liquids and gases; good elasticity under compression; contact angle of the pores are the favorable features of GDE [49]. The most widely available carbon cloth-based GDL are ELAT [50], Avcarb [51] and CeTech [52], where carbon paper-based GDL are Avcarb [53], Toray [54], Freudenberg [55], Sigracet [56] and Spectracarb [57]. Carbon cloth has an ordered arrangement of fibers, larger pores, high porosity, and permeability, resulting in reduced mass transportation resistance and accelerating effective CO_2 removal [58]. Carbon paper, on the other hand, has a packed high denser structure and could be utilized to improve back diffusion of water by retaining a hydraulic pressure at the cathode [59]. A study on structural multiplicity and acclimatization dependence of DMFC reported that, using carbon cloth as the anode GDL and carbon paper as cathode GDL for a DMFC outperformed all other configurations [60]. Metal foams, metal meshes, and sintered metals are used as alternative GDLs in DMFCs [61,62]. Studies reported that the metal foam-enhanced passive DMFC performed better in terms of oxygen transportation and consequently cell performance [63].

3. MEA Degradation and Mitigation Strategies in DMFC

The MEA has often been termed the heart of DMFC. An insight into the degradation mechanism of the MEA can apparently increase the reliability of the DMFC stack. As a consequence, an assessment of the degradation in the membrane and gas diffusion electrode of a DMFC stack is mandatory. A comprehensive and systematic review is a promising way of representing such degradation, as those events can be classified as the primary, secondary, and tertiary consequences which cause the degradation of the stack components. This article logically assesses the combinations of the undesired events that can potentially lead to the undesired state based on numerous research articles reported. The primary consequences are those unsought or most intricate causes, the secondary consequence is the intermediate cause, and the tertiary consequences are at the bottom. In the present work, an assessment is performed on those degradation mechanisms that can lead to the failure of MEA components of DMFC. Nevertheless, similar assessments have been already performed by several researchers that largely focus on hydrogen-based PEMFCs [64–67]. Though the degradation of DMFC is similar to hydrogen-based PEMFCs, there are significant alterations that have to be assessed. The following sections elaborate on the degradation mechanism in the MEA components of DMFC.

3.1. Electrolyte Membrane

The primary constituent of the membrane degradation is the methanol crossover, followed by various other constituents such as CO poisoning, membrane thickness, methanol concentration, and its impurities, membrane assembly defects, reaction parameters, thermal and mechanical stability, and radical (OH*/HOO*/ROO*).

3.1.1. Methanol Crossover

In a liquid-fed DMFC, methanol in excess is provided to the anode of the MEA. It is desired to have all or for most of the methanol diffuse into the anode for the reaction. A phenomenon follows called "methanol crossover (MCO)" where a certain amount of the methanol diffuses over the membrane from the anode to the cathode [68]. The MCO in general is defined using Fick's law for diffusion across a polymer membrane as given in Equation (4) [69] and its schematic representation is given in Figure 4 [70].

$$J = -D\frac{dC}{dZ} \tag{4}$$

where J is the methanol flux, D is the coefficient of diffusion, C is the methanol concentration and Z is the location within the membrane.

The Inherent water diffusion characteristic of the membrane and methanol solubility in water triggers MCO [71]. Consequently, merely a few elements of methanol oxidize at the anode layer, where the rest of the elements are diffused through the membrane resulting in MCO [72]. This transportation is primarily by the diffusion and electro-osmotic drag (EOD) mechanism [73]. The mass transportation diffusion mechanism is fundamental during MCO, and it is proportional to methanol feed concentration. EOD is generated when proton transfer drags several methanol molecules. The EOD contribution is proportional in MCO, i.e., it increases with increase in amount of methanol fraction at the membrane-electrode (anode) interface and the current generated by the cell [71].

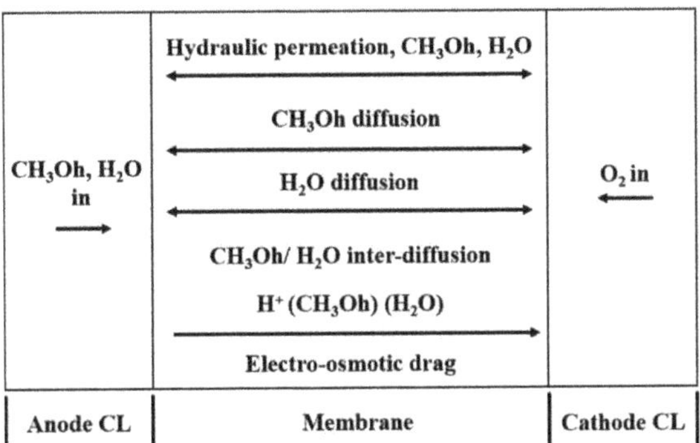

Figure 4. Schematic representation of methanol crossover in DMFC [70].

The MCO occurs mostly owing to the nature of the Nafion membrane, as Nafion is made up of hydrophilic side chains that comprise ionic sulfonic acid ($-SO_3H$) groups, which clustered in conjunction to produce ionic channels. While the water flow through ionic channels aids in the transport of protons, allowing for superior proton conductivity, it allows for the passage of methanol across the membrane. This aliphatic polymer structure of Nafion facilitates the development of larger ionic channels, resulting in enhanced methanol permeability [72].

Studies reported that the MCO mostly relies on the operating temperature [74–76] and the method of delivering the fuel [77]. It is given that the rate of fuel/oxidant delivered, and their concentration will substantially impact MCO [78]. The linear dependency of current density on temperature and methanol concentration was reported in numerous studies [79–83]. Figure 5 [84] represents cell current density versus methanol crossover current density as a function of temperature and methanol concentration, respectively. While the precise co-relation between permeability and MCO is complicated, certain variables such as the thickness of membrane [85] and methanol feed concentration [86] impact crossover in definite directions, i.e., for greater methanol concentration or thinner membrane, the crossover rates are higher [87]. In addition to permeability, the co-efficient of diffusion and solubility properties are often reported [88]. It is essential to measure these concepts and connect them to the MCO fluxes of operational DMFCs at this phase. This MCO leads to lower theoretical open-circuit voltage (OCV) of ~0.7 to 0.8 V vs. ~1.2 V and low power density of DMFC [89]. MCO produces a mixed potential at the cathode, and consequently decreases the methanol consumption in the anode, thereby decreasing the voltage of the DMFC around 400 mV [90].

The most amount of the MCO to the cathode will be electrochemically oxidized. This leads to the decrease in the performance of the cell and is subjected to the consumption of the cathode reactants [91]. In prevailing liquid feed DMFC, dilute methanol (6.41% to 9.61 vol% CH_3OH or 2 to 3 mol) is provided to the anode side of MEA. In such instances, the methanol diffuses to the anode and reacts incompletely.

The residual methanol leaves the CL and thus diffuses through the membrane to the cathode. The deficit of fuel decreases efficiency, which consequently reduces the cathode performance. The methanol loss at the anode side initiates the concentration at the surface of the catalyst to deterioration in the vicinity of the membrane accelerates the mass transportation resistance [92,93].

The hydrophilic properties of Nafion are enhanced by the concentration of hydrophilic domains and as a result, DMFCs elements are easily transported over perfluoro sulfonic acid membranes (mostly water and methanol). This consequently influences MCO from the anode to the cathode, which is primarily accomplished by the diffusion via the water-filled channels inside the Nafion structure and active transportation with protons and associated water solvent molecules during EOD [94]. A rigid polymer matrix [95] and a small free-volume membrane [96] could potentially bring down the electro-osmosis methanol diffusion. At the cathode, the MCO is oxidized to CO_2 and H_2O, lowering fuel efficiency and leading to cathode depolarization [97]. It is postulated that MCO reduces the efficiency of DMFC by 35%. Although the loss of oxygen from the cathode to the anode has an adverse effect on the efficiency of DMFC, it may be overlooked in contrast to the impact of the MCO. On contrary, mass transfer of N and CO through the membrane has no substantial effect on DMFC performance [95].

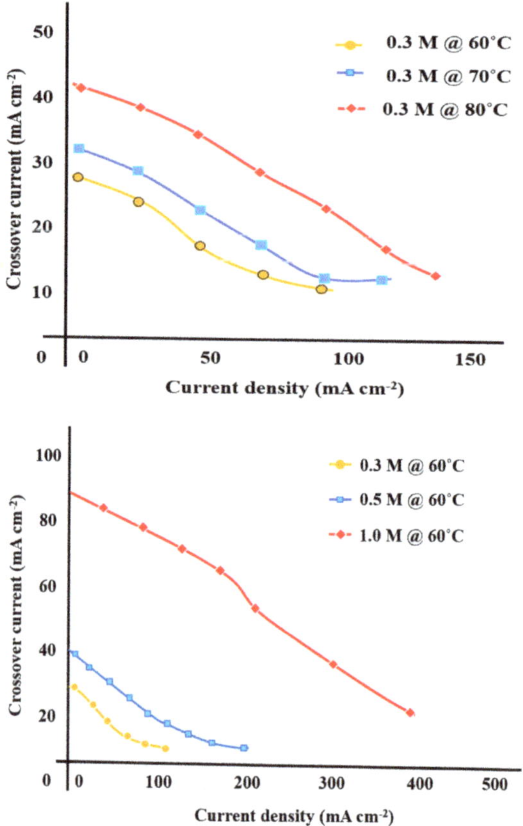

Figure 5. Cell current density versus methanol crossover current density as a function of temperature a methanol concentration [84].

It is evident from studies that the MCO is a significant factor that leads to the degradation of the membrane component. From the findings of various experiments reported, higher methanol concentration and substantial membrane thinning was detected, because the Nafion is soluble in methanol. This results in increased MCO and intense mixed potential formation on the cathode. Thus, the incompetence of the membrane to function as an efficient alcohol barrier decreases the performance of the cell as a result of mixed potential at the cathode [98–101]. Figure 6 [102] illustrates a graph comparing methanol crossover rates at OCV condition and in the presence of 1 and 5 M methanol concentration for some of the tested membranes including the commercial Nafion® 115. This measurement was performed by employing an online gas chromatograph to analyze the CO_2 at the cathode. In the sequences E-730, F-1850, and Nafion®-115, the crossover increased.

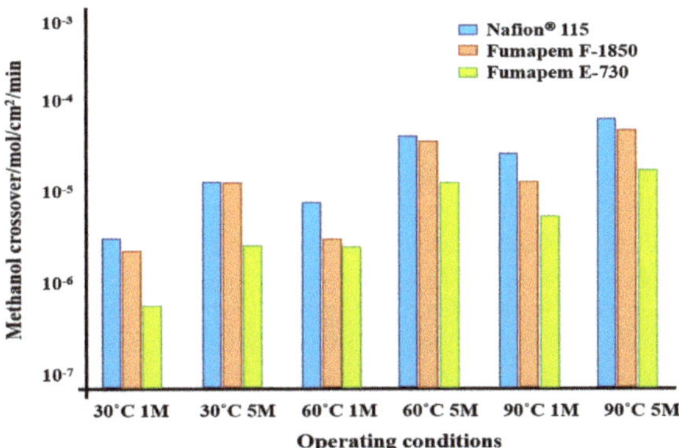

Figure 6. Methanol crossover rate of different membranes under different operating conditions (temperature and methanol concentration) [102].

Numerous studies have examined DMFC performance utilizing anion exchange membranes (AEMs) rather than the more typical proton-based Nafion membrane, with the vast majority reporting higher Ohmic resistance and lower MCO [103–105]. While AEM-DMFCs have a greater OCV and marginally improved initial performance compared to Nafion-based DMFCs, the resilience of such membranes often degrades in contrast to Nafion-based membranes, culminating in quick performance degradation [106–108]. The quaternized polyaromatics-based AEM polymers such as quaternized polyether ketone and quaternized polysulfone have recently received considerable attention, because of their outstanding conductivity properties, good mechanical and chemical stability [109,110]. However, synthesis of quaternized polyaromatics embraces hazardous compounds such as chloromethylation of benzene rings, which are highly carcinogenic and cytotoxic. Furthermore, the cost for tailoring polymers of quaternized polyaromatics is relatively expensive, which stands as the primary impediment to widespread commercialization [111,112]. Table 1 [113] provides the snapshot of electrochemical characteristics of different membranes used for DMFC.

Table 1. Electrochemical characteristics of different membranes used for DMFC [113].

Membrane Acronym	E-730	F-1850	Nafion® 115
Type of polymer	SPEEK	PFSA	PFSA
Equivalent weight (g mol^{-1})	700	1800	1100
Membrane thickness (μm)	30	120	125
Crossover current (mA cm^{-2})	48	100	195
Maximum power density @ 90 °C (mW cm^{-2})	77	38	64

Various studies have been conducted to control MCO [114,115]. The methanol concentration gradient through the Nafion membrane has high impacts over MCO; however, this approach is said to increase flooding and result in lower energy density [116]. Given that, the methanol tolerant catalysts cannot directly control the MCO rate, though it is capable of reducing the detrimental consequences of it [117–119].

3.1.2. Stack Assembly Effects

At high operating temperatures, the membrane degradation rate accelerates owing to the development of pinholes in the membrane in conjunction with cathode degradation and delamination [120]. If any flaw or pinhole appears on the membrane, the reactants blend under the catalytic effect that forms reaction intermediates such as CO, resulting in the hot-spot formation and, in turn, the pinhole dimensions or structural deformation [121]. In extreme circumstances, this could contribute to the explosion of cells, or structural flaws are caused by the following reasons: (i) When the membrane is stabbed by fibers or particles in the MEA during manufacturing and processor during stack assembly [122]; (ii) When the membrane is stabbed or lacerated by MEA or bipolar plate edges under stress produced by the stack bolts [123]; (iii) During long-term operation of the stack, peroxide corrosion makes the membrane thinner, and which leads to pinhole formation, that are more likely to be appeared in membrane drying cases [124]. Despite the above-mentioned reasons, membrane pinhole development can occur under extremely transient circumstances with a high number of humidity cycles. Because of the substantial volumetric expansion of ionomeric membranes under high humidity and in the presence of liquid water, variations in RH cause high mechanical stresses in the membrane, which contribute to membrane pinholes through in the lack of chemical degradation [122]. However, the pinhole formation triggered by short-term mechanical damage or steady long-term corrosion is not known as their effects on performance are not apparent; hence, it is imperative to develop approaches for perceiving membrane pinholes formation to circumvent the potential hazard of explosion or constant performance reduction, and subsequently the reduction in lifespan.

Studies have shown that stresses developed within the system could damage the membrane. These developed stresses are said to have various origins. The first is the initial stresses that arises within the MEA during stack assembly [125]. Numerous studies on optimizing the assembly procedure (such as bolt assembly), especially for the homogenization of the arising mechanical stresses have been reported [126]. A study on the mechanical response of membrane subjected to various clamping conditions observed the formation of pinhole at the center of the membrane which is due to the extreme form of non-uniformity in this constraint configuration [127]. One study observed that clamping stress is conducive to the durability and integrity of the membrane that, however, depends on the design specifications [128]. Evaluating various stack compression methods by considering the vibration effects in the clamping systems of the stack can offer better perceptions of optimal stack assembly procedure. Supplementary efforts are yet to be established to deliver models that combine the physical and electrochemical process concerning assembling and structural issues viz. stress, displacement (deformation), damage (cracks, delamination).

3.1.3. Thermal/Mechanical Stability

In general, Nafion membrane is volatile on operating the stack at temperatures range of above 100 °C. Better cell performance of the cell for Nafion-based membranes is achievable with the operating temperature range of 25 °C to 90 °C. Over 90 °C, the current density is limited as temperature increased, because of the membrane degradation [129]. In recent times, significant attention has been paid to Polybenzimidazole (PBI)-based membranes because of their high thermal stability, which is attributed to their tolerance when operating at higher temperatures (i.e., >100 °C) than Nafion membranes. To achieve maximum power output, it is desirable to increase the operating temperature; PBI membranes are able to endure while operating at temperatures in excess of 100 °C with a vaporized feed [130].

3.1.4. Methanol Concentration and Impurities

The other factors leading to the damage of the membranes are the impurities present in the methanol. DMFCs utilizing various types of commercial methanol con numerous compositions of impurities. The water content in the humidified methane encompasses cationic impurities namely Na^+, Cl^- and Mg^{2+}, that possibly contaminates the membrane surface [131]. The proton loss is more acute with low-valent cations, which is due to the decreased affinity of the sulfonic acid groups present in the Nafion membrane [132]. It was reported that that the MEA substantially degraded with an increase in the corrosion of austenitic stainless steel by the supply of various metal solution concentrations into the fuel stream of DMFC [133]. In addition to impurities with the methanol, the concentration of the fuel adversely affects the MEA performances. Optimum methanol concentration fall in the range of 1–2 M, which entails water dilution and subsequently far lower energy density. The higher alcohol concentration worsens the influence of permeated fuel on the performance of the membrane due to the MCO effect [134]. A study stated that the membrane becomes highly porous and the degree of membrane swelling is increased when there is an increase in alcohol concentration [135]. In general, the pure polar solvent has a greater membrane absorption than the pure non-polar solvent. When water is blended with a solvent, fortunately, the degree of swelling is negatively proportionate to the polarity of the solvent [136].

3.1.5. Membrane Thickness

Membrane thickness in a DMFC influences water crossover as well as water management, which thereby causes membrane degradation. A study reported that when the passive DMFC is worked with low methanol concentration and with a thicker membrane shows better performance at low current densities than at high current densities [137]. The thinnest sPEEK membrane was proven to have good DMFC performance, though also showing low Faradays efficiency (i.e., high methanol permeation with low Ohmic losses). On the other hand, the thickest membrane showed superior qualities concerning methanol permeation [138].

3.1.6. Reactive Oxygen Species

The formation of reactive oxygen species, such as the hydroxyl radical (OH), and hydroperoxyl/peroxyl radicals (HOO/ROO) can also cause the degradation of the membrane. Given that, the Nafion membranes are prone to the high absorption of reactive oxygen species, which leads to a high swelling ratio rate of membranes and fail to properly hold the contact between membrane and catalyst [139]. In the case of the PFSA membrane, the formation of reactive oxygen species also opens up the path for methanol permeability, which leads to high MCO [140].

3.1.7. Mitigation Strategies for Membrane Degradation

The contemporary reliance on Nafion membrane use for DMFC is significant due to its high conductivity and mechanical and chemical stability. The degradation of Nafion membranes, on the other hand, represents a considerable drawback. Hence, exploring an alternate membrane or increasing the performance of the Nafion is critical. Numerous literature has been reported to address the membrane degradation issue. Concerning the methanol crossover reduction, it can be accomplished by utilizing cross-linking procedures or adding up nanosized inorganic fillers within the membrane to improve the tortuosity path, as well as by fine-tuning the ion exchange capacity (IEC) [141]. The alternative approach contemplates the change of the chemical characteristics of the polymer network adjoining the ionic groups to modify the degree of dissociation and the degree of interpenetrated networks [142,143]. These approaches can reduce the level of methanol permeation permeability while keeping the proton conductivity at suitable levels; however, apparently, the cathode is less polarized in the presence of lower methanol permeation [144]. This corresponds to lower overpotentials for the oxygen reduction reaction. This strategy is further assisted by using methanol-tolerant cathode electro-catalysts with high activity for oxygen reduction [145]. FuMA-Tech created a new class of fluorinated membranes for use in DMFCs [146], which exhibited ion exchange capacities of 0.4 and 0.5 meq g^{-1}, that is nominally equated to equivalent weights of 1800 and 2300 g mol^{-1}, respectively. These results differed greatly from those of conventional PFSA membranes, such as Nafion$^{®}$ 115, with an equivalent weight of 1100 g mol^{-1} and an IEC of 0.91 meq g^{-1}. These blend membranes were designed to restrict methanol crossing, while the cast membrane thickness of 30–50 m enabled minimizing the membrane area resistance and cost of material as contrasted to Nafion$^{®}$ 115, which has a thickness of 125 m.

A relationship (reciprocal of the product of the area-specific resistance and the crossover) between DMFC power density and membrane selectivity is reported in a study [102]. This equation could be used to estimate DMFC performance based on fundamental membrane parameters in the presence of identical catalyst-loading, mechanical, and interfacial features. As explained above, selectivity is connected to intrinsic membrane features. Consequently, determining membrane selectivity does not always necessitate testing in DMFCs. Conductivity, thickness, and methanol permeation properties may all be used to calculate selectivity. If the electrode parameters are known, membrane selectivity can provide an indicator of DMFC performance at low temperatures, according to our research.

Using reinforced composite membrane is another strategy to improve membrane durability in DMFCs. The membrane with silane grafted on graphene oxide-treated mordenite with graphene oxide shows improved durability than conventional Nafion [147]. A thin Nafion membrane comprised of highly surface-functionalized sulfonated silica-coated polyvinylidene fluoride (S-SiO$_2$@PVDF) nanofiber mat and methanol-resistant chitosan. The proposed membrane showed three times greater wet tensile strength (25.2 MPa) and 1.6 times greater elongation (83.5%) than those of the pure Nafion membrane. More vitally, the improved ionic conductivity, reduced methanol permeability, and extremely limited swelling were attained for the composite membrane. These results show that the production of such a mechanically robust membrane with improved proton conductivity, and methanol resistance is a great structural design system. Likewise, in theory, decreasing the molecular weight or increasing the content of silane coupling agent coating could help enhance the conductivity of nanofibers. Similar studies were reported using quaternized chitosan reinforced with surface-functionalized PVDF electrospun nanofibers [148], Polyvinyl alcohol (PVA) reinforced composite membranes [149], poly(styrene sulfonic acid)-grafted poly(vinylidene fluoride) reinforced composites [150], Graphene quantum dot reinforced hyperbranched polyamide membrane [151], to improve the DMFC membrane durability.

3.2. Electrocatalyst Degradation

The DMFC widespread hindrance is highly influenced by the degradation of the catalyst layer which is accelerated by the following reasons: Catalyst poisoning; Pt/Ru agglomeration; Pt/Ru delamination; Pt/Ru dissolution; Ru leaching from Pt/Ru surface; Surface oxide formation and the membrane dissolution.

3.2.1. Catalytic Poisoning

CO poisoning, the most severe catalyst deactivation proc, is a critical concern, particularly in direct methanol fuel cells (DMFCs). Figure 7 [152] schematically illustrates the CO formation on the surface of the Pt catalyst.

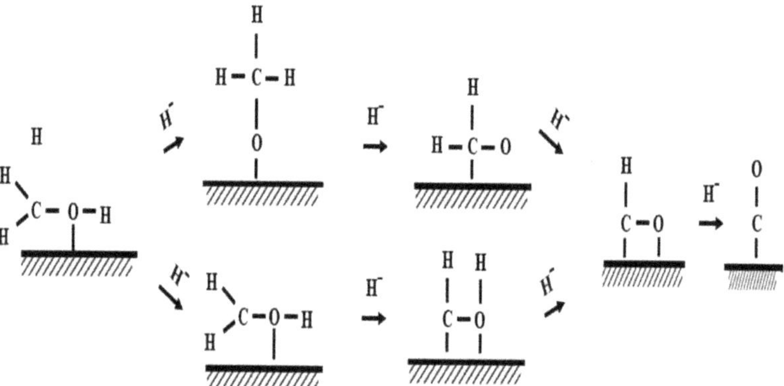

Figure 7. CO formation on the surface of Pt catalyst [152].

The intermediates of methanol electro-oxidation impede alcohol oxidation and absorb CO molecules on the electrocatalyst surface which then leads to poisoning the electrocatalyst and hinders the electro-oxidation kinetics [153]. This consequence lead to the use of alloys. Ruthenium as a potentially promising circumvent the CO poisoning [154]. In addition, transition metal carbides have benefits in terms of poison resistance; for instance, tungsten carbide (WC) has unique character traits such as high electrical conductivity, acid resistance, relatively low cost, and resistance to CO poisoning during methanol electro-oxidation [155].

3.2.2. Pt/Ru Agglomeration

The overall growth in particle size during methanol oxidation is described by agglomeration. The mechanism varies on the particles' distance and support characteristics. Given that the specific particles are in the nearby vicinity or linked to each other, thus producing the larger sized group, the catalyst material diffused and increase together throughout potential cycling. This further formed as a layer with the membrane and MPL on either side and degrades on constant operation [156]. The mechanism of Pt agglomeration at the cathode side of DMFC is given in Figure 8 [157]. Constant advancements in catalyst ink development have been attained such as decreasing the Pt/Ru loading and substituting Pt-black with Pt-supported carbon black [144]. The carbon support typically comprises several carbon black particles, namely, Vulcan (20 μm) or Ketjen black (50 μm). When the particles turn out to be coupled together by fusion bonding in the ink solution, then the aggregates produce agglomerates by attractive van der Waals forces [158].

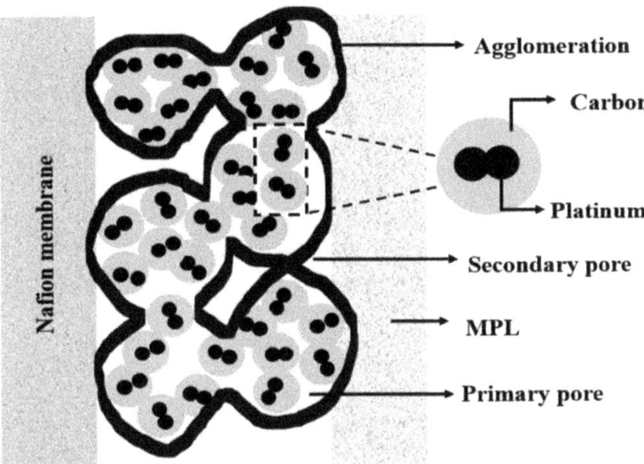

Figure 8. Pt agglomeration at cathode catalyst of DMFC [157].

In a study, aggregates of various forms were modelled as particle clusters and distributed randomly in space to develop an agglomerate composition of the catalyst layer. It was observed that when aggregates overlapped, a new spot of agglomeration was randomly formed [159].

3.2.3. Pt/Ru Dissolution

Conventional Pt-Ru-based electrocatalysts are reported to comprehend a substantial quantity of unalloyed Ru in the form of intermetallic phases or segregated oxides, which are responsible for the dissolution of Ru under collapsed pseudocapacitive currents [160]. Pt or Ru dissolution from the Pt–Ru anode catalyst by potentials greater than 0.5 V vs. DHE, observed by migration and accumulation to the cathode can decrease the activity of both anode and cathode catalysts and a deterioration of cell performance [161]. Figure 9 [162] illustrates the Pt dissolution mechanism in a typical DMFC. In a study, it is reported that if the Pt/Ru atomic ratio is unchanged or decreased, it also can lead to Ru losses, relying on the comparative amount of Pt and Ru lost. The standard Pt/Ru atomic ratio of the Pt-Ru catalyst is 1:1, hence, a drop in Ru concentration in the catalyst leads to a decline in MOR. Nevertheless, based on the degree of alloying, the Pt–Ru catalyst's MOR activity would increase after Ru loss [163].

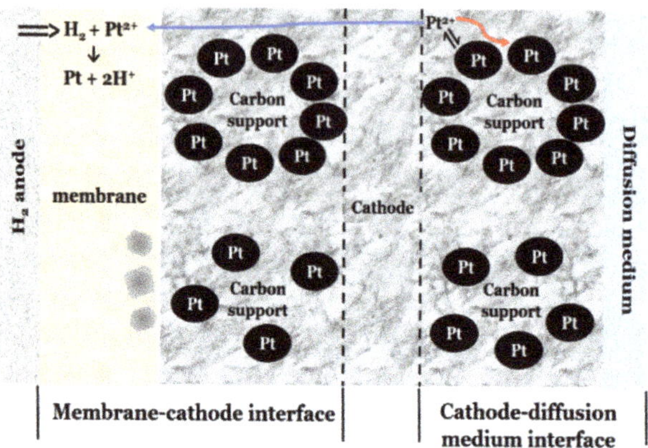

Figure 9. Dissolution of Pt from the surface of carbon-supported catalyst [162].

3.2.4. Pt/Ru Delamination

Due to the low binding energy between the catalyst-coated membrane and the CLs, delamination of CLs often occurs throughout repetitive operations. The bonding force supplied by the direct pressing approach of CCM cannot withstand the swelling of the membrane during the operation of the DMFC, resulting in delamination [164]. A study reported that the isotropic membrane swelling (i.e., expansion coefficient difference between the CL and membrane) enhances the variance in swelling ratio with catalyst encompasses a Nafion binder, which would most likely increase electrode layer delamination [165].

3.2.5. Ru Leaching from Pt/Ru Catalyst

In the absence of Ru near the Pt catalyst, the firmly bonded CO molecule generated following reactant transfer will degrade the Pt surface. As a result, it is critical to avoid certain operating circumstances that lead to Ru leaching [166]. Additionally, the Ru leached from the anode will be deposited on the Nafion membrane, causing membrane fouling. Finally, the anode's leached Ru can be deposited at the cathode, lowering the cathode catalyst's total oxygen reduction activity. Though the existence of Ru is crucial for MOR, when the anode potential achieves 0.5 V vs. RHE, the Ru (in oxidized form) particles would be receptive to leaching [167]. By quantifying the amount of Ru oxide from various sections of the DMFC, we can ascertain that the potential distribution in the cell is not uniform even in a single cell arrangement under typical operating circumstances. Because of the comparatively large anodic potentials, this condition may also result in preferential Ru leaching from the methanol inflow area [168].

3.2.6. Surface Oxidation Formation and Practical Growth

The high binding energies of both CO and oxygen-containing species such as surface oxides or adsorbed OH groups induce the quite often substantiated poisoning of pure platinum utilized as an anodic catalyst [169]. Intermediate species may also poison the catalyst and deteriorate the fuel cell performance. For instance, methanol will partially oxidize to form intermediate species such as formic acid, methyl formate, and formaldehyde [170]. Pt surface oxidation at the cathode is investigated in numerous studies using the oxide reduction peak of cyclic-voltammetry measurements, and it is determined that the electrocatalytic activity for oxygen reduction decreases, which contributes to degradation in the catalyst layer [171]. The substantial, oxidized Ru percentage in the anode catalyst was demonstrated to exert a major influence in the development of particles at the anode side and was found to be disseminated throughout the cathode in its oxidized form and, therefore, can have a considerable impact on the oxygen reduction activity (ORR) [172].

3.2.7. Carbon Corrosion

Carbon black (specifically Vulcan XC-72) is the most frequently used backing layer for DMFCs. These are typically made using pyrolyzing hydrocarbons. The high surface area ($\sim$250 m^2 g^{-1} for Vulcan XC-72), low cost, and effortless accessibility of carbon black help decrease the total cost of the cell [173]. While widely used as a catalyst–support, CBs still endure complications such as (i) the presence of organo-sulfur impurities [174] and (ii) deep micropores or recesses which deceits the catalyst nanoparticles making them inaccessible to reactants consequently leading to reduced catalytic activity. Under the acidic environment of a conventional DMFC, thermochemical stability is essential, and its deficiency results in corrosion of the carbon support and dissolution of the catalyst layer [175]. Carbon black, on the other hand, is mostly made up of planar graphite carbon and amorphous carbon, both of which have a lot of dangling bonds and flaws. The dangling bonds quickly generate surface oxides, leading to a faster corrosion rate during electrochemical oxidation [176]. A study reported that agglomeration of catalysts which is triggered by reactant starvation is correlated to carbon support corrosion [177].

3.2.8. Mitigation Strategies for Catalyst Degradation

Among various degradation mechanisms, the Pt/Ru agglomeration and dissolution were found to be the most significant contributors to catalyst degradation in DMFCs. Therefore, considerable attention has been paid to reducing the catalysts agglomeration and dissolution. Pt/Ru supported on TiO$_2$ embedded carbon nanofibers (Pt-Ru/TECNF), was reported as a highly active catalyst for methanol oxidation, which demonstrated reduced agglomeration compared to the conventional Pt catalysts supported on carbon [178,179]. A study reported that using ethanol solvent as catalyst ink for anode catalyst increased the interaction between Pt particles and ionomer, resulting in reduced agglomeration [180]. A similar study reported that using N-methyl pyrrolidone and dimethyl sulfoxide as catalyst ink enables the reduction of catalysts agglomeration [181]. Polyaniline-Silica (PANI-SiO$_2$) nanocomposite was created as a support for improving the performance of Pt/Ni electrocatalysts, to improve catalyst stability. The Pt/Ni/SiO$_2$-PANI electrocatalyst demonstrated exceptional catalytic activity. PANI-SiO$_2$, as an organic–inorganic hybrid catalyst support, significantly increased the stability and CO poisoning tolerance of the resultant electrocatalyst, according to experimental and theoretical data [182]. A thin, permeable silicon oxide (SiO$_x$) nanomembrane encapsulates a well-defined Pt thin film (SiO$_x$/Pt), is used as a catalyst-coated membrane to improve durability. The proposed catalyst demonstrated exceptional CO tolerance and highly active methanol oxidation, which also shows an improved lifespan compared to conventional catalyst [183]. TiO$_2$-Fe$_2$O$_3$@SiO$_2$-incorporated graphene oxide nanohybrid prepared by the hydrothermal method was used as a catalyst for DMFC. The nanohybrid showed greater stability with 91.58% retaining the initial current density later on 5000 s in the life span current–time curve [184]. Nickel–palladium supported onto mesostructured silica nanoparticles (NiPd–MSN) was used as an electrocatalyst for DMFC. The results of the study showed greater stability toward oxidation with 61% current retention and superior tolerance to the carbonaceous species accumulation compared with other electrocatalysts [185]. Poly(3,4-ethyl) (PEDOT) backed with carbon-supported Pt is used as anode catalysts for DMFC methanol oxidation which exhibits high mass activity and superior stability after 500 durability cycles, which is greater than those of commercial Pt/C catalyst [186]. A study demonstrated a dual-template approach to produce well-defined cage-bell nanostructures containing Pt core and a mesoporous PtM (M $\frac{1}{4}$ Co, Ni) bimetallic shell (Pt@mPtM (M $\frac{1}{4}$ Co, Ni) CB). The unique nanostructure and bimetallic properties of Pt@mPtM (M $\frac{1}{4}$ Co, Ni) CBs showed higher catalytic activity, superior durability, and greater CO tolerance for the methanol oxidation reaction than commercial Pt/C [187].

3.3. Gas Diffusion Layer

GDL is the most often overlooked cell component subjected to degradation in the DMFC system. This is attributed to the fact that the GDL degradation is much limited by the factors such as cell potential, porosity, and the effect due to the temperature.

3.3.1. Cell Potential

The diffusion layers are usually made of a carbon cloth or carbon paper that contributes a substantial role to the species transportation and structural integrity of MEA for PEM and any alcohol fuel cells [188]. The carbon base provides DMFCs with fairly good electrical conductivity between the catalyst and current collecting plates [189]. In general, GDL is around 100–400 μm thick and porous, which permits gas diffusion to the catalyst. Thinner layers are normally better as they possess nominal electrical resistance and let fuel and oxidants effortlessly pass through [190]. Nevertheless, carbon corrosion occurs at different voltage rates under several fuel cell operating conditions; however, the severity is low compared to a H_2-based PEM fuel cell as the nominal operating voltage is low for DMFC [191].

3.3.2. Porosity

GDL porosity can also be one of the parameters that can impact its durability and performance. The GDL is frequently wet-proofed with a hydrophobic material such as Teflon$^\circledR$ (PTFE). The hydrophobic material permits excess water rejection, thereby inhibiting flooding [192,193]. The primary physical parameters influencing GDL degradation are the gas permeability and the pore size diameter [194]. A study reported that the optimal pore size diameter to be around 25–40 μm, and greater than that would result in excess flooding, which degrades the GDL; however, increased porosity increases the current density [195]. The Teflon$^\circledR$ content and GDL thickness thus proven to have a larger impact on GDL degradation just than the porosity [196].

3.3.3. Operating Temperature

Limited water diffusion through the cathode GDL leads to flooding, and too much water diffusion can lead to cathode active layer and the polymer membrane dry triggering excessive cell resistance, which is called Ohmic polarization [197]. An increase in cell temperature than the standard defined level (60–100 °C) tends to dry the GDL, which consequently results in degradation [198]. On the other hand, low operating temperature is affected due to the progressive damage in carbon fiber and due to the change in the structure of the microporous layer [199]. Nevertheless, at higher operating temperatures, inherent mechanical characteristics of the GDL were drastically affected by the temperature-dependent parameters of the PTFE and epoxy resin resulting in a considerable reduction in resistance to GDL compression than found at lower operating temperatures [200].

3.3.4. Mitigation Strategies of GDL Degradation

The critical aspect impeding the output performance of conventional DMFCs is the minimum efficiency of the mass transport of oxygen, which is often a result of water flooding. Currently, researchers are largely determined on the water back diffusion from the cathode to the anode, as this can potentially solve the flooding at the cathode and decrease methanol crossover. A new hybrid catalyst layer (CL) was described in a study where relatively hydrophobic and hydrophilic CLs were integrated to form a hybrid CL [201]. The results of the study showed that the hydrophilic and hydrophobic control can efficiently generate a better distribution of methanol and water concentration. A three-dimensional graphene framework was applied to manufacture cathode MPL for improving water management [202]. The results indicated that the performance and stability were improved remarkably. From the literature review, water back diffusion improvement is an efficient approach for water management, inhibiting cathode flooding and reducing the MCO from the anode to the cathode. Though substantial improvements have been attained using the aforesaid techniques, the oxygen mass transport within the catalyst layer is however a challenge. A trilaminar-catalytic layered GDL design could accelerate the water back diffusion and encourage oxygen mass transportation. In a study, a trilaminar-catalytic layer comprises an inner, middle, and outer layer that is used for DMFC [203]. The middle layer has lower porosity compared to the inner and outer layers. This produces a water pressure gradient between the inner and middle layers and an oxygen concentration gradient between the outer and middle layers. Thus, the trilaminar-catalytic layered MEA can enhance water back diffusion from the cathode to the anode, as well as oxygen mass transportation, by creating beneficial gradients. An optimized MPL design is crucial to mitigate methanol crossover and improve DMFC performance. A study on the effects of MPL design in a DMFC indicated that anode MPL decreases the methanol concentration and liquid saturation in the anode CL. Cathode MPL improves the water back-flow from the cathode to the anode and Hydrophobic anode CL improves the water back-flow from the cathode to anode [204]. A study on the effect of porosity of the copper-fibre sintered felt (CFSF) demonstrated that GDL with a super-hydrophobic pattern that has a porosity of 60% attains the best performance compared to those of 50% and 70% porous, since it facilitates water removal when the water balance coefficient (WBC) is high [205].

4. Influence of Water Flooding in MEA Degradation

Cathode flooding of DMFC is a key contributor to the recoverable operational losses in DMFCs. Cathode flooding is categorized as catalyst flooding and backing flooding. Catalyst flooding is associated with several ORR active sites, the catalyst thickness, pore size, and its distribution [206]. The backing flooding is associated with the hydrophobicity and porosity of the backing layer [207]. Figure 10 [208] depicts the schematic representation of the water flooding mechanism in DMFC. The effect of cathode flooding in DMFC is considerably hazardous than a PEMFC, because of the aqueous methanol feed in the anode [209]. Almost 80% of the overall water content of DMFC cathode originates from the anode side, primarily through diffusion and electro-osmotic drag processes. This mostly degrades the initial fuel cell efficiency and also leads to excessive voltage drop rates throughout prolonged operation [210].

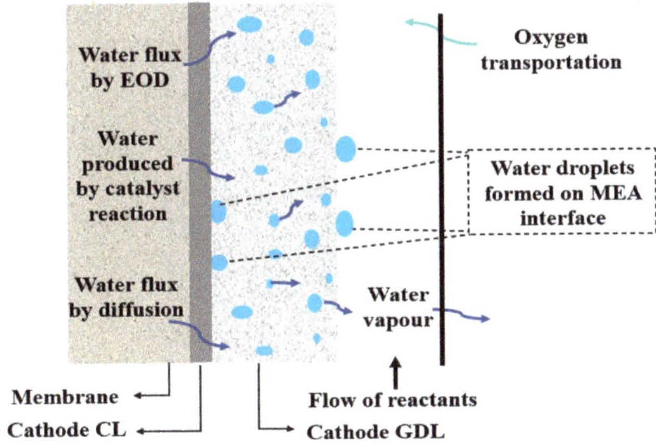

Figure 10. Water flooding mechanism in DMFC [208].

In general, DMFCs provide stable performance during the initial phase of a durability test. However, when the operating period is expanded, voltage loss is accelerated owing to water accumulation [211]. Water flooding intensifies with time and is known to effectively halt the operation of DMFC after only a few hundred hours of operation, particularly while catalyst-coated membrane (CCM)-type MEAs are employed. The following performance recovery of a DMFC can be achieved by cathode drying over longer durations, that could be as extensive as a few days [212]. Water flooding in a DMFC is commonly ascribed to a lack of hydrophobicity in the cathode GDL/MPL due to the dispersion of the poly-tetrafluoroethylene (PTFE) additive [213]. However, emphasizing the loss of GDL hydrophobic character as the sole reason over such an early phase of cathode flooding acceleration led to a misinterpretation of the precise accelerated rate of structural changes of GDL and potentially trigger an overlook of the significant contribution by morphological changes in the CCL [214]. In a study, using the Sessile-drop method, it is observed that decreasing hydrophobicity leads to a small contact angle of water droplets in the GDL fibers [215]. Despite the irreversible performance reduction, there is a chance that the physical degradation of electrodes, especially the cathode, will have implications for water drainage characteristics during a prolonged test, potentially increasing the water flooding conflict [216]. However, the number of lite is limited with linked water flooding behavior to morphological alterations in the CCL during protracted DMFC operation.

5. Conclusions

The DMFCs have the potential to play an important role in the future, specifically in replacing the Li-ion-based batteries for able and military applications. Inadequate reliability can potentially impede the commercialization of DMFCs. As a consequence, a reliability assessment can provide more insight into the component's attributes. Therefore, the present assessment emphasized the general degradation mechanism of MEA components of DMFCs. Incidentally, the durability of the MEA components needs to be circumvented for these systems to penetrate the market; specifically, the long-term durability of these systems should range from 3000 to 5000 operating hours. Aside from the durability of MEA components, operational methods and design can have a substantial influence on the durability characteristics of MEA, which might be a prerequisite for MEA robustness. This critical assessment can improve the reliability and, subsequently, complement the market penetration at a faster rate through a structured procedure which can be potentially useful for DMFC research. In this work, the reliability analysis is also carried out with the basic structured procedure, while excluding the system modelling and quantitative/qualitative analysis.

The authors believe that a systematic root cause analysis or a fault tree analysis (FTA) method of the present literature can help DMFC researchers and manufacturers to gain a holistic insight into the durability mechanism in a simple yet effective manner.

Author Contributions: Conceptualization, A.J., D.K.M. and N.M.K.; resources, A.J. and N.M.K.; data curation, A.J. and N.M.K.; writing—original draft preparation, A.J., D.K.M. and N.M.K.; writing—review and editing, A.J. and N.M.K.; visualization, D.K.M.; supervision, N.M.K.; project administration, A.J. and N.M.K. All authors have read and agreed to the published version of the manuscript.

Funding: This research received no external funding.

Institutional Review Board Statement: Not applicable.

Informed Consent Statement: Not applicable.

Data Availability Statement: Not applicable.

Acknowledgments: The authors would like to acknowledge the technical support from SusSo Foundation Cares, a Non-profitable Organization (NPO), India.

Conflicts of Interest: The authors declare no conflict of interest.

References

1. Chakraborty, S.; Kumar, N.M.; Jayakumar, A.; Dash, S.K.; Elangovan, D. Selected Aspects of Sustainable Mobility Reveals Implementable Approaches and Conceivable Actions. *Sustainability* **2021**, *13*, 12918. [CrossRef]
2. Jayakumar, A. An Assessment on Polymer Electrolyte Membrane Fuel Cell Stack Components. In *Applied Physical Chemistry with Multidisciplinary Approaches*; Apple Academic Press: Cambridge, MA, USA, 2018; ISBN 978-1-315-16941-5.

3. Sun, C.; Negro, E.; Vezzù, K.; Pagot, G.; Cavinato, G.; Nale, A.; Herve Bang, Y.; Di Noto, V. Hybrid Inorganic-Organic Proton-Conducting Membranes Based on SPEEK Doped with WO3 Nanoparticles for Application in Vanadium Redox Flow Batteries. *Electrochim. Acta* **2019**, *309*, 311–325. [CrossRef]
4. Garraín, D.; Banacloche, S.; Ferreira-Aparicio, P.; Martínez-Chaparro, A.; Lechón, Y. Sustainability Indicators for the Manufacturing and Use of a Fuel Cell Prototype and Hydrogen Storage for Portable Uses. *Energies* **2021**, *14*, 6558. [CrossRef]
5. Cost-Effective Iron-Based Aqueous Redox Flow Batteries for Large-Scale Energy Storage Application: A Review. *J. Power Sources* **2021**, *493*, 229445. [CrossRef]
6. Wu, F.; Maier, J.; Yu, Y. Guidelines and Trends for Next-Generation Rechargeable Lithium and Lithium-Ion Batteries. *Chem. Soc. Rev.* **2020**, *49*, 1569–1614. [CrossRef]
7. Staffell, I.; Scamman, D.; Abad, A.V.; Balcombe, P.; Dodds, P.E.; Ekins, P.; Shah, N.; Ward, K.R. The Role of Hydrogen and Fuel Cells in the Global Energy System. *Energy Environ. Sci.* **2019**, *12*, 463–491. [CrossRef]
8. Halme, A.; Selkäinaho, J.; Noponen, T.; Kohonen, A. An Alternative Concept for DMFC—Combined Electrolyzer and H2 PEMFC. *Int. J. Hydrogen Energy* **2016**, *41*, 2154–2164. [CrossRef]
9. Munjewar, S.S.; Thombre, S.B.; Patil, A.P. Passive Direct Alcohol Fuel Cell Using Methanol and 2-Propanol Mixture as a Fuel. *Ionics* **2019**, *25*, 2231–2241. [CrossRef]
10. Cheng, Y.; Zhang, J.; Lu, S.; Jiang, S.P. Significantly Enhanced Performance of Direct Methanol Fuel Cells at Elevated Temperatures. *J. Power Sources* **2020**, *450*, 227620. [CrossRef]
11. Dao, D.V.; Adilbish, G.; Le, T.D.; Nguyen, T.T.D.; Lee, I.-H.; Yu, Y.-T. Au@CeO2 Nanoparticles Supported Pt/C Electrocatalyst to Improve the Removal of CO in Methanol Oxidation Reaction. *J. Catal.* **2019**, *377*, 589–599. [CrossRef]
12. Badwal, S.P.S.; Giddey, S.; Kulkarni, A.; Goel, J.; Basu, S. Direct Ethanol Fuel Cells for Transport and Stationary Applications—A Comprehensive Review. *Appl. Energy* **2015**, *145*, 80–103. [CrossRef]
13. Pan, Z.; Bi, Y.; An, L. Mathematical Modeling of Direct Ethylene Glycol Fuel Cells Incorporating the Effect of the Competitive Adsorption. *Appl. Therm. Eng.* **2019**, *147*, 1115–1124. [CrossRef]
14. Chu, Y.H.; Shul, Y.G. Combinatorial Investigation of Pt–Ru–Sn Alloys as an Anode Electrocatalysts for Direct Alcohol Fuel Cells. *Int. J. Hydrogen Energy* **2010**, *35*, 11261–11270. [CrossRef]
15. Achmad, F.; Kamarudin, S.K.; Daud, W.R.W.; Majlan, E.H. Passive Direct Methanol Fuel Cells for Portable Electronic Devices. *Appl. Energy* **2011**, *88*, 1681–1689. [CrossRef]
16. Bahrami, H.; Faghri, A. Review and Advances of Direct Methanol Fuel Cells: Part II: Modeling and Numerical Simulation. *J. Power Sources* **2013**, *230*, 303–320. [CrossRef]
17. Mallick, R.K.; Thombre, S.B.; Shrivastava, N.K. Vapor Feed Direct Methanol Fuel Cells (DMFCs): A Review. *Renew. Sustain. Energy Rev.* **2016**, *56*, 51–74. [CrossRef]
18. Falcão, D.S.; Oliveira, V.B.; Rangel, C.M.; Pinto, A.M.F.R. Review on Micro-Direct Methanol Fuel Cells. *Renew. Sustain. Energy Rev.* **2014**, *34*, 58–70. [CrossRef]
19. Karimi, M.B.; Mohammadi, F.; Hooshyari, K. Recent Approaches to Improve Nafion Performance for Fuel Cell Applications: A Review. *Int. J. Hydrogen Energy* **2019**, *44*, 28919–28938. [CrossRef]
20. Lufrano, F.; Baglio, V.; Staiti, P.; Antonucci, V.; Arico', A.S. Performance Analysis of Polymer Electrolyte Membranes for Direct Methanol Fuel Cells. *J. Power Sources* **2013**, *243*, 519–534. [CrossRef]
21. Pethaiah, S.S.; Arunkumar, J.; Ramos, M.; Al-Jumaily, A.; Manivannan, N. The Impact of Anode Design on Fuel Crossover of Direct Ethanol Fuel Cell. *Bull. Mater. Sci.* **2016**, *39*, 273–278. [CrossRef]
22. Kumar, P.; Dutta, K.; Das, S.; Kundu, P.P. An Overview of Unsolved Deficiencies of Direct Methanol Fuel Cell Technology: Factors and Parameters Affecting Its Widespread Use. *Int. J. Energy Res.* **2014**, *38*, 1367–1390. [CrossRef]
23. Park, J.-Y.; Park, K.-Y.; Kim, K.B.; Na, Y.; Cho, H.; Kim, J.-H. Influence and Mitigation Methods of Reaction Intermediates on Cell Performance in Direct Methanol Fuel Cell System. *J. Power Sources* **2011**, *196*, 5446–5452. [CrossRef]
24. Bresciani, F.; Rabissi, C.; Zago, M.; Gazdzicki, P.; Schulze, M.; Guétaz, L.; Escribano, S.; Bonde, J.L.; Marchesi, R.; Casalegno, A. A Combined In-Situ and Post-Mortem Investigation on Local Permanent Degradation in a Direct Methanol Fuel Cell. *J. Power Sources* **2016**, *306*, 49–61. [CrossRef]
25. Kreuer, K.-D. Ion Conducting Membranes for Fuel Cells and Other Electrochemical Devices. *Chem. Mater.* **2014**, *26*, 361–380. [CrossRef]
26. Ahmad, H.; Kamarudin, S.K.; Hasran, U.A.; Daud, W.R.W. Overview of Hybrid Membranes for Direct-Methanol Fuel-Cell Applications. *Int. J. Hydrogen Energy* **2010**, *35*, 2160–2175. [CrossRef]
27. Kim, D.S.; Guiver, M.; McGrath, J.; Pivovar, B.; Kim, Y.S. Molecular Design Aspect of Sulfonated Polymers for Direct Methanol Fuel Cells. *ECS Trans.* **2010**, *33*, 711. [CrossRef]
28. Sharma, S. *Membranes for Low Temperature Fuel Cells: New Concepts, Single-Cell Studies and Applications*; De Gruyter: Berlin, Germany, 2019; ISBN 978-3-11-064732-7.
29. Lin, H.-L.; Wang, S.-H. Nafion/Poly(Vinyl Alcohol) Nano-Fiber Composite and Nafion/Poly(Vinyl Alcohol) Blend Membranes for Direct Methanol Fuel Cells. *J. Membr. Sci.* **2014**, *452*, 253–262. [CrossRef]
30. Choi, J.S.; Sohn, J.-Y.; Shin, J. A Comparative Study on EB-Radiation Deterioration of Nafion Membrane in Water and Isopropanol Solvents. *Energies* **2015**, *8*, 5370–5380. [CrossRef]

31. Aydın Ünal, F.; Erduran, V.; Timuralp, C.; Şen, F. 15—Fabrication and Properties of Polymer Electrolyte Membranes (PEM) for Direct Methanol Fuel Cell Application. In *Nanomaterials for Direct Alcohol Fuel Cells*; Şen, F., Ed.; Micro and Nano Technologies; Elsevier: Amsterdam, The Netherlands, 2021; pp. 283–302. ISBN 978-0-12-821713-9.
32. Ercelik, M.; Ozden, A.; Devrim, Y.; Colpan, C.O. Investigation of Nafion Based Composite Membranes on the Performance of DMFCs. *Int. J. Hydrogen Energy* **2017**, *42*, 2658–2668. [CrossRef]
33. Wang, J.; Liao, J.; Yang, L.; Zhang, S.; Huang, X.; Ji, J. Highly Compatible Acid–Base Blend Membranes Based on Sulfonated Poly(Ether Ether Ketone) and Poly(Ether Ether Ketone-Alt-Benzimidazole) for Fuel Cells Application. *J. Membr. Sci.* **2012**, *415–416*, 644–653. [CrossRef]
34. Liu, C.; Wang, X.; Xu, J.; Wang, C.; Chen, H.; Liu, W.; Chen, Z.; Du, X.; Wang, S.; Wang, Z. PEMs with High Proton Conductivity and Excellent Methanol Resistance Based on Sulfonated Poly (Aryl Ether Ketone Sulfone) Containing Comb-Shaped Structures for DMFCs Applications. *Int. J. Hydrogen Energy* **2020**, *45*, 945–957. [CrossRef]
35. Dutta, K.; Das, S.; Kundu, P.P. Synthesis, Preparation, and Performance of Blends and Composites of π-Conjugated Polymers and Their Copolymers in DMFCs. *Polym. Rev.* **2015**, *55*, 630–677. [CrossRef]
36. Kludský, M.; Vopička, O.; Matějka, P.; Hovorka, Š.; Friess, K. Nafion® Modified with Primary Amines: Chemical Structure, Sorption Properties and Pervaporative Separation of Methanol-Dimethyl Carbonate Mixtures. *Eur. Polym. J.* **2018**, *99*, 268–276. [CrossRef]
37. Klose, C.; Breitwieser, M.; Vierrath, S.; Klingele, M.; Cho, H.; Büchler, A.; Kerres, J.; Thiele, S. Electrospun Sulfonated Poly(Ether Ketone) Nanofibers as Proton Conductive Reinforcement for Durable Nafion Composite Membranes. *J. Power Sources* **2017**, *361*, 237–242. [CrossRef]
38. Pérez, L.C.; Brandão, L.; Sousa, J.M.; Mendes, A. Segmented Polymer Electrolyte Membrane Fuel Cells—A Review. *Renew. Sustain. Energy Rev.* **2011**, *15*, 169–185. [CrossRef]
39. Prokop, M.; Kodym, R.; Bystron, T.; Drakselova, M.; Paidar, M.; Bouzek, K. Degradation Kinetics of Pt during High-Temperature PEM Fuel Cell Operation Part II: Dissolution Kinetics of Pt Incorporated in a Catalyst Layer of a Gas-Diffusion Electrode. *Electrochim. Acta* **2020**, *333*, 135509. [CrossRef]
40. Kamarudin, S.K.; Hashim, N. Materials, Morphologies and Structures of MEAs in DMFCs. *Renew. Sustain. Energy Rev.* **2012**, *16*, 2494–2515. [CrossRef]
41. Xinyao, Y.; Zhongqing, J.; Yuedong, M. Preparation of Anodes for DMFC by Co-Sputtering of Platinum and Ruthenium. *Plasma Sci. Technol.* **2010**, *12*, 224–229. [CrossRef]
42. Dutta, K.; Das, S.; Rana, D.; Kundu, P.P. Enhancements of Catalyst Distribution and Functioning Upon Utilization of Conducting Polymers as Supporting Matrices in DMFCs: A Review. *Polym. Rev.* **2015**, *55*, 1–56. [CrossRef]
43. Gulaboski, R.; Mirceski, V.; Komorsky-Lovric, S.; Lovric, M. Three-Phase Electrodes: Simple and Efficient Tool for Analysis of Ion Transfer Processes across Liquid-Liquid Interface—Twenty Years On. *J Solid State Electrochem.* **2020**, *24*, 2575–2583. [CrossRef]
44. Li, Q.; Wang, T.; Havas, D.; Zhang, H.; Xu, P.; Han, J.; Cho, J.; Wu, G. High-Performance Direct Methanol Fuel Cells with Precious-Metal-Free Cathode. *Adv. Sci.* **2016**, *3*, 1600140. [CrossRef] [PubMed]
45. Sgroi, M.F.; Zedde, F.; Barbera, O.; Stassi, A.; Sebastián, D.; Lufrano, F.; Baglio, V.; Aricò, A.S.; Bonde, J.L.; Schuster, M. Cost Analysis of Direct Methanol Fuel Cell Stacks for Mass Production. *Energies* **2016**, *9*, 1008. [CrossRef]
46. Samad, S.; Loh, K.S.; Wong, W.Y.; Lee, T.K.; Sunarso, J.; Chong, S.T.; Wan Daud, W.R. Carbon and Non-Carbon Support Materials for Platinum-Based Catalysts in Fuel Cells. *Int. J. Hydrogen Energy* **2018**, *43*, 7823–7854. [CrossRef]
47. Khotseng, L. *Oxygen Reduction Reaction*; IntechOpen: Gujarat, India, 2018; ISBN 978-1-78984-813-7.
48. Yan, X.H.; Zhao, T.S.; An, L.; Zhao, G.; Zeng, L. A Crack-Free and Super-Hydrophobic Cathode Micro-Porous Layer for Direct Methanol Fuel Cells. *Appl. Energy* **2015**, *138*, 331–336. [CrossRef]
49. Abdelkareem, M.A.; Sayed, E.T.; Nakagawa, N. Significance of Diffusion Layers on the Performance of Liquid and Vapor Feed Passive Direct Methanol Fuel Cells. *Energy* **2020**, *209*, 118492. [CrossRef]
50. Ivanova, N.A.; Alekseeva, O.K.; Fateev, V.N.; Shapir, B.L.; Spasov, D.D.; Nikitin, S.M.; Presnyakov, M.Y.; Kolobylina, N.N.; Soloviev, M.A.; Mikhalev, A.I.; et al. Activity and Durability of Electrocatalytic Layers with Low Platinum Loading Prepared by Magnetron Sputtering onto Gas Diffusion Electrodes. *Int. J. Hydrogen Energy* **2019**, *44*, 29529–29536. [CrossRef]
51. Netzeband, C.; Arlt, T.; Wippermann, K.; Lehnert, W.; Manke, I. Three-Dimensional Multiscale Analysis of Degradation of Nano- and Micro-Structure in Direct Methanol Fuel Cell Electrodes after Methanol Starvation. *J. Power Sources* **2016**, *327*, 481–487. [CrossRef]
52. Nagy, K.A.; Tóth, I.Y.; Ballai, G.; Varga, Á.T.; Szenti, I.; Sebők, D.; Kopniczky, J.; Hopp, B.; Kukovecz, Á. Wetting and Evaporation on a Carbon Cloth Type Gas Diffusion Layer for Passive Direct Alcohol Fuel Cells. *J. Mol. Liq.* **2020**, *304*, 112698. [CrossRef]
53. Poplavsky, V.V.; Dorozhko, A.V.; Matys, V.G. Ion-Beam Formation of Electrocatalysts for Fuel Cells with Polymer Membrane Electrolyte. *J. Synch. Investig.* **2017**, *11*, 326–332. [CrossRef]
54. Liu, G.; Li, X.; Wang, M.; Wang, M.; Kim, J.Y.; Woo, J.Y.; Wang, X.; Lee, J.K. A Study on Anode Diffusion Layer for Performance Enhancement of a Direct Methanol Fuel Cell. *Energy Convers. Manag.* **2016**, *126*, 697–703. [CrossRef]
55. Falcão, D.S.; Silva, R.A.; Rangel, C.M.; Pinto, A.M.F.R. Performance of an Active Micro Direct Methanol Fuel Cell Using Reduced Catalyst Loading MEAs. *Energies* **2017**, *10*, 1683. [CrossRef]
56. Hsieh, S.-S.; Hung, L.-C.; Liu, C.-C.; Huang, C.-F. Analyses of Electrochemical Impedance Spectroscopy and Cyclic Voltammetry in Micro-Direct Methanol Fuel Cell Stacks. *Int. J. Energy Res.* **2016**, *40*, 2162–2175. [CrossRef]

57. Rashed, M.K.; Mohd Salleh, M.A.; Abdulbari, H.A.; Shah Ismail, M.H.; Izhar, S. The Effects of Electrode and Catalyst Selection on Microfluidic Fuel Cell Performance. *ChemBioEng Rev.* **2015**, *2*, 356–372. [CrossRef]

58. Gauthier, E.; Benziger, J.B. Gas Management and Multiphase Flow in Direct Alcohol Fuel Cells. *Electrochim. Acta* **2014**, *128*, 238–247. [CrossRef]

59. Xue, R.; Zhang, Y.; Liu, X. A Novel Cathode Gas Diffusion Layer for Water Management of Passive μ-DMFC. *Energy* **2017**, *139*, 535–541. [CrossRef]

60. Yuan, W.; Tang, Y.; Yang, X.; Liu, B.; Wan, Z. Structural Diversity and Orientation Dependence of a Liquid-Fed Passive Air-Breathing Direct Methanol Fuel Cell. *Int. J. Hydrogen Energy* **2012**, *37*, 9298–9313. [CrossRef]

61. Yi, P.; Peng, L.; Lai, X.; Li, M.; Ni, J. Investigation of Sintered Stainless Steel Fiber Felt as Gas Diffusion Layer in Proton Exchange Membrane Fuel Cells. *Int. J. Hydrogen Energy* **2012**, *37*, 11334–11344. [CrossRef]

62. Zhao, Z.; Zhang, F.; Zhang, Y.; Zhang, D. Performance Optimization of MDMFC with Foamed Stainless Steel Cathode Current Collector. *Energies* **2021**, *14*, 6608. [CrossRef]

63. Zhang, J.; Zhu, Y.L.; Qi, G.; Li, J.Y. Current Development of Key Materials for Low Temperature Fuel Cells. *Key Eng. Mater.* **2017**, *727*, 670–677. [CrossRef]

64. Madheswaran, D.K.; Jayakumar, A. Recent Advancements on Non-Platinum Based Catalyst Electrode Material for Polymer Electrolyte Membrane Fuel Cells: A Mini Techno-Economic Review. *Bull. Mater. Sci.* **2021**, *44*, 287. [CrossRef]

65. Yao, D.; Jao, T.-C.; Zhang, W.; Xu, L.; Xing, L.; Ma, Q.; Xu, Q.; Li, H.; Pasupathi, S.; Su, H. In-Situ Diagnosis on Performance Degradation of High Temperature Polymer Electrolyte Membrane Fuel Cell by Examining Its Electrochemical Properties under Operation. *Int. J. Hydrogen Energy* **2018**, *43*, 21006–21016. [CrossRef]

66. Han, J.; Han, J.; Yu, S. Experimental Analysis of Performance Degradation of 3-Cell PEMFC Stack under Dynamic Load Cycle. *Int. J. Hydrogen Energy* **2020**, *45*, 13045–13054. [CrossRef]

67. Chu, T.; Zhang, R.; Wang, Y.; Ou, M.; Xie, M.; Shao, H.; Yang, D.; Li, B.; Ming, P.; Zhang, C. Performance Degradation and Process Engineering of the 10 KW Proton Exchange Membrane Fuel Cell Stack. *Energy* **2021**, *219*, 119623. [CrossRef]

68. Casalegno, A.; Bresciani, F.; Zago, M.; Marchesi, R. Experimental Investigation of Methanol Crossover Evolution during Direct Methanol Fuel Cell Degradation Tests. *J. Power Sources* **2014**, *249*, 103–109. [CrossRef]

69. Shrivastava, N.K.; Chadge, R.B.; Bankar, S.L. Modelling and Simulation of Passive Feed Direct Methanol Fuel Cell. *Int. J. Energy Technol. Policy* **2017**, *13*, 4–18. [CrossRef]

70. Direct Methanol Fuel Cell Methanol and Water Crossover. Available online: https://www.wpclipart.com/science/how_things_work/Direct_Methanol_Fuel_Cell_Methanol_and_Water_Crossover.png.html (accessed on 13 November 2021).

71. Xu, C.; Faghri, A.; Li, X.; Ward, T. Methanol and Water Crossover in a Passive Liquid-Feed Direct Methanol Fuel Cell. *Int. J. Hydrogen Energy* **2010**, *35*, 1769–1777. [CrossRef]

72. Sharifi, S.; Rahimi, R.; Mohebbi-Kalhori, D.; Colpan, C.O. Numerical Investigation of Methanol Crossover through the Membrane in a Direct Methanol Fuel Cell. *Iran. J. Hydrog. Fuel Cell* **2018**, *5*, 21–33. [CrossRef]

73. Chi, N.T.Q.; Bae, B.; Kim, D. Electro-Osmotic Drag Effect on the Methanol Permeation for Sulfonated Poly(Ether Ether Ketone) and Nafion117 Membranes. *J. Nanosci. Nanotechnol.* **2013**, *13*, 7529–7534. [CrossRef]

74. García-Nieto, D.; Barragán, V.M. A Comparative Study of the Electro-Osmotic Behavior of Cation and Anion Exchange Membranes in Alcohol-Water Media. *Electrochim. Acta* **2015**, *154*, 166–176. [CrossRef]

75. Endo, N.; Ogawa, Y.; Ukai, K.; Kakihana, Y.; Higa, M. DMFC Performance of Polymer Electrolyte Membranes Prepared from a Graft-Copolymer Consisting of a Polysulfone Main Chain and Styrene Sulfonic Acid Side Chains. *Energies* **2016**, *9*, 658. [CrossRef]

76. Huang, K.-L.; Liao, Y.-H.; Chen, S.-J. Effects of Operating Parameters on Gas-Phase PAH Emissions from a Direct Methanol Fuel Cell. *Aerosol Air Qual. Res.* **2019**, *19*, 2196–2204. [CrossRef]

77. Yuan, Z.; Chuai, W.; Guo, Z.; Tu, Z.; Kong, F. The Self-Adaptive Fuel Supply Mechanism in Micro DMFC Based on the Microvalve. *Micromachines* **2019**, *10*, 353. [CrossRef] [PubMed]

78. Araya, S.S.; Andreasen, S.J.; Kær, S.K. Experimental Characterization of the Poisoning Effects of Methanol-Based Reformate Impurities on a PBI-Based High Temperature PEM Fuel Cell. *Energies* **2012**, *5*, 4251. [CrossRef]

79. Bogolowski, N.; Drillet, J.-F. Appropriate Balance between Methanol Yield and Power Density in Portable Direct Methanol Fuel Cell. *Chem. Eng. J.* **2015**, *270*, 91–100. [CrossRef]

80. Ji, F.; Yang, L.; Sun, H.; Wang, S.; Li, H.; Jiang, L.; Sun, G. A Novel Method for Analysis and Prediction of Methanol Mass Transfer in Direct Methanol Fuel Cell. *Energy Convers. Manag.* **2017**, *154*, 482–490. [CrossRef]

81. Ramesh, V.; Krishnamurthy, B. Modeling the Transient Temperature Distribution in a Direct Methanol Fuel Cell. *J. Electroanal. Chem.* **2018**, *809*, 1–7. [CrossRef]

82. Govindarasu, R.; Somasundaram, S. Studies on Influence of Cell Temperature in Direct Methanol Fuel Cell Operation. *Processes* **2020**, *8*, 353. [CrossRef]

83. Andoh, S.; Fujita, A.; Miitsu, T.; Kuroda, Y.; Mitsushima, S. Practical and Reliable Methanol Concentration Sensor for Direct Methanol Fuel Cells. *Electrochemistry* **2021**, *89*, 250–255. [CrossRef]

84. Tsen, W.-C. Composite Proton Exchange Membranes Based on Chitosan and Phosphotungstic Acid Immobilized One-Dimensional Attapulgite for Direct Methanol Fuel Cells. *Nanomaterials* **2020**, *10*, 1641. [CrossRef]

85. Sudaroli, B.M.; Kolar, A.K. An Experimental Study on the Effect of Membrane Thickness and PTFE (Polytetrafluoroethylene) Loading on Methanol Crossover in Direct Methanol Fuel Cell. *Energy* **2016**, *98*, 204–214. [CrossRef]

86. Gwak, G.; Lee, K.; Ferekh, S.; Lee, S.; Ju, H. Analyzing the Effects of Fluctuating Methanol Feed Concentration in Active-Type Direct Methanol Fuel Cell (DMFC) Systems. *Int. J. Hydrogen Energy* **2015**, *40*, 5396–5407. [CrossRef]
87. Zhang, C.; Yue, X.; Yang, Y.; Lu, N.; Zhang, S.; Wang, G. Thin and Methanol-Resistant Reinforced Composite Membrane Based on Semi-Crystalline Poly (Ether Ether Ketone) for Fuel Cell Applications. *J. Power Sources* **2020**, *450*, 227664. [CrossRef]
88. Wippermann, K.; Klafki, K.; Kulikovsky, A.A. In Situ Measurement of the Oxygen Diffusion Coefficient in the Cathode Catalyst Layer of a Direct Methanol Fuel Cell. *Electrochim. Acta* **2014**, *141*, 212–215. [CrossRef]
89. Gago, A.S.; Esquivel, J.-P.; Sabaté, N.; Santander, J.; Alonso-Vante, N. Comprehensive Characterization and Understanding of Micro-Fuel Cells Operating at High Methanol Concentrations. *Beilstein J. Nanotechnol.* **2015**, *6*, 2000–2006. [CrossRef] [PubMed]
90. Zago, M.; Bisello, A.; Baricci, A.; Rabissi, C.; Brightman, E.; Hinds, G.; Casalegno, A. On the Actual Cathode Mixed Potential in Direct Methanol Fuel Cells. *J. Power Sources* **2016**, *325*, 714–722. [CrossRef]
91. Majidi, P.; Altarawneh, R.M.; Ryan, N.D.W.; Pickup, P.G. Determination of the Efficiency of Methanol Oxidation in a Direct Methanol Fuel Cell. *Electrochim. Acta* **2016**, *199*, 210–217. [CrossRef]
92. Zhao, X.; Yuan, W.; Wu, Q.; Sun, H.; Luo, Z.; Fu, H. High-Temperature Passive Direct Methanol Fuel Cells Operating with Concentrated Fuels. *J. Power Sources* **2015**, *273*, 517–521. [CrossRef]
93. Wan, N. High Performance Direct Methanol Fuel Cell with Thin Electrolyte Membrane. *J. Power Sources* **2017**, *354*, 167–171. [CrossRef]
94. Xue, Y.; Chan, S. Layer-by-Layer Self-Assembly of CHI/PVS–Nafion Composite Membrane for Reduced Methanol Crossover and Enhanced DMFC Performance. *Int. J. Hydrogen Energy* **2015**, *40*, 1877–1885. [CrossRef]
95. Rambabu, G.; Bhat, S.D. Carbon-Polymer Nanocomposite Membranes as Electrolytes for Direct Methanol Fuel Cells. In *Membrane Technology*; CRC Press: Boca Raton, FL, USA, 2018; ISBN 978-1-315-10566-6.
96. Yee, R.S.L.; Zhang, K.; Ladewig, B.P. The Effects of Sulfonated Poly(Ether Ether Ketone) Ion Exchange Preparation Conditions on Membrane Properties. *Membranes* **2013**, *3*, 182–195. [CrossRef] [PubMed]
97. Iannaci, A.; Mecheri, B.; D'Epifanio, A.; Licoccia, S. Sulfated Zirconium Oxide as Electrode and Electrolyte Additive for Direct Methanol Fuel Cell Applications. *Int. J. Hydrogen Energy* **2014**, *39*, 11241–11249. [CrossRef]
98. Yang, C.-C. Fabrication and Characterization of Poly(Vinyl Alcohol)/Montmorillonite/Poly(Styrene Sulfonic Acid) Proton-Conducting Composite Membranes for Direct Methanol Fuel Cells. *Int. J. Hydrogen Energy* **2011**, *36*, 4419–4431. [CrossRef]
99. Maiti, J.; Kakati, N.; Lee, S.H.; Jee, S.H.; Viswanathan, B.; Yoon, Y.S. Where Do Poly(Vinyl Alcohol) Based Membranes Stand in Relation to Nafion® for Direct Methanol Fuel Cell Applications? *J. Power Sources* **2012**, *216*, 48–66. [CrossRef]
100. Zhong, S.; Cui, X.; Gao, Y.; Liu, W.; Dou, S. Fabrication and Properties of Poly(Vinyl Alcohol)-Based Polymer Electrolyte Membranes for Direct Methanol Fuel Cell Applications. *Int. J. Hydrogen Energy* **2014**, *39*, 17857–17864. [CrossRef]
101. Sidharthan, K.A.; Joseph, S. Preparation and Characterization of Polyvinyl Alcohol Based Nanocomposite Membrane for Direct Methanol Fuel Cell. *AIP Conf. Proc.* **2019**, *2162*, 020143. [CrossRef]
102. Aricò, A.S.; Sebastian, D.; Schuster, M.; Bauer, B.; D'Urso, C.; Lufrano, F.; Baglio, V. Selectivity of Direct Methanol Fuel Cell Membranes. *Membranes* **2015**, *5*, 793–809. [CrossRef] [PubMed]
103. Katzfuß, A.; Poynton, S.; Varcoe, J.; Gogel, V.; Storr, U.; Kerres, J. Methylated Polybenzimidazole and Its Application as a Blend Component in Covalently Cross-Linked Anion-Exchange Membranes for DMFC. *J. Membr. Sci.* **2014**, *465*, 129–137. [CrossRef]
104. Alam, T.M.; Hibbs, M.R. Characterization of Heterogeneous Solvent Diffusion Environments in Anion Exchange Membranes. *Macromolecules* **2014**, *47*, 1073–1084. [CrossRef]
105. Xu, W.; Zhao, Y.; Yuan, Z.; Li, X.; Zhang, H.; Vankelecom, I.F.J. Highly Stable Anion Exchange Membranes with Internal Cross-Linking Networks. *Adv. Funct. Mater.* **2015**, *25*, 2583–2589. [CrossRef]
106. Verjulio, R.W.; Santander, J.; Ma, J.; Alonso-Vante, N. Selective CoSe2/C Cathode Catalyst for Passive Air-Breathing Alkaline Anion Exchange Membrane μ-Direct Methanol Fuel Cell (AEM-MDMFC). *Int. J. Hydrogen Energy* **2016**, *41*, 19595–19600. [CrossRef]
107. Benvenuti, T.; García-Gabaldón, M.; Ortega, E.M.; Rodrigues, M.A.S.; Bernardes, A.M.; Pérez-Herranz, V.; Zoppas-Ferreira, J. Influence of the Co-Ions on the Transport of Sulfate through Anion Exchange Membranes. *J. Membr. Sci.* **2017**, *542*, 320–328. [CrossRef]
108. Higa, M.; Mehdizadeh, S.; Feng, S.; Endo, N.; Kakihana, Y. Cell Performance of Direct Methanol Alkaline Fuel Cell (DMAFC) Using Anion Exchange Membranes Prepared from PVA-Based Block Copolymer. *J. Membr. Sci.* **2020**, *597*, 117618. [CrossRef]
109. Hu, X.; Huang, Y.; Liu, L.; Ju, Q.; Zhou, X.; Qiao, X.; Zheng, Z.; Li, N. Piperidinium Functionalized Aryl Ether-Free Polyaromatics as Anion Exchange Membrane for Water Electrolysers: Performance and Durability. *J. Membr. Sci.* **2021**, *621*, 118964. [CrossRef]
110. Zakaria, Z.; Kamarudin, S.K. A Review of Quaternized Polyvinyl Alcohol as an Alternative Polymeric Membrane in DMFCs and DEFCs. *Int. J. Energy Res.* **2020**, *44*, 6223–6239. [CrossRef]
111. Mustain, W.E.; Chatenet, M.; Page, M.; Kim, Y.S. Durability Challenges of Anion Exchange Membrane Fuel Cells. *Energy Environ. Sci.* **2020**, *13*, 2805–2838. [CrossRef]
112. Chen, N.; Wang, H.H.; Kim, S.P.; Kim, H.M.; Lee, W.H.; Hu, C.; Bae, J.Y.; Sim, E.S.; Chung, Y.-C.; Jang, J.-H.; et al. Poly(Fluorenyl Aryl Piperidinium) Membranes and Ionomers for Anion Exchange Membrane Fuel Cells. *Nat Commun.* **2021**, *12*, 2367. [CrossRef] [PubMed]
113. Electrochemical Characteristics of Different Membranes Used for DMFC. Available online: https://www.fumatech.com/EN/Membranes/Fuel-cells/Products%2Bfumapem/index.html (accessed on 13 November 2021).

114. Chen, X.; Li, T.; Shen, J.; Hu, Z. From Structures, Packaging to Application: A System-Level Review for Micro Direct Methanol Fuel Cell. *Renew. Sustain. Energy Rev.* **2017**, *80*, 669–678. [CrossRef]
115. Xia, Z.; Zhang, X.; Sun, H.; Wang, S.; Sun, G. Recent Advances in Multi-Scale Design and Construction of Materials for Direct Methanol Fuel Cells. *Nano Energy* **2019**, *65*, 104048. [CrossRef]
116. Yan, X.H.; Gao, P.; Zhao, G.; Shi, L.; Xu, J.B.; Zhao, T.S. Transport of Highly Concentrated Fuel in Direct Methanol Fuel Cells. *Appl. Therm. Eng.* **2017**, *126*, 290–295. [CrossRef]
117. Kim, J.; Jang, J.-S.; Peck, D.-H.; Lee, B.; Yoon, S.-H.; Jung, D.-H. Methanol-Tolerant Platinum-Palladium Catalyst Supported on Nitrogen-Doped Carbon Nanofiber for High Concentration Direct Methanol Fuel Cells. *Nanomaterials* **2016**, *6*, 148. [CrossRef]
118. Lo Vecchio, C.; Sebastián, D.; Lázaro, M.J.; Aricò, A.S.; Baglio, V. Methanol-Tolerant M–N–C Catalysts for Oxygen Reduction Reactions in Acidic Media and Their Application in Direct Methanol Fuel Cells. *Catalysts* **2018**, *8*, 650. [CrossRef]
119. Elangovan, A.; Xu, J.; Sekar, A.; Rajendran, S.; Liu, B.; Li, J. Platinum Deposited Nitrogen-Doped Vertically Aligned Carbon Nanofibers as Methanol Tolerant Catalyst for Oxygen Reduction Reaction with Improved Durability. *Appl. Nano* **2021**, *2*, 303–318. [CrossRef]
120. Prapainainar, P.; Du, Z.; Theampetch, A.; Prapainainar, C.; Kongkachuichay, P.; Holmes, S.M. Properties and DMFC Performance of Nafion/Mordenite Composite Membrane Fabricated by Solution-Casting Method with Different Solvent Ratio. *Energy* **2020**, *190*, 116451. [CrossRef]
121. Divya, K.; Sri Abirami Saraswathi, M.S.; Rana, D.; Alwarappan, S.; Nagendran, A. Custom-Made Sulfonated Poly (Ether Sulfone) Nanocomposite Proton Exchange Membranes Using Exfoliated Molybdenum Disulfide Nanosheets for DMFC Applications. *Polymer* **2018**, *147*, 48–55. [CrossRef]
122. Barbera, O.; Stassi, A.; Sebastian, D.; Bonde, J.L.; Giacoppo, G.; D'Urso, C.; Baglio, V.; Aricò, A.S. Simple and Functional Direct Methanol Fuel Cell Stack Designs for Application in Portable and Auxiliary Power Units. *Int. J. Hydrogen Energy* **2016**, *41*, 12320–12329. [CrossRef]
123. Wang, L.; Yuan, Z.; Wen, F.; Cheng, Y.; Zhang, Y.; Wang, G. A Bipolar Passive DMFC Stack for Portable Applications. *Energy* **2018**, *144*, 587–593. [CrossRef]
124. Sombatmankhong, K.; Yunus, K.; Fisher, A.C. Electrocogeneration of Hydrogen Peroxide: Confocal and Potentiostatic Investigations of Hydrogen Peroxide Formation in a Direct Methanol Fuel Cell. *J. Power Sources* **2013**, *240*, 219–231. [CrossRef]
125. Li, Y.; Liang, L.; Liu, C.; Li, Y.; Xing, W.; Sun, J. Self-Healing Proton-Exchange Membranes Composed of Nafion–Poly(Vinyl Alcohol) Complexes for Durable Direct Methanol Fuel Cells. *Adv. Mater.* **2018**, *30*, 1707146. [CrossRef]
126. Wang, A.; Yuan, W.; Huang, S.; Tang, Y.; Chen, Y. Structural Effects of Expanded Metal Mesh Used as a Flow Field for a Passive Direct Methanol Fuel Cell. *Appl. Energy* **2017**, *208*, 184–194. [CrossRef]
127. Sakurai, Y.; Kawai, A. Mechanical Stress Effect on Ionic Conductivity of Perflurosulfonic Acid (PFSA) Film by Photolithography. *J. Photopolym. Sci. Technol.* **2012**, *25*, 723–727. [CrossRef]
128. Yuan, Z.Y.; Zhang, Y.F.; Fu, W.T.; Li, Z.P.; Liu, X.W. Investigation of the Direct Methanol Fuel Cell with Novel Assembly Method. *Fuel Cells* **2013**, *13*, 794–803. [CrossRef]
129. Ince, A.C.; Karaoglan, M.U.; Glüsen, A.; Colpan, C.O.; Müller, M.; Stolten, D. Semiempirical Thermodynamic Modeling of a Direct Methanol Fuel Cell System. *Int. J. Energy Res.* **2019**, *43*, 3601–3615. [CrossRef]
130. Li, L.-Y.; Yu, B.-C.; Shih, C.-M.; Lue, S.J. Polybenzimidazole Membranes for Direct Methanol Fuel Cell: Acid-Doped or Alkali-Doped? *J. Power Sources* **2015**, *287*, 386–395. [CrossRef]
131. Kimiaie, N.; Trappmann, C.; Janßen, H.; Hehemann, M.; Echsler, H.; Müller, M. Influence of Contamination with Inorganic Impurities on the Durability of a 1 KW DMFC System. *Fuel Cells* **2014**, *14*, 64–75. [CrossRef]
132. Nicotera, I.; Simari, C.; Enotiadis, A. 2—Nafion-Based Cation-Exchange Membranes for Direct Methanol Fuel Cells. In *Direct Methanol Fuel Cell Technology*; Dutta, K., Ed.; Elsevier: Amsterdam, The Netherlands, 2020; pp. 13–36. ISBN 978-0-12-819158-3.
133. Wang, L.; Kang, B.; Gao, N.; Du, X.; Jia, L.; Sun, J. Corrosion Behaviour of Austenitic Stainless Steel as a Function of Methanol Concentration for Direct Methanol Fuel Cell Bipolar Plate. *J. Power Sources* **2014**, *253*, 332–341. [CrossRef]
134. Simari, C.; Nicotera, I.; Aricò, A.S.; Baglio, V.; Lufrano, F. New Insights into Properties of Methanol Transport in Sulfonated Polysulfone Composite Membranes for Direct Methanol Fuel Cells. *Polymers* **2021**, *13*, 1386. [CrossRef]
135. Lufrano, E.; Simari, C.; Lo Vecchio, C.; Aricò, A.S.; Baglio, V.; Nicotera, I. Barrier Properties of Sulfonated Polysulfone/Layered Double Hydroxides Nanocomposite Membrane for Direct Methanol Fuel Cell Operating at High Methanol Concentrations. *Int. J. Hydrogen Energy* **2020**, *45*, 20647–20658. [CrossRef]
136. Li, J.; Fan, K.; Cai, W.; Ma, L.; Xu, G.; Xu, S.; Ma, L.; Cheng, H. An In-Situ Nano-Scale Swelling-Filling Strategy to Improve Overall Performance of Nafion Membrane for Direct Methanol Fuel Cell Application. *J. Power Sources* **2016**, *332*, 37–41. [CrossRef]
137. Li, X.; Miao, Z.; Marten, L.; Blankenau, I. Experimental Measurements of Fuel and Water Crossover in an Active DMFC. *Int. J. Hydrogen Energy* **2021**, *46*, 4437–4446. [CrossRef]
138. Mollá, S.; Compañ, V. Polymer Blends of SPEEK for DMFC Application at Intermediate Temperatures. *Int. J. Hydrogen Energy* **2014**, *39*, 5121–5136. [CrossRef]
139. Krishnan, N.N.; Henkensmeier, D.; Jang, J.H.; Kim, H.-J. Nanocomposite Membranes for Polymer Electrolyte Fuel Cells. *Macromol. Mater. Eng.* **2014**, *299*, 1031–1041. [CrossRef]
140. Zatoń, M.; Rozière, J.; Jones, D.J. Current Understanding of Chemical Degradation Mechanisms of Perfluorosulfonic Acid Membranes and Their Mitigation Strategies: A Review. *Sustain. Energy Fuels* **2017**, *1*, 409–438. [CrossRef]

141. Bunlengsuwan, P.; Paradee, N.; Sirivat, A. Influence of Sulfonated Graphene Oxide on Sulfonated Polysulfone Membrane for Direct Methanol Fuel Cell. *Polym.-Plast. Technol. Eng.* **2017**, *56*, 1695–1703. [CrossRef]
142. Dutta, K.; Kumar, P.; Das, S.; Kundu, P.P. Utilization of Conducting Polymers in Fabricating Polymer Electrolyte Membranes for Application in Direct Methanol Fuel Cells. *Polym. Rev.* **2014**, *54*, 1–32. [CrossRef]
143. Rosli, N.A.H.; Loh, K.S.; Wong, W.Y.; Yunus, R.M.; Lee, T.K.; Ahmad, A.; Chong, S.T. Review of Chitosan-Based Polymers as Proton Exchange Membranes and Roles of Chitosan-Supported Ionic Liquids. *Int. J. Mol. Sci.* **2020**, *21*, 632. [CrossRef]
144. Aricò, A.S.; Stassi, A.; D'Urso, C.; Sebastián, D.; Baglio, V. Synthesis of Pd3Co1@Pt/C Core-Shell Catalysts for Methanol-Tolerant Cathodes of Direct Methanol Fuel Cells. *Chem.-A Eur. J.* **2014**, *20*, 10679–10684. [CrossRef]
145. Choi, B.; Nam, W.-H.; Chung, D.Y.; Park, I.-S.; Yoo, S.J.; Song, J.C.; Sung, Y.-E. Enhanced Methanol Tolerance of Highly Pd Rich Pd-Pt Cathode Electrocatalysts in Direct Methanol Fuel Cells. *Electrochim. Acta* **2015**, *164*, 235–242. [CrossRef]
146. Baglio, V.; Stassi, A.; Barbera, O.; Giacoppo, G.; Sebastian, D.; D'Urso, C.; Schuster, M.; Bauer, B.; Bonde, J.L.; Aricò, A.S. Direct Methanol Fuel Cell Stack for Auxiliary Power Units Applications Based on Fumapem® F-1850 Membrane. *Int. J. Hydrogen Energy* **2017**, *42*, 26889–26896. [CrossRef]
147. Prapainainar, P.; Pattanapisutkun, N.; Prapainainar, C.; Kongkachuichay, P. Incorporating Graphene Oxide to Improve the Performance of Nafion-Mordenite Composite Membranes for a Direct Methanol Fuel Cell. *Int. J. Hydrogen Energy* **2019**, *44*, 362–378. [CrossRef]
148. Liu, G.; Tsen, W.-C.; Jang, S.-C.; Hu, F.; Zhong, F.; Zhang, B.; Wang, J.; Liu, H.; Wang, G.; Wen, S.; et al. Composite Membranes from Quaternized Chitosan Reinforced with Surface-Functionalized PVDF Electrospun Nanofibers for Alkaline Direct Methanol Fuel Cells. *J. Membr. Sci.* **2020**, *611*, 118242. [CrossRef]
149. Mollá, S.; Compañ, V. Polyvinyl Alcohol Nanofiber Reinforced Nafion Membranes for Fuel Cell Applications. *J. Membr. Sci.* **2011**, *372*, 191–200. [CrossRef]
150. Li, H.-Y.; Lee, Y.-Y.; Lai, J.-Y.; Liu, Y.-L. Composite Membranes of Nafion and Poly(Styrene Sulfonic Acid)-Grafted Poly(Vinylidene Fluoride) Electrospun Nanofiber Mats for Fuel Cells. *J. Membr. Sci.* **2014**, *466*, 238–245. [CrossRef]
151. Xu, G.; Wu, Z.; Xie, Z.; Wei, Z.; Li, J.; Qu, K.; Li, Y.; Cai, W. Graphene Quantum Dot Reinforced Hyperbranched Polyamide Proton Exchange Membrane for Direct Methanol Fuel Cell. *Int. J. Hydrogen Energy* **2021**, *46*, 9782–9789. [CrossRef]
152. Zhang, X.; Zhang, M.; Deng, Y.; Xu, M.; Artiglia, L.; Wen, W.; Gao, R.; Chen, B.; Yao, S.; Zhang, X.; et al. A Stable Low-Temperature H2-Production Catalyst by Crowding Pt on α-MoC. *Nature* **2021**, *589*, 396–401. [CrossRef]
153. Ramli, Z.A.C.; Kamarudin, S.K. Platinum-Based Catalysts on Various Carbon Supports and Conducting Polymers for Direct Methanol Fuel Cell Applications: A Review. *Nanoscale Res. Lett.* **2018**, *13*, 1–25. [CrossRef]
154. Sahin, O.; Kivrak, H. A Comparative Study of Electrochemical Methods on Pt–Ru DMFC Anode Catalysts: The Effect of Ru Addition. *Int. J. Hydrogen Energy* **2013**, *38*, 901–909. [CrossRef]
155. Nie, M.; Du, S.; Li, Q.; Hummel, M.; Gu, Z.; Lu, S. Tungsten Carbide as Supports for Trimetallic AuPdPt Electrocatalysts for Methanol Oxidation. *J. Electrochem. Soc.* **2020**, *167*, 044510. [CrossRef]
156. Bresciani, F.; Rabissi, C.; Casalegno, A.; Zago, M.; Marchesi, R. Experimental Investigation on DMFC Temporary Degradation. *Int. J. Hydrogen Energy* **2014**, *39*, 21647–21656. [CrossRef]
157. Yaldagard, M.; Jahanshahi, M.; Seghatoleslami, N. Carbonaceous Nanostructured Support Materials for Low Temperature Fuel Cell Electrocatalysts—A Review. *World J. Nano Sci. Eng.* **2013**, *3*, 33. [CrossRef]
158. Yang, Z.; Kim, C.; Hirata, S.; Fujigaya, T.; Nakashima, N. Facile Enhancement in CO-Tolerance of a Polymer-Coated Pt Electrocatalyst Supported on Carbon Black: Comparison between Vulcan and Ketjenblack. *ACS Appl. Mater. Interfaces* **2015**, *7*, 15885–15891. [CrossRef]
159. Motsoeneng, R.G.; Modibedi, R.M.; Mathe, M.K.; Khotseng, L.E.; Ozoemena, K.I. The Synthesis of PdPt/Carbon Paper via Surface Limited Redox Replacement Reactions for Oxygen Reduction Reaction. *Int. J. Hydrogen Energy* **2015**, *40*, 16734–16744. [CrossRef]
160. Xie, J.; Zhang, Q.; Gu, L.; Xu, S.; Wang, P.; Liu, J.; Ding, Y.; Yao, Y.F.; Nan, C.; Zhao, M.; et al. Ruthenium–Platinum Core–Shell Nanocatalysts with Substantially Enhanced Activity and Durability towards Methanol Oxidation. *Nano Energy* **2016**, *21*, 247–257. [CrossRef]
161. Jovanovič, P.; Šelih, V.S.; Šala, M.; Hočevar, S.; Ruiz-Zepeda, F.; Hodnik, N.; Bele, M.; Gaberšček, M. Potentiodynamic Dissolution Study of PtRu/C Electrocatalyst in the Presence of Methanol. *Electrochim. Acta* **2016**, *211*, 851–859. [CrossRef]
162. Kovtunenko, V.A.; Karpenko-Jereb, L. Lifetime of Catalyst under Voltage Cycling in Polymer Electrolyte Fuel Cell Due to Platinum Oxidation and Dissolution. *Technologies* **2021**, *9*, 80. [CrossRef]
163. Han, M.; Zeng, J.; Xia, J.; Liao, S. Effect of Thermal Treatment on Structural Change of Anode Electrocatalysts for Direct Methanol Fuel Cells. *Particuology* **2014**, *15*, 45–50. [CrossRef]
164. Sharma, R.; Gyergyek, S.; Lund, P.B.; Andersen, S.M. Recovery of Pt and Ru from Spent Low-Temperature Polymer Electrolyte Membrane Fuel Cell Electrodes and Recycling of Pt by Direct Redeposition of the Dissolved Precursor on Carbon. *ACS Appl. Energy Mater.* **2021**, *4*, 6842–6852. [CrossRef]
165. Zhang, C.; Yue, X.; Luan, J.; Lu, N.; Mu, Y.; Zhang, S.; Wang, G. Reinforced Poly(Ether Ether Ketone)/Nafion Composite Membrane with Highly Improved Proton Conductivity for High Concentration Direct Methanol Fuel Cells. *ACS Appl. Energy Mater.* **2020**, *3*, 7180–7190. [CrossRef]
166. Schoekel, A.; Melke, J.; Bruns, M.; Wippermann, K.; Kuppler, F.; Roth, C. Quantitative Study of Ruthenium Cross-over in Direct Methanol Fuel Cells during Early Operation Hours. *J. Power Sources* **2016**, *301*, 210–218. [CrossRef]

167. Lee, K.-S.; Jeon, T.-Y.; Yoo, S.J.; Park, I.-S.; Cho, Y.-H.; Kang, S.H.; Choi, K.H.; Sung, Y.-E. Effect of PtRu Alloying Degree on Electrocatalytic Activities and Stabilities. *Appl. Catal. B Environ.* **2011**, *102*, 334–342. [CrossRef]

168. Arlt, T.; Manke, I.; Wippermann, K.; Riesemeier, H.; Mergel, J.; Banhart, J. Investigation of the Local Catalyst Distribution in an Aged Direct Methanol Fuel Cell MEA by Means of Differential Synchrotron X-Ray Absorption Edge Imaging with High Energy Resolution. *J. Power Sources* **2013**, *221*, 210–216. [CrossRef]

169. Sun, Y.; Zhou, Y.; Zhu, C.; Hu, L.; Han, M.; Wang, A.; Huang, H.; Liu, Y.; Kang, Z. A Pt–Co$_3$O$_4$–CD Electrocatalyst with Enhanced Electrocatalytic Performance and Resistance to CO Poisoning Achieved by Carbon Dots and Co$_3$O$_4$ for Direct Methanol Fuel Cells. *Nanoscale* **2017**, *9*, 5467–5474. [CrossRef] [PubMed]

170. Abdelkareem, M.A.; Masdar, M.S.; Tsujiguchi, T.; Nakagawa, N.; Sayed, E.T.; Barakat, N.A.M. Elimination of Toxic Products Formation in Vapor-Feed Passive DMFC Operated by Absolute Methanol Using Air Cathode Filter. *Chem. Eng. J.* **2014**, *240*, 38–44. [CrossRef]

171. Chung, D.Y.; Lee, K.-J.; Sung, Y.-E. Methanol Electro-Oxidation on the Pt Surface: Revisiting the Cyclic Voltammetry Interpretation. *J. Phys. Chem. C* **2016**, *120*, 9028–9035. [CrossRef]

172. Park, S.-H.; Joo, S.-J.; Kim, H.-S. An Investigation into Methanol Oxidation Reactions and CO, OH Adsorption on Pt-Ru-Mo Catalysts for a Direct Methanol Fuel Cell. *J. Electrochem. Soc.* **2014**, *161*, F405. [CrossRef]

173. Siller-Ceniceros, A.A.; Sánchez-Castro, M.E.; Morales-Acosta, D.; Torres-Lubian, J.R.; Martínez, G.E.; Rodríguez-Varela, F.J. Innovative Functionalization of Vulcan XC-72 with Ru Organometallic Complex: Significant Enhancement in Catalytic Activity of Pt/C Electrocatalyst for the Methanol Oxidation Reaction (MOR). *Appl. Catal. B Environ.* **2017**, *209*, 455–467. [CrossRef]

174. Karuppanan, K.K.; Panthalingal, M.K.; Biji, P. Chapter 26—Nanoscale, Catalyst Support Materials for Proton-Exchange Membrane Fuel Cells. In *Handbook of Nanomaterials for Industrial Applications*; Hussain, C.M., Ed.; Micro and Nano Technologies; Elsevier: Amsterdam, The Netherlands, 2018; pp. 468–495. ISBN 978-0-12-813351-4.

175. Baruah, B.; Deb, P. Performance and Application of Carbon-Based Electrocatalysts in Direct Methanol Fuel Cell. *Mater. Adv.* **2021**, *2*, 5344–5364. [CrossRef]

176. Sundqvist, B. Carbon under Pressure. *Phys. Rep.* **2021**, *909*, 1–73. [CrossRef]

177. Zainoodin, A.M.; Kamarudin, S.K.; Masdar, M.S.; Daud, W.R.W.; Mohamad, A.B.; Sahari, J. Investigation of MEA Degradation in a Passive Direct Methanol Fuel Cell under Different Modes of Operation. *Appl. Energy* **2014**, *135*, 364–372. [CrossRef]

178. Tsukagoshi, Y.; Ishitobi, H.; Nakagawa, N. Improved Performance of Direct Methanol Fuel Cells with the Porous Catalyst Layer Using Highly-Active Nanofiber Catalyst. *Carbon Resour. Convers.* **2018**, *1*, 61–72. [CrossRef]

179. Abdullah, N.; Kamarudin, S.K.; Shyuan, L.K. Novel Anodic Catalyst Support for Direct Methanol Fuel Cell: Characterizations and Single-Cell Performances. *Nanoscale Res. Lett.* **2018**, *13*, 1–13. [CrossRef]

180. Vu, T.H.T.; Nguyen, M.D.; Mai, A.T.N. Influence of Solvents on the Electroactivity of PtAl/RGO Catalyst Inks and Anode in Direct Ethanol Fuel Cell. *J. Chem.* **2021**, *2021*, 6649089. [CrossRef]

181. So, M.; Ohnishi, T.; Park, K.; Ono, M.; Tsuge, Y.; Inoue, G. The Effect of Solvent and Ionomer on Agglomeration in Fuel Cell Catalyst Inks: Simulation by the Discrete Element Method. *Int. J. Hydrogen Energy* **2019**, *44*, 28984–28995. [CrossRef]

182. Lashkenari, M.S.; Ghorbani, M.; Silakhori, N.; Karimi-Maleh, H. Enhanced Electrochemical Performance and Stability of Pt/Ni Electrocatalyst Supported on SiO$_2$-PANI Nanocomposite: A Combined Experimental and Theoretical Study. *Mater. Chem. Phys.* **2021**, *262*, 124290. [CrossRef]

183. Robinson, J.E.; Labrador, N.Y.; Chen, H.; Sartor, B.E.; Esposito, D.V. Silicon Oxide-Encapsulated Platinum Thin Films as Highly Active Electrocatalysts for Carbon Monoxide and Methanol Oxidation. *ACS Catal.* **2018**, *8*, 11423–11434. [CrossRef]

184. Yusoff, N.; Kumar, S.V.; Rameshkumar, P.; Pandikumar, A.; Shahid, M.M.; Rahman, M.A.; Huang, N.M. A Facile Preparation of Titanium Dioxide-Iron Oxide@Silicon Dioxide Incorporated Reduced Graphene Oxide Nanohybrid for Electrooxidation of Methanol in Alkaline Medium. *Electrochim. Acta* **2016**, *192*, 167–176. [CrossRef]

185. Mansor, M.; Timmiati, S.N.; Wong, W.Y.; Mohd Zainoodin, A.; Lim, K.L.; Kamarudin, S.K. NiPd Supported on Mesostructured Silica Nanoparticle as Efficient Anode Electrocatalyst for Methanol Electrooxidation in Alkaline Media. *Catalysts* **2020**, *10*, 1235. [CrossRef]

186. Niyaz, M.; Sawut, N.; Jamal, R.; Abdiryim, T.; Helil, Z.; Liu, H.; Xie, S.; Song, Y. Preparation of PEDOT-Modified Double-Layered Hollow Carbon Spheres as Pt Catalyst Support for Methanol Oxidation. *Int. J. Hydrogen Energy* **2021**, *46*, 31623–31633. [CrossRef]

187. Yin, S.; Wang, Z.; Li, C.; Yu, H.; Deng, K.; Xu, Y.; Li, X.; Wang, L.; Wang, H. Mesoporous Pt@PtM (M = Co, Ni) Cage-Bell Nanostructures toward Methanol Electro-Oxidation. *Nanoscale Adv.* **2020**, *2*, 1084–1089. [CrossRef]

188. Jayakumar, A.; Sethu, S.P.; Ramos, M.; Robertson, J.; Al-Jumaily, A. A Technical Review on Gas Diffusion, Mechanism and Medium of PEM Fuel Cell. *Ionics* **2015**, *21*, 1–18. [CrossRef]

189. Deng, H.; Zhang, Y.; Zheng, X.; Li, Y.; Zhang, X.; Liu, X. A CNT (Carbon Nanotube) Paper as Cathode Gas Diffusion Electrode for Water Management of Passive μ-DMFC (Micro-Direct Methanol Fuel Cell) with Highly Concentrated Methanol. *Energy* **2015**, *82*, 236–241. [CrossRef]

190. Chen, Y.; Ke, Y.; Xia, Y.; Cho, C. Investigation on Mechanical Properties of a Carbon Paper Gas Diffusion Layer through a 3-D Nonlinear and Orthotropic Constitutive Model. *Energies* **2021**, *14*, 6341. [CrossRef]

191. Randrianarizafy, B.; Schott, P.; Gerard, M.; Bultel, Y. Modelling Carbon Corrosion during a PEMFC Startup: Simulation of Mitigation Strategies. *Energies* **2020**, *13*, 2338. [CrossRef]

192. Jayakumar, A. A Comprehensive Assessment on the Durability of Gas Diffusion Electrode Materials in PEM Fuel Cell Stack. *Front. Energy* **2019**, *13*, 325–338. [CrossRef]
193. Pethaiah, S.S.; Sadasivuni, K.K.; Jayakumar, A.; Ponnamma, D.; Tiwary, C.S.; Sasikumar, G. Methanol Electrolysis for Hydrogen Production Using Polymer Electrolyte Membrane: A Mini-Review. *Energies* **2020**, *13*, 5879. [CrossRef]
194. Kothekar, K.P.; Shrivastava, N.K.; Thombre, S.B. 11—Gas Diffusion Layers for Direct Methanol Fuel Cells. In *Direct Methanol Fuel Cell Technology*; Dutta, K., Ed.; Elsevier: Amsterdam, The Netherlands, 2020; pp. 317–339. ISBN 978-0-12-819158-3.
195. Hsieh, S.-S.; Chen, I.-C.; Hwong, C.-H. Development of a Four-Cell DMFC Stack with Air-Breathing Cathode and Dendrite Flow Field. *Int. J. Energy Res.* **2014**, *38*, 1693–1711. [CrossRef]
196. Hwang, C.M.; Ishida, M.; Ito, H.; Maeda, T.; Nakano, A.; Hasegawa, Y.; Yokoi, N.; Kato, A.; Yoshida, T. Influence of Properties of Gas Diffusion Layers on the Performance of Polymer Electrolyte-Based Unitized Reversible Fuel Cells. *Int. J. Hydrogen Energy* **2011**, *36*, 1740–1753. [CrossRef]
197. Liu, G.; Li, X.; Wang, H.; Liu, X.; Chen, M.; Woo, J.Y.; Kim, J.Y.; Wang, X.; Lee, J.K. Design of 3-Electrode System for in Situ Monitoring Direct Methanol Fuel Cells during Long-Time Running Test at High Temperature. *Appl. Energy* **2017**, *197*, 163–168. [CrossRef]
198. García-Salaberri, P.A.; Vera, M. On the Effect of Operating Conditions in Liquid-Feed Direct Methanol Fuel Cells: A Multiphysics Modeling Approach. *Energy* **2016**, *113*, 1265–1287. [CrossRef]
199. Kim, Y.-S.; Peck, D.-H.; Kim, S.-K.; Jung, D.-H.; Lim, S.; Kim, S.-H. Effects of the Microstructure and Powder Compositions of a Micro-Porous Layer for the Anode on the Performance of High Concentration Methanol Fuel Cell. *Int. J. Hydrogen Energy* **2013**, *38*, 7159–7168. [CrossRef]
200. He, Y.-L.; Miao, Z.; Zhao, T.-S.; Yang, W.-W. Numerical Study of the Effect of the GDL Structure on Water Crossover in a Direct Methanol Fuel Cell. *Int. J. Hydrogen Energy* **2012**, *37*, 4422–4438. [CrossRef]
201. Lee, K.; Ferekh, S.; Jo, A.; Lee, S.; Ju, H. Effects of Hybrid Catalyst Layer Design on Methanol and Water Transport in a Direct Methanol Fuel Cell. *Electrochim. Acta* **2015**, *177*, 209–216. [CrossRef]
202. Yuan, W.; Zhang, X.; Hou, C.; Zhang, Y.; Wang, H.; Liu, X. Enhanced Water Management via the Optimization of Cathode Microporous Layer Using 3D Graphene Frameworks for Direct Methanol Fuel Cell. *J. Power Sources* **2020**, *451*, 227800. [CrossRef]
203. Deng, H.; Zhou, J.; Zhang, Y. A Trilaminar-Catalytic Layered MEA Structure for a Passive Micro-Direct Methanol Fuel Cell. *Micromachines* **2021**, *12*, 381. [CrossRef]
204. Ko, J.; Kang, K.; Park, S.; Kim, W.-G.; Lee, S.-H.; Ju, H. Effect of Design of Multilayer Electrodes in Direct Methanol Fuel Cells (DMFCs). *Int. J. Hydrogen Energy* **2014**, *39*, 1571–1579. [CrossRef]
205. Yuan, W.; Han, F.; Chen, Y.; Chen, W.; Hu, J.; Tang, Y. Enhanced Water Management and Fuel Efficiency of a Fully Passive Direct Methanol Fuel Cell with Super-Hydrophilic/ -Hydrophobic Cathode Porous Flow-Field. *J. Electrochem. Energy Convers. Storage* **2018**, *15*, 031003. [CrossRef]
206. Chan, D.-S.; Hsueh, K.-L. A Transient Model for Fuel Cell Cathode-Water Propagation Behavior inside a Cathode after a Step Potential. *Energies* **2010**, *3*, 920–939. [CrossRef]
207. Wu, Q.X.; Zhao, T.S.; Yang, W.W. Effect of the Cathode Gas Diffusion Layer on the Water Transport Behavior and the Performance of Passive Direct Methanol Fuel Cells Operating with Neat Methanol. *Int. J. Heat Mass Transf.* **2011**, *54*, 1132–1143. [CrossRef]
208. Shrestha, S. Fuel Cell: Direct Methanol Fuel Cell (DMFC). Available online: https://energyandclimatechangeinfo.blogspot.com/2021/06/direct-methanol-fuel-cell-dmfc-uses.html (accessed on 12 February 2021).
209. Kim, J.H.; Lee, G.G.; Kim, W.T. Comparison of Liquid Water Dynamics in Bent Gas Channels of a Polymer Electrolyte Membrane Fuel Cell with Different Channel Cross Sections in a Channel Flooding Situation. *Energies* **2017**, *10*, 748. [CrossRef]
210. Li, X.; Faghri, A.; Xu, C. Water Management of the DMFC Passively Fed with a High-Concentration Methanol Solution. *Int. J. Hydrogen Energy* **2010**, *35*, 8690–8698. [CrossRef]
211. Wang, Z.; Zhang, X.; Nie, L.; Zhang, Y.; Liu, X. Elimination of Water Flooding of Cathode Current Collector of Micro Passive Direct Methanol Fuel Cell by Superhydrophilic Surface Treatment. *Appl. Energy* **2014**, *126*, 107–112. [CrossRef]
212. Sun, J.; Zhang, G.; Guo, T.; Che, G.; Jiao, K.; Huang, X. Effect of Anisotropy in Cathode Diffusion Layers on Direct Methanol Fuel Cell. *Appl. Therm. Eng.* **2020**, *165*, 114589. [CrossRef]
213. Zhang, J.; Feng, L.; Cai, W.; Liu, C.; Xing, W. The Function of Hydrophobic Cathodic Backing Layers for High Energy Passive Direct Methanol Fuel Cell. *J. Power Sources* **2011**, *196*, 9510–9515. [CrossRef]
214. Fu, X.; Ni, H.; Zhang, F.; Yang, Y.; Shao, X.; Huang, Z.; Zhang, R.; Liu, Q.; Hu, S. Polypyrrole Nanowires as a Cathode Microporous Layer for Direct Methanol Fuel Cell to Enhance Oxygen Transport. *Int. J. Energy Res.* **2021**, *45*, 3375–3384. [CrossRef]
215. Hutzenlaub, T.; Paust, N.; Zengerle, R.; Ziegler, C. The Effect of Wetting Properties on the Bubble Dynamics in the Flow Field of Direct Methanol Fuel Cells. *Meet. Abstr.* **2010**, *3*, 177. [CrossRef]
216. Yuan, W.; Hou, C.; Zhang, X.; Zhong, S.; Luo, Z.; Mo, D.; Zhang, Y.; Liu, X. Constructing a Cathode Catalyst Layer of a Passive Direct Methanol Fuel Cell with Highly Hydrophilic Carbon Aerogel for Improved Water Management. *ACS Appl. Mater. Interfaces* **2019**, *11*, 37626–37634. [CrossRef]

 *sustainability*

Article

Increased Digital Resource Consumption in Higher Educational Institutions and the Artificial Intelligence Role in Informing Decisions Related to Student Performance

Anjeela Jokhan [1,†], Aneesh A. Chand [2,*,†], Vineet Singh [2,†] and Kabir A. Mamun [2,*]

1   Permanent Secretary for Ministry of Education, Heritage and Arts, Suva, Fiji; anjeela.jokhan@gmail.com
2   School of Information Technology, Engineering, Mathematics and Physics (STEMP), Suva, Fiji; vineet.singh@usp.ac.fj
*   Correspondence: aneeshamitesh@gmail.com (A.A.C.); kabir.mamun@usp.ac.fj (K.A.M.)
†   Current address: Faculty of Science, Technology and Environment, University of the South Pacific, Suva, Fiji.

Abstract: As education is an essential enabler in achieving Sustainable Development Goals (SDGs), it should "ensure inclusive, equitable quality education, and promote lifelong learning opportunities for all". One of the frameworks for SDG 4 is to propose the concepts of "equitable quality education". To attain and work in the context of SDG 4, artificial intelligence (AI) is a booming technology, which is gaining interest in understanding student behavior and assessing student performance. AI holds great potential for improving education as it has started to develop innovative teaching and learning approaches in education to create better learning. To provide better education, data analytics is critical. AI and machine learning approaches provide rapid solutions with high accuracy. This paper presents an AI-based analytics tool created to predict student performance in a first-year Information Technology literacy course at The University of the South Pacific (USP). A Random Forest based classification model was developed which predicted the performance of the student in week 6 with an accuracy value of 97.03%, sensitivity value of 95.26%, specificity value of 98.8%, precision value of 98.86%, Matthews correlation coefficient value of 94% and Area Under the ROC Curve value of 99%. Hence, such a method is very useful in predicting student performance early in their courses of allowing for early intervention. During the COVID-19 outbreak, the experimental findings demonstrate that the suggested prediction model satisfies the required accuracy, precision, and recall factors for forecasting the behavioural elements of teaching and e-learning for students in virtual education systems.

Keywords: SDGs; artificial intelligence; higher education institutions; machine learning; data classification; early warning system; student performance prediction

 check for updates

Citation: Jokhan, A.; Chand, A.A.; Singh, V.; Mamun, K.A. Increased Digital Resource Consumption in Higher Educational Institutions and the Artificial Intelligence Role in Informing Decisions Related to Student Performance. *Sustainability* **2022**, *14*, 2377. https://doi.org/10.3390/su14042377

Academic Editors: Nallapaneni Manoj Kumar, Grigorios L. Kyriakopoulos and Luis Jesús Belmonte-Ureña

Received: 3 December 2021
Accepted: 11 January 2022
Published: 18 February 2022

Publisher's Note: MDPI stays neutral with regard to jurisdictional claims in published maps and institutional affiliations.

1. Introduction

Artificial intelligence (AI), connectivity (the Internet of Things), information digitisation, additive manufacturing (such as 3D printing), virtual or augmented reality, machine learning, blockchain, robotics, quantum computing, and synthetic biology are all examples of areas where the digital revolution can help to facilitate Sustainable Development Goals (SDGs) [1,2]. Similarly, the digital transformation will fundamentally affect many aspects of global communities and economies, resulting in a shift in how the sustainability paradigm is interpreted. Digitalization is a key driver of disruptive, multiscalar change, not just a "tool" for resolving sustainability concerns. Working with digital revolution is already reshaping leisure, work, education, behaviour, and governance. Generally, these contributions can boost labour, energy, resource, and carbon productivity, as well as cut production costs, improve service access, and dematerialise production [2]. Rapid increase in digital resources have influenced the education sector to achieve the SDGs.

The SDGs Agenda of the United Nations endorsed by global leaders in 2015 include climate change mitigation, poverty eradication, and universal access to education [1]. Today in all functions achieving SDGs is very important, and in achieving these SDGs, the role of education is crucial. SDGs are indivisible and encompass economic, social, and environmental dimensions [1]. Equitable quality education (EQE) is one of the major challenges faced by most academic institutes around the globe. SDG 4 talks about the concepts of "equitable quality education". Quality education is thought to lead to a more sustainable world. The major criterion for Education for Sustainable Development (ESD) is to provide a culture that assists students in completing their academics while also providing greater opportunity for addressing problems. Despite these goals, it is questionable what education for sustainability intends to do. The following are some of the historical policy statements [2]:

- The United Nations Education, Scientific, and Cultural Organization (UNESCO), in collaboration with the United Nations Environment Programme (UNEP), hosted the world's first intergovernmental conference on environmental education from 14 to 26 October 1977 in Tbilisi, Georgia (USSR).
- The Earth Summit in Rio de Janeiro in 1992 saw the launch of Education for Sustainable Development (ESD): The United Nations Conference on Environment and Development (Rio Summit, Earth Summit) and Agenda 21's Chapter 36 on Education, Training, and Public Awareness consolidated international discussions on the critical role of education, training, and public awareness in achieving sustainable development.
- In 2002, during the World Summit on Sustainable Development, the Decade for ESD was announced: A proposal for the Decade of Education for Sustainable Development (ESD) was included in the Johannesburg Plan of Implementation. At its 57th session in December 2002, the United Nations General Assembly passed a resolution declaring the UN Decade of Education for Sustainable Development (DESD) to begin in January 2005.
- In 2014, the announcement of the Global Action Programme (GAP) for ESD was introduced during the UNESCO World Conference on ESD.
- In 2015 the World Education Forum in Incheon, Korea R, emphasised the importance of education as a primary driver of development and achievement of the SDGs.

Therefore, with the above argument it is also debatable if these efforts have succeeded in changing curricula and teaching methods to be more sustainable. The concept and understanding of sustainability is crucial to the development of acceptable educational pedagogies, their implementation, and their capacity to provide what they are created for.

EQE is one of the key drivers of a country's economic prosperity and supports sustainability. Exponential growth in enrolment in Higher Education Institutions (HEI) has been observed in the past twenty years [2], as a result of the perceived importance of further education in career development and opportunities. In contrast to the students historically entering tertiary education, a shift in student demographics has been recorded with an increasingly heterogeneous student population taking multi-modal course deliveries [3]. With the increase in student numbers, the demand for state-of-the-art services and resources from the learners has also escalated [4].

Competition amongst the HEI is stiff, as they strive to attract students to take their programmes. With the growing consumerism in higher education, criteria amongst students to choose HEI is more complex, considering factors such as the delivery of the service, reputation and likelihood of getting a better career amongst the traditional socio-economic factors [5]. While student enrolment is dependent on the reputation and attractions on offer, student satisfaction and success are the impetus that drive student retention. Thus, factors that lead to student success are given increasing importance.

Various measures of student success and attainment are used by the HEIs, which include the use of cross-sectional and longitudinal data measuring student progress, completion rates for courses and programmes, to the success of their alumni [6]. Universities strive to maximize successful completion of courses and programmes with student support services, tools and technologies that have been shown to enhance student learning. This warrants the use of new and innovative pedagogies to captivate interest and maximize the potential of the learners.

Today, education relies heavily on ICT, with new tools in the field of higher education [7–9]. For instance, distance learning is not a challenge anymore with off-campus students accessing learning resources using e-learning and m-learning tools [10–14]. In addition to this, AI is gaining interest as it can be used to execute tasks normally associated with human intelligence [15,16], such as social networking applications for learning, speech recognition, learning management systems (LMS), decision-making, cloud learning services, visual perception, mobile learning applications and translating languages [8,9]. Currently, most universities are livestreaming lectures and offering full courses online. Massive Open Online Courses (MOOCs) have made higher education courses from some of the world's most prominent universities available to anybody with a reasonable Internet connection anywhere in the world [17]. Virtual reality will increasingly allow students to participate in field trips and obtain practical experience without ever leaving the classroom or their homes. Through Internet platforms like as *chegg.com*, students have access to "personal" instructors 24 h a day, from anywhere on the globe [18,19]. Textbooks, school libraries, and even consolidated campus attendance are all on the decline.

In conjunction with SDG 4, which aims to "provide inclusive and equitable quality education and encourage life-long learning opportunities for everyone", the digital revolution in education will undoubtedly enhance access to high-quality education throughout the world [1]. However, in order to do this, the essential broadband and energy infrastructure must be supplied simultaneously in poor countries and rural places. The rapid digital revolution of education will have an influence on our cities' structure and social connections. The necessity for centralised campuses and accompanying infrastructure will shrink as education is increasingly given remotely, allowing students to learn from home, either individually or via "virtual classes".

Student performance is one of the most essential elements for any learning institution [20]. Student enrolment and attendance records, as well as their examination results, are the most conventional form of data mining (DM) in higher education institutions [11,20–22]. In this age of big data, education data mining (EDM) is an interdisciplinary field where machine learning, statistics, DM, psycho-pedagogy, information retrieval, cognitive psychology and recommended systems methods and techniques are used in various educational data sets to resolve educational issues [23]. This phenomenon surrounding the EDM can be better explained in Figure 1. To date, little has been done in EDM using AI in education in the developing world. In the current dynamic status of EDM, numerous studies have been carried out in relation to different typologies of DM in educational environment [23–25].

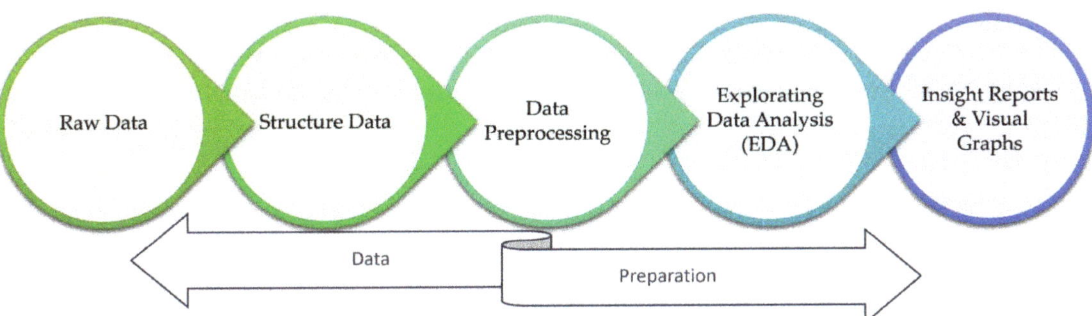

Figure 1. Steps in data mining.

Common representative classifications are as follows:
- Analysis and Visualization of Data,

- Providing Feedback for Supporting Instructors,
- Recommendations for Students,
- Predicting Student's Performance,
- Student Modelling, and
- Social Network Analysis.

The use of AI in the educational environment is imperative because it can contribute significantly to the improvement of the teaching and learning processes, as well as encourage the process of knowledge construction [15–17]. Based on the results of a report on the sustainability of higher education and TEL [26], when identifying the necessary conditions for technology to assist and not obstruct teaching and learning, we need to be very careful.

This research is designed to model an AI based predictor for student performance in a higher education online course and the significant contributions of the paper can be recounted as follows:

- A framework for an AI based student performance predictor is proposed,
- Digital resources are used in informing decisions related to student performance,
- Al prediction for student performance is designed and analyzed for a first-year IT literacy course at The University of the South Pacific (USP).

In this work, the main focus was to achieve better accuracy when compared with previous research [20] and the early prediction of student performance by employing AI in EDM. A Random Forest (RF) classifier model is applied to the data set and an accuracy of 97.03% was achieved at week 6.

The paper is organized as follows. Section 2 summarizes the literature with the current direction, the role of digital learning, and the involvement of AI in HEI. Section 3 provides the design and architecture of the developed model (i.e., intelligent Early Warning System (iEWS)). The methodology used to predict student performance is presented in Section 4. Section 5 provides the results and discussion, and the conclusion and research suggestions are provided in Section 6.

2. Types of Early Warning Systems

A substantial body of research shows that progressive trends for students are significant contributors to student performance in online learning. There are several methods associated with EWS. One of the most common techniques is the use of statistical analysis to predict performance. Until recently, statistical approaches were largely applied in educational institutes to understand potential student pass/fail and dropout rates. More recently, different approaches have been combined to show better performance in EDM. Different predictive techniques are used in order to have a better prediction rate. Different classification methods are applied for given data sets. Figure 2 depicts the graphical representation for a list of the common methods used for EWS for student performance prediction.

2.1. The Evolution of EWS in Higher Education

The topic of predictive algorithms is often regarded as the most relevant field of study within the data analytics discipline. EWS is widely used in various fields of study and has impacted the education sector [20,27,28] more recently. One of the prime reasons for applying EWS is that universities use it to track student progress and recognize students at risk of failing a course or dropping out of a course or programme [11,29]. Various techniques are being proposed, applied and tested, and there are many advanced tools available in the literature which have better-predicting accuracy in the field of EDM [30,31]. One of the pre-processing algorithms of EDM is known as Clustering 32. Interestingly, DM is one of the most popular techniques which is widely applied in education to analyse student performance [25,32]. EWS has been used widely in secondary schools in the United States for many years. It has been used to track student success in schools and to identify measures that predict the likelihood of dropping out of school [33,34]. The features and variables that were collected for EWS were based on demographic/historical data, ongoing test results and the use of LMS. Once the EWS identifies an at-risk student, the teacher has the option of providing corrective measures which includes indicating different alert signals on a student's Moodle page and alert message via e-mail messages or text messages [29,30]. Additionally, students were allowed to get a referral to an academic advisor to address the problem faced in a particular course.

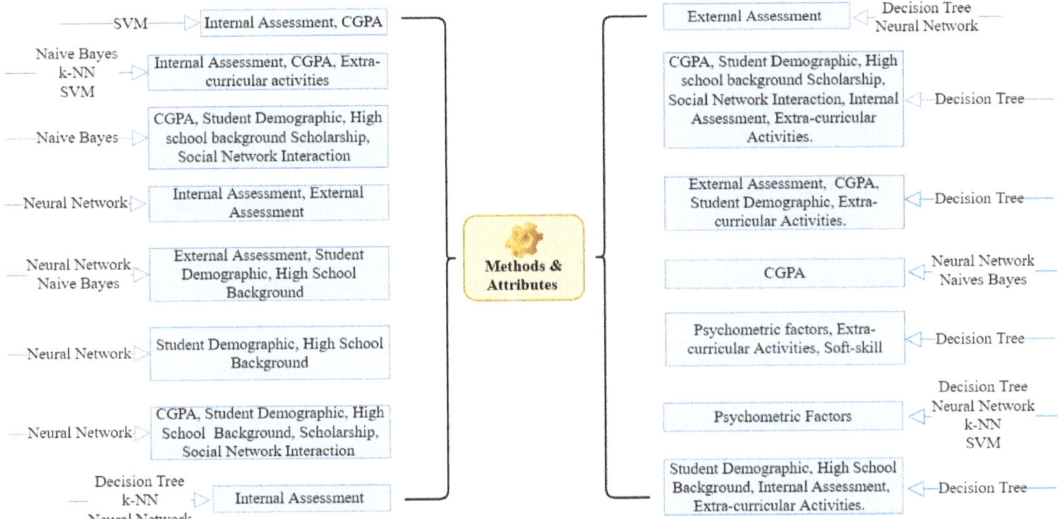

Figure 2. Lists of the common methods and attributes used in EWS as predictive tools.

There has been an increase in different types of prediction models used in learning analytics. According to [35], analytical researchers are trying to predict with better accuracy and employing different classification tools to compare accuracies. The common classification algorithms, such as EM, C4.5, Naive Bayes Classifier, Support Vector Machines, K-nearest neighbor [29], neural network models [36], and decision tree methods [37] are also employed.

In most cases, the analysis is performed to predict whether a student will pass or fail a course based on the binary response variable 'pass/fail'. Principally, one of the fundamentals and keen methodologies usually applied in predictive models is that the analysis is usually performed on a single course rather than used for several courses. As a systemic approach, model features and response variables are used in classifying at-risk students but for a prediction model [29], the beginning of the semester is too early to identify at-risk students. It is often difficult to contrast the studies and identify which study has obtained the most accurate results.

Azcona and Casey [37], argue that a single course analysis is more efficient in terms of accuracy. This may be because each course is structured differently and, therefore, the feature will be not the same for classification in different courses. In a similar study by Ognjanovic et al. [38], it was evident that predictive models could be applied to multiple courses. However, they noted that the inherent differences in disciplines caused specific variables to be strong for some courses and weak for other courses. Hence, the nature of the course should be considered before selecting variables for an early warning system.

2.2. AI in Early Warning Systems

The involvement of AI in previous years has attracted several controversial remarks [15,16]. The use of AI in computing power, DM and Big Data technologies appears to be a more advanced tool in predicting with better accuracy [15,32]. As mentioned earlier, AI used a better classification tool to predict the accuracy of any EDM. Ognjanovic et al. [38] and Andriessen et al. [39] both examined AI methods used in learning platforms and the relationship between education and AI, respectively. To add to this, academic performance in game-based learning strategies was studied by Stojanovska et al. [40]. They also studied flip teaching techniques, and video conferencing sessions by mining personality traits, learning style and satisfaction. Basavaraju et al. [41] proposed a study by supervised learning to use the android app. Table 1 shows the different research carried out in the field of AI and DM methods and their accuracies. EMD study was carried out where student's behavioural features were used to model the system. The system yielded 22.1% accuracy, and later, using an ensemble method, they noticed there was an increase in the accuracy of 25.8% [42].

A Deep Neural Network (DNN) was used to analyze student performance in Keras library. They used online data sets and achieved 83.4% accuracy, and the quality of the classifier was measured by Cost Function and Accuracy [43]. In 2017, a Recurrent Neural Network (RNN) was implemented to predict the students' performance for logged data from 108 students. The predicting feature used was log data of an LMS and the results revealed a 90% accuracy [44]. A review was carried out on predicting student performance using DM methods and showed that the results of Neural Network and Decision Tree had achieved an accuracy of 98% and 91%, respectively [45]. A prediction model was developed using an Artificial Neural Network (ANN). The work was designed to predict the Cumulative Grade Point Average of students. The academic datasets were modelled in one of the universities in Bangladesh. They performed a compassion test with the predicted and original grades. The highest accuracy of 99.98% and Root Mean Square Error of the work was 0.176546 [46].

Table 1. Research carried out in the field of AI and DM methods with its accuracy.

Method	Feature	Accuracy (%)	Ref.
DNN	External assessment, Student Demographic, High school background	74	[47]
	Student Demographic, High school background	72	[48]
	CGPA, Student Demographic, High school background, Scholarship, Social network interaction	71	[49]
	CGPA	75	[50]
	External assessment	97	[51]
	Psychometric factors	69	[52]
	Internal assessments	81	[53]
SVM	Internal assessment, CGPA	80	[54]
	Internal assessment, CGPA, Extra-curricular activities	80	[55]
	Psychometric factors	83	[56]
Decision Tree	Psychometric factors, Extra-curricular activities, soft skills	88	[57]
	External assessment, CGPA, Student Demographic, Extra-curricular activities	90	[58]
	Internal assessment, Student Demographic, Extra-curricular activities	90	[59]
	Internal assessment, CGPA, Extra-curricular activities	73	[55]
	CGPA, Student Demographic, High school background, Scholarship, Social network interaction	73	[49]
	CGPA	91	[50]
	External assessment	85	[60]
	Psychometric factors	65	[51]
	Internal assessments	76	[24]

3. Design and Architecture of Intelligent Early Warning System (iEWS) Model

This study retrieved complete online interaction data for undergraduate students of a fully online first year course, Communication Information Literacy, at the USP for one semester. The USP uses Moodle LMS where all the online, face-to-face and blended courses are hosted. Moodle requires user authentication to access the registered courses for a particular student and detailed interactions for each student for the course are recorded in the Moodle database including system login, logout, material access, assignment submission, discussion forum activities, score package records, quiz activities and numerous other activities and resource data. All these data are stored on individual activity/resource table and all other interactions in the course are stored in a log table.

An EWS (Student Alert Moodle Plugin) developed by the Faculty of Science, Technology and Environment at the USP was implemented in week 4 of the semester in the course [11]. The data from the EWS plugin were used to extract features to develop iEWS predictor. The architecture is shown in Figure 3 and the process flow is discussed in Table 2.

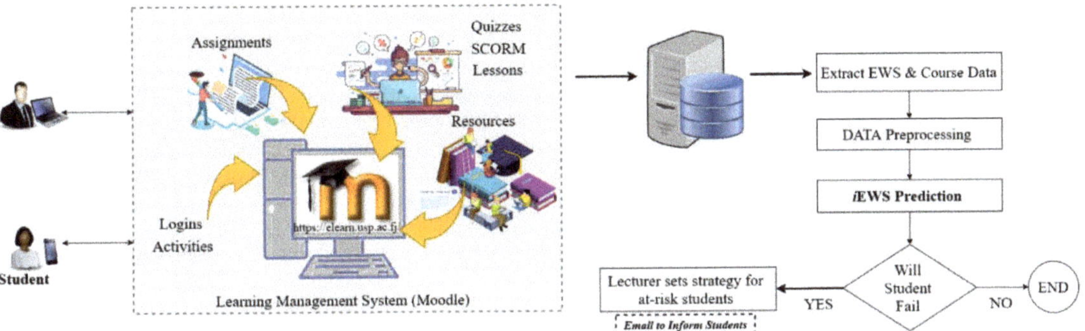

Figure 3. iEWS Architecture.

Table 2. iEWS process flow.

1	Students and teachers interact with the course activities.
2	All interactions are recorded on Moodle Database.
3	EWS data are calculated using Moodle DB and recorded in EWS DB.
4	EWS data are extracted, and data prepressing is done (Data cleaning and EWS features are extracted).
5	EWS features are used to develop the iEWS predictor.
6	The iEWS predictor is tested with the test data.
7	If iEWS predicts a student to fail, then teacher sets strategies for these students.

4. Methodology

This study discusses the proposed predictor called iEWS, which uses students' EWS data based on online course login, interaction and completion for as early as Week 6 to predict if the student will pass or fail a course. The following sections discuss the dataset, data cleaning and extraction of features, statistical measures and validation scheme used to measure the performance and RF classifier used for prediction.

4.1. Dataset

On implementation of EWS in week 4, the completion rates, interaction rates and average logins per week increased. Completion rate is based on the number of the course activities mentioned earlier that were completed by the students in each week. EWS data collection started in week 4, after the EWS was implemented for weekly/fortnightly intervals. In this research, a total of 1523 student data-sets were used, in which 1271 students passed (positive samples) and 252 students failed (negative samples).

4.2. Features

The following attributes from EWS plugin were used for this study:

- AvgCompRate—average percentage of online activities completed by students each week,
- AvgLogin—average number of logins by students each week.
- CourseworkScore—the coursework marks for Weeks 6, 8 and 10.

4.3. Reducing the Imbalance between Classes

After investigating the dataset, it was clear that the number of positive samples (students passed) was much bigger than the negative samples (students failed). This clearly resulted in a high-class imbalance of dataset.

The k-nearest neighbour technique was employed to reduce the imbalance of the dataset (i.e., between samples and classes) to remove redundant positive samples. Euclidean distance between all the samples in the dataset was calculated. Firstly, the cut-off was set by dividing the number of positive instances and negative instances (1271/252) which equals to a ratio of 5.04, thus K = 5 was set. This implies that there was a removal of a positive sample if there existed at least a positive sample within five nearest neighbours. After initial filtering, imbalance classes still remained, therefore, the K value was continuously increased until both the sets were approximately similar in size. This method eventually reduced the initial positive samples of 1271 to 256 with a threshold value of 29 (k = 29), which implies a positive sample was removed if at least one negative sample existed within the 29 nearest neighbours. The negative instances were not changed and remained at 252. The final dataset after filtering (filtered negative samples and positive samples) was used to carry out 6-, 8-, 10- fold cross-validation and assess the predictor's performance.

4.4. Tool

MATLAB® software was used to carry out data pre-processing, feature extraction, reducing the imbalance between classes, splitting the data set into "N" folds of approximately equal sample size with similar positive and negative counts, creating a Weka data format (ARFF) file for Weka Classifiers. Weka was developed by University of Waikato in New Zealand, for classification and performance assessment [61,62].

The code was written in Java to train and test a set of classifiers provided by Weka for which performance assessment was carried out for different "N" folds. Net beans IDE was used for Java code and Weka.jar library downloaded from (http://www.cs.waikato.ac.nz/ml/weka/snapshots/weka_snapshots.html (accessed on 15 October 2021)) and referenced in the java project to access and run the Weka classifiers required [62]. Different classifiers were used to train and test to finalize the best classifier for iEWS predictor, based on the performance from each of the classifiers stated below.

4.5. Classifier

C4.5 (J48) is an algorithm used to generate a decision tree for classification of different applications [63]. PART is a partial decision tree algorithm, developed from C4.5 and RIPPER algorithms [64]. A decision table represents conditional logic with a list of tasks depicting business rules that can be used with the same number of conditions, which makes it different from the decision tree [65]. One Rule (OneR) is a simple classification algorithm that creates one rule for predictor in the data and then selects the rule with minimum error rate [66]. Decision stump consists of one level of Decision Tree, and uses only a single attribute for splitting [67]. Logistic regression is a statistical model, which uses a logistic function to model and predict the probability of an outcome that can have two values or binary classes [68]. Sequential Minimal Optimization (SMO) algorithm is based on the Support Vector Machine (SVM) solving quadratic programming (QP) problem, which arises during the training of SVM [69]. Multilayer perceptrons (MLP) is one type of neural network, which has a similar structure as a single layer perceptron, with one or many hidden layers and two phases [70].

4.6. RF

RF and decision trees are well known and used for the supervised learning model with associated learning algorithms that analyze data used for classification and regression analysis. It has been used in many other similar studies [24,48–50,55,57–60] and it gives high accuracy as shown in Figure 3. RF is an ensemble approach that includes a lot of trees for decision. The growing level of trees in a candidate feature set is calculated by an optimal law. The candidate feature set is a random subset of all features, which is distinct at each tree level. The RF grouping is an ensemble identification, corresponding to a new approach consisting not just one but several classifiers as well. In reality, hundreds of classifiers are built into RF grouping, and their selections are commonly combined by plurality vote. The concept remains that sometimes the combination of ensemble classifiers are more reliable than any of the ensembles [71,72], evicting conflicts among subsets of features. The RF classification is, therefore, commonly used for remotely sensed imagery processing. The common element in all of these procedures is that for the k-th tree, a random vector $\varnothing_k$ is generated independent of the prior random vectors $\varnothing_k \ldots \varnothing_{k-1}$, but with the same distribution; the tree is grown using the training set and $\varnothing_k$, resulting in a classifier $h(\varnothing_k)$ where x is an input vector [71]. The genetic expression to predict a class of an observation is obtained by:

$$H(x) = argmax_y \sum_{i-1}^{k} I(h_i(X, \theta_k) = Y) \tag{1}$$

where, $argmax_y$ represent the Y maximize value of $\sum_{i-1}^{k} I(h_i(X, \theta_k) = Y)$ which is the output variable, $I(h_i(X, \theta_k))$ is the indicator function, and $h_i(X, \theta_k)$ is a single decision tree.

The classifier comprises various trees which are uniformly assembled by pseudo-randomly selecting subsets of feature vector components, that is, trees are assembled in randomly picked subspaces that preserve the maximum precision of training data and increase the accuracy of generalization as it increases in complexity [73].

4.7. Statistical Measures

To evaluate the performance of the proposed predictor and compare with the existing predictors, few measures such as sensitivity (Sn), specificity (Sp), accuracy (Acc), precision (Pre) and Matthews correlation coefficient (MCC) were employed in this work.

On the other hand, specificity assesses the proportion of correctly identified number of students failed. A specificity of 1 demonstrates an accurate predictor which is able to predict negative instance of the dataset (number of students failed) whereas a specificity equal to 0 shows that the predictor is unable to identify the number of students failed. The metric for specificity is defined as:

$$Sensitivity = \frac{P_+}{P_+ + P_-} \tag{2}$$

where, P_+ is number of students passed predicted correctly and P_- represents the number of students passed incorrectly classified by the predictor

On the other hand, specificity assesses the proportion of correctly identified number of students failed. A specificity of 1 demonstrates an accurate predictor which is able to predict a negative instance of the dataset (number of students failed), whereas specificity equal to 0 shows that the predictor is unable to identify the number of students failed. The metric for specificity is defined as:

$$Specificity = \frac{F_+}{F_+ + F_-} \tag{3}$$

where, F_+ is the number of students failed predicted correctly and F_- represents the number of incorrectly classified students failed by the predictor.

For a predictor to correctly distinguish between positive samples and negative samples, the accuracy of the predictor is evaluated. A predictor with an accuracy equal to 1 shows an accurate predictor, whereas a zero accuracy means the predictor is completely incorrect. Accuracy is calculated as:

$$Accuracy = \frac{P_+ + F_+}{P + F} \tag{4}$$

where P and F are the total numbers of passed and failed students, respectively.

Precision is another assessment measure of the predictor, defined as the ratio of the number of correctly identify students passed over sum of correctly classified passed and failed students.

$$Precision = \frac{P_+}{P_+ + F_+} \tag{5}$$

The final statistical measure used in this paper is the Matthews correlation coefficient (MCC). It shows the value of the correlation coefficient between predicted and observed instances. The MCC metric is calculated as:

$$MCC = \frac{(F_+ \times P_+) - (F_- \times P_-)}{\sqrt{(P_+ + P_-)(P_+ + F_-)(F_- + P_-)(F_+ + F_-)}} \tag{6}$$

A best predictor is the one that achieves high performance in the five statistical measures discussed. However, it should perform better at least in some of the measures compared to the existing predictors. A predictor that is unable to predict passed or failed students correctly cannot be used for prediction.

4.8. Validation Scheme

The effectiveness of a new predictor needs to be assessed with a validation method. Two of the most commonly used ones are the jackknife and n-fold validation scheme [23,73]. In the validation phase, an independent test set has to be used to assess the predictor. The jackknife validation is less arbitrary than the n-fold cross-validation and provides unique results for a dataset. As per the literature [74,75], the same validation scheme (n-fold cross-validation) technique was used in this study. The n-fold cross-validation technique was carried out in the following steps listed in Table 3 and shown in Figure 4.

Table 3. Steps for cross-validation approach.

1	Split pre-processed data set into n folds of approximately equal sample size with similar positive and negative samples in each.
2	Separate one of the folds as an independent test set and use the other n-1 folds as training data.
3	Train the model with training data and adjust the parameters of the predictor
4	Use the independent test set (2) to validate the predictor by computing all the statistical measures
5	Repeat steps 1 to 4 for other folds until n folds for validation and calculate the average of each statistical measure for n-folds and record the result

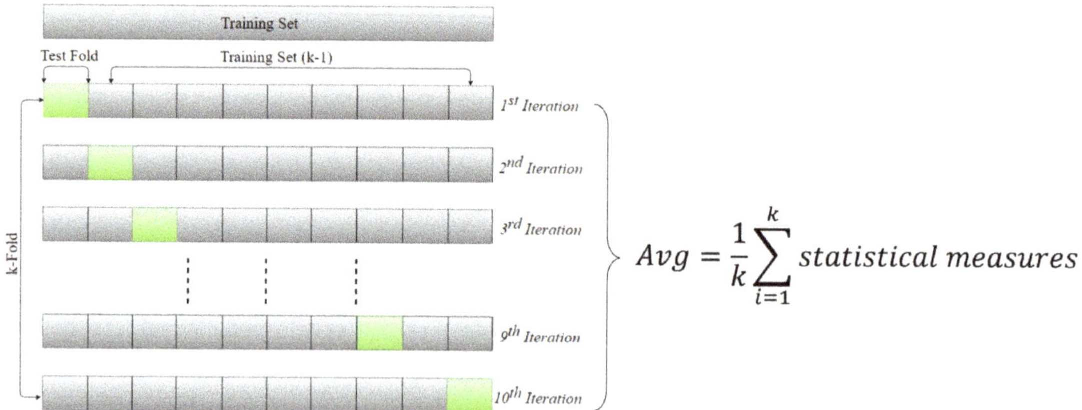

$$Avg = \frac{1}{k}\sum_{i=1}^{k} statistical\ measures$$

Figure 4. n-fold cross-validation technique.

In this study, 6-, 8- and 10-fold cross-validations was conducted to assess iEWS predictor and recorded the result.

5. Results and Discussion

In order to verify the performance of any proposed predictor, it has to be assessed using different measures. The five statistical metrics: accuracy, sensitivity, specificity, precision and Matthews correlation coefficient, which are normally used, were used in this study [29,36,37,48,70]. This section presents the results of the proposed predictor.

5.1. Comparison with Statistical Analysis

In the previous study [20], a statistical model was developed with an accuracy of 60.8%. It is worth noting that the same dataset was used to develop an iEWS predictor and the accuracy was compared [20]. In comparison to the old EWS model, this new iEWS predicted the accuracy of 97%, which is an improvement of at least 36.2%. Accuracy of prediction was 97% in week 6, 98% in week 8 and 98.4% in week 10.

Furthermore, the main advantage of the proposed iEWS is that it can predict whether a student can pass or fail so the corrective measures can be taken as early as possible. The model was able to identify and predict the student's performance just by analysing the three attributes (i.e., avgcomprate, avglogin, and courseworkscore). It is worth noting that out of nine different classification tools, RF predicted the best performance (accuracy) with the given attributes. Therefore, weeks 6, 8, and 10 datasets are employed to develop the model. It was seen that week 6 showed very promising results for which the sensitivity, specificity, precision, accuracy, MCC for iEWS for 6-, 8- and 10-fold cross-validation trials were calculated.

5.2. iEWS Prediction with RF

The aim of Moodle-based EWS is to monitor the learning progress of students in a course and to identify at-risk students as early as possible so teachers can implement strategies to assist those students. The early prediction in week 6 (which very high accuracy) of the semester by the proposed iEWS shows a promising tool that can be used by HEIs to intervene and assist the more vulnerable students. This prediction uses significant features of average completion rate, average login frequency and coursework from EWS plugin in this first year IT course.

The effective use of RF classifier in EWS also contributes to the outcome. In short, the combination of EWS data + RF classifier play a significant role in predicting whether students pass or fail the course. The results for Week 6 are given in Figure 5 with three different folds. A huge improvement in accuracy for proposed iEWS by at least 36.2% is seen over the statistical model in [20]. It is also observed that iEWS predictor recorded high sensitivity, specificity, precision and MCC, implying its great performance. The promising results show the ability of the proposed iEWS predictor to correctly identify students passing and failing the course as early as week 6 of the semester. Consequently, using an RF-based model has the potential to accelerate educational development, and the efficiency of education may be shown to increase dramatically. By effectively and efficiently using RF methods in the context of teaching and learning, education will be transformed, radically altering teaching, learning, and research. Educators that use digital tools will acquire a better knowledge of how their students are developing with their studies, allowing them to intervene early and increase student performance and retention.

It is worth noting that the features and classifier used for this study may not work for other courses as the online presence and activities differ in courses. The more online activities a course would have, the better the ability for prediction, as the activities will contribute to the completion rate and coursework of EWS. A similar study was carried out to predict at-risk students in a course using standards-based grading where they created a specific course predictive model to identify at-risk student in week 5 [31]. The common tool used in this study was SVM, K-NN and Naive Bayes classifier. The Naïve Bayes classifier had the best results among the seven testing models. The different accuracy of the prediction model used are showed in Figure 6.

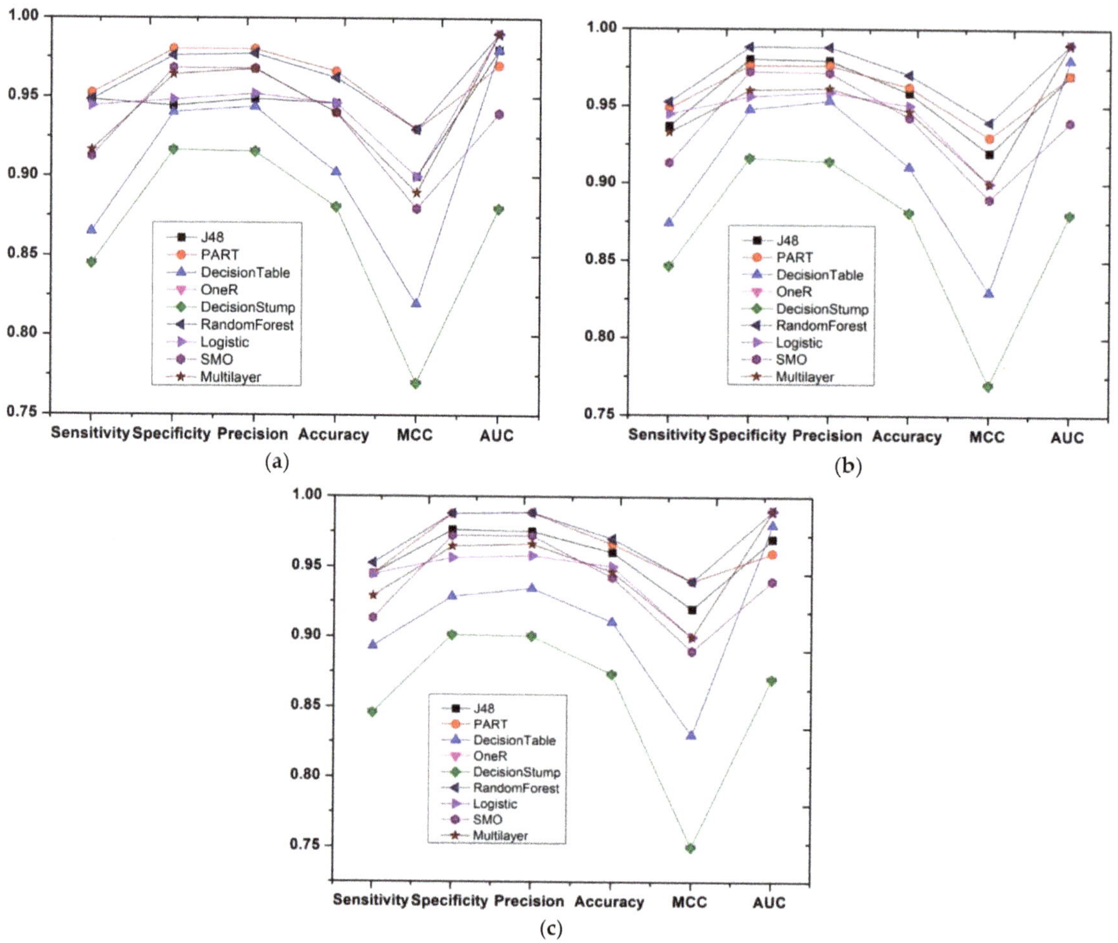

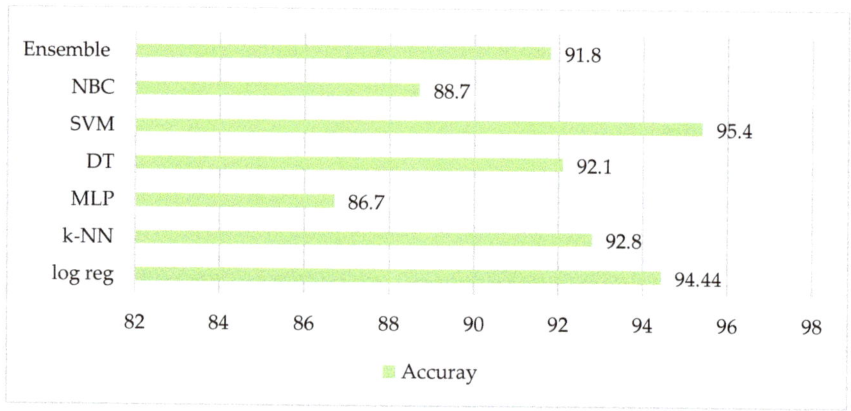

Figure 5. iEWS prediction for (**a**) 6-fold, (**b**) 8-fold, and (**c**) 10-fold.

Figure 6. Different level of accuracy of the prediction model [29].

In most cases, the EWS report relied on midterm grades [35]. At this point, it is often too late into the term and students either cannot cope or drop out of the course. This has been one of the drawbacks of EWS. For this reason, improving the accuracy of EWS and predicting performance much earlier is of great importance. In iEWS, RF classification is used, which predicted more accurately and early in the semester. In this study, since the EWS was introduced in the course in week 4, the earliest prediction could be made in week 6. However, if EWS is engaged in a course much earlier, detection could be even sooner.

As discussed earlier, the proposed model is able to predicate the students' performance as early as week 6 of the semester, with an accuracy of 97.03%. Furthermore, most literature studies propose self-developed models to predict the student performance, but they have failed to mention how early in the semester the prediction of student performances were made. However, the proposed model enabled sustainability in education by providing a iEWS for students as well as for educators. It also saves energy, time, and resources while predicating the students' performance as early as possible.

6. Conclusions

The tendency of students to procrastinate and fall under at-risk categories is often reported by numerous academics as a significant factor that negatively influences student success in higher education blended courses, making its prediction a very useful task for universities and students alike. In this context, this research conducts a different approach, i.e., an AI-based predictor that can predict students' performance as early as possible in the era of the SDGs from a systems perspective. The use of ICT tools contributes to an excellent learning environment among students and learning pedagogies. Such tools were heavily involved in the current education system which uplifted and connected the whole society.

In this work, an AI approach is applied to the same model, the RF classifier model was developed with week 6 EWS data and an accuracy of 97.03% was achieved. An AI platform is designed with LMS and EWS, and the RF classifier is applied with respective sensitivity, specificity and precision. All methods appeared to be sensitive to the increment in the number of classes. RF, with an accuracy of 97.03%, showed a better performance using categorical features compared to other classification methods (see Figure 5). When comparing the accuracy of prediction of student performance using iEWS with that determined through a statistical analysis, it proved higher by more than 35%. In future, this work can be expanded by using a different predictive method and feature vectors of different lengths from different courses. Moreover, different hybrid feature vectors can be created using pre-education grade, students' submission, logins, gender, location of origin, and social interaction behaviour to examine the effect of various time-related indicators on the EWS and at-risk student's predication as earliest possible.

This research objective process can help all those involved in education and sustainability collaborate more effectively, allowing educational institutions to develop a clear vision of what sustainability means to them, and work towards transforming individuals, groups, organisations, communities, and systems by developing the skills needed to transition to a more sustainable future. One of the most significant effects of digitisation in the coming decades will undoubtedly be in the field of virtual education. The development and delivery of course content and curricula will be drastically altered as a result of the digitisation of education. Curricula will need to reflect this digitally capable culture to ensure pupils remain engaged in learning, given the increased digital awareness and competency of students, even those as young as pre-school age. Curricula with more flexibility, standardisation, and even globalisation have the potential to promote equitability and give more options.

In broad terms, sustainability in education is an attempt to reconcile growing a quality learning environment with socio-economic objectives. The framework established to address the requirement to contextualise the function of ESD helps both educators and students to see the wider picture and grasp the role of education in sustainable development. During the COVID-19 outbreak, these experimental findings demonstrate that the suggested prediction model satisfies the required accuracy, precision, and recall factors for forecasting the behavioural elements of teaching and e-learning for students in virtual education systems. Its phases should be thought of as conceptual, since greater specificity will be heavily influenced by the setting, institutional capability, challenge, timing, and resources available to the educational redesign process.

Author Contributions: Conceptualization, A.J.; methodology, A.A.C.; software, V.S.; validation, A.A.C. and V.S.; formal analysis, V.S.; investigation, A.A.C. and K.A.M.; resources, A.J.; data curation, A.J.; writing—original draft preparation, A.A.C., V.S. and K.A.M.; writing—review and editing, A.A.C., A.J. and V.S.; visualization, A.J.; supervision, A.J.; project administration, A.J.; funding acquisition, A.J. and K.A.M. All authors have read and agreed to the published version of the manuscript.

Funding: This research project is funded by The University of the South Pacific (USP) under the Pacific Islands Forum Secretariat grant.

Institutional Review Board Statement: Not applicable.

Informed Consent Statement: Not applicable.

Data Availability Statement: Data is contained within the article.

Acknowledgments: Authors would like to acknowledge the support from the School of Information Technology, Engineering, Mathematics and Physics (STEMP) and Research Office.

Conflicts of Interest: The authors declare no conflict of interest.

Abbreviations

The following abbreviations are used in this manuscript:

ANN	Artificial Neural Network
DM	Data Mining
DNN	Deep Neural Network
EQE	Equitable Quality Education
ESD	Education for Sustainable Development
EWS	Early Warning System
HEI	Higher Education Institutions
iEWS	Intelligent Early Warning System
KNN	K-nearest neighbors
LMS	Learning Management Systems
RF	Random Forest
SDGs	Sustainable Development Goals
SVM	Support Vector Machine

References

1. United Nations. Sustainable Development Goals. Available online: http://www.undp.org/content/undp/en/home/sustainable-development (accessed on 15 October 2021).
2. Messner, D.; Nakicenovic, N.; Zimm, C.; Clarke, G.; Rockström, J.; Aguiar, A.P.; Boza-Kiss, B.; Campagnolo, L.; Chabay, I.; Collste, D.; et al. *The Digital Revolution and Sustainable Development: Opportunities and Challenges-Report Prepared by The World in 2050 Initiative*; International Institute for Applied Systems Analysis (IIASA): Laxenburg, Austria, 2019.
3. Kioupi, V.; Voulvoulis, N. Education for sustainable development: A systemic framework for connecting the SDGs to educational outcomes. *Sustainability* **2019**, *11*, 6104. [CrossRef]
4. Jayaprakash, S.M.; Moody, E.W.; Lauría, E.J.; Regan, J.R.; Baron, J.D. Early alert of academically at-risk students: An open source analytics initiative. *J. Learn. Anal.* **2014**, *1*, 6–47. [CrossRef]
5. Petruzzellis, L.; Romanazzi, S. Educational value: How students choose university: Evidence from an Italian university. *Int. J. Educ. Manag.* **2010**, *24*, 139–158. [CrossRef]
6. Young, M.F.; Slota, S.; Cutter, A.B.; Jalette, G.; Mullin, G.; Lai, B.; Simeoni, Z.; Tran, M.; Yukhymenko, M. Our princess is in another castle: A review of trends in serious gaming for education. *Rev. Educ. Res.* **2012**, *82*, 61–89. [CrossRef]
7. Islam, A.A.; Mok, M.M.C.; Gu, X.; Spector, J.; Hai-Leng, C. ICT in higher education: An exploration of practices in Malaysian universities. *IEEE Access* **2019**, *7*, 16892–16908. [CrossRef]
8. Tseng, H.; Yi, X.; Yeh, H.T. Learning-related soft skills among online business students in higher education: Grade level and managerial role differences in self-regulation, motivation, and social skill. *Comput. Hum. Behav.* **2019**, *95*, 179–186. [CrossRef]

9. Sharma, B.; Nand, R.; Naseem, M.; Reddy, E.V. Effectiveness of online presence in a blended higher learning environment in the Pacific. *Stud. High. Educ.* **2020**, *45*, 1547–1565. [CrossRef]
10. Rodrigues, H.; Almeida, F.; Figueiredo, V.; Lopes, S. Mapping key concepts of e-learning and educational-systematic review through published papers. In Proceedings of the 11th Annual International Conference of Education, Research and Innovation, Seville, Spain, 12–14 November 2018; pp. 8949–8952.
11. Chand, A.A.; Lal, P.P.; Chand, K.K. Remote learning and online teaching in Fiji during COVID-19: The challenges and opportunities. *Int. J. Surg.* **2021**, *92*, 106019. [CrossRef] [PubMed]
12. Sharma, B.N.; Reddy, P.; Reddy, E.; Narayan, S.S.; Singh, V.; Kumar, R.; Naicker, R.; Prasad, R. Use of Mobile Devices for Learning and Student Support in the Pacific Region. In *Handbook of Mobile Teaching and Learning*; Zhang, Y., Cristol, D., Eds.; Springer: Berlin/Heidelberg, Germany, 2019.
13. Sharma, B.; Kumar, R.; Rao, V.; Finiasi, R.; Chand, S.; Singh, V.; Naicker, R. A mobile learning journey in Pacific education. In *Mobile Learning in Higher Education in the Asia-Pacific Region*; Springer: Singapore, 2017; pp. 581–605.
14. Singh, V. Android based student learning system. In Proceedings of the 2015 2nd Asia-Pacific World Congress on Computer Science and Engineering (APWC on CSE), Suva, Fiji, 2–4 December 2015; IEEE: New York, NY, USA; pp. 1–9.
15. Denning, P.J.; Lewis, T.G. Intelligence May Not Be Computable: A hierarchy of artificial intelligence machines ranked by their learning power shows their abilities–and their limits. *Am. Sci.* **2019**, *107*, 346–350. [CrossRef]
16. Zhao, Y.; Liu, G. How do teachers face educational changes in artificial intelligence era. *Adv. Soc. Sci. Educ. Humanit. Res.* **2018**, *300*, 47–50.
17. Zawacki-Richter, O.; Marín, V.I.; Bond, M.; Gouverneur, F. Systematic review of research on artificial intelligence applications in higher education–where are the educators? *Int. J. Educ. Technol. High. Educ.* **2019**, *16*, 39. [CrossRef]
18. Kumar, N.M.; Krishna, P.R.; Pagadala, P.K.; Kumar, N.S. Use of smart glasses in education-a study. In Proceedings of the 2018 2nd International Conference on I-SMAC (IoT in Social, Mobile, Analytics and Cloud)(I-SMAC) I-SMAC (IoT in Social, Mobile, Analytics and Cloud)(I-SMAC), Coimbatore, India, 30–31 August 2018; IEEE: New York, NY, USA; pp. 56–59.
19. Kumar, N.M.; Singh, N.K.; Peddiny, V.K. Wearable smart glass: Features, applications, current progress and challenges. In Proceedings of the 2018 Second International Conference on Green Computing and Internet of Things (ICGCIoT), Bangalore, India, 16–18 August 2018; IEEE: New York, NY, USA; pp. 577–582.
20. Jokhan, A.; Sharma, B.; Singh, S. Early warning system as a predictor for student performance in higher education blended courses. *Stud. High. Educ.* **2019**, *44*, 1900–1911. [CrossRef]
21. Pierrakeas, C.; Koutsonikos, G.; Lipitakis, A.D.; Kotsiantis, S.; Xenos, M.; Gravvanis, G.A. The variability of the reasons for student dropout in distance learning and the prediction of dropout-prone students. In *Machine Learning Paradigms*; Springer: Cham, Switzerland, 2020; pp. 91–111.
22. Hochachka, W.M.; Caruana, R.; Fink, D.; Munson, A.R.T.; Riedewald, M.; Sorokina, D.; Kelling, S. Data-mining discovery of pattern and process in ecological systems. *J. Wildl. Manag.* **2007**, *71*, 2427–2437. [CrossRef]
23. Dutt, A.; Ismail, M.A.; Herawan, T. A systematic review on educational data mining. *IEEE Access* **2017**, *5*, 15991–16005. [CrossRef]
24. Romero, C.; Ventura, S.; Espejo, P.G.; Hervás, C. Data mining algorithms to classify students. In Proceedings of the Educational Data Mining 2008, The 1st International Conference on Educational Data Mining, Montreal, QC, Canada, 20–21 June 2008.
25. Buenaño-Fernandez, D.; Villegas-CH, W.; Luján-Mora, S. The use of tools of data mining to decision making in engineering education—A systematic mapping study. *Comput. Appl. Eng. Educ.* **2019**, *27*, 744–758. [CrossRef]
26. Buenaño-Fernández, D.; Gil, D.; Luján-Mora, S. Application of machine learning in predicting performance for computer engineering students: A case study. *Sustainability* **2019**, *11*, 2833. [CrossRef]
27. Avramidis, E.; Skidmore, D. Reappraising learning support in higher education. *Res. Post-Compuls. Educ.* **2004**, *9*, 63–82. [CrossRef]
28. Wolff, A.; Zdrahal, Z.; Nikolov, A.; Pantucek, M. Improving retention: Predicting at-risk students by analysing clicking behaviour in a virtual learning environment. In Proceedings of the Third International Conference on Learning Analytics and Knowledge, Leuven, Belgium, 8–13 April 2013; pp. 145–149.
29. Marbouti, F.; Diefes-Dux, H.A.; Madhavan, K. Models for early prediction of at-risk students in a course using standards-based grading. *Comput. Educ.* **2016**, *103*, 1–15. [CrossRef]
30. Howard, E.; Meehan, M.; Parnell, A. Contrasting prediction methods for early warning systems at undergraduate level. *Internet High. Educ.* **2018**, *37*, 66–75. [CrossRef]
31. Hu, Y.H.; Lo, C.L.; Shih, S.P. Developing early warning systems to predict students' online learning performance. *Comput. Hum. Behav.* **2014**, *36*, 469–478. [CrossRef]
32. Romero, C.; Ventura, S. Educational data mining: A review of the state of the art. *IEEE Trans. Syst. Man Cybern. Part C Appl. Rev.* **2010**, *40*, 601–618. [CrossRef]
33. Casillas, A.; Robbins, S.; Allen, J.; Kuo, Y.L.; Hanson, M.A.; Schmeiser, C. Predicting early academic failure in high school from prior academic achievement, psychosocial characteristics, and behavior. *J. Educ. Psychol.* **2012**, *104*, 407. [CrossRef]
34. Agaoglu, M. Predicting instructor performance using data mining techniques in higher education. *IEEE Access* **2016**, *4*, 2379–2387. [CrossRef]
35. Gašević, D.; Dawson, S.; Rogers, T.; Gasevic, D. Learning analytics should not promote one size fits all: The effects of instructional conditions in predicting academic success. *Internet High. Educ.* **2016**, *28*, 68–84. [CrossRef]

36. Calvo-Flores, M.D.; Galindo, E.G.; Jiménez, M.P.; Piñeiro, O.P. Predicting students' marks from Moodle logs using neural network models. *Curr. Dev. Technol.-Assist. Educ.* **2006**, *1*, 586–590.
37. Azcona, D.; Casey, K. Micro-analytics for student performance prediction. *Int. J. Comput. Sci. Softw. Eng.* **2015**, *4*, 218–223.
38. Ognjanovic, I.; Gasevic, D.; Dawson, S. Using institutional data to predict student course selections in higher education. *Internet High. Educ.* **2016**, *29*, 49–62. [CrossRef]
39. Andriessen, J.; Sandberg, J. Where is education heading and how about AI. *Int. J. Artif. Intell. Educ.* **1999**, *10*, 130–150.
40. Vasileva-Stojanovska, T.; Malinovski, T.; Vasileva, M.; Jovevski, D.; Trajkovik, V. Impact of satisfaction, personality and learning style on educational outcomes in a blended learning environment. *Learn. Individ. Differ.* **2015**, *38*, 127–135. [CrossRef]
41. Basavaraju, P.; Varde, A.S. Supervised learning techniques in mobile device apps for Androids. *ACM SIGKDD Explor. Newsl.* **2017**, *18*, 18–29. [CrossRef]
42. Amrieh, E.A.; Hamtini, T.; Aljarah, I. Mining educational data to predict student's academic performance using ensemble methods. *Int. J. Database Theory Appl.* **2016**, *9*, 119–136. [CrossRef]
43. Bendangnuksung, P.P. Students' performance prediction using deep neural network. *Int. J. Appl. Eng. Res.* **2018**, *13*, 1171–1176.
44. Okubo, F.; Yamashita, T.; Shimada, A.; Ogata, H. A neural network approach for students' performance prediction. In Proceedings of the Seventh International Learning Analytics & Knowledge Conference, Vancouver, BC, Canada, 13–17 March 2017; pp. 598–599.
45. Shahiri, A.M.; Husain, W. A review on predicting student's performance using data mining techniques. *Procedia Comput. Sci.* **2015**, *72*, 414–422. [CrossRef]
46. Sikder, M.F.; Uddin, M.J.; Halder, S. Predicting students yearly performance using neural network: A case study of BSMRSTU. In Proceedings of the 2016 5th International Conference on Informatics, Electronics and Vision (ICIEV), Dhaka, Bangladesh, 13–14 May 2016; IEEE: New York, NY, USA; pp. 524–529.
47. Oladokun, V.O.; Adebanjo, A.T.; Charles-Owaba, O.E. Predicting students academic performance using artificial neural network: A case study of an engineering course. *Bull. Educ. Res.* **2008**, *40*, 157–164.
48. Ramesh, V.A.M.A.N.A.N.; Parkavi, P.; Ramar, K. Predicting student performance: A statistical and data mining approach. *Int. J. Comput. Appl.* **2013**, *63*, 35–39. [CrossRef]
49. Osmanbegovic, E.; Suljic, M. Data mining approach for predicting student performance. *Econ. Rev. J. Econ. Bus.* **2012**, *10*, 3–12.
50. Jishan, S.T.; Rashu, R.I.; Haque, N.; Rahman, R.M. Improving accuracy of students' final grade prediction model using optimal equal width binning and synthetic minority over-sampling technique. *Decis. Anal.* **2015**, *2*, 112. [CrossRef]
51. Arsad, P.M.; Buniyamin, N. A neural network students' performance prediction model (NNSPPM). In Proceedings of the 2013 IEEE International Conference on Smart Instrumentation, Measurement and Applications (ICSIMA), Kuala Lumpur, Malaysia, 25–27 November 2013; IEEE: New York, NY, USA; pp. 1–5.
52. Gray, G.; McGuinness, C.; Owende, P. An application of classification models to predict learner progression in tertiary education. In Proceedings of the 2014 IEEE International Advance Computing Conference (IACC), Gurgaon, India, 21–22 February 2014; IEEE: New York, NY, USA; pp. 549–554.
53. Wang, T.; Mitrovic, A. Using neural networks to predict student's performance. In Proceedings of the International Conference on Computers in Education, Auckland, New Zealand, 3–6 December 2002; IEEE: New York, NY, USA; pp. 969–973.
54. Hämäläinen, W.; Vinni, M. Comparison of machine learning methods for intelligent tutoring systems. In Proceedings of the International Conference on Intelligent Tutoring Systems, Jhongli, Taiwan, 26–30 June 2006; Springer: Berlin/Heidelberg, Germany; pp. 525–534.
55. Mayilvaganan, M.; Kalpanadevi, D. Comparison of classification techniques for predicting the performance of students academic environment. In Proceedings of the 2014 International Conference on Communication and Network Technologies, Sivakasi, India, 18–19 December 2014; IEEE: New York, NY, USA; pp. 113–118.
56. Sembiring, S.; Zarlis, M.; Hartama, D.; Ramliana, S.; Wani, E. Prediction of student academic performance by an application of data mining techniques. In Proceedings of the International Conference on Management and Artificial Intelligence IPEDR, Bali, Indonesia, 1–3 April 2011; Volume 6, pp. 110–114.
57. Mishra, T.; Kumar, D.; Gupta, S. Mining students' data for prediction performance. In Proceedings of the 2014 Fourth International Conference on Advanced Computing & Communication Technologies, Rohtak, India, 8–9 February 2014; pp. 255–262.
58. Natek, S.; Zwilling, M. Student data mining solution–knowledge management system related to higher education institutions. *Expert Syst. Appl.* **2014**, *41*, 6400–6407. [CrossRef]
59. Naren, J. Application of data mining in educational database for predicting behavioural patterns of the students. *Int. J. Comput. Sci. Inf. Technol.* **2014**, *5*, 4649–4652.
60. Bunkar, K.; Singh, U.K.; Pandya, B.; Bunkar, R. Data mining: Prediction for performance improvement of graduate students using classification. In Proceedings of the 2012 Ninth International Conference on Wireless and Optical Communications Networks (WOCN), Indore, India, 20-22 September 2012; pp. 1–5.
61. Holmes, G.; Donkin, A.; Witten, I.H. Weka: A machine learning workbench. In *ANZIIS'94-Australian New Zealnd Intelligent Information Systems*; IEEE: New York, NY, USA, 1994; pp. 357–361.
62. Witten, I.H.; Frank, E.; Trigg, L.E.; Hall, M.A.; Holmes, G.; Cunningham, S.J. Weka: Practical Machine Learning Tools and Techniques with Java Implementations. In Proceedings of the ICONIP/ANZIIS/ANNES'99 International Workshop "Future Directions for Intelligent Systems and Information Sciences", Dunedin, New Zealand, 22–23 November 1999; pp. 192–196.

63. Bhargava, N.; Sharma, G.; Bhargava, R.; Mathuria, M. Decision tree analysis on j48 algorithm for data mining. *Int. J. Adv. Res. Comput. Sci. Softw. Eng.* **2013**, *3*, 1114–1119.
64. Ali, S.; Smith, K.A. On learning algorithm selection for classification. *Appl. Soft Comput.* **2006**, *6*, 119–138. [CrossRef]
65. Qian, Y.; Liang, J.; Li, D.; Zhang, H.; Dang, C. Measures for evaluating the decision performance of a decision table in rough set theory. *Inf. Sci.* **2008**, *178*, 181–202. [CrossRef]
66. Muda, Z.; Yassin, W.; Sulaiman, M.N.; Udzir, N.I. Intrusion detection based on k-means clustering and OneR classification. In Proceedings of the 2011 7th International Conference on Information Assurance and Security (IAS), Melacca, Malaysia, 5–8 December 2011; IEEE: New York, NY, USA; pp. 192–197.
67. Ayinde, A.Q.; Adetunji, A.B.; Bello, M.; Odeniyi, O.A. Performance Evaluation of Naive Bayes and Decision Stump Algorithms in Mining Students' Educational Data. *Int. J. Comput. Sci. Issues* **2013**, *10*, 147.
68. Peng, C.Y.J.; Lee, K.L.; Ingersoll, G.M. An introduction to logistic regression analysis and reporting. *J. Educ. Res.* **2002**, *96*, 3–14. [CrossRef]
69. Platt, J. *Sequential Minimal Optimization: A Fast Algorithm for Training Support Vector Machines*; Microsoft: Washington, DC, USA, 1998; Available online: https://web.iitd.ac.in/~{}\{\}sumeet/tr-98-14.pdf (accessed on 15 October 2021).
70. Gardner, M.W.; Dorling, S.R. Artificial neural networks (the multilayer perceptron)—A review of applications in the atmospheric sciences. *Atmos. Environ.* **1998**, *32*, 2627–2636. [CrossRef]
71. Breiman, L. Random forests. *Mach. Learn.* **2001**, *45*, 5–32. [CrossRef]
72. Chan, J.C.W.; Paelinckx, D. Evaluation of Random Forest and Adaboost tree-based ensemble classification and spectral band selection for ecotope mapping using airborne hyperspectral imagery. *Remote Sens. Environ.* **2008**, *112*, 2999–3011. [CrossRef]
73. Ho, T.K. The random subspace method for constructing decision forests. *IEEE Trans. Pattern Anal. Mach. Intell.* **1998**, *20*, 832–844.
74. Xue, Y.; Chen, H.; Jin, C.; Sun, Z.; Yao, X. NBA-Palm: Prediction of palmitoylation site implemented in Naive Bayes algorithm. *BMC Bioinform.* **2006**, *7*, 458. [CrossRef]
75. Paliwal, M.; Kumar, U.A. Neural networks and statistical techniques: A review of applications. *Expert Syst. Appl.* **2009**, *36*, 2–17. [CrossRef]

 *sustainability*

Article

Portfolio Analysis of Clean Energy Vehicles in Japan Considering Copper Recycling

Jun Osawa

Department Management Systems, College of Informatics and Human Communication, Kanazawa Institute of Technology, Nonoichi 921-8501, Japan; j-osawa@neptune.kanazawa-it.ac.jp

Abstract: Several countries are moving toward carbon neutrality to mitigate climate change. The introduction of clean energy vehicles (CEVs) is a measure to offset the adverse effects of global warming. However, each CEV has its strengths and weaknesses. An optimal CEV portfolio must be formulated to create effective policies that promote innovative technologies and introduce them into the market. CEVs also consume more copper than gasoline vehicles. Copper is associated with supply risks, which most previous conventional studies have failed to address. Therefore, this study proposes a novel CEV optimization model for sustainable consumption of copper resources through recycling along with reduction of CO_2 emissions. This study aims to analyze the optimal portfolio for domestic passenger vehicles and the assumed effects of copper recycling and usage reduction. For this analysis, this study set up scenarios for the recycling rate of copper contained in end-of-life vehicles and the reduction rate of copper used in newly sold vehicles. Our simulation results showed that increased recycling rates and reduced use of copper are necessary for the diffusion of battery electric vehicles. Furthermore, the simulation results indicated that if these improvements are not implemented, the deployment of fuel cell vehicles needs to be accelerated.

Keywords: automotive industry; circular economy; clean energy vehicle; copper; global warming; optimization; recycling

Citation: Osawa, J. Portfolio Analysis of Clean Energy Vehicles in Japan Considering Copper Recycling. *Sustainability* **2023**, *15*, 2113. https://doi.org/10.3390/su15032113

Academic Editors: Nallapaneni Manoj Kumar, Subrata Hait, Anshu Priya and Varsha Bohra

Received: 11 December 2022
Revised: 11 January 2023
Accepted: 19 January 2023
Published: 22 January 2023

1. Introduction

Several countries are moving toward carbon neutrality to mitigate the negative impacts of greenhouse gas (GHG) emissions and climate change. The Japanese government aims for carbon neutrality and supports the development of various new technologies that can reduce CO_2 emissions. The transportation sector contributed to approximately 17.7% of all domestic CO_2 emissions in 2020 [1]. Hence, there is growing demand for measures to reduce these emissions.

A remedial measure under consideration is the introduction of clean energy vehicles (CEVs), such as hybrid electric vehicles (HEVs), plug-in hybrid electric vehicles (PHEVs), battery electric vehicles (BEVs), and fuel cell vehicles (FCVs). However, each type of CEV has strengths and weaknesses. In addition to CO_2 emission rates, they also differ in terms of purchase costs, fuel costs, additional infrastructure development costs, and other factors. Thus, widespread adoption of a single type of vehicle is not necessarily optimal. Although a certain vehicle type may be superior in one aspect, it may be inferior in other characteristics. Therefore, to create an effective policy to introduce CEVs, it is necessary to consider an optimal portfolio combining gasoline vehicles (GVs) and CEVs, along with multiple indicators such as CO_2 emissions, costs, and other metrics. The portfolio of vehicles means the composition of the number of vehicles sold by vehicle type.

Ichinohe and Endo [2] and Yeh et al. [3] have used the linear programming model (MARKAL energy system model) to calculate passenger vehicle portfolios that minimize the total energy system cost under the constraint of the target CO_2 emission rates in Japan and the United States, respectively. Yamada and Hondo [4] set fuel and vehicle purchase costs as constraints to construct a domestic passenger vehicle portfolio generating the lowest achievable life cycle CO_2. The authors examined CO_2 emissions from the vehicle production and disposal stages in addition to the energy production stage (well-to-tank) and the driving stage (tank-to-wheel). Arimori and Nakano [5] and Romejko and Nakano [6] set the total of vehicle purchase, fuel, and infrastructure costs as the objective function with oil consumption and CO_2 emissions as constraints when calculating the optimal portfolios for passenger vehicles, trucks, and buses in Japan and Poland, respectively. The authors focused on energy security based on oil dependence in the transportation sector. These previous studies [4–6] have also used the linear programming model.

The focus of previous studies remained on the energy sources used by vehicles and their CO_2 emissions. Nevertheless, these studies did not consider the sustainability of metal resources, such as copper. Part compositions differ between GVs and CEVs. Consequently, the types and amounts of metal resources required to produce GVs and CEVs also differ. For example, large amounts of copper are used in BEV lithium-ion batteries and motor windings. Therefore, the demand for these metal resources is expected to increase in the future. In other words, there is a trade-off between reduction in CO_2 emissions and consumption of metal resources. Furthermore, the primary supply of copper and other metals remains uncertain due to the reserve-to-production ratios and uneven distribution of ore-producing countries and regions.

Kato et al. [7] have used linear programming to calculate optimal passenger vehicle portfolios in six regions (OECD North America, OECD Europe, OECD Pacific, China, India, and All Other). The authors analyzed the case for minimizing the consumption of copper under the constraints of the CO_2 emissions reduction targets. The authors focused on copper demand but did not consider copper supply availability, including recycling. The global demand for copper is expected to increase as CEVs become more widely used, and copper supply shortages may become a real problem. Primary copper resource supply is insecure, hence copper sustainability must be considered in terms of secondary resource supply from recycling. Thus, it is necessary to develop an optimization model that integrates costs, CO_2 emissions, and metal supply and demand.

While not examining the optimal portfolio of GVs and CEVs, several studies have forecasted the demand for metal resources and estimated the amount of recycling. Habib et al. [8] used an S-curve to estimate the number of HEVs, PHEVs, and BEVs and future demand for metal resources, such as copper and rare metals, in five global regions. Similarly, Li et al. [9] and Shi [10] used the Bass model to estimate the future demand for rare-earth elements and cobalt, respectively, in China. Yano et al. [11] estimated the number of end-of-life vehicles (ELVs) among passenger HEVs and the recycling potential of rare earth elements in Japan, while Li et al. [12,13] estimated ELVs in China and the amount of recyclable nonferrous metals and plastic, respectively. However, these studies focused on specific vehicle type and did not consider the impact of recycling on supply constraints. The authors focused only on metal resources and did not consider other factors such as CO_2 emissions and costs. Each CEV type should be considered for implementation from multiple perspectives.

These findings demonstrate the need for strategies and policies that consider metal resources recycling and supply constraints in addition to other factors such as CO_2 emissions during CEV promotion. In this study, we focus on copper since the demand is expected to increase with the spread of CEVs in the future. It is also widely used as a base metal in industries other than the automotive industry, which could lead to a supply-demand crunch. Moreover, the estimated reserves-to-production ratio is relatively short, at approximately 40 years [14].

For these reasons, we propose a new model that optimizes portfolios by considering copper demand and supply, including copper recycling, social system costs (vehicle purchase, energy consumption, and infrastructure costs), and CO_2 emissions in the process of introducing CEVs. This study aims to analyze the optimal portfolio for domestic passenger vehicles and the assumed effects of copper recycling and usage reduction. Passenger GVs, HEVs, PHEVs, BEVs, and FCVs were considered in this study.

Section 2 describes the innovative optimization model. Section 3 elaborates on the various preconditions for the optimization model. Section 4 presents the portfolio calculation results generated by the optimization model and explains the differences between portfolios and the amount of recycled and consumed copper due to changes in recycling and usage reduction rates. In addition, the quantitative analysis results for the effects of copper recycling and usage reduction are presented by comparing the social system costs in each case. Section 5 presents the conclusions of this study and discusses future prospects based on the discoveries made herein.

2. Optimization Model

2.1. Framework

In this study, we developed an optimization model that considers costs, CO_2 emissions, and copper supply and demand, including recycling. Figure 1 shows an overview of the optimization model constructed herein. The input consisted of the characteristics of each vehicle type and other prerequisite data. Furthermore, the impact could be analyzed by inputting the recycling and usage reduction rates for each case. The output was the optimal number of new vehicles of each type sold in each target year. The number of vehicles owned, the number of ELVs, CO_2 emissions, amount of copper consumed and recycled, and social system costs were calculated for each optimal portfolio.

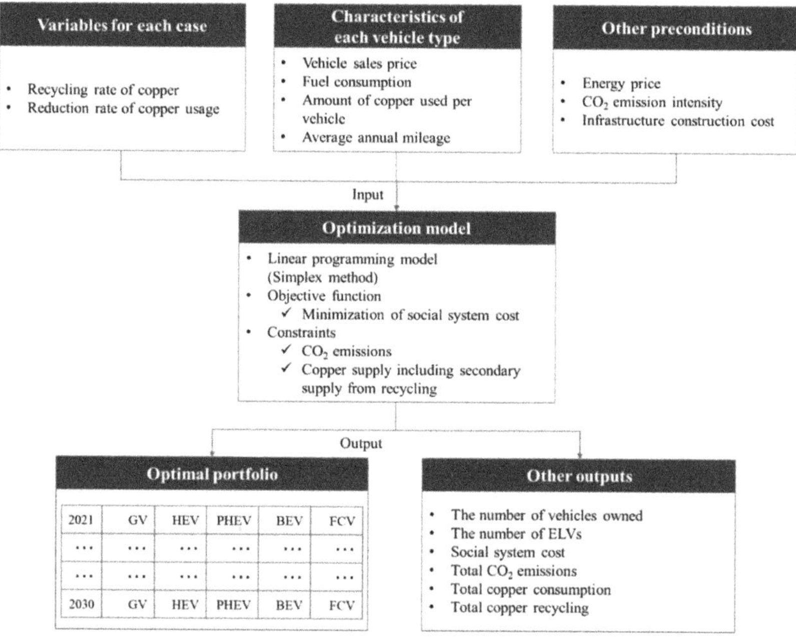

Figure 1. Overview of the CEV portfolio optimization model.

The objective function was set for the social system cost. The optimization problem was then solved using the constraint method with CO_2 emissions and the supply of copper as the constraint conditions. Based on the foregoing parameters, "optimal" in this study refers to the following: the social system cost was the lowest when the CO_2 emission reduction target was achieved and the demand for copper was lower than or equal to its supply. In the present study, the target area was Japan, while the target years were set between 2021 and 2030. The mathematical formulae of the model are explained below.

Compared with previous studies, this study added new factors to assess copper demand and supply, including recycling. Specifically, copper supply constraint was added to the optimization model, and copper consumption per vehicle, copper recycling rate, and usage reduction rate were added to the inputs.

2.2. Objective Function

The objective function is defined as the social system cost, which is the sum of the vehicle purchase cost, energy consumption cost associated with fuel use while driving, and cost of building new infrastructure, such as charging stations. The calculation was performed using Equation (1).

$$min\ SC_t(X_{it})$$

$$SC_t(X_{it}) = \sum_i X_{it} P_{it} + \sum_i \sum_j S_{it}(X_{it}) FC_{it} EP_{jt} AM + \sum_i X_{it} IC_{it} \tag{1}$$

$$S_{it}(X_{it}) = X_{it} + S_{it-1} - \sum_k ELV_{ikt-1} \tag{2}$$

where i is the vehicle type [GV, HEV, PHEV, BEV, FCV]; j is the energy source [gasoline, electricity, hydrogen]; t is the target year [2021–2030]; k is the vehicle age; SC_t is the social system cost in year t [Yen]; X_{it} is the sale of a new vehicle type i in year t [Units]; S_{it} is the number of vehicle types i owned in year t [Units]; P_{it} is the sales price of vehicle type i in year t [Yen]; FC_{it} is the average fuel consumption of vehicle type i in year t [MJ/km]; EP_{jt} is the price of energy j in year t [Yen/MJ]; AM is the annual average mileage [km]; IC_{it} is the additional infrastructure construction cost for each vehicle type i in year t [Yen/Unit], and ELV_{ikt} is the number of ELVs for each vehicle type i with k years of age in year t [Units].

The Weibull distribution was used to calculate the number of ELVs. The cumulative retirement probability based on the Weibull distribution is expressed in Equation (3). The survival probability is 1 minus the distribution function in Equation (3). Also, the shape parameter m and scale parameter η were set to be 2.59 and 14.44, respectively, from Osawa [15]. These parameters were estimated to minimize the sum of squares of the errors between the estimated survival probability and the actual survival rate in 2020. The actual survival rate in 2020 was calculated from the number of units sold in the year of registration and the number of units remaining in 2020 according to the existing literature [16]. The number of units remaining represents the number of units sold in each registration year that had not been retired by 2020. In addition, ELVs can be calculated using Equation (4).

$$W_k = 1 - exp\left[-(k/\eta)^m\right] \tag{3}$$

$$ELV_{ikt} = X_{ikt}\{W_{kt} - W_{kt-1}\} \tag{4}$$

where m is shape parameter ($m > 0$); η is scale parameter ($\eta > 0$); W_{kt} is the cumulative retirement probability of vehicle with k years of age in year t [%], and X_{ikt} is the sales of new vehicle type i in year t-k [Units].

2.3. Constraint Conditions

2.3.1. CO$_2$ Emissions

CO$_2$ emissions were set as the constraint in this study, as shown in Equations (5) and (6). The CO$_2$ emissions reduction target was set based on the Sixth Strategic Energy Plan of the Agency for Natural Resources and Energy [17].

$$TC_t(X_{it}) \leq TC_{t0}(X_{it0})(1 - EG_t) \tag{5}$$

$$TC_t(X_{it}) = \sum_i S_{it}(X_{it}) FC_{it} CU_{it} AM \tag{6}$$

where t^0 is the base year [2013]; TC_t are the CO$_2$ emissions in year t [t-CO$_2$]; CU_{it} is the CO$_2$ emission intensity of vehicle type i in year t [g-CO$_2$/MJ], and EG_t is the CO$_2$ emission reduction rate target in year t [%].

2.3.2. Copper Supply

Copper supply was set as a constraint in this study. Specifically, the constraint was set such that the demand for copper was less than or equal to the supply, as shown in Equations (7) and (8). The formula for estimating the amount of copper recycled was established with reference to previous research [18].

$$\sum_i X_{it} BC_{it} \leq PS_t + TR_t \tag{7}$$

$$TR_t = \sum_i \sum_k ELV_{ikt} BC_{ikt} VR_t BR_t \tag{8}$$

where BC_{it} is the copper consumption per vehicle type i in year t [g/Unit]; PS_t is the primary supply of copper in year t [kg]; TR_t is the total amount of copper recycled in year t [kg]; VR_t is the recovery rate of ELVs in year t [%], and BR_t is the recycle rate of copper in year t [%].

3. Preconditions

3.1. New Vehicle Sales and Vehicle Sales Price

The total number of new passenger vehicles sold after 2021 was assumed to be proportional to Japan's population. The number of new passenger vehicles sold between 2021 and 2030 was set based on reference materials [19,20]. Furthermore, the sales volume of each vehicle type in the past (1993–2020), the premise for calculating the number of ELVs, was set based on the existing literature [19].

According to Kimura [21], the GV sales price is set at 1.8 million yen. The differences in part composition between GVs and HEVs, PHEVs, and BEVs and their amounts were also set based on previous studies [21–23]. Then, the sales prices of HEVs, PHEVs, and BEVs were set by multiplying the amount of increase or decrease in parts by the producer price freight rate and commercial margin rate [24] and adding or subtracting it to the GV sales price. For FCVs, the sales price was set by multiplying the shipment value in the literature [25] by the producer price freight rate and commercial margin rate [24]. For years with insufficient data, such as 2026–2029, prices were estimated by assuming a change equivalent to the average annual change in the preceding and following years. Figure 2 presents the results of the preceding estimates.

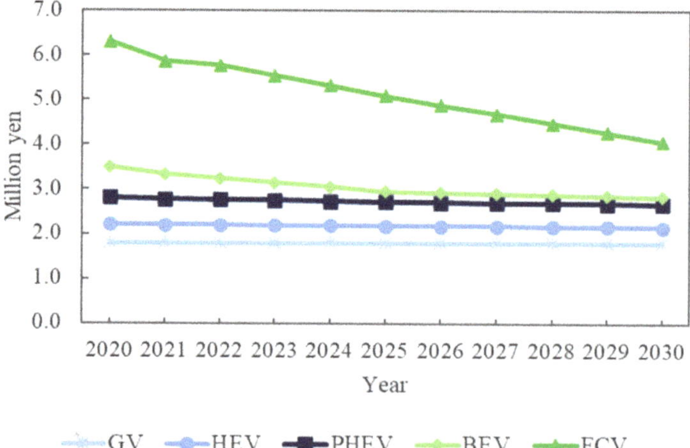

Figure 2. Trends in sale prices for each type of vehicle. GV: gasoline vehicles; HEV: hybrid electric vehicles; PHEV: plug-in hybrid electric vehicles; BEV: battery electric vehicles; FCV: fuel cell vehicles.

3.2. Fuel Consumption

The fuel consumption rates for passenger vehicles sold in 2013 and 2020 were derived from the Ministry of Land, Infrastructure, Transport, and Tourism [26]. The fuel consumption of passenger vehicles sold in 2030 was based on the standard fuel consumption value reported by the Automobile Fuel Economy Standards Subcommittee [27]. CEV fuel consumption was determined by multiplying the foregoing values by the ratio of each CEV to GV fuel consumption per km tank-to-wheel [28]. Figure 3 shows the changes in fuel consumption for each vehicle type. The fuel economy between 2021 and 2029 was estimated assuming improvement as per the average change per year between 2020 and 2030.

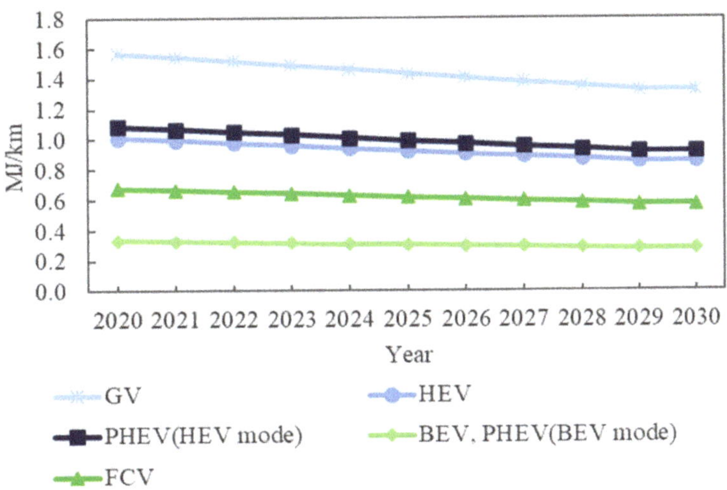

Figure 3. Trends in fuel consumption. GV: gasoline vehicles; HEV: hybrid electric vehicles; PHEV: plug-in hybrid electric vehicles; BEV: battery electric vehicles; FCV: fuel cell vehicles.

Furthermore, the average fuel consumption of owned vehicles in the target year was calculated from the weighted average of the fuel consumption of sold vehicles in the target year and the average fuel consumption of owned vehicles in the previous year.

3.3. Energy Price

The energy sources were gasoline for the GVs, HEVs, and PHEVs; electricity for the BEVs and PHEVs; hydrogen for the FCVs.

According to previous reports [5,29], the price of gasoline was calculated by adding petroleum, coal, and volatile oil taxes and refining and distribution margins to the crude oil price. Crude oil price was set according to the case reported in the Sustainable Development of the IEA's World Energy Outlook [30]. The electricity prices were set according to the Central Research Institute of Electric Power Industry [31]. The hydrogen supply price was set in accordance with the Agency for Natural Resources and Energy [32].

Figure 4 shows the price changes for each energy source. Energy prices between 2021 and 2029 were estimated based on the assumption that they will change by the average amount of change per year between 2020 and 2030.

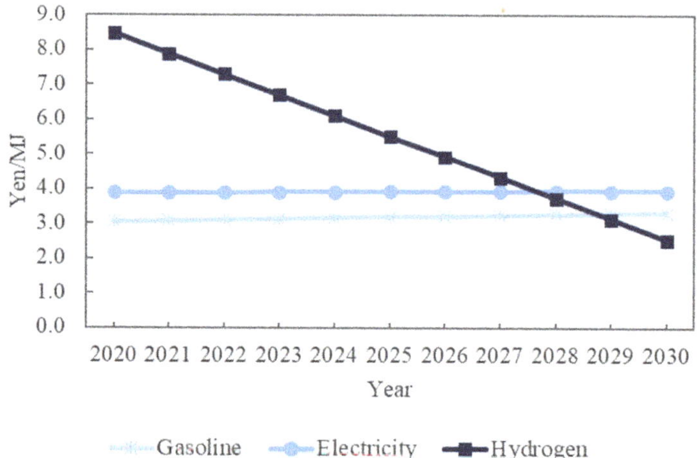

Figure 4. Trends in energy prices.

3.4. CO_2 Emission Intensity

This study considers CO_2 emissions during both the automobile driving and energy source production stages. The CO_2 emission intensity at the vehicle driving stage (tank-to-wheel) was obtained from a previous study [33]. According to the PHEV, electricity is used when the vehicle is operated in BEV mode, while gasoline is used when the vehicle is operated in HEV mode. Furthermore, CO_2 emission intensity at the energy production stage (well-to-tank) was estimated as previously described [17,28,34–37]. The CO_2 emission intensity at the well-to-wheel was estimated from the sum of well-to-tank and tank-to-wheel CO_2 emission intensities (Figure 5).

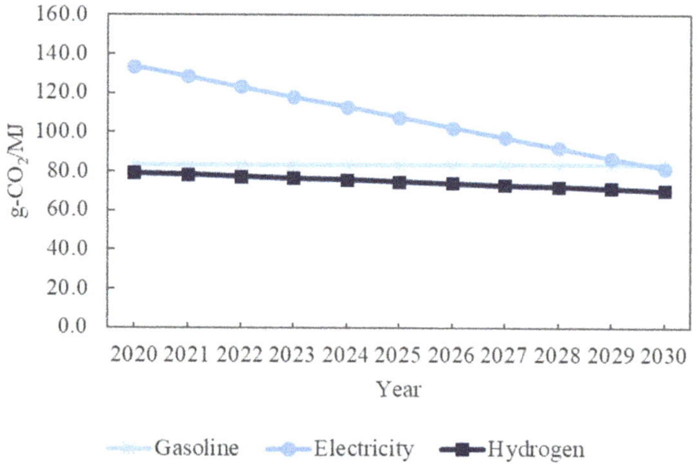

Figure 5. Trends in CO_2 emission intensity.

3.5. Other Preconditions

The construction cost per charging station for PHEVs and BEVs was set based on previous studies [5]. For PHEVs and BEVs, it was assumed that one regular charger must be installed per vehicle, in addition to a fast charger shared with other vehicles. The prices of the fast and regular chargers were assumed to remain the same from year to year. The construction cost per hydrogen station for FCVs was set based on previous research [38]. The infrastructure number required per vehicle was set assuming that supply facilities were sufficient to accommodate the number of GVs. The infrastructure required per vehicle was determined by dividing the number of gas stations in 2020 [39] by the number of GVs owned [40]. Here, it was assumed that new infrastructure will be constructed in response to additional BEV, PHEV, and FCV sales, owing to the insufficient charging and hydrogen stations at this time.

The amount of copper used per vehicle was estimated based on previous findings [41,42], as shown in Figure 6. Copper consumption is expected to increase in electric power and other industries. Therefore, it is unlikely that the passenger vehicle sector share of total primary supply will significantly increase. Primary copper supply to the passenger vehicle sector after 2021 was assumed to be equal to the demand for passenger vehicle sales in 2019.

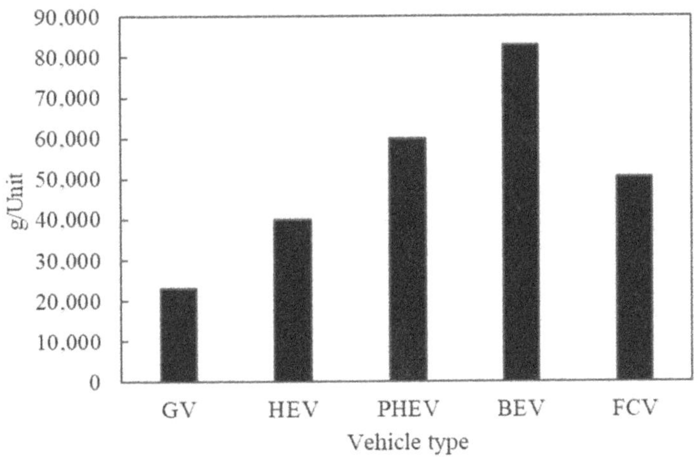

Figure 6. Copper consumption for each vehicle type. GV: gasoline vehicles; HEV: hybrid electric vehicles; PHEV: plug-in hybrid electric vehicles; BEV: battery electric vehicles; FCV: fuel cell vehicles.

The average annual mileage was estimated at 9601 km based on data from the Ministry of Land, Infrastructure, Transport, and Tourism [43]. The ratios of the driving modes to the PHEV mileage were set to 50% for BEV driving and 50% for HEV driving in accordance with the Central Research Institute of Electric Power Industry [44].

Not all ELVs are delivered to domestic recyclers, some are exported. Therefore, the average recovery rate from 2011 to 2020 was estimated at approximately 71% based on previous data [45].

In the Sixth Strategic Energy Plan of the Government of Japan [17], Japan's policy aims for 46% or 50% reduction of GHG emissions by FY2030 compared with the FY2013 level. Although this target is not specific to the transportation sector, the CO_2 emission reduction target for 2030 was set to 50% below the 2013 CO_2 emissions in this study. The CO_2 emission reduction target between 2021 and 2029 was estimated assuming a change equal to the average change per year between 2020 and 2030.

4. Simulation Results

4.1. Scenarios for Copper Recycling and Usage

In the future, copper recycling and reduced copper usage will become important issues owing to the tight supply and demand of copper. Therefore, in this study, cases were set up from two perspectives, namely the recycling rate of copper contained in ELVs and the reduction rate of copper use in newly sold vehicles (Table 1). The recycling rate at present is set at 50% based on a previous study [46]. The next section presents the results of the portfolio calculations for each case and their comparison.

Table 1. Copper recycling rate and usage reduction rate in each case.

Scenario		Recycling Rate in 2030	Usage Reduction Rate for Each Year by 2030
Free of copper constraints (free case)		-	-
Base case		50.0%	0.0%
Recycling rate improvement case	Low	70.0%	0.0%
	High	95.0%	0.0%
Usage reduction case	Low	50.0%	0.5%
	High	50.0%	1.0%

4.2. Optimal Portfolio Calculation

Figure 7 shows the sales volume composition ratio of each vehicle type in 2030 for each case.

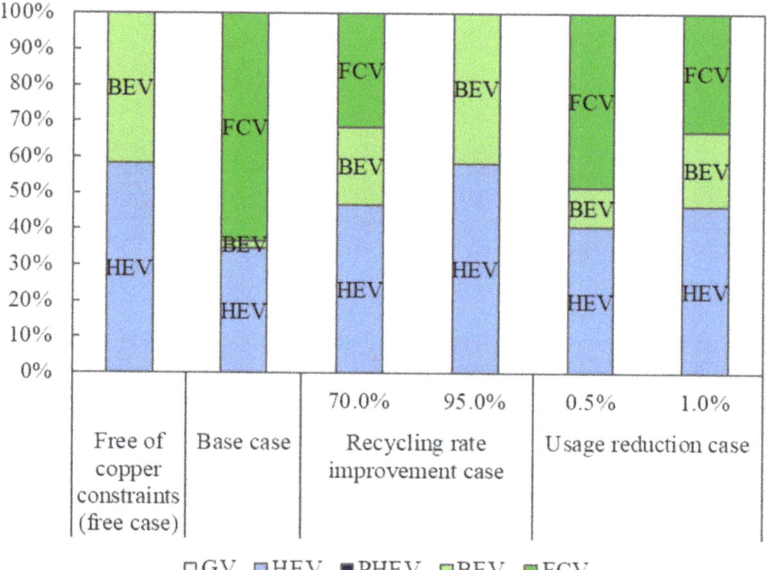

Figure 7. Sales share of each vehicle type in 2030. GV: gasoline vehicles; HEV: hybrid electric vehicles; PHEV: plug-in hybrid electric vehicles; BEV: battery electric vehicles; FCV: fuel cell vehicles.

First, without considering copper resource constraints (free case), HEVs and BEVs accounted for approximately 58% and 42%, respectively, whereas if the current copper recycling rate continues (base case), HEVs, BEVs, and FCVs accounted for approximately 35%, 2%, and 64%, respectively. Compared with the free case, the base case showed greater diffusion for FCVs than for BEVs. This result can be attributed to the fact that copper constraints made it difficult to promote BEVs that use more copper. It was also assumed that FCVs were selected to achieve the CO_2 emission constraint, since they are the next most environmentally friendly vehicle type after BEVs.

Next, comparing the base case with cases where the recycling rate improves to 70% and 95% showed that the major vehicle type shifts from FCVs to BEVs and HEVs as the recycling rate improves. This result is attributed to an increase in the amount of recycled copper and total supply of copper available for use in passenger vehicles. As shown in Figure 8, the amount of copper recycled per year is expected to increase from 2020 to 2030. Even in the base case, recycling volumes are expected to increase owing to an increase in the number of scrapped CEVs with high copper content. Comparing the improvement case of the recycling rate with the base case, the amount of recycled copper is expected to expand more significantly. In cases where recycling rates improve to 70% and 95%, the amount of copper recycled in 2030 is, respectively, more than 1.5 and 2 times as much as that in 2020 (Figure 8). As a result, the potential for widespread BEV use, which requires more copper, will expand. It is also assumed that HEVs were selected for their good balance of environmental performance, copper resource consumption, and sales price.

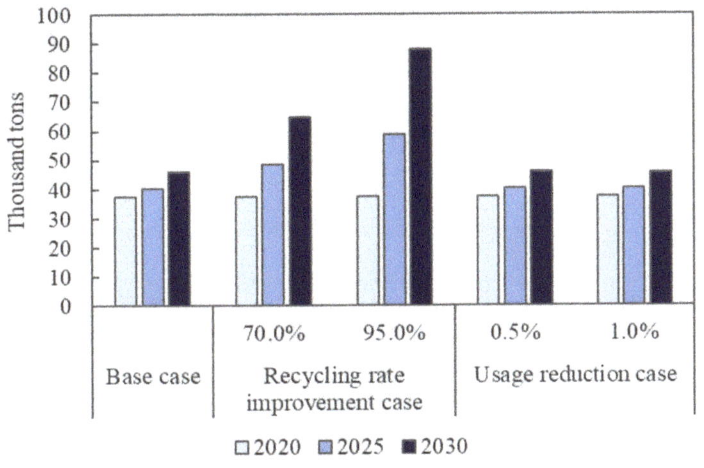

Figure 8. Amount of copper recycled from end-of-life vehicles.

Comparing the usage reduction case and base case showed a shift in major vehicle types from FCVs to BEVs and HEVs, as the copper use reduction rate increases. Reduced copper usage per CEV has led to the deployment of more BEVs, with almost the same amount of total copper consumption as in the base case (Figure 9). Furthermore, comparing portfolio changes with the usage reduction case and recycling rate improvement case showed similar results, although the HEV, BEV, and FCV compositions were different. In addition, reducing copper use by 1% per year had almost the same effect as increasing the recycling rate from 50% to 70%.

Then, as shown in Figure 10, HEVs accounted for a large share of the total number of vehicles owned in all cases. In the 2020s, a price gap remained between HEVs, BEVs, and FCVs, and, as a result, HEVs played an important role. Comparing the cases where the recycling rate improves to 70% and 95% with the base case showed that as recycling rate improves, BEVs, rather than FCVs, accounted for a larger share of the total. In the base case, BEVs and FCVs accounted for approximately 5% and 12%, respectively, of the total number of vehicles owned by 2030. However, the BEV share increased to approximately 10% and 12% when the recycling rate improved to 70% and 95%, respectively. In addition, the usage reduction case showed similar results to the recycling rate improvement case.

For these results, as copper recycling rate improves, or copper use decreases, HEVs and BEVs are expected to play a greater role by 2030. Conversely, if these improvements are not implemented, the spread of FCVs must be accelerated. Furthermore, in all cases, PHEVs did not exhibit significant diffusion, presumably since HEVs are preferred when social system costs are considered, while BEVs are preferred when CO_2 emissions are considered, resulting in a lesser selection of PHEVs.

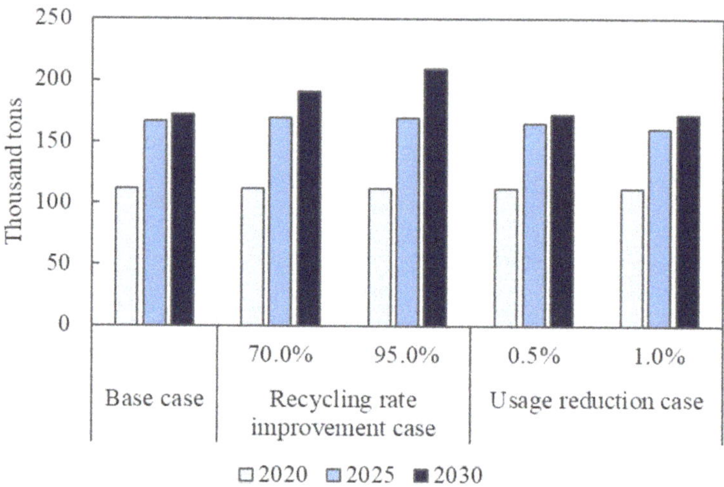

Figure 9. Total copper consumption from new vehicle sales.

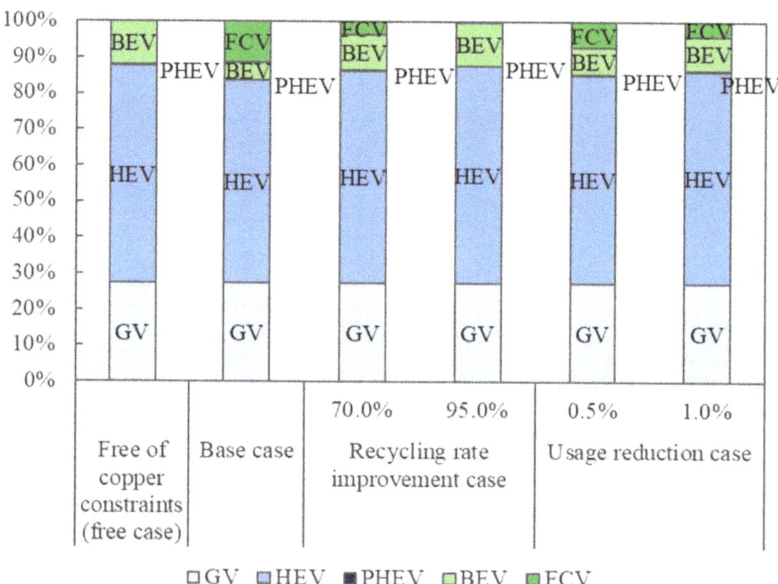

Figure 10. Share of vehicles owned for each vehicle type in 2030. GV: gasoline vehicles; HEV: hybrid electric vehicles; PHEV: plug-in hybrid electric vehicles; BEV: battery electric vehicles; FCV: fuel cell vehicles.

Compared with previous studies [5,7], the number of FCVs sold increased due to copper supply constraints. On the other hand, BEVs accounted for a large share, as in the previous study, when the copper supply-demand situation was mitigated by increasing the copper recycling rate and decreasing the amount of copper used per vehicle. However, except for the case where the recycling rate improves to 95%, the diffusion of FCVs as well as BEVs was important. Metal resource management during the vehicle's life cycle is essential. In addition, the ratio of HEVs increased compared with previous studies. These results were attributed to reducing social system costs while meeting the constraints of copper supply and CO_2 emission reduction targets. However, it should be noted that there are differences in the prerequisite data, such as CO_2 emission reduction targets, between the previous study and this study.

4.3. Evaluating the Impact of Copper Resource Constraints on Social System Costs

This section examines how changes in copper recycling rate and use per vehicle affect social system costs.

Figure 11 shows the average annual social system cost for each case from 2021 to 2030. The green line represents the social system cost for the free case, where copper resource constraints are not considered, while the blue bars represent the remaining five cases, where copper resource constraints are considered.

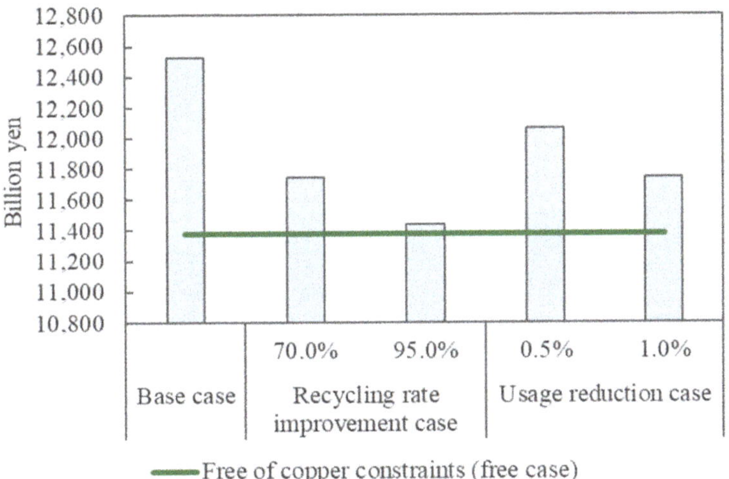

Figure 11. Average annual social system costs in each case.

The social system cost in the free case was approximately 11 trillion yen. On the other hand, the base case showed that additional costs of approximately 1.2 trillion yen would be required. Compared with the base case, improving the recycling rate from 50% to 70% or reducing the amount used by 1% would reduce social system costs by approximately 800 billion yen, and improving the recycling rate to 95% would reduce social system costs by approximately 1.1 trillion yen.

Even reducing copper use by 0.5% would reduce the social system cost by approximately 450 billion yen, compared with the base case, since the price of FCVs and their infrastructure construction costs remain high in the 2020s. Therefore, improving recycling rate and reducing material use can have a significant effect on society.

Currently, the Japanese government provides subsidies to consumers purchasing CEVs, construction of infrastructure, such as charging stations, development of next-generation batteries and motors, and development of hydrogen supply chain technologies. For example, the Green Innovation Fund has allocated up to approximately 150 billion yen to the research and development of new batteries for higher performance, resource conservation, promotion of recycling, and reduction of GHG emissions during manufacturing [47]. Given the impact on society, it is important to further promote the development of vehicle recycling systems and alternative technologies.

5. Conclusions

The introduction of CEVs in the transportation sector is being promoted to achieve carbon neutrality. Each CEV has different characteristics. It is necessary to consider an optimal portfolio based on multiple perspectives to introduce CEVs effectively. Similar previous studies have not adequately factored in the copper recycling and supply constraints. Hence, in this study, a novel portfolio optimization model was developed to consider copper demand and supply, including copper recycling, social system costs and CO_2 emissions. The study also aimed to analyze the optimal portfolio for domestic passenger vehicles and the assumed effects of copper recycling and usage reduction.

The present model was used to calculate the most economically rational passenger vehicle portfolio to achieve CO_2 emission and copper supply constraints in Japan. In addition, a scenario analysis was conducted by varying copper recycling and usage reduction rates. The following conclusions were drawn:

- When copper resource constraints were not considered, HEVs and BEVs accounted for a large share of the total sales. On the other hand, when the current recycling rate of copper resources continues (base case), a greater diffusion of FCVs was observed compared with BEVs;

- Comparing the cases where the recycling rate improves to 70% and 95% with the base case, the major vehicle types shift from FCVs to BEVs as the recycling rate improves. This result was attributed to the increase in the amount of recycled copper and total supply of copper available for use in passenger vehicles. Furthermore, when copper usage was reduced by 0.5% or 1% per year, similar results were observed as in the case of recycling rate improvement. Thus, changes in copper recycling and usage reduction rates will change the types of vehicles with widespread use;

- Increased recycling rates and reduced usage could reduce social system costs by approximately 450–1100 billion yen, compared with the base case. Improving the recycling rate and reducing the amount of material are beneficial to society. Therefore, it is important to invest in vehicle recycling systems and develop alternative technologies to reduce the amount of copper used. Furthermore, to better promote recycling, it is necessary to strengthen the cooperation between the arterial and venous side of the automotive industry.

Furthermore, compared with previous studies, the results showed that the share of FCVs increased due to copper supply constraints and that significant improvements in the recycling rate or the amount of copper used per vehicle were necessary to promote BEVs. The comparison results indicated the importance of metal resource management during the vehicle's life cycle. These conclusions are the results of an optimization model that focuses on supply, including recycling, and the demand for copper resources in response to increasing CEV production and use.

The optimization model used in this study provides valuable insights for analyzing the conditions for the introduction of CEVs to achieve CO_2 emission reduction targets. However, it should be noted that there are uncertainties in the prerequisite data such as fuel, infrastructure, and vehicle costs.

Future work should extend the present model to analyze how portfolios are affected by changes in copper prices, energy prices, and CO_2 emission intensity. Future research should also consider metal resources other than copper and areas outside of Japan. In addition, since this study did not consider specific recycling methods or alternative technologies, future studies should endeavor to design three "R" (reuse, reduce, and recycle) scenarios based on these latest developments.

Funding: This research received no external funding.

Data Availability Statement: The data presented in this study are available on reasonable request from the corresponding author.

Conflicts of Interest: There are no conflicts of interest to declare.

Abbreviations and Symbols

Abbreviation/Symbol	Definition
BEVs	battery electric vehicles
CEVs	clean energy vehicles
ELVs	end-of-life vehicles
FCVs	fuel cell vehicles
GHG	greenhouse gas
GVs	gasoline vehicles
HEVs	hybrid electric vehicles
PHEVs	plug-in hybrid electric vehicles
i	vehicle type [GV, HEV, PHEV, BEV, FCV]
j	energy source [gasoline, electricity, hydrogen]
k	vehicle age
m	shape parameter of the Weibull distribution ($m > 0$)
η	scale parameter of the Weibull distribution ($\eta > 0$)
t	target year [2021–2030]
t^0	base year [2013]
AM	annual average mileage [km]
BC	copper consumption per vehicle [g/Unit]
BR	recycle rate of copper [%]
CU	CO_2 emission intensity of vehicle [g-CO_2/MJ]
EG	CO_2 emission reduction rate target in year t [%]
ELV	number of ELVs [Units]
EP	price of energy [Yen/MJ]
FC	average fuel consumption of a vehicle [MJ/km]
IC	additional infrastructure construction cost [Yen/Unit]
P	sales price of a vehicle [Yen]
PS	primary supply of copper [kg]
S	number of vehicles owned [Units]
SC	social system cost [Yen]
TC	CO_2 emissions [t-CO_2]
TR	total amount of copper recycled [kg]
VR	recovery rate of ELVs [%]
W	cumulative retirement probability of vehicle [%]
X	new vehicle sales [Units]

References

1. Ministry of Land, Infrastructure, Transport and Tourism. CO_2 Emissions in the Transport Sector. Available online: https://www.mlit.go.jp/sogoseisaku/environment/sosei_environment_tk_000007.html (accessed on 5 July 2022). (In Japanese)
2. Ichinohe, M.; Endo, E. Analysis of the vehicle mix in the passenger-car sector in Japan for CO_2 emissions reduction by a MARKAL model. *Appl. Energy* **2006**, *83*, 1047–1061. [CrossRef]
3. Yeh, S.; Farrell, A.; Plevin, R.; Sanstad, A.; Weyant, J. Optimizing U.S. Mitigation Strategies for the Light-Duty Transportation Sector: What We Learn from a Bottom-Up Model. *Environ. Sci. Technol.* **2008**, *42*, 8202–8210. [CrossRef] [PubMed]
4. Yamada, S.; Hondo, H. Optimal Product-Replacement Model considering Lifetime Distribution: Analysis of Introduction of Green Cars for Carbon Dioxide Emission Reduction. *Energy Resour.* **2009**, *30*, 9–15. (In Japanese) [CrossRef]
5. Arimori, Y.; Nakano, M. Portfolio Optimization for Clean Energy Vehicles in Japan. *Trans. JSME Ser. C* **2012**, *78*, 2571–2582. (In Japanese) [CrossRef]
6. Romejko, K.; Nakano, M. Portfolio analysis of alternative fuel vehicles considering technological advancement, energy security and policy. *J. Cleaner Prod.* **2017**, *142*, 39–49. [CrossRef]
7. Kato, K.; Nonaka, T.; Nakano, M. Optimization Model for Global Portfolio of Clean Energy Vehicles Considering Metal Resource. *Trans. JSME Ser. C* **2013**, *79*, 77–89. (In Japanese) [CrossRef]
8. Habib, K.; Hansdóttir, S.T.; Habib, H. Critical metals for electromobility: Global demand scenarios for passenger vehicles, 2015–2050. *Resour. Conserv. Recycl.* **2020**, *154*, 104603. [CrossRef]
9. Li, X.; Ge, J.; Chen, W.; Wang, P. Scenarios of rare earth elements demand driven by automotive electrification in China: 2018–2030. *Resour. Conserv. Recycl.* **2019**, *145*, 322–331. [CrossRef]

10. Shi, Q. Cobalt demand for automotive electrification in China: Scenario analysis based on the Bass model. *Front. Energy Res.* **2022**, *10*, 699. [CrossRef]

11. Yano, J.; Muroi, T.; Sakai, S.I. Rare earth element recovery potentials from end-of-life hybrid electric vehicle components in 2010–2030. *J. Mater. Cycles Waste Manag.* **2016**, *18*, 655–664. [CrossRef]

12. Li, Y.; Fujikawa, K.; Wang, J.; Li, X.; Ju, Y.; Chen, C. The Potential and Trend of End-Of-Life Passenger Vehicles Recycling in China. *Sustainability* **2020**, *12*, 1455. [CrossRef]

13. Li, Y.; Huang, S.; Liu, Y.; Ju, Y. Recycling Potential of Plastic Resources from End-of-Life Passenger Vehicles in China. *Int. J. Environ. Res. Public Health* **2021**, *18*, 10285. [CrossRef] [PubMed]

14. Japan Oil, Gas and Metals National Corporation. Mineral Resources Material Flow 2020. 2021. Available online: https://mric.jogmec.go.jp/wp-content/uploads/2021/06/material_flow2020.pdf (accessed on 13 May 2022). (In Japanese)

15. Osawa, J. Evaluation of the recycling potential of rare metals in the passenger vehicle sector in Japan. In Proceedings of the 2023 13th International Conference on Future Environment and Energy (ICFEE 2023), Tokyo, Japan, 13–15 January 2023.

16. Automobile Inspection & Registration Information Association. Table of Average Vehicle Age by Vehicle Type. 2020. Available online: https://www.airia.or.jp/publish/file/r5c6pv000000u7a6-att/02_syarei02.pdf (accessed on 30 May 2022). (In Japanese)

17. Agency for Natural Resources and Energy. 6th Strategic Energy Plan. 2021. Available online: https://www.enecho.meti.go.jp/en/category/others/basic_plan/pdf/6th_outline.pdf (accessed on 12 May 2022).

18. Karai, T.; Nakano, M.; Kimura, F. Estimation on Sustainability for Copper Domestic Supply by Considering Recycling Flow: A Case Study of Automobile, Electric Appliances and Construction Industries. *Trans. JSME Ser. C* **2010**, *76*, 3744–3751. (In Japanese) [CrossRef]

19. Japan Automobile Manufacturers Association. The Motor Industry of Japan 2021. 2021. Available online: https://www.jama.or.jp/english/publications/MIoJ2021_e.pdf (accessed on 12 May 2022).

20. National Institute of Population and Social Security Research. Future Estimated Population of Japan (Estimated in 2017). 2017. Available online: https://www.ipss.go.jp/pp-zenkoku/j/zenkoku2017/pp_zenkoku2017.asp (accessed on 12 May 2022).

21. Kimura, S. Impact Analysis on Next-generation Vehicles to Chubu Economy, by using Chubu-region Multi-regional Input Output Table. *Input-Output Anal.* **2019**, *26*, 80–98. (In Japanese) [CrossRef]

22. Fuji Keizai. *Future Outlook for Energy and Large-Scale Rechargeable Batteries and Materials 2020: Electric Vehicle and Automotive Battery Sector Edition*; Fuji Keizai: Tokyo, Japan, 2020. (In Japanese)

23. Fuji Keizai. *In-Depth Analysis and Research on HEV and EV Related Markets for 2020*; Fuji Keizai: Tokyo, Japan, 2020. (In Japanese)

24. Ministry of Land, Infrastructure, Transport and Tourism. Input-Output Table Focusing on the Transportation Sector for 2011. 2015. Available online: https://www.mlit.go.jp/k-toukei/unyubumonwotyuushintoshitasangyourenkanhyou_h23Excel.html (accessed on 10 June 2022). (In Japanese)

25. Fuji Keizai. *Future Prospects for Fuel Cell Related Technologies and Markets for 2020*; Fuji Keizai: Tokyo, Japan, 2021. (In Japanese)

26. Ministry of Land, Infrastructure, Transport and Tourism. List of Automobile Fuel Consumption. 2022. Available online: https://www.mlit.go.jp/jidosha/jidosha_fr10_000051.html (accessed on 12 May 2022). (In Japanese)

27. Automobile Fuel Economy Standards Subcommittee. Summary Material. 2019. Available online: https://www.mlit.go.jp/common/001294895.pdf (accessed on 12 May 2022). (In Japanese)

28. The Japan Automobile Research Institute. Analysis Report of the Overall Efficiency and GHG Emissions. 2011. Available online: https://pdf4pro.com/amp/cdn/ghg-jari-or-jp-1f227e.pdf (accessed on 13 May 2022). (In Japanese)

29. Yanagisawa, A. Structure of Passing on the Price of Crude Oil to the Price of Gasoline in Japan. 2011. Available online: https://eneken.ieej.or.jp/data/4158.pdf (accessed on 13 May 2022). (In Japanese)

30. International Energy Agency. World Energy Outlook 2021. 2021. Available online: https://iea.blob.core.windows.net/assets/88dec0c7-3a11-4d3b-99dc-8323ebfb388b/WorldEnergyOutlook2021.pdf (accessed on 13 May 2022).

31. Central Research Institute of Electric Power Industry. Study of the Economy, Industry, and Energy Supply-Demand Structure in Japan through FY2030. 2021. Available online: https://criepi.denken.or.jp/jp/serc/source/pdf/Y20506.pdf (accessed on 5 July 2022). (In Japanese)

32. Agency for Natural Resources and Energy. Promotion of Investment in the Hydrogen/Ammonia Supply Chain and Measures to Expand Its Demand. 2022. Available online: https://www.meti.go.jp/shingikai/enecho/shoene_shinene/suiso_seisaku/pdf/002_01_00.pdf (accessed on 13 May 2022). (In Japanese)

33. Ministry of the Environment. Calculation Method and Emission Factor List. 2020. Available online: https://ghg-santeikohyo.env.go.jp/files/calc/itiran_2020_rev.pdf (accessed on 13 May 2022). (In Japanese)

34. Agency for Natural Resources and Energy. Greenhouse Gas Emissions in the Life Cycle of Biomass Power Generation. 2020. Available online: https://www.meti.go.jp/shingikai/enecho/shoene_shinene/shin_energy/biomass_sus_wg/pdf/008_03_00.pdf (accessed on 13 May 2022). (In Japanese)

35. Agency for Natural Resources and Energy. About Future Renewable Energy Policy. 2021. Available online: https://www.meti.go.jp/shingikai/enecho/denryoku_gas/saisei_kano/pdf/025_01_00.pdf (accessed on 13 May 2022). (In Japanese)

36. Central Research Institute of Electric Power Industry. Comprehensive Assessment of Life Cycle CO_2 Emissions from Power Generation Technologies in Japan. 2016. Available online: https://criepi.denken.or.jp/hokokusho/pb/reportDownload?reportNoUkCode=Y06&tenpuTypeCode=30&seqNo=1&reportId=8713 (accessed on 13 May 2022). (In Japanese)

37. Mizuho Research & Technologies. Evaluation of Greenhouse Gas Emissions of Hydrogen in the Life Cycle. 2017. Available online: https://www.mizuho-ir.co.jp/seminar/info/2017/pdf/bif0131_02mizuho.pdf (accessed on 13 May 2022). (In Japanese)
38. Agency for Natural Resources and Energy. Current Status of FCV and Hydrogen Station Business. 2021. Available online: https://www.meti.go.jp/shingikai/energy_environment/suiso_nenryo/pdf/024_01_00.pdf (accessed on 29 June 2022). (In Japanese)
39. Agency for Natural Resources and Energy. Changes in the Number of Gasoline Dealers and Gas Stations. 2021. Available online: https://www.enecho.meti.go.jp/category/resources_and_fuel/distribution/hinnkakuhou/data/2021_07_30_01.pdf (accessed on 13 May 2022). (In Japanese)
40. Ministry of Land, Infrastructure, Transport and Tourism. Number of Vehicles Owned by Vehicle Type. 2021. Available online: https://www.mlit.go.jp/common/001405880.pdf (accessed on 13 May 2022). (In Japanese)
41. International Copper Association. The Electric Vehicle Market and Copper Demand. 2017. Available online: https://copperalliance.org/wp-content/uploads/2017/06/2017.06-E-Mobility-Factsheet-1.pdf (accessed on 16 June 2022).
42. Toyota Motor Corporation. The MIRAI Life Cycle Assessment for Communication. 2015. Available online: https://global.toyota/pages/global_toyota/sustainability/esg/challenge2050/challenge2/life_cycle_assessment_report_en.pdf (accessed on 13 May 2022).
43. Ministry of Land, Infrastructure, Transport and Tourism. Annual Report of Automobile Transportation Statistics. 2021. Available online: https://www.e-stat.go.jp/stat-search/files?page=1&layout=datalist&toukei=00600330&tstat=000001078083&cycle=8&year=20201&month=0&result_back=1&tclass1val=0 (accessed on 13 May 2022). (In Japanese)
44. Central Research Institute of Electric Power Industry. Evaluation of GHG Emissions from the Manufacture and Operation of Electric Vehicles and Internal Combustion Engine Vehicles—Comparative Analysis according to the Ratio of Thermal Power Generation for Business Use. 2021. Available online: https://criepi.denken.or.jp/jp/serc/source/pdf/Y21503.pdf (accessed on 15 July 2022). (In Japanese)
45. Japan Automobile Recycling Promotion Center. Automotive Recycling Data Book 2020. 2021. Available online: https://www.jarc.or.jp/data/databook/ (accessed on 30 May 2022). (In Japanese)
46. Yoshimura, A.; Matsuno, Y. Dynamic Material Flow Analysis of Copper and Copper Alloy in Global Scale—Forecast of In-Use Stock and Inputs and the Estimation of Scrap Recovery Potential. *J. Jpn. Inst. Metals* **2018**, *82*, 8–17. (In Japanese) [CrossRef]
47. Ministry of Economy, Trade and Industry. Plans for Research and Development and Social Implementation Related to the "Development of Next Generation Storage Batteries and Next Generation Motors" Project. 2021. Available online: https://www.meti.go.jp/policy/energy_environment/global_warming/gifund/pdf/gif_12_randd.pdf (accessed on 29 June 2022). (In Japanese)

 *sustainability*

Article

Development, Critical Evaluation, and Proposed Framework: End-of-Life Vehicle Recycling in India

Zambri Harun [1], Altaf Hossain Molla [1,*], Mohd Radzi Abu Mansor [1] and Rozmi Ismail [2]

[1] Department of Mechanical and Manufacturing Engineering, Faculty of Engineering and Built Environment, Universiti Kebangsaan Malaysia (UKM), Bangi 43600, Malaysia
[2] Faculty of Social Sciences and Humanities, Universiti Kebangsaan Malaysia (UKM), Bangi 43600, Malaysia
* Correspondence: altafhossain1994@gmail.com; Tel.: +60-1123-307-859

Abstract: Over the last couple of decades, the automobile sector in India has seen dramatic growth, following the phenomenal booming of engenders rapid proliferation of end-of-life vehicles (ELVs). Therefore, efficient and sustainable handling of ELVs is paramount. India has been striving to establish a practical regulatory framework to handle ELVs sustainably. This study explores India's current ELV recycling system to promote sustainable development. Subsequently, this article evaluates the present ELV recycling system to determine the existing issues in ELV recycling to prevent failure and enhance and standardize the processes involved in ELV recycling to achieve the optimum standard for product and process quality. This paper proposes pragmatic frameworks and offers recommendations for setting up an efficient ELV recycling system to resolve current issues and expedite sustainable development. This study has been performed through a mixed-method approach; a literature and policy review accompanied by detailed structured interviews with major stakeholders and industrial visits. This investigation reveals that India's ELV recycling system is at the embryonic stage and struggling against numerous inherent impediments. However, the proposed frameworks, together with practical recommendations, provide a paradigm for expediting materials recycling from ELVs and resolving perennial issues. This research may assist the government of India in implementing any upcoming regulatory and legal framework.

Keywords: end-of-life vehicle (ELV); recycling; framework; sustainable development

check for
updates

Citation: Harun, Z.; Molla, A.H.; Mansor, M.R.A.; Ismail, R. Development, Critical Evaluation, and Proposed Framework: End-of-Life Vehicle Recycling in India. *Sustainability* **2022**, *14*, 15441. https://doi.org/10.3390/su142215441

Academic Editor: Paolo Rosa

Received: 26 September 2022
Accepted: 3 November 2022
Published: 21 November 2022

Publisher's Note: MDPI stays neutral with regard to jurisdictional claims in published maps and institutional affiliations.

1. Introduction

An end-of-life vehicle (ELV) is a vehicle that must be disposed of since it is no longer functional, owing to a mechanical issue [1]. ELV recycling has been particularly resonant in recent decades [2], as ELV incorporates a plethora of precious metallic and non-metallic substances [3–5]; these are economically significant and industrially essential [6,7]. As well as having substantial salvage value, it also encompasses noxious pollutants that impose threats to society in a myriad of aspects, to a substantial degree [8–10]. Hence, addressing ELVs in an effective and sustainable way is imperative. India has witnessed a phenomenal expansion of the automobile industry, rapid economic prosperity, and extensive industrialization, prompting an overwhelming demand for vehicles. Consequently, the number of cars on the road is proliferating, eventually contributing directly to the corresponding ELVs. In 2021–22, India manufactured 22.9 million vehicles encompassing passenger, commercial, two-wheeler, three-wheeler, and quadricycle vehicles; India's domestic and import automobile market is snowballing significantly to 17.5 million domestic sales and 5.6 million exports, respectively [11]. Table 1 enumerates vehicle production in India from 2016 to 2022. Figure 1 depicts India's domestic sales and export of automobiles from 2016 to 2022.

Table 1. Automobile production in India from 2016 to 2022 in millions [10].

Category	2016–17	2017–18	2018–19	2019–20	2020–21	2021–22
Passenger vehicles	3.8	4.02	4.028	3.42	3	3.6
Commercial vehicles	0.81	0.895	1.1	0.75	0.624	0.8
Three-wheeler	0.78	1.02	1.27	1.1	0.614	0.75
Two-wheeler	19.9	23.15	24.49	21	18.34	17.7
Quadricycle	0.00158	0.001713	0.005388	0.006	0.003	0.004
Total	25.3	29.09	30.91	26.35	22.65	22.93

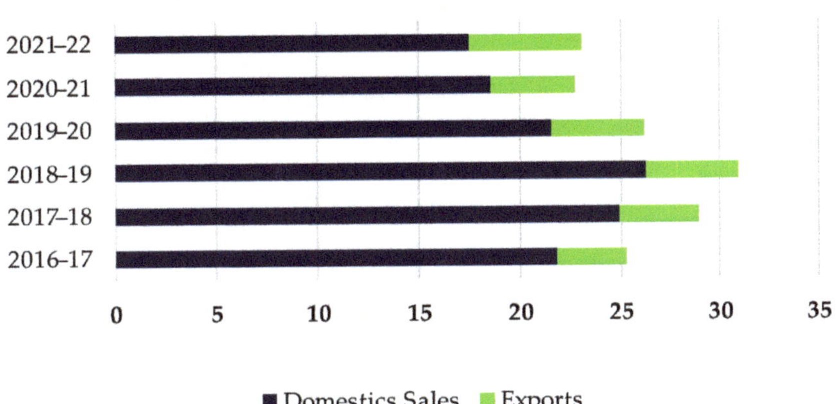

Figure 1. Domestics sales and export of automobiles in India from 2016 to 2022.

India strives to supply virgin materials to its automobile sector as the natural resources are depleting dramatically due to extensive extraction activities [12]. Scrap vehicles contain many distinct materials that are economically viable and industrially valuable; it also incorporates a series of potentially detrimental substances and harmful components that potentially causes environmental pollution [13,14] Through an effective and sustainable regulatory framework, ELVs can be recycled and reused; consequently, they can act as a secondary source of materials to occupy the raw materials' deficiencies in industries [14–16], and the incorporated deleterious substances that can contribute to pollution can be disposed of effectively to mitigate the environmental impact [17–19]. India's current ELV recycling system is exceptionally unorganized, and informality in this sector is prominent [20]. The absence of practical and effective frameworks for recycling ELVs in India thwarts the development and degenerates environmental issues by contributing to greenhouse gases emission and land pollution [21]. The material extraction and the recycling rate are limited and below the standard; only 7 % of aluminum and 76% of iron recovered from ELVs have been successfully recycled [22]. The recycling industries are not well-equipped with advanced technologies that enable them to recover and recycle certain economically valuable and rare materials, resulting in losses of these critical metals [23].

The strict law, accompanied by a stringent policy and practical guidelines, is required to establish an effective ELV recycling system to improve the materials recovery rate, reduce energy consumption, mitigate greenhouse gas emissions, and eventually ensure road and vehicle safety [24]. Countries such as China, Japan, the EU, and the USA have well-defined laws and policies regarding ELV recycling that enables a well-established and well-built ELV recycling system [25,26]. Still, India does not have stringent regulations relating to ELV recycling, even though India has taken specific rudimentary initiatives. India's ELV recycling system is predominant in the informal sectors, which are the units engaged in

ELV recycling with the primary objective of maximizing economic benefits [22]. Most ELVs are not dismantled environmentally soundly in their practices as they disregard any standard guidelines. The final product of ELV recycling, automotive shredder residue (ASR), is disposed of without proper treatment, even though ASR is metal-rich and contains significant salvage value [10,27]. Additionally, little is known about their operations. In this regard, well-defined laws and policies are required to set up an environmentally sound framework and procedure to promote sustainability in ELV recycling [28,29].

Conspicuously observed, researchers have performed an insignificant number of studies on ELV recycling in India to elucidate the ELV recycling system's current status and develop a comprehensive understanding of how the ELV recycling system works in different parts of India. This development, in a sense, is instrumental in enhancing the material extraction rate from ELV by preserving the environment and conserving natural resources [30]. It will educate and make individuals aware of ELV recycling promoting sustainable development for a thriving and prosperous future. A limited number of evaluative studies have been performed on India's ELV recycling system to address the perennial and persistent challenges in ELV handling, enhance and standardize the process, prevent failure, and promote sustainability [12]. This creates a knowledge gap that necessitates a deeper investigation into this issue. Although few research studies have proposed a framework for ELV recycling, these studies concentrate on one ELV recycling hub in a state in India. Hence, the lack of a comprehensive and inclusive study that investigates the significant number of ELV recycling hubs across India before proposing a framework is evident in the literature, which creates a severe lack of consideration [22]. The absence of practical and beneficial suggestions for future strategies to resolve the persistent issues in ELV recycling from experts and research scholars is quite apparent.

Numerous research and knowledge gaps have been identified in the ELV recycling system in India. This study is inspired in response to the challenges that India's ELV system is struggling against. This study investigates the current ELV recycling system in formal as well as informal ELV recycling sectors in India, along with shedding light on laws and policies regarding ELV handling in India to develop a comprehensive understanding of how the country's formal and informal ELV recycling sectors operate and stringency of India's rules and policies regarding ELV recycling. This paper critically evaluates the present ELV recycling system to determine the existing issues in ELV recycling to prevent failure and enhance and standardize the processes involving ELV recycling to achieve the optimum standard for product and process quality. This study also suggests a pragmatic framework and makes recommendations for setting up an efficient ELV recycling system to resolve the current issues and expedite sustainable development. This research adopts an explorative method to meet the aims of this study, as field investigations, together with the interviews with stakeholders, are instrumental in unveiling the ground-level practices and persistent issues and recording the attitudes and aspirations of stakeholders. The suggested frameworks, along with offered recommendations, may facilitate the stakeholders involved in ELV recycling to enhance the materials recovery and recycling rates from ELVs. This study may be significant in India's ELV recycling and have a profound contribution to recycling sectors, society, and the environmental aspect as this study emphasizes resource recovery and sustainability. This research may assist the government of India in implementing any upcoming regulatory and legal framework and eliminate the informality from the ELV recycling sector.

2. Methodology

This study delves into the current scenario of ELV recycling systems in India by adopting an explorative and integrated bottom-up approach. The research is based on the original primary as well as secondary research. Through primary research, this study has obtained valuable information, data, and insights and recorded the current practices to meet the aims of the study, while secondary research involves gathering the data and information from previously conducted studies in different countries to provide a foundational understanding of this research. Both qualitative and quantitative approaches have been employed in primary research.

2.1. Research Design and Sampling

This research aims to elucidate the current status of the ELV recycling system in India, critically evaluate India's present ELV recycling sector, propose pragmatic frameworks, and offer recommendations for future strategies for setting up an efficient ELV recycling system based on Indian values. This paper concentrates on five influential Indian automobile sectors, namely, Kolkata, Chennai, Mumbai, Delhi, and Jamshedpur, to understand the current practices of ELV handling in India in a broader aspect. This research employs a mixed-method approach incorporating quantitative and qualitative techniques to achieve its objectives. The concerned key stakeholders, formal as well as informal, encompassing manufacturing industries, automobile organizations, environmental protection boards, formal sector recyclers, transport department authorities, ELV traders, dismantlers, third-party vendors, workers, and academicians have all interacted with a structured questionnaire reviewed and approved by prominent research scholars in the ELV recycling field as well as government authorized agencies. There have been interactions with a total of 537 selected individuals. Figure 2 shows the interviews of all key stakeholders involved in ELV recycling.

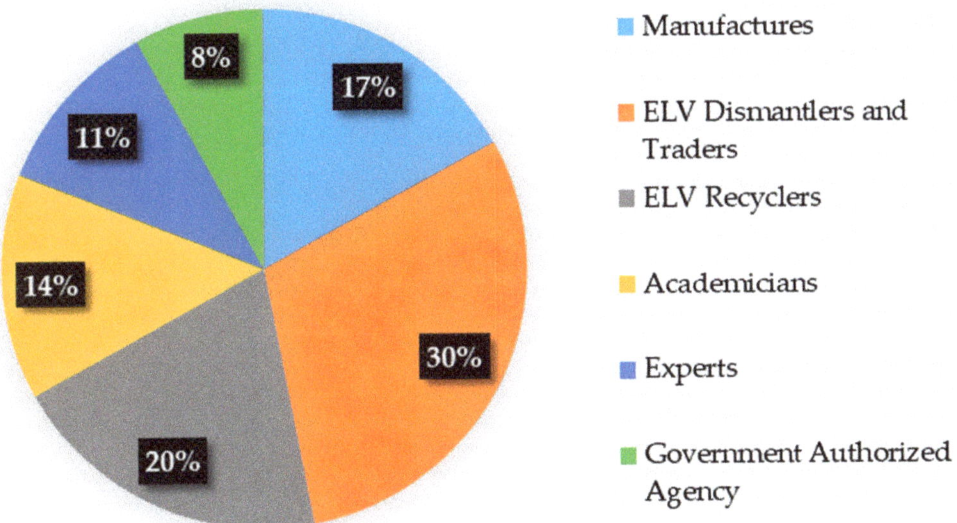

Figure 2. Interview composition involving stakeholders.

Figure 3 depicts the years of working experience of the stakeholders in the ELV recycling sector. The primary aim of these interviews was to apprehend and understand the status of the ELV recycling system, existing persistent problems in ELV recycling concerning the law and policy, technology, skills, and information of the ELV recycling system, since the comprehensive knowledge of these critical elements is of paramount of importance to critically evaluate the present ELV recycling system in India and, consequently, address the perennial issues regarding ELV recycling system in India.

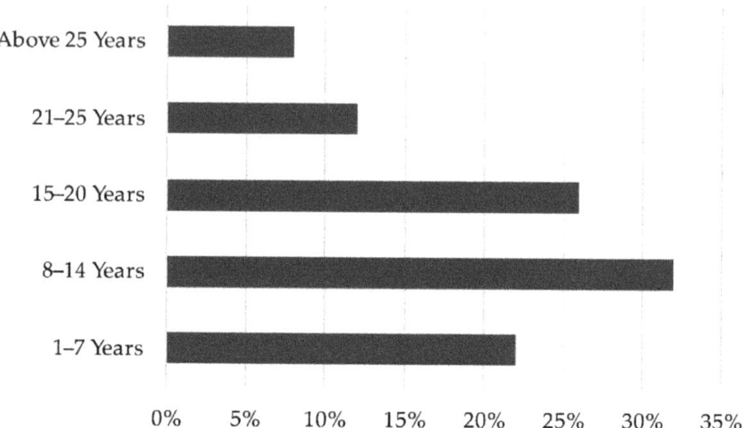

Figure 3. Years of working experience in the ELV recycling sector.

2.2. Qualitative Data

Through systematic interviews and focus group discussions (FGDs) with all crucial stakeholders involved in the ELV recycling system, guided by a structured questionnaire, coupled with a thorough field investigation of the five mentioned automobile hubs, respectively, we have obtained insights, information, data, and experience regarding the current practice of ELV recycling in different automobile hubs in India in a broader aspect. The obtained information and data are organized systematically for analysis. This research has performed thematic and narrative analyses to interpret and analyse the qualitative data. Based on this knowledge, information, and expertise, this investigation addresses the current issues pertinent to ELV handling through critical analysis to prevent failure and enhance and standardize the processes involving ELV recycling to achieve the optimum standard for product and process quality. Qualitative research is instrumental in exploring current practices and issues and offering recommendations [31].

2.3. Quantitative Data

Quantitative data was instrumental in corroborating these research findings. We have collected quantitative data through a standard questionnaire recommended and approved by preeminent research scholars in this contemporary epoch by administering the questionnaire to the crucial stakeholders involved in ELV recycling. The selected people (based on their expertise and insights) in five automobile hubs have responded to the questionnaire. The obtained data and information through quantitative research are organized systematically for analysis. This paper has conducted a descriptive analysis to interpret the quantitative data. The acquired data buttresses findings, assists in critical analysis, and proposes frameworks for ELV recycling in this paper. This data is imperative for our ongoing evaluative project about India's ELV recycling system, and this article is a part of that project. Quantitative research is essential for generating objective data [32]. Figure 4 represents the general methodology of this study.

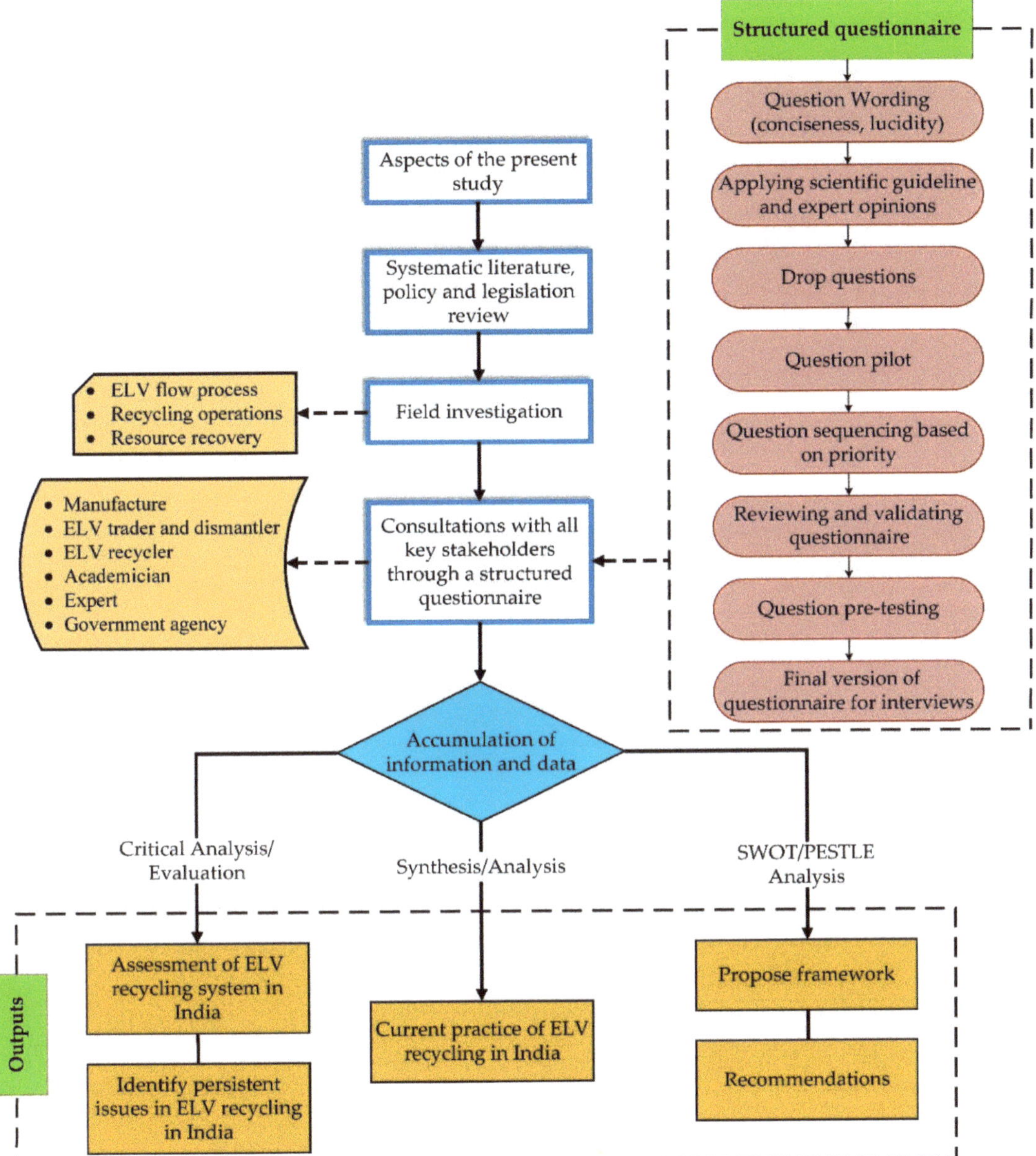

Figure 4. The study's general methodology.

2.4. Secondary Research

Secondary research was indispensable to developing a fundamental understanding and knowledge of the ELV system in India and the other countries that already have a well-established ELV recycling system. Through an extensive literature survey, we have developed the foundation of this study. India's laws and policies regarding end-of-life vehicles have been studied from the literature to provide clear information regarding government directives towards automobile dismantling procedures. This study also proposes pragmatic frameworks for setting up an efficient ELV recycling system in India based on Indian values to encounter the persistent problems in ELV handling [33,34]. An extensive literature review and a regulation review have been conducted to perform Strength-Weakness-Opportunity-Threat (SWOT) and Political-Economic-Social-Technological-Legal-Environmental (PESTLE) analysis; this is instrumental in proposing an efficient ELV system and design, re-built and implementing a new policy regarding ELV handling. Literature reviews regarding the advanced countries with efficient ELV systems, together with insights acquired from practical interactions with key stakeholders involved in ELV recycling, this article makes pragmatic recommendations to enhance the materials recycling rates and expedite sustainable development.

3. Status Quo of End-of-Life Vehicle Recycling in India

3.1. Law and Policy

ELVs are an essential secondary source of materials [35–37]; in contrast, they contribute significantly to environmental pollution as they contain harmful substances [38,39]. The requirement of stringent regulations and a practical regulatory framework is instrumental to the handling and properly disposing of ELVs [40–42]. At this moment, India does not have strict rules regarding ELV recycling. In contrast, the developed and advanced nations, namely Japan, the EU, and the USA, already have well-established robust regulations that compel the last owner to recycle the scrap vehicle [21,43,44]; consequently, through a regulatory framework, they dispose of scrap vehicles in environmentally sound ways and recover higher values from ELVs [45]. Even though the existing policies and regulations governing the ELV recycling system and the environment are lenient and limited, the Indian government understood the need for comprehensive laws to handle ELV recycling, waste recycling, protect the environment, and promote sustainable development. Hence, the Indian government has taken a few initiatives, but these have not sufficed. The legislated statutes and policies the Indian government has taken so far are listed in Table 2.

Table 2. The list of recycling and environmental protection regulations in India.

Law	Year	Vision/Mission
Water (Prevention and Control of Pollution) Act	1974	• To control pollution and protect the environment
Air (Prevention and Control of Pollution) Act	1981	• To increase the quality of air and mitigate the pollution level
Environmental (Protection) Act	1986	• To protect the environment
The Central Motor Vehicles Act	1988	• To provide a regulatory framework for the registration and deregistration of vehicles.
The Hazardous Wastes (Management and Handling) Rules	1989	• To dispose of the wastes in environmentally sound ways.
The Municipal Solid Waste (Management and Handling)	2000	• To dispose and recycle solid waste in cities
The Ozone Depleting Substances (Regulation and Control) Rules	2000	• To protect the environment
The Batteries (Management and Handling) Rules	2001	• To recycle batteries and promote sustainability
The National Environment Policy	2006	• To legalize and boost the informal sector for collecting and recycling various materials
The National Action Plan for Climate Change	2009	• It aims to find solutions to environmental problems in India
The e-waste (Management and Handling) Rules	2011	• To recycle and dispose of the e-waste
The Plastic Waste (Management and Handling) Rules	2011	• To recycle plastics and control the pollution
Hazardous and Other Wastes (Management and Transboundary Movement) Rules	2016	• To protect the environment and enhance recycling
Solid Waste Management Rules	2016	• To recycle solid waste and dispose properly
Plastic Waste Management Rules (Amendment)	2016	• To recycle plastics and control the pollution

With the compliance of the provided guidelines and provisions by the United Nations (UN) to enhance the recycling of materials, promote sustainability in recycling, and protect the environment, the collaboration of the Automotive Research Association of India (ARAI), Society for Automobile Association (SIAM), and Automotive Industry Standards Committee (AISC), has prepared the Automotive Industry Standards for ELVs (AIS 129), the guidelines and the provisions for the ELV recycling in environmental sounding ways. These guidelines and conditions have been discussed in detail in the following sections.

The Indian government (through the Ministry of Road, Transport, and Highway, MORTH), in the Union Budget for 2021–2022, has initiated Vehicle Scrapping Policy. This policy includes legislation about vehicle fitness tests, conditions, the vehicle deregistration process, taxes, registration fees, and guidelines for vehicle manufacturers. In this regard, compulsory testing of vehicles will be most likely to begin from June 2023, although the successful implementation of this regulation is dependent on information authority.

3.1.1. Registered Vehicle Scrapping Facility (RVSF)

In Automotive Industry Standards for ELVs (AIS 129), the guidelines and the provisions for ELV recycling in environmentally sounding ways, introduced by the collaboration of the Automotive Research Association of India (ARAI), Society for Automobile Association (SIAM), and Automotive Industry Standards Committee (AISC), Registered Vehicle Scrapping Facility (RVSF) defines as that any organization has a "Registration for Vehicle Scrapping" issued by the authority for conducting the proper operations of ELV recycling such as depolluting, dismantling, shredding, and other functions. The RVSF aims to dispose of unfit and abandoned vehicles environmentally friendly way. It assesses the fitness of the vehicle through an automated testing facility, and if a vehicle fails the fitness test, it will be recommended for scrapping. The RVSF has been developed to endeavor to formalize the current informality-dominated ELV recycling sector and encourage vehicle owners to recycle abandoned vehicles at RVSF.

Terms and Conditions for RVSF

Automotive Industry Standards for ELVs (AIS 129) lucidly mentions the terms and conditions to meet the eligibility criteria for RVSF. The crucial eligibility criteria are as below:

I. Ownership of the RVSF is open to any legitimate entity, whether it is an individual, an organization, a society, or a trust.
II. The entity must have a valid PAN card, valid GST registration, and a certificate of incorporation.
III. The entity must have the permission letter for setting up an RVSF from the state or union territory authority where it is to be located.
IV. The entity shall commit to meeting the basic technical requirements for ELV recycling operations specified by the Central Pollution Control Board.
V. The entity should have sufficient human resources to conduct ELV recycling operations in environmentally sound ways.
VI. The entity should possess ISO 9001 (quality management system)/ISO 14001 (environmental certification)/ISO 45001 (occupational health and safety) quality certifications, a permission letter to operate from State Pollution Control Board, and necessary cyber security certifications.
VII. The entity should adhere to provisions for labor and other laws and regulations.

Minimum Technical Requirements for RVSF

Automotive Industry Standards for ELVs (AIS 129) conspicuously specifies the technical requirements to meet the eligibility criteria for RVSF. The minimum technical requirements for an RVSF are as follows.

A. RVSF should have an impervious surface, like concrete for spillage collection, decanters, and degreaser operations.

B.   RVSF should have appropriate and sufficient storage for recovered spare components, like tires.

C.   RVSF should have suitable and proper containers and storage tanks to store batteries, filters/ plastic circuit board (PCB)/ plastic circuit Polycyclohexylenedimethylene terephthalate (PCT)-containing condensers, and other fluids.

D.   RVSF should have equipment and facility for removing batteries and fluid tanks, neutralizing explosive substances recovered from ELVs, and removing catalysts.

E.   RVSF should have equipment for recovering the metal parts from ELVs made of Copper, Aluminum, Magnesium, and other heavy metals in a sustainable way to recycle for further uses.

F.   Recovering significant plastic components are recommended; the recovery of glass from ELVs should be performed effectively.

Automotive Industry Standards for ELVs (AIS 129) recommends a list of equipment that RVSF should possess for different operations of ELV recycling. Table 3 enumerates the recommended equipment for RVSF.

Table 3. Recommended equipment for RVSF.

Operation	Equipment
Pre-treatments	• Vehicle Lift • Pry bar • Center Punch • Wheel Popper • Hydraulic Tube Cutter • Pneumatic Air Gun • Windshield removal tool
Depolluting	• Air-conditioning gas recovery unit • Exhaust gases neutralization system • Filter Wrench • Airbag deployment unit • Piercing equipment • Suction equipment • Bleeding system • Container • Draining Tray
Dismantling	• A set of screwdrivers • Spanner • Mallet • Ratchet • Cutter • Cutting Plier • Windscreen Cutter
Shredding	• Shredder • Bailing press unit

3.2. ELVs Flow

India's ELV recycling system is at the nascent stage but fledging, developing across megacities and automobile industrial hubs. India's ELV recycling sector has been divided into formal and informal sectors. Formal recycling sectors are authorized and registered recycling centers that operate environmentally friendly by following proper guidelines and regulations provided by the authority. Whereas informal recycling sectors are unauthorized and unregistered recycling centers consisting of individuals, groups, and small businesses driven by economic gain, their operations are not environmentally friendly and don't follow standard guidelines.

The informal sector dominates the current ELV recycling system, even though they operate with significantly higher material recycling and recovery rate and play an instrumental role. Certain operations are not environmentally sound, harming the environment without a regulatory framework. A few formal sectors have been evolving recently; their functions are more environmentally, socially, and economically sustainable. They cannot operate beyond the scope of a regulatory framework; they are also required to compete with the informal sector regarding materials recycling and recovery. Figure 5 depicts the current practice of the ELV recycling system in India, both formal and informal sectors.

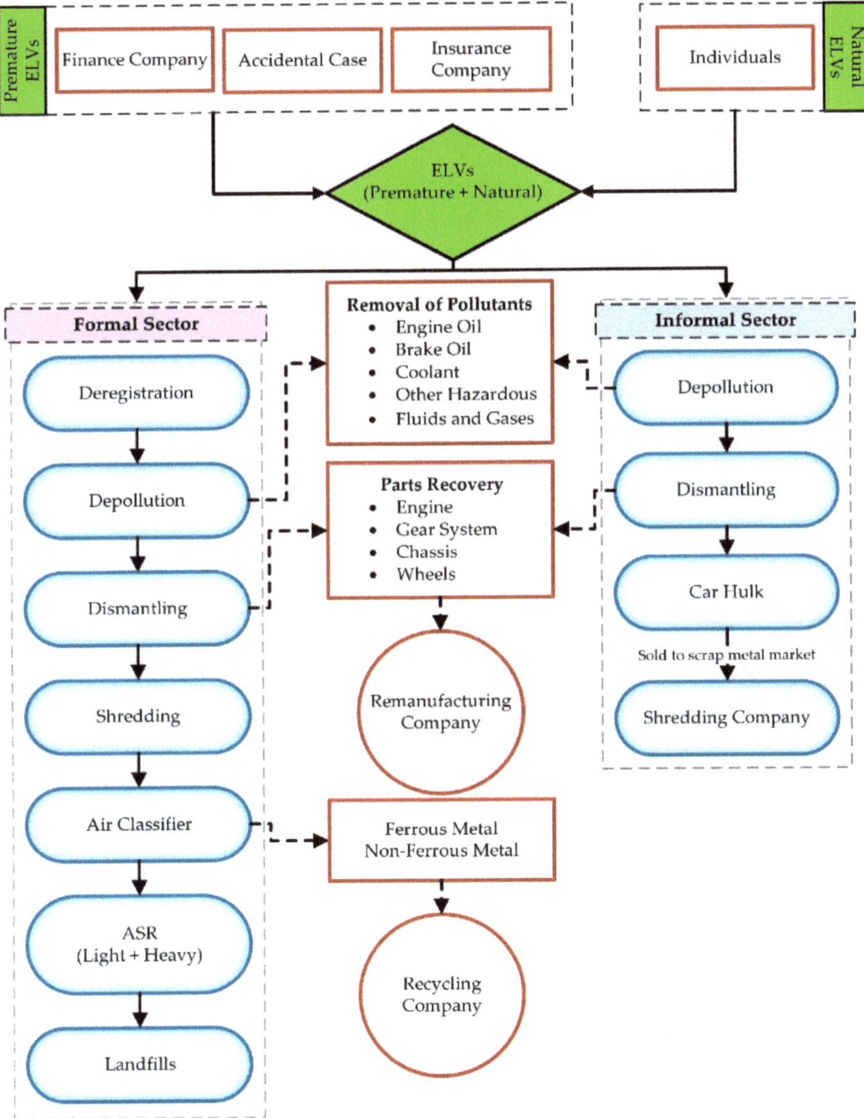

Figure 5. The current practice of ELV recycling system in India, both formal and informal sectors.

In the informal sector, there is no vehicle deregistration procedure; ELVs directly undergo a perfunctory depollution operation to remove the pollutants from ELVs. However, the removed pollutants are not disposed of in environmentally sound ways. Through the dismantling procedure, economically valuable parts such as the engine, gear system, chassis, and wheels are recovered. The recovered parts and components from ELV are sold at the scrap metal market. After that, a shredding company purchases the left car hulk as the informal sector lacks the facility to process the car hulk to recover further values. Whereas the formal sector operates with guidelines provided by authorities; ELVs undergo proper documentation, followed by the depollution operation, dismantling procedure for parts recovery, shredding operation for further value recovery, and eventually disposal of ASR to landfills, as there is no facility for ASR treatment. Even though formal sector operations are environmentally sound, India has remarkably few formal sectors.

Table 4 details depollution operations in India's formal and informal ELV recycling sectors. The formal ELV recycling sector performs few specific depollution operations sustainably, whereas the informal sector disposes of a few basic hazardous components from ELVs. Airbags contain harmful gases. Neither the formal nor informal ELV recycling sector disposes of airbags in an environmentally sound way, which contributes to the greenhouse effect and eventually threatens society and the environment.

Table 4. Depolluting operations both in formal and informal sectors.

Depollution Operations	Formal ELV Recycling Sector Managing Sustainably (✔)/ Not (✘)	Informal ELV Recycling Sector Managing Sustainably (✔)/ Not (✘)
Before Lifting the Vehicles		
Removal of battery	✔	✔
Remove the fuel filter cap and oil filler	✔	✘
Removal of tires and wheels	✔	✔
Removal of parts containing mercury	✔	✘
Air-conditioning gases	✘	✘
Lifting the vehicles on the depollution frame		
Engine oil	✔	✔
Transmission oil	✔	✘
Coolant	✔	✘
Brake fluids	✔	✘
Removal of catalysts	✘	✘
Washer bottle	✔	✘
Brake/clutch reservoirs	✘	✘
Power steering reservoirs	✔	✔
Fuel tank	✔	✘
After removing from the de-pollution frame		
Deploy airbags and other pyrotechnics on-site (if equipped and capable of carrying out this operation).	✘	✘
Eliminate any airbags and pyrotechnics (if fitted and unable to be used in situ).	✘	✘

4. Critical Evaluation

India's ELV recycling sector is now operating with various issues, albeit flourishing and proliferating mostly informal sectors [22,46]. We have observed a myriad of challenges that the ELV recycling system in India is experiencing and striving against. These challenges impede the development of the ELV sector in the techno-socio-economic aspect. For combatting climate change, mitigating greenhouse gas emissions, expediting sustainable development, and promoting energy-efficient vehicles, the government of India has been endeavoring to create a pragmatic framework for setting up an efficient ELV management system by enacting acts and making regulations, even though certain rudimentary causative factors have identified, analyzed, and elucidated in details, can be effective in enabling authorities in evolving effective ELV management systems not only in India but also in other developing and developed nations. Figure 6 graphically depicts critical issues in ELV recycling in India.

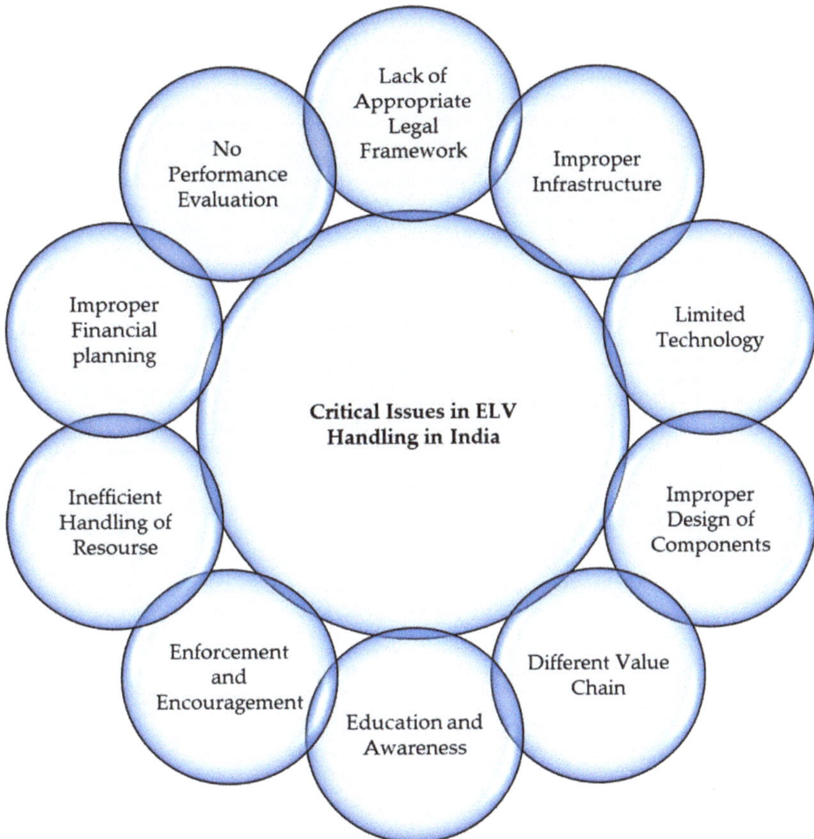

Figure 6. Critical issues in ELV handling in India.

4.1. Lack of Appropriate Legal Framework

India does not have a comprehensive and precise legal framework for recycling ELV sustainably other than that mentioned in Table 3. Existing policies and regulations are ineffectual in promoting recycling and managing a sheer deluge of ELVs proliferating every day [47]. Starting from the collection of ELVs, followed by a series of operations, deregistration, de-pollution, dismantling, shredding, ASR treatment, and throwing residue into landfills, in every stage, the laxity of legislation, regulation, and lack of proper guidelines are pervasive. Whereas countries such as China, Japan, the EU, and the USA have well-defined regulations [28,48,49], in terms of law, ELV recycling is mandatory for all citizens after a certain period, which is called the retirement of vehicles; the criterion of retirement varies from to country. Due to the apparent governance vacuum, ELV dismantlers and traders disregard the negative consequences of their activity and emphasize only economic gains from recycling. As a consequence of current practice, instead of expediting sustainable development and the material recovery rate, it is causing environmental degradation and preventing advancement in the material recovery process. The government is fully accountable for forming legislation, goals, coherent strategy, code of practice, ELV regulation, and priorities to contribute to the progress of ELV management system implementation.

4.2. Improper Infrastructure

Proper and effective infrastructure is instrumental for ELV recycling [50]. Improper and ineffective infrastructure is another critical impediment to the advancement of India's ELV recycling system [51]. As the informal sector dominates India's ELV recycling system, they don't have an advanced technique to handle ELV recycling; they only recover certain economically valuable components from ELVs and dispose of other crucial and scarce materials. Hence, these materials are not recycled. No ELV dismantler deals with different vehicles, light, medium, and heavy vehicles. As India's ELV recycling system is marked by low investment, this low investment creates poor infrastructure. However, India's recently developed few formal recycling sectors are well-equipped and have proper infrastructure for ELV recycling operations. As the informal sector is predominant in India and plays an imperative role in recycling, the Indian government should assist the informal sector in having better infrastructure for ELV recycling operations.

4.3. Limited Technology

ELV treatment encompasses many convoluted and cumbersome processes, sub-suming, dismantling, segregation, shredding, subsequent repair, reuse, recycling, re-manufacturing, residue management, and final disposition [52,53]. As ELV dismantlers operate with rudimentary technologies, the recycling rate, recovering rate, and materials extraction rate are comparatively poor to developed nations. ASR contains various critical and scarce materials [10]. Without recovering materials from ASR, the materials recovery and recycling rate cannot be increased; there is no facility for ASR treatment for materials recovery in the informal sector. As a consequence, the materials-rich ASR is disposed of in landfills. Hence, well-trained, consummate technicians and advanced and cutting-edge technologies are required to carry out ELV operations to maximize efficiency and enhance the rate of materials extraction. In order to improve the recovery and materials extraction rate, the authorities should look for appropriate material extraction methods in the local context so that the recycling and recovery rates are on par with developed nations.

4.4. Improper Design of Components

Inadequately designed vehicle component is the significant causative factor towards the poor performance of the ELV recycling system [54,55]. It has been observed and studied that, in present ELVs, the vehicle's components are not designed for recycling and remanufacturing, which makes the segregation, dismantling, and other subsequent operations onerous, arduous, challenging, and expensive [56]. Additionally, many component manufacturers adopt the short-term vision for product design to make their products economically inexpensive; hence, these poorly designed components cannot be recycled for future uses. Adapting the proper design method for components recycling can enhance the recovery and recycling and make the ELVs recycling operation substantially easier [57]. Automotive manufacturers and distributors should code elements and materials to make the reuse and recovery of materials more accessible.

4.5. Different Value Chain

Despite the fact that the automobile industry and ELV recycling system are potentially intertwined [58], they are not working together and collaborating. For the dismantling and other subsequent operations, the ELV dismantler needs accurate and valuable information from the manufacturers. This information can make ELV handling and parts recovery more effective [59]. In contrast, the manufacturers are also not aware of the capabilities and technology of ELV dismantlers. The cooperation between the automobile industry and the ELV sector can enhance the recycling rate and increase the economic gains for both. The manufacturer can buy excellent quality recycled components from ELV recyclers for future use. Information about the element shared by the manufacturer can make ELV processing easier and promote sustainability.

4.6. Education and Awareness

Public awareness and education are instrumental in developing and implementing an effective ELV recycling system [60]. Indian people are generally oblivious and unconcerned about the environmental aspects [51]. Almost everyone pollutes the environment but, ironically, refuses to help clean it up; many people are only concerned with their economic interests, regardless of environmental protection. Already, Indians are inhaling toxic air and drinking polluted water; if the Indian government does not emphasize public awareness and environmental education, the environmental degradation issue will become more severe. Promoting sustainable environmental education and raising public consciousness among Indians are long-term goals that must begin by taking initiatives from the government from the primary schooling of children.

4.7. Enforcement and Encouragement

Enforcement and encouragement are conducive to establishing an efficient ELV recycling system [34,61]. In a regulatory legal framework, enforcement compels vehicle owners to transfer their abandoned vehicles to the ELV processing organization through a deregistration process, award, and concession to inspire them to recycle their abandoned cars. In India's ELV recycling system, the absence of enforcement and motivation is prevalent. There is no compulsion to recycle the vehicle after the specified age and no award, which prompts the vehicle owners to recycle their obsolete cars. The Indian government is also quite reluctant to invest money in encouraging and empowering research and development programs on ELV recycling.

4.8. Improper Handling of Resources

Appropriate handling of resources is paramount to sustainable development, the circular economy, and recycling and energy recovery [62–64]. The Indian government should take the initiative to establish an independent management body to deploy and allocate resources efficiently and efficaciously, taking into account the availability of resources, competency, responsibility, and infrastructure. India's improper handling and mismanagement of resources have been observed and investigated [51].

4.9. Improper Financial Planning

Proper financial planning and the appropriate budget allocation are expedient for developing an effective ELV management system [65–67]. In India, the lack of appropriate financial planning for ELV recycling, especially for supporting the informal sector in having better infrastructure, is quite evident; even though Indian authorities invest a significant portion of their budget into sustainable development, proper ELVs recycling, and development of India as the hub for energy-efficient vehicles, but not in accurate and appropriate ways, the government attempts are futile. Now is the time for the authority to firmly establish an effective cost management system for proper planning and controlling of system costs, as well as to make provisions and allocate the necessary budget for the development of ELV management systems.

4.10. No Performance Evaluation

Indubitably, performance evaluation, together with the development of an inspection center, enhances the performance of a system towards specific, well-defined objectives [68]. Currently, no firmly established assessment system in India assesses the employee, process, method, equipment, tools, and other subsequent factors. The lack of performance assessment and inspection prompts, especially the informal recyclers, to disregard environmental impacts to maximize profit. An effective strategy should be developed for periodically monitoring the system, maintaining the documentation of records, and promoting through incentives, tax relaxation, and awards.

5. Proposed Frameworks

5.1. ELV Recycling System

Through critical analysis, identified issues and challenges in the ELV recycling system in India need to resolve by following a practical, systematic, and efficient framework that would not only enhance the material extraction rate but also promote sustainability in ELV recycling. Figure 7 represents a practical, efficient, and systematic framework for ELV recycling based on Indian values. This framework delineates directives lucidly at every stage of ELV recycling; it begins with the collection process of ELVs and terminates by proposing an efficient approach to handling the ASR. In India, there are no laws that govern the compulsory recycling of abandoned vehicles, proper collection process of ELV, and deregistration process. For the depollution process, the ELV recycler removes certain harmful fluids but not all potential pollutants, and removed pollutants are not disposed of properly; the suggested framework delineates the proper depollution process to protect the environment and promote sustainability by proposing a safe center for the disposal of stripped detrimental substances. The government should develop the proposed safe center to dispose of hazardous substances. The dismantling process is unorganized and chaotic; this framework shows the directives toward a systematic and organized dismantling process that can significantly enhance the materials recovery process. Presently, India has no ASR processing system to improve the recycling and recovery rate. The proposed framework has delved into this issue and taken into account this, and this study has emphasized ASR treatment and offered a systemic and effective ASR processing system.

The collaboration among the government, manufacturers, and ELV recyclers is instrumental in establishing a practical framework for ELV recycling. Still, there is no information-sharing facility among them. No fund management system supports informal and formal recyclers in depolluting, collecting ELVs, sustainable ASR disposal, and other operations. As this proposed framework delineates a fund management system where the authority should allocate funds to assist the stakeholders in collecting ELVs, deregistration process, depollution process, dismantling process and ASR treatment, and monitoring stakeholder's operations. This framework introduces an information management system and a fund management system. These management systems would help each other by providing the required and valuable information needed to establish a practical framework. The fund management would enable the ELV recycling sector to improve the infrastructure to boost the recovery and recycling rate and promote sustainability in ELV recycling.

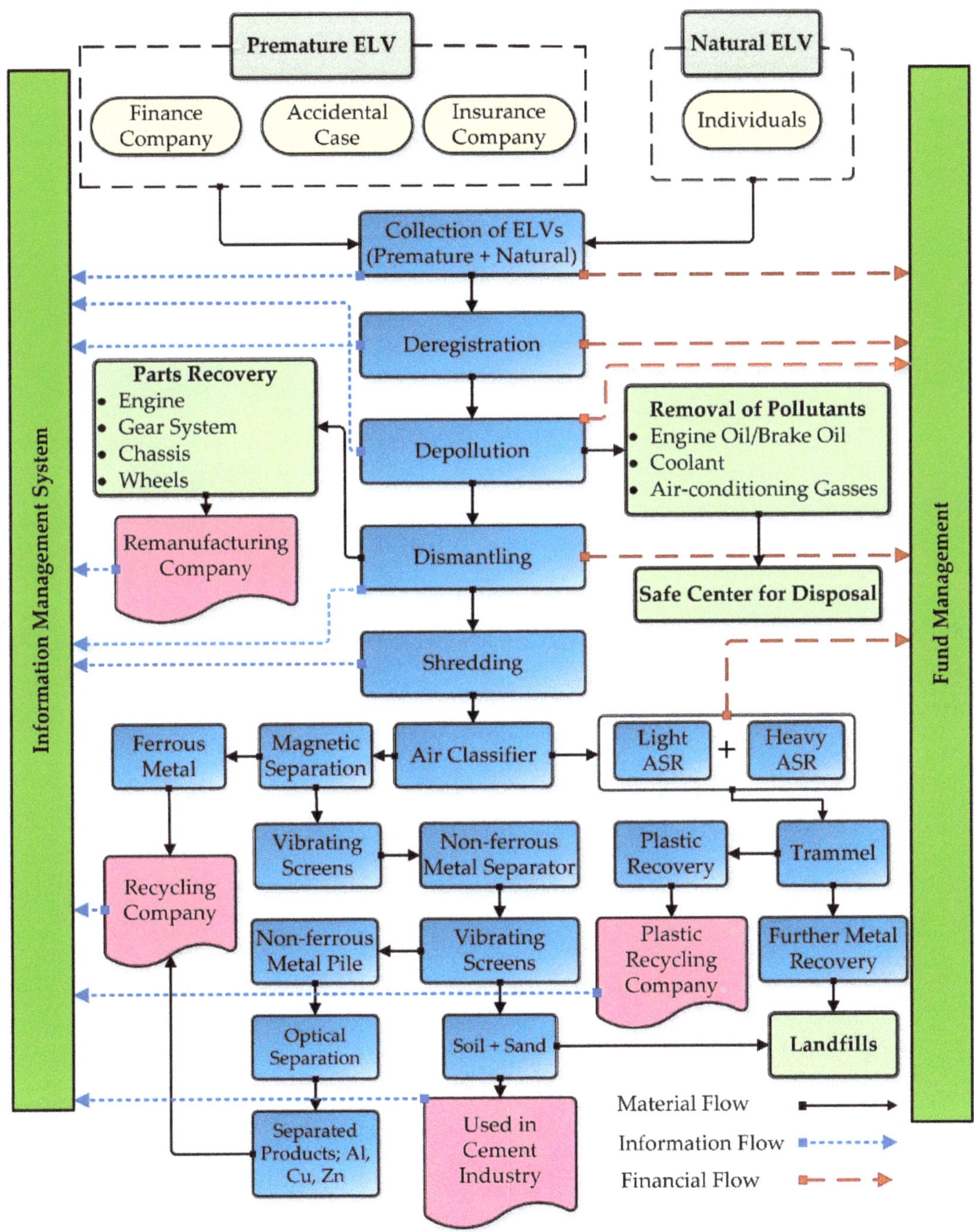

Figure 7. Proposed framework for ELV recycling system in India.

5.2. Policy Framework

Currently, there is no comprehensive and complete set of rules and regulations to govern the ELV recycling system in India. The ELV recycling system is complex and convoluted due to the numerous stages involved. To effectively handle and regulate the ELV system, every step involved in ELV recycling requires well-defined directives [69]. Firstly, there is no comprehensive regulation regarding the declaration of the ELV, a fitness test of the vehicle, and the deregistration process; the government should provide clear directives and guidelines about identifying the ELV, a fitness test of the vehicle, and the deregistration process. Secondly, for recycling, individuals voluntarily direct their abandoned vehicles to ELV recyclers; no existing law regulates the collection of ELV; authority should enact the regulation that would compel the car owner to recycle their cars after a specified age. Thirdly, no authorized management system monitors and accesses the de-pollution process of ELV, dismantling operation, shredding, and ASR treatment; hence informal recyclers perform these operations with disregard for environmental impacts; consequently, it causes environmental degradation [70]. The government should form an agency to access and monitor ELV recycling processes. Fourthly, the Indian ELV recycling sector is dominated by the informal sector. They do not follow the environmentally sound procedure for ELV recycling operations. They do not have the facility and technology to operate sustainably; hence, the materials recycling and recovery rates are comparatively low. The government should provide guidelines regarding materials recycling and recovery.

As mentioned above, the lack of regulations regarding declaration, deregistration, collection of ELVs, depolluting, dismantling, shredding, and disposal of the final product, ASR, hinders the sustainable development of ELV recycling in India. The policy and legislation should address the existing regulatory vacuum to make the Indian ELV recycling system viable in a sustainable manner. The Ministry of Transport, Human Resources, Natural Resources and Environment, Domestic Trade, Cooperation and Consumerism, and Finance must work together to control and monitor the sustainable management of ELV recycling.

Objectives of the Policy on ELV

The objectives of policy and regulation regarding ELV recycling in India should be emphasized on

- 10R-strategies (Refuse, Rethink, Reduce, Reuse, Repair, Refurbish, Remanufacture, Repurpose, Recycle, Recovery) to promote sustainability in ELV recycling in India.
- Promoting circular economy in India.
- Mitigation of pollution levels and improving environmental quality.
- Efficient use of natural resources.
- Reduce open dumping for ELV.
- Implementation of a solid waste management system.
- Enhance the resilience factors and productivity in ELV recycling.
- Development of proper infrastructure for ELV recycling.

6. Recommendations

The ELV recycling system in India is beset by a plethora of rudimentary issues, challenges, and threats, impeding the development of the ELV recycling sector in India [51]. Appropriate directives and initiatives by authorities are instrumental in the thriving and flourishing ELV recycling sector sustainably. Based on our observations and analysis in the previous section, we have identified a host of critical and persistent issues in ELV recycling. Without resolving these perennial issues in ELV recycling, further development and enhancement are elusive. In this section, we have made practical recommendations for efficiently encountering the problems identified in recycling for the long-term viability of ELV recycling. Figure 8 represents critical issues and corresponding recommendations regarding the ELV recycling system in India.

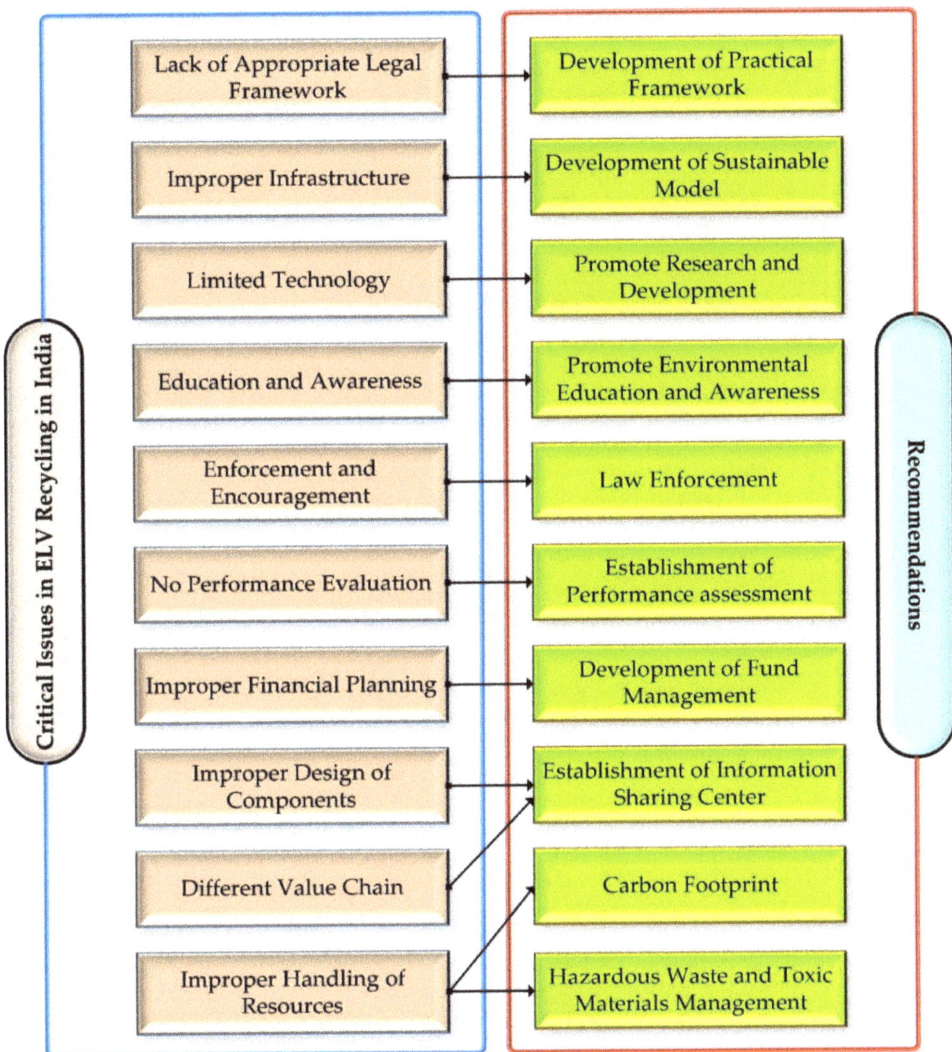

Figure 8. Critical issues and corresponding recommendations regarding ELV recycling system in India.

6.1. Development of Practical Framework

Developing a practical framework is the first crucial element toward sustainability [71,72]; authorities cannot provide appropriate directions to the ELV recycling sector without a realistic framework. In the Indian ELV sector, informality is prevalent because of the absence of a practical framework from authority. The Indian government should develop a pragmatic framework based on Indian values for ELV recycling to eliminate the informalities and malpractices and promote sustainability in ELV recycling. This practical framework will assist the ELV recycler by providing appropriate guidelines regarding ELV recycling operations, inevitably enhancing the material extraction rate.

6.2. Establishment of Information Sharing Center

The lack of collaboration among ELV recyclers, manufacturers, and authorities is thwarting the prosperity of ELV recycling; hence an information-sharing center is required to establish. The information-sharing center will preserve all data shared by key stakeholders, manufacturers, government authorities, and car owners. This system will have numerous interfaces for all key stakeholders, manufacturers, car owners, and management. They will upload the components, operations, material flow information, and information about hazardous waste. This system will assist all crucial parties involved in recycling by providing accurate and authentic pieces of information. This system will have a profound impact on the ELV recycling system.

6.3. Development of a Sustainable Model

By considering socio-economic-environment aspects and the long-term viability of ELV recycling, the government of India should develop a sustainable model for ELV recycling, which will not only enhance the recycling and recovery rate but also preserve the environment [73,74]. By adopting the 4Rs (reuse, recycle, remanufacturing, repair) standardization, authorities can create a sustainable model which will guide the automotive sector toward obtaining the highest possible standard of product and operation, safeguarding the environment, and consumer safety. This model will cherish business sustainability and increase competitiveness.

6.4. Promote Research and Development

ELV recycling involves numerous convoluted processes. Technology plays an imperative role in development and advancement [75]. Because of poor technology, India's ELV recycling performance is limited. In order to enhance the materials extraction and recovery rate, the ELV recycling sector requires advanced technology. The Indian government, together with vehicle manufacturers, should allocate funds to promote research and development to establish an effective recycling system. This allocated fund should be given to the government and vehicle manufacturer research and development institute to promote sustainability in ELV recycling.

6.5. Development of Fund Management

The treatment of abandoned vehicles for recycling is associated with a significant cost as many processes are involved. The car manufacturer or car owner can cover this expense for operations of ELV for recycling; still, the government has yet to define who is accountable for the expense of recycling. The Indian authority should establish fund management for ELV recycling, which will be responsible for the relevant expense of recycling. It encompasses various expenses, including recycling fees, subsidies, rewards, costs associated with waste management, and social awareness. The vehicle manufacturer, the government, and the vehicle owner should all contribute to this fund. Different countries have different policies regarding the fees for recycling vehicles accordingly to their economic and social aspects. In countries like India, vehicle manufacturers and car owners should bear the entire cost of recycling in equal proportion. Whereas, the government should monitor and audit this fund. This fund should be utilized to promote green technology for the long-term viability of ELV recycling. This fund should be used to dispose of hazardous and toxic waste properly.

6.6. Performance Assessment

Performance evaluation is crucial in enhancing the recycled product quality and operations involved in recycling. India has yet to establish a proper performance evaluation process to assess the recycled product's quality, operation, environmental impact, and waste management. Performance evaluation, together with appropriate inspection, will assist the Indian government in achieving the well-defined objectives of ELV recycling.

6.7. Promote Environmental Education and Awareness

Long-term goals, such as promoting sustainable environmental education and raising public awareness among Indians, must begin with children. Indians are generally unaware and unconcerned about environmental issues. It is expected that public awareness programs and training for organization members and stakeholders will improve general society's level of knowledge and awareness of ELVs and practitioners' competency and skills in performing ELVs processes.

6.8. Carbon Footprint Application

Numerous distinct power tools are used in the ELV recycling system during the operations of ELVs; this equipment requires a lot of energy and produces a significant quantity of greenhouse gases [76,77], which is detrimental to the environment. The application of carbon footprint, which measures the total amount of greenhouse gas emissions, can assist the ELV recycler in evaluating energy consumption and reducing energy consumption.

6.9. Law Enforcement

The relevant ELV recycling laws are paramount to properly governing and regulating the Indian ELV recycling sector. The Indian government has yet to enact the appropriate and suitable rules and regulations in different aspects, incorporating environment, transport, use of resources, waste disposal, and business model. The quality of the environment in India has become a critical concern and has been deepening significantly because of the absence of proper regulation.

6.10. Hazardous Waste and Toxic Materials Management

The detrimental automotive shredder residue, toxic materials, poisonous electronic waste, waste fluid, oil, battery, and other unwanted debris should not be disposed of randomly. To meet this demand, we can construct a hazardous waste and toxic materials processing center in each major city, as Indian people's environmental awareness is low. Hazardous waste should be disposed of at a hazardous waste and toxic materials processing center. The hazardous waste and toxic materials processing center should deal with toxic electronic waste and hazardous and toxic scrap automobile waste. It will keep the materials used in ELV shells as clear as possible.

7. Discussions and Future Implications

This study delineates the status quo of the ELV recycling system holistically in India, which is a combination of current law and policy analysis and ELV flow in formal and informal sectors, respectively. This research summarizes the relevant laws regarding recycling and environmental protection and thoroughly sheds light on recent drafting about ELV recycling and RVSF. This analysis reveals that India does not yet have rigorous regulations governing the recycling of ELVs, and lax and insufficient laws and regulations govern the current ELV recycling sector. As this research demonstrates the existing vacuum of regulation to regulate the ELV recycling sector, which may draw the attention of Indian authorities to devise and enact an appropriate set of rules to regulate and monitor the ELV recycling sector sustainably and systematically.

This investigation reveals the ELV recycling practice in both formal and informal sectors and finds that India's ELV recycling system is still in its infancy and the informal sector dominates the current ELV recycling system. The ELV dismantling operations in the informal sector are not environmentally friendly, contributing significantly to environmental degradation. The materials recovery and recycling rates are below standard in the informal sector compared to other developed nations. The Indian government should develop an appropriate and practical framework for ELV recycling based on Indian values and provide this framework to stakeholders to upgrade their operations and practices.

This research critically evaluates the current ELV recycling system in India to address the persisting perennial critical issues that the ELV recycling system is encountering.

These challenges hinder the development of the current ELV recycling system. The identified issues in the critical evaluations section may assist and encourage the authority to appropriate counter measurement to promote sustainability in the ELV recycling sector.

Aside from critical evaluations, this paper proposes pragmatic and efficient frameworks based on Indian values that would enhance materials recovery and recycling rates, promote sustainability in ELV recycling, and provide appropriate directions in terms of law and policy. The proposed frameworks in this study may assist the stakeholders in enhancing materials recovery and recycling rates and cherishing sustainability in the ELV recycling sector. At the same time, these frameworks provide lucid directives in terms of law and policy for the authority to devise and enact an appropriate set of regulations to regulate the ELV recycling sector sustainably.

Eventually, this study has provided practical and valuable recommendations for effectively addressing the highlighted issues for the long-term sustainability of ELV recycling. These recommendations may assist the government of India in setting up an efficient ELV recycling sector by addressing persistent issues in the ELV recycling sector. The findings of this research may encourage the authorities and stakeholders to adopt and implement the proposed frameworks to enhance the recycling rate while respecting the environment. The outcomes of this study can be useful for other countries in the Indian subcontinent.

8. Research Limitations

This research has a few potential limitations.

- This research has investigated the ELV recycling system in India, and the frameworks in this study are proposed based on Indian values; therefore, the outcomes of this study might be limited to the Indian subcontinent. Therefore, other developing countries may not benefit from the results of this investigation.
- This investigation has been performed in five well-known automotive sectors: Mumbai, Kolkata, Chennai, Jamshedpur, and Delhi. This research could have provided a more holistic ELV recycling practice in India by Incorporating additional automobile hubs.
- Aside from field investigation, literature, and policy review, there have been interactions with 537 chosen people, including all key stakeholders, academicians, and government officials. This study could have provided more accurate outcomes with a higher number of interactions.
- Very little research has been performed on ELV recycling in India. The lack of literature about ELV recycling in India initially made it challenging to develop the fundamental basis of this research.

9. Future Prospects and Suggestions

This study emphasizes the development, critical evaluation, and proposed framework for ELV recycling in India. As well as that, this study suggests specific vital topics that require in-depth analysis to advance the ELV recycling system in India.

- Materials recycling and recovery rates are imperative in ELV recycling. Measuring and comparing the materials extraction rates between formal and informal sectors requires further exploration.
- A carbon emission assessment is instrumental for monitoring the air quality around the automotive hubs. Assessing and comparing the carbon emission from formal and informal sectors requires in-depth study.
- ELV contains several hazardous materials. Appropriate disposal methods for these detrimental substances require further thorough investigation.
- Recycling industries depend on manufacturing industries. Hence, exploring the synergy between recycling and manufacturing industries needs to be evaluated thoroughly.

10. Conclusions

This study has delved deeper into India's ELV recycling system and elucidated the current practice of ELV recycling in formal and informal sectors in India in a greater context to develop a profound understanding that is instrumental in expanding the ELV sector by obtaining the best potential values from ELVs. This study's findings reveal that ELV recycling in India is at the embryonic stage and struggling with a host of challenges in different aspects. Currently, there is no comprehensive and complete set of rules and regulations to govern the ELV recycling system in India. The vacuum of appropriate policy engenders a prevalence of informality in India's ELV recycling sector.

After reflecting on current practice, this research has critically evaluated the existing ELV recycling system in India and identified the impediments thwarting the development of the ELV recycling system in India. The lack of practical frameworks, lax laws and policies, and inappropriate technology are the critical deterrent to sustainability in recycling in India.

Identification of hindrances was the first step to encountering the existing impediments. This study has proposed pragmatic frameworks and made practical recommendations to approach the identified issues and expedite sustainable development. The proposed frameworks and practical suggestions are made based on Indian values to assist in developing an efficient ELV recycling system in India to cherish sustainable growth and expand the ELV sector without affecting the environment and conserving natural resources. This research may encourage informal ELV recyclers to adopt the sustainable ELV recycling model, which will assist in formalizing the Indian ELV recycling sector and reducing landfills. This study may increase the public awareness level of sustainable ELV recycling. The proposed framework is likely to assist the government of India in implementing any upcoming regulatory and legal framework regarding ELV recycling. These frameworks and recommendations are also suitably applicable to other developing nations in the Indian subcontinent.

Author Contributions: Each author's contributions are listed as follows: Conceptualization, A.H.M.; Methodology, A.H.M.; Software, A.H.M.; Validation, A.H.M.; Formal Analysis, A.H.M.; Investigation, A.H.M.; Resources, Z.H. and A.H.M.; Data Curation, A.H.M.; Writing—Original Draft Preparation, A.H.M.; Writing—Review and Editing, Z.H. and R.I.; Visualization, A.H.M. and Z.H.; Supervision, Z.H.; Project Administration, Z.H.; Funding Acquisition, Z.H. and M.R.A.M. All authors have read and agreed to the published version of the manuscript.

Funding: This research was supported by the Transdisciplinary Research Grant Scheme TRGS/1/2020/UKM/02/1/1 and UKM internal grant GUP-2018-012.

Institutional Review Board Statement: The study was conducted according to the guidelines of the International Conference of Harmonization Good Clinical Practice, and approved by the Research Ethics Committee of The National University of Malaysia (UKM PPI/111/8/JEP-2021-595, 11 November 2021).

Informed Consent Statement: Not applicable.

Data Availability Statement: The data presented in this study are available in this published article.

Acknowledgments: The authors would like to thank the anonymous reviewers for their help.

Conflicts of Interest: The authors declare that they have no known competing financial interests or personal relationships that could have appeared to influence the work reported in this paper.

References

1. Korica, P.; Cirman, A.; Žgajnar Gotvajn, A. Comparison of end-of-life vehicles management in 31 European countries: A LMDI analysis. *Waste Manag. Res. J. Sustain. Circ. Econ.* **2022**, *40*, 1156–1166. [CrossRef]
2. Zhou, F.; Lim, M.K.; He, Y.; Lin, Y.; Chen, S. End-of-life vehicle (ELV) recycling management: Improving performance using an ISM approach. *J. Clean. Prod.* **2019**, *228*, 231–243. [CrossRef]
3. Karagoz, S.; Aydin, N.; Simic, V. End-of-life vehicle management: A comprehensive review. *J. Mater. Cycles Waste Manag.* **2020**, *22*, 416–442. [CrossRef]
4. Petronijević, V.; Đorđević, A.; Stefanović, M.; Arsovski, S.; Krivokapić, Z.; Mišić, M. Energy Recovery through End-of-Life Vehicles Recycling in Developing Countries. *Sustainability* **2020**, *12*, 8764. [CrossRef]

5. Abdollahifar, M.; Doose, S.; Cavers, H.; Kwade, A. Graphite Recycling from End-of-Life Lithium-Ion Batteries: Processes and Applications. *Adv. Mater. Technol.* **2022**. *early view.* [CrossRef]

6. Crespo, M.S.; González, M.V.G.; Peiró, L.T. Prospects on end of life electric vehicle batteries through 2050 in Catalonia. *Resour. Conserv. Recycl.* **2022**, *180*, 106133. [CrossRef]

7. Go, T.; Wahab, D.; Rahman, M.; Ramli, R.; Azhari, C. Disassemblability of end-of-life vehicle: A critical review of evaluation methods. *J. Clean. Prod.* **2011**, *19*, 1536–1546. [CrossRef]

8. Rovinaru, F.I.; Rovinaru, M.D.; Rus, A.V. The Economic and Ecological Impacts of Dismantling End-of-Life Vehicles in Romania. *Sustainability* **2019**, *11*, 6446. [CrossRef]

9. Nakano, K.; Shibahara, N. Comparative assessment on greenhouse gas emissions of end-of-life vehicles recycling methods. *J. Mater. Cycles Waste Manag.* **2015**, *19*, 505–515. [CrossRef]

10. Williams, K.S.; Khodier, A. Meeting EU ELV targets: Pilot-scale pyrolysis automotive shredder residue investigation of PAHs, PCBs and environmental contaminants in the solid residue products. *Waste Manag.* **2020**, *105*, 233–239. [CrossRef]

11. Society of Indian Automobile Manufactures. Available online: https://www.siam.in/statistics.aspx?mpgid=8&pgidtrail=9 (accessed on 5 May 2022).

12. Arora, N.; Bakshi, S.K.; Bhattacharjya, S. Framework for sustainable management of end-of-life vehicles management in India. *J. Mater. Cycles Waste Manag.* **2018**, *21*, 79–97. [CrossRef]

13. Zhang, L.; Ji, K.; Liu, W.; Cui, X.; Liu, Y.; Cui, Z. Collaborative approach for environmental and economic optimization based on life cycle assessment of end-of-life vehicles' dismantling in China. *J. Clean. Prod.* **2020**, *276*, 124288. [CrossRef]

14. Numfor, S.A.; Takahashi, Y.; Matsubae, K. Energy recovery from end-of-life vehicle recycling in Cameroon: A system dynamics approach. *J. Clean. Prod.* **2022**, *361*, 132090. [CrossRef]

15. Jang, Y.-C.; Choi, K.; Jeong, J.-H.; Kim, H.; Kim, J.-G. Recycling and Material-Flow Analysis of End-of-Life Vehicles towards Resource Circulation in South Korea. *Sustainability* **2022**, *14*, 1270. [CrossRef]

16. Mallampati, S.R.; Lee, B.H.; Mitoma, Y.; Simion, C. Sustainable recovery of precious metals from end-of-life vehicles shredder residue by a novel hybrid ball-milling and nanoparticles enabled froth flotation process. *J. Clean. Prod.* **2018**, *171*, 66–75. [CrossRef]

17. Ghosh, S.K.; Ghosh, S.K.; Baidya, R. Circular Economy in India: Reduce, Reuse, and Recycle Through Policy Framework. In *Circular Economy: Recent Trends in Global Perspective*; Springer: Singapore, 2021; pp. 183–217. [CrossRef]

18. Sato, F.E.K.; Furubayashi, T.; Nakata, T. Application of energy and CO_2 reduction assessments for end-of-life vehicles recycling in Japan. *Appl. Energy* **2019**, *237*, 779–794. [CrossRef]

19. Numfor, S.; Omosa, G.; Zhang, Z.; Matsubae, K. A Review of Challenges and Opportunities for End-of-Life Vehicle Recycling in Developing Countries and Emerging Economies: A SWOT Analysis. *Sustainability* **2021**, *13*, 4918. [CrossRef]

20. Mohan, T.V.K.; Amit, R.K. Dismantlers' dilemma in end-of-life vehicle recycling markets: A system dynamics model. *Ann. Oper. Res.* **2018**, *290*, 591–619. [CrossRef]

21. Chen, Y.; Ding, Z.; Liu, J.; Ma, J. Life cycle assessment of end-of-life vehicle recycling in China: A comparative study of environmental burden and benefit. *Int. J. Environ. Stud.* **2019**, *76*, 1019–1040. [CrossRef]

22. Sharma, L.; Pandey, S. Recovery of resources from end-of-life passenger cars in the informal sector in India. *Sustain. Prod. Consum.* **2020**, *24*, 1–11. [CrossRef]

23. Andersson, M.; Söderman, M.L.; Sandén, B.A. Challenges of recycling multiple scarce metals: The case of Swedish ELV and WEEE recycling. *Resour. Policy* **2019**, *63*, 101403. [CrossRef]

24. Khan, S.A.R.; Godil, D.I.; Thomas, G.; Tanveer, M.; Zia-Ul-Haq, H.M.; Mahmood, H. The Decision-Making Analysis on End-of-Life Vehicle Recycling and Remanufacturing under Extended Producer Responsibility Policy. *Sustainability* **2021**, *13*, 11215. [CrossRef]

25. Yu, J.; Wang, S.; Serrona, K.R.B. Comparative Analysis of ELV Recycling Policies in the European Union, Japan and China. *Investig. Linguisticae* **2019**, *XLIII*, 34–56. [CrossRef]

26. Zhang, L.; Lu, Q.; Yuan, W.; Jiang, S.; Wu, H. Characterizing end-of-life household vehicles' generations in China: Spatial-temporal patterns and resource potentials. *Resour. Conserv. Recycl.* **2022**, *177*, 105979. [CrossRef]

27. Khodier, A.; Williams, K.; Dallison, N. Challenges around automotive shredder residue production and disposal. *Waste Manag.* **2018**, *73*, 566–573. [CrossRef]

28. Soo, V.K.; Doolan, M.; Compston, P.; Duflou, J.R.; Peeters, J.; Umeda, Y. The influence of end-of-life regulation on vehicle material circularity: A comparison of Europe, Japan, Australia and the US. *Resour. Conserv. Recycl.* **2021**, *168*, 105294. [CrossRef]

29. Rosa, P.; Terzi, S. Improving end of life vehicle's management practices: An economic assessment through system dynamics. *J. Clean. Prod.* **2018**, *184*, 520–536. [CrossRef]

30. Yi, S.; Lee, H. Economic analysis to promote the resource circulation of end-of-life vehicles in Korea. *Waste Manag.* **2021**, *120*, 659–666. [CrossRef]

31. Li, Y.; Liu, Y.; Chen, Y.; Huang, S.; Ju, Y. Estimation of end-of-life electric vehicle generation and analysis of the status and prospects of power battery recycling in China. *Waste Manag. Res.* **2022**, *40*, 1424–1432. [CrossRef]

32. Sitinjak, C.; Ismail, R.; Bantu, E.; Fajar, R.; Samuel, K. The understanding of the social determinants factors of public acceptance towards the end of life vehicles. *Cogent Eng.* **2022**, *9*, 2088640. [CrossRef]

33. Wang, J.; Sun, L.; Fujii, M.; Li, Y.; Huang, Y.; Murakami, S.; Daigo, I.; Pan, W.; Li, Z. Institutional, Technology, and Policies of End-of-Life Vehicle Recycling Industry and Its Indication on the Circular Economy- Comparative Analysis Between China and Japan. *Front. Sustain. Food Syst.* **2021**, *2*, 13. [CrossRef]

34. Yu, Z.; Tianshan, M.; Rehman, S.A.; Sharif, A.; Janjua, L. Evolutionary game of end-of-life vehicle recycling groups under government regulation. *Clean Technol. Environ. Policy* **2020**, 1–12. [CrossRef]
35. Yu, Z.; Khan, S.A.R.; Zia-Ul-Haq, H.M.; Tanveer, M.; Sajid, M.J.; Ahmed, S. A Bibliometric Analysis of End-of-Life Vehicles Related Research: Exploring a Path to Environmental Sustainability. *Sustainability* **2022**, *14*, 8484. [CrossRef]
36. Sanz, V.M.; Serrano, A.M.; Schlummer, M. A mini-review of the physical recycling methods for plastic parts in end-of-life vehicles. *Waste Manag. Res.* **2022**, *40*, 0734242X221094917. [CrossRef]
37. Swain, B.; Park, J.R.; Lee, C.G. Industrial recycling of end-of-life vehicle windshield glass by mechanical beneficiation and complete recovery of polyvinyl butyral. *J. Clean. Prod.* **2022**, *334*, 130192. [CrossRef]
38. Edun, A.; Hachem-Vermette, C. Energy and environmental impact of recycled end of life tires applied in building envelopes. *J. Build. Eng.* **2021**, *39*, 102242. [CrossRef]
39. Zhang, Z.; Malik, M.Z.; Khan, A.; Ali, N.; Malik, S.; Bilal, M. Environmental impacts of hazardous waste, and management strategies to reconcile circular economy and eco-sustainability. *Sci. Total Environ.* **2022**, *807*, 150856. [CrossRef]
40. Zimon, D.; Madzík, P. Standardized management systems and risk management in the supply chain. *Int. J. Qual. Reliab. Manag.* **2020**, *37*, 305–327. [CrossRef]
41. Farahani, S.; Otieno, W.; Barah, M. Environmentally friendly disposition decisions for end-of-life electrical and electronic products: The case of computer remanufacture. *J. Clean. Prod.* **2019**, *224*, 25–39. [CrossRef]
42. Wang, L.; Chen, M. Policies and perspective on end-of-life vehicles in China. *J. Clean. Prod.* **2013**, *44*, 168–176. [CrossRef]
43. D'Adamo, I.; Gastaldi, M.; Rosa, P. Recycling of end-of-life vehicles: Assessing trends and performances in Europe. *Technol. Forecast. Soc. Change* **2020**, *152*, 119887. [CrossRef]
44. Li, Y.; Fujikawa, K.; Wang, J.; Li, X.; Ju, Y.; Chen, C. The Potential and Trend of End-Of-Life Passenger Vehicles Recycling in China. *Sustainability* **2020**, *12*, 1455. [CrossRef]
45. Blume, T.; Walther, M. The End-of-life Vehicle Ordinance in the German automotive industry—Corporate sense making illustrated. *J. Clean. Prod.* **2013**, *56*, 29–38. [CrossRef]
46. Govindan, K.; Shankar, M.; Kannan, D. Application of fuzzy analytic network process for barrier evaluation in automotive parts remanufacturing towards cleaner production—A study in an Indian scenario. *J. Clean. Prod.* **2016**, *114*, 199–213. [CrossRef]
47. Luthra, S.; Kumar, V.; Kumar, S.; Haleem, A. Barriers to implement green supply chain management in automobile industry using interpretive structural modeling technique: An Indian perspective. *J. Ind. Eng. Manag.* **2011**, *4*, 231–257. [CrossRef]
48. Hu, S.; Wen, Z. Monetary evaluation of end-of-life vehicle treatment from a social perspective for different scenarios in China. *J. Clean. Prod.* **2017**, *159*, 257–270. [CrossRef]
49. Li, Y.; Liu, Y.; Chen, Y.; Huang, S.; Ju, Y. Projection of end-of-life vehicle population and recyclable metal resources: Provincial-level gaps in China. *Sustain. Prod. Consum.* **2022**, *31*, 818–827. [CrossRef]
50. Al-Quradaghi, S.; Zheng, Q.P.; Betancourt-Torcat, A.; Elkamel, A. Optimization Model for Sustainable End-of-Life Vehicle Processing and Recycling. *Sustainability* **2022**, *14*, 3551. [CrossRef]
51. Molla, A.H.; Shams, H.; Harun, Z.; Ab Rahman, M.N.; Hishamuddin, H. An Assessment of Drivers and Barriers to Implementation of Circular Economy in the End-of-Life Vehicle Recycling Sector in India. *Sustainability* **2022**, *14*, 13084. [CrossRef]
52. Cossu, R.; Lai, T. Automotive shredder residue (ASR) management: An overview. *Waste Manag.* **2015**, *45*, 143–151. [CrossRef]
53. Tarrar, M.; Despeisse, M.; Johansson, B. Driving vehicle dismantling forward—A combined literature and empirical study. *J. Clean. Prod.* **2021**, *295*, 126410. [CrossRef]
54. Sinha, A.; Mondal, S.; Boone, T.; Ganeshan, R. Analysis of issues controlling the feasibility of automobile remanufacturing business in India. *Int. J. Serv. Oper. Manag.* **2017**, *26*, 459–475. [CrossRef]
55. Nag, U.; Sharma, S.K.; Govindan, K. Investigating drivers of circular supply chain with product-service system in automotive firms of an emerging economy. *J. Clean. Prod.* **2021**, *319*, 128629. [CrossRef]
56. Wahab, D.; Blanco-Davis, E.; Ariffin, A.; Wang, J. A review on the applicability of remanufacturing in extending the life cycle of marine or offshore components and structures. *Ocean Eng.* **2018**, *169*, 125–133. [CrossRef]
57. Mayyas, A.; Omar, M.; Hayajneh, M.; Mayyas, A.R. Vehicle's lightweight design vs. electrification from life cycle assessment perspective. *J. Clean. Prod.* **2017**, *167*, 687–701. [CrossRef]
58. Ropi, N.M.; Hishamuddin, H.; Wahab, D.A. Analysis of the Supply Chain Disruption Risks in the Malaysian Automotive Remanufacturing Industry-a Case Study. *Int. J. Integr. Eng.* **2020**, *12*, 1–11. [CrossRef]
59. Zhao, Q.; Chen, M. A comparison of ELV recycling system in China and Japan and China's strategies. *Resour. Conserv. Recycl.* **2011**, *57*, 15–21. [CrossRef]
60. Sitinjak, C.; Ismail, R.; Tahir, Z.; Fajar, R.; Simanullang, W.F.; Bantu, E.; Samuel, K.; Rose, R.A.C.; Yazid, M.R.M.; Harun, Z. Acceptance of ELV Management: The Role of Social Influence, Knowledge, Attitude, Institutional Trust, and Health Issues. *Sustainability* **2022**, *14*, 10201. [CrossRef]
61. Kaya, D.I.; Pintossi, N.; Dane, G. An Empirical Analysis of Driving Factors and Policy Enablers of Heritage Adaptive Reuse within the Circular Economy Framework. *Sustainability* **2021**, *13*, 2479. [CrossRef]
62. Tan, J.; Tan, F.J.; Ramakrishna, S. Transitioning to a Circular Economy: A Systematic Review of Its Drivers and Barriers. *Sustainability* **2022**, *14*, 1757. [CrossRef]
63. Grafström, J.; Aasma, S. Breaking circular economy barriers. *J. Clean. Prod.* **2021**, *292*, 126002. [CrossRef]

64. Kayikci, Y.; Kazancoglu, Y.; Lafci, C.; Gozacan, N. Exploring barriers to smart and sustainable circular economy: The case of an automotive eco-cluster. *J. Clean. Prod.* **2021**, *314*, 127920. [CrossRef]

65. Lindahl, P.; Robèrt, K.-H.; Ny, H.; Broman, G. Strategic sustainability considerations in materials management. *J. Clean. Prod.* **2014**, *64*, 98–103. [CrossRef]

66. Mamat, T.N.A.R.; Saman, M.Z.M.; Sharif, S.; Simic, V. Key success factors in establishing end-of-life vehicle management system: A primer for Malaysia. *J. Clean. Prod.* **2016**, *135*, 1289–1297. [CrossRef]

67. Modoi, O.-C.; Mihai, F.-C. E-Waste and End-of-Life Vehicles Management and Circular Economy Initiatives in Romania. *Energies* **2022**, *15*, 1120. [CrossRef]

68. Mamat, T.N.A.R.; Saman, M.Z.M.; Sharif, S.; Simic, V.; Wahab, D.A. Development of a performance evaluation tool for end-of-life vehicle management system implementation using the analytic hierarchy process. *Waste Manag. Res.* **2018**, *36*, 1210–1222. [CrossRef] [PubMed]

69. Harun, Z.; Wan Mustafa, W.M.S.; Abd Wahab, D.; Abu Mansor, M.R.; Saibani, N.; Ismail, R.; Mohd Ali, H.; Hashim, N.; Mohd Paisal, M. An Analysis of End-of-Life Vehicle Policy Implementation in Malaysia from the Perspectives of Laws and Public Perception. *Jurnal Kejuruteraan* **2021**, *33*, 709–718. [CrossRef]

70. Darom, N.A.M.; Hishamuddin, H.; Ramli, R.; Nopiah, Z.M.; Sarker, R.A. Investigation of Disruption Management Practices and Environmental Impact on Malaysian Automotive Supply Chains: A Case Study Approach. *Jurnal Kejuruteraan* **2020**, *32*, 341–348. [CrossRef]

71. Melo, A.C.S.; Braga, A.E.; Leite, C.D.P.; Bastos, L.D.S.L.; Nunes, D.R.D.L. Frameworks for reverse logistics and sustainable design integration under a sustainability perspective: A systematic literature review. *Res. Eng. Des.* **2020**, *32*, 225–243. [CrossRef]

72. Kaviani, M.A.; Tavana, M.; Kumar, A.; Michnik, J.; Niknam, R.; de Campos, E.A.R. An integrated framework for evaluating the barriers to successful implementation of reverse logistics in the automotive industry. *J. Clean. Prod.* **2020**, *272*, 122714. [CrossRef]

73. Pigosso, D.C.A.; De, M.; Pieroni, P.; Kravchenko, M.; Awan, U.; Sroufe, R.; Bozan, K. Designing Value Chains for Industry 4.0 and a Circular Economy: A Review of the Literature. *Sustainability* **2022**, *14*, 7084. [CrossRef]

74. Hina, M.; Chauhan, C.; Kaur, P.; Kraus, S.; Dhir, A. Drivers and barriers of circular economy business models: Where we are now, and where we are heading. *J. Clean. Prod.* **2022**, *333*, 130049. [CrossRef]

75. Bockholt, M.T.; Kristensen, J.H.; Colli, M.; Jensen, P.M.; Wæhrens, B.V. Exploring factors affecting the financial performance of end-of-life take-back program in a discrete manufacturing context. *J. Clean. Prod.* **2020**, *258*, 120916. [CrossRef]

76. Liu, Z.; Adams, M.; Cote, R.P.; Chen, Q.; Wu, R.; Wen, Z.; Liu, W.; Dong, L. How does circular economy respond to greenhouse gas emissions reduction: An analysis of Chinese plastic recycling industries. *Renew. Sustain. Energy Rev.* **2018**, *91*, 1162–1169. [CrossRef]

77. Harun, Z.; Molla, A.H.; Khan, A.; Zainal Safuan, U.; Azman, M.S.S.; Hashim, H. End-of-Life Vehicle (ELV) Emission Evaluation using IoT in Malaysia. In Proceedings of the 2022 9th International Conference on Information Technology, Computer, and Electrical Engineering (ICITACEE), Semarang, Indonesia, 25–26 August 2022. [CrossRef]

 *sustainability*

Article

Development of a Reverse Logistics Modeling for End-of-Life Lithium-Ion Batteries and Its Impact on Recycling Viability—A Case Study to Support End-of-Life Electric Vehicle Battery Strategy in Canada

Giovanna Gonzales-Calienes *[iD], Ben Yu and Farid Bensebaa

Energy, Mining and Environment Research Centre, National Research Council Canada, 1200 Montreal Road, Ottawa, ON K1A 0R6, Canada
* Correspondence: giovanna.gonzalescalienes@nrc-cnrc.gc.ca; Tel.: +1-613-9931700

Abstract: The deployment of a sustainable recycling network for electric vehicle batteries requires the development of an infrastructure to collect and deliver batteries to several locations from which they can be transported to companies for repurposing or recycling. This infrastructure is still not yet developed in North America, and consequently, spent electric vehicle batteries in Canada are dispersed throughout the country. The purpose of this reverse logistics study is to develop a spatial modeling framework to identify the optimal locations of battery pack dismantling hubs and recycling processing facilities in Canada and quantify the environmental and economic impacts of the supporting infrastructure network for electric vehicle lithium-ion battery end-of-life management. The model integrates the geographic information system, material flow analysis for estimating the availability of spent battery stocks, and the life cycle assessment approach to assess the environmental impact. To minimize the costs and greenhouse gas emission intensity, three regional recycling clusters, including dismantling hubs, recycling processing, and scrap metal smelting facilities, were identified. These three clusters will have the capacity to satisfy the annual flow of disposed batteries. The Quebec–Maritimes cluster presents the lowest payload distance, life-cycle carbon footprint, and truck transportation costs than the Ontario and British Columbia–Prairies clusters. Access to end-of-life batteries not only makes the battery supply chain circular, but also provides incentives for establishing recycling facilities. The average costs and carbon intensity of recycled cathode raw materials are CAD 1.29/kg of the spent battery pack and 0.7 kg CO_{2e}/kg of the spent battery pack, respectively, which were estimated based on the optimization of the transportation distances.

Keywords: end-of-life lithium-ion battery; battery recycling; geographic information system; material flow analysis; GHG emissions; transportation network optimization; reverse logistics

Citation: Gonzales-Calienes, G.; Yu, B.; Bensebaa, F. Development of a Reverse Logistics Modeling for End-of-Life Lithium-Ion Batteries and Its Impact on Recycling Viability—A Case Study to Support End-of-Life Electric Vehicle Battery Strategy in Canada. *Sustainability* **2022**, *14*, 15321. https://doi.org/10.3390/su142215321

Academic Editors: Nallapaneni Manoj Kumar, Subrata Hait, Anshu Priya and Varsha Bohra

Received: 8 September 2022
Accepted: 11 November 2022
Published: 18 November 2022

Publisher's Note: MDPI stays neutral with regard to jurisdictional claims in published maps and institutional affiliations.

1. Introduction

Greenhouse gas (GHG) emission reduction is a crucial target to fight climate change, and accelerating decarbonization in diverse economic sectors is a net-zero emissions pathway to achieve it. Several national governments worldwide have announced net-zero GHG emission pledges (in law, proposed, and policy documents), which commit to developing long-term, low GHG emission strategies in line with limiting temperature increases to 1.5 °C above preindustrial levels by 2050. Traditionally, transportation has been heavily reliant on fossil fuels, which accounted for more than 90% of the transport sector energy needs in 2020. Electrification plays a central role in decarbonizing light- and heavy-duty vehicles, which rely on policies to promote electric mobility (battery- or fuel cell-powered electric vehicles) [1].

The electrification of road transportation entails challenges related to battery supply chain sustainability and end-of-life (EoL) management. Technology advancements in lithium-ion batteries (LIBs) have become the main drivers for accelerating the development of the electric vehicle (EV) market. The raw material inputs associated with LIBs, such as lithium, nickel, cobalt, and graphite, present a high level of criticality with the associated high costs and environmental impacts [2–4]. EoL management of spent EV LIBs is an important aspect of the closed-loop life cycle approach for EV LIBs. Reusing and recycling LIBs will reduce the dependence on mining and refining of the critical materials in LIBs, decrease the negative environmental impact, and generate local socioeconomic benefits. Before the repurposing and recycling phases of EV LIBs, LIBs must be collected and dismantled, and key valuable components must be transported to repurposing or recycling facilities.

The transportation of spent EV LIBs is a critical activity that influences the environmental impact and costs of EoL management. A current review of state-of-art of spent EV LIBs transportation emphasizes the importance of the strategic design of a reverse logistics network of spent EV LIBs [5–7]. Wilson and Goffnett [8] describe the reverse logistics in supply chain management as the flow of materials in the reverse direction to the main flow. In the case of spent batteries, reverse logistics deal with the collection of EoL LIBs from the end users and the transport of such batteries to downstream facilities including remanufacturing, reusing, and recycling [5]. Some studies suggest that optimized facility siting to minimize distances decreases transportation costs and emissions [9–13]. Additionally, it is implied that a reduction in transportation distances with a design of local dismantling facilities also reduces safety risks [10,14,15] and reduces unnecessary transportation costs when spent batteries are sent to recycling facilities and the remaining can be reused [14,15].

There is limited research on optimization research on EoL infrastructure siting and the role of transportation of spent batteries for EoL processing stages. This research is critical to evaluate the viability of EoL recycling, such as the optimal locations of EoL management facilities, while assessing the environmental and cost impacts of collection and transportation. The recycling viability of building EoL management facilities should be evaluated with a view to different factors, such as recycled material amounts and prices, transport distances, financial metrics, policies, and types of technology [16].

Table 1 provides key information on the relevant studies in this field. Some studies on EV LIBs reverse logistics have been focused on modeling the optimization of siting locations of EoL facilities. Wang et al. [13] designed an optimal recycling network in China assuming transportation costs and carbon taxes. Hoyer, et al. [17] developed an optimization model for technology and capacity planning in recycling networks, where the strategy of setting up all the recycling plants at the beginning of the planning horizon presented the higher net present value. Tadaros, et al. [18] developed a reverse supply chain model for used LIBs in Sweden. The model is defined as an unlinked discrete multiperiod problem without taking into account any inventories. Hendrickson, et al. [19] used geographic information system (GIS) and life cycle assessment (LCA) to assess the environmental impacts of used battery supply logistics beyond GHG emissions, highlighting the impact of transportation on human health damages in California and pointing out the importance of an optimized system design for siting new facilities. Although Rallo, et al. [20] did not use an optimization modeling approach, their research work analyzed the economic viability of centralized and decentralized dismantling scenarios in Europe and concluded that a decentralized facility scenario can reduce logistics costs and CO_{2e} emissions rather than a centralized configuration of dismantling facilities. Nguyen-Tien, et al. [21] investigated the impact of travel distance of transporting spent batteries on the business and economic viability of building EoL recycling facilities.

Table 1. Studies that report a reverse logistics optimization for spent EV LIBs and the environmental and economic impacts of transporting spent batteries. Adapted from [5,7].

Studies	EoL Stage	Modeling Approach	Decision Variable	Spent Battery Transportation Impact	Geographical Scope	Assumptions
[13]	Collection, recycling, disposal	Mixed-integer linear programing	Optimal facility location	Economic and environmental	China	Travel distances are estimated as straight-line distances between collection and recycling facilities. Baseline emissions not reported.
[17]	Collection, recycling	Mixed-integer linear programing	Optimal investment plan	Economic	Germany	Transportation cost decreases in a decentralized collection facility design.
[18]	Inspection, recycling	Mixed-integer linear programing	Optimal facility location and allocation of demand zones to each facility	Economic	Sweden	Assumes different transport modes to optimize transportation cost.
[19]	Collection, dismantling, recycling	GIS and LCA	Optimal facility location	Environmental	California	Transportation cost value is not specified.
[20]	Dismantling	Economic and environmental assessment	-	Economic and environmental	Europe	Germany as centralized scenario and Spain as decentralized network
[21]	Collection recycling	Material flow analysis Geospatial supply chain model Economic and environmental assessment	Optimal facility location	Economic and environmental	UK	Recycling demand distributed equally between collection sites. Transportation costs assumed from EverBatt model [22,23]

With few peer-reviewed publications on the optimization of reverse logistics for spent EV LIBs, our work aims at addressing three relevant gaps of the current state of reverse logistics research as highlighted above, i.e., the forecasted demand for LIB recycling, optimization of EoL management facilities, and economic and environmental assessments. For that purpose, this study developed a detailed spatial model for reverse logistics in Canada by integrating material flow analysis, geographical information system tools, transportation costs, and life cycle approach for GHG emissions. The contribution of this study is to include new spatial modeling parameters and criteria for spent battery supply chains as follows: (i) local demographical, geographical, and socioeconomic factors to geolocate future spent battery collection sites and allocate recycling demand, (ii) siting location criteria and routing optimization procedures applied to regional recycling clusters for future collection sites, dismantling hubs, and recycling facilities, and (iii) recovering materials from the whole battery pack, i.e., the battery cell and the balance-of-system components.

With the fast growth of LIB demand, EV LIB recycling will play a critical role in the sustainability of the electrification of the transport sector in Canada and elsewhere. The 2030 Emissions Reduction Plan presents the goal to reach a new climate target of cutting emissions by 40% below 2005 levels by 2030 in order to achieve net-zero emissions by 2050. Transportation accounted for 25% of total GHG emissions in 2019, and its carbon footprint has risen 16% in the last 17 years. A key milestone is to reach 100% zero-emission electric vehicles (ZEVs) in total light-duty vehicle (passenger cars and light commercial vehicles) sales by 2035 [24]. A ZEV is considered a battery electric vehicle (BEV), plug-in hybrid electric vehicle (PHEV), and fuel cell electric vehicle (FCV). With no commercial deployment for FCVs, BEVs and PHEVs will continue to dominate the ZEV market, and it is expected that it will continue to grow in order to reach Canada's net-zero targets.

Hence, this study aims to understand how the transportation of spent EV LIBs has an influence on the recycling viability in Canada. The purpose of this study is to develop a reverse logistics modeling framework for EoL lithium-ion batteries, combining Geographical Information System (GIS), Material Flow Analysis (MFA), and Life Cycle Assessment (LCA) tools and propose a case study, which provides for the first time a detailed framework for recycling clusters of spent EV LIBs in Canada. The case study proposes an optimized transportation network that minimizes transportation distances to identify the best siting locations for dismantling hubs and recycling facilities and allocates battery mass for recycling while evaluating the associated costs and environmental impacts. Furthermore, the overall life-cycle GHG emissions and costs of EV LIBs at the EoL phase are completed by including life-cycle GHG emissions and the associated costs of transporting spent batteries along the reverse logistics network system. The overall life-cycle GHG emissions of recycling processes to recover EV LIB cathode materials can be compared with the EV LIB cathode production from virgin materials.

2. Methods

A spatial model of a transportation network for spent batteries was developed to identify and optimize siting of dismantling hubs and recycling processing facilities in Canada to minimize the costs and environmental impacts of spent battery transportation by using GIS.

This spatial modeling was integrated with MFA and LCA methods and estimated the quantity of spent EV LIBs viable for recycling by 2040, the transportation distance, and the potential GHG emissions and costs related to transportation of spent batteries from collection centers to EoL processing facilities.

In this study, we focused on the reverse logistics of battery recycling clusters. Figure 1 depicts the overall reverse logistics network considered for one recycling cluster. This included three types of actors or activities: collection, dismantling, and recycling.

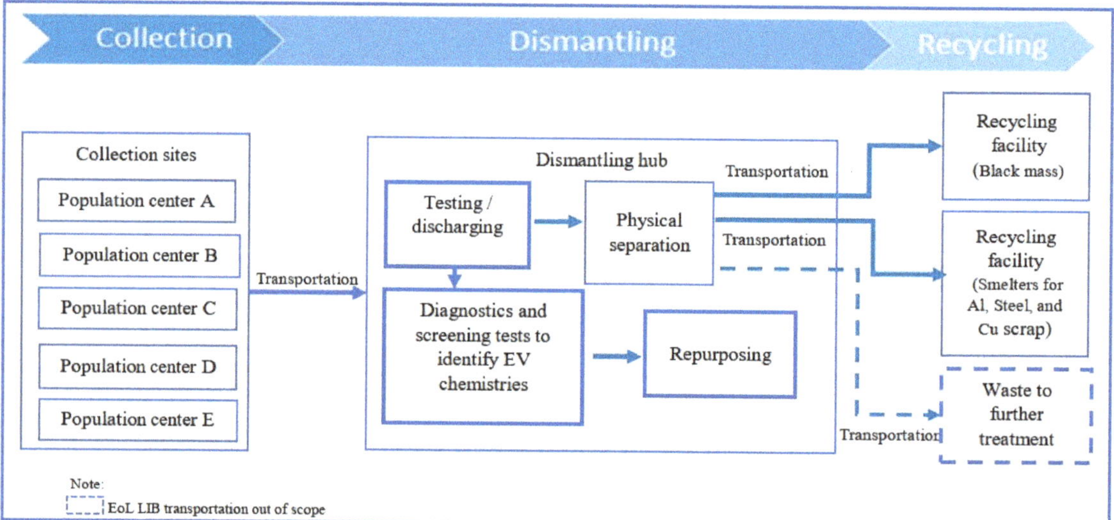

Figure 1. High-level representation of spent EV LIB packs in reverse logistics network in a recycling cluster, which has three main EoL management facilities, predefined collection sites located in major population centers. Recycling demand is allocated to each facility then spent batteries are transported to dismantling hubs to be diverted to repurposing and recycling processes. The battery pack is separated into black mass (battery cell materials) and other metals before being sent to recycling facilities.

End-of-life EV LIBs are collected and transported to dismantling hubs for discharging and testing processes to assess suitability for either repurposing or recycling [25]. Spent EV LIBs are transported to a recycling facility after physical separation (dismantling). Our model did not include the transportation of batteries from EV users to the collection points, which is a common practice in the literature [26].

Because only battery cells should be shipped to a centralized battery recycling facility for raw materials recovery [19], the main pathways identified for the spent battery stream (Figure 1) started with collection of these spent LIB packs that are transported to dismantling hubs to be physically separated by shredding. Second-life applications, such as repurposing of end-of-life batteries, is an optional pathway that would potentially modify the distribution of the batteries before recycling, and this would have implications on the facility optimization process described in this study. Hence, the repurposing pathway was considered in the current study by estimating an annual percentage of the total EV battery packs to be repurposed.

After dismantling the battery pack, the battery cell components, such as anode, cathode, carbon black, and binder, are crushed to produce a black mass (filter cake). The cathode chemistry will affect the relative Li, Ni, Mn, Co, and graphite contents of this black mass. Plastics and electrolytes from the battery cell are sent to further waste treatments. The battery cell's nonhazardous components, such as aluminum from the cathode current collector, positive terminal assembly, and cell container and copper from the anode current collector and negative terminal assembly, may be separated and transported to smelters depending on whether pyrometallurgical or hydrometallurgical recycling processes are used. In the case of a pyrometallurgical process, the nickel and cobalt in the black mass and copper can be recovered, but aluminum in the black mass is lost. In the case of a hydrometallurgical process, the black mass does not contain aluminum and copper. The final destinations of the battery cells (as black mass) are centralized recycling processing facilities to recover metal sulfates (pyrometallurgical process) or cathode materials (hydrometallurgical process) to be reused in cathode and battery cell production, respectively.

The balance-of-system components (BOS), often referred to as battery management systems, module and battery pack terminals, heat conductors, and module and pack enclosures, are separated from the battery cell and are also dismantled, sorted, and shredded to separate metals, such as aluminum, copper, and steel scraps. These metals are sold and transported by truck to a few dedicated smelters across Canada that process them as secondary material inputs. Transportation to smelting facilities was included in the scope of this study. However, other BOS components, such as plastics, electronic scraps, used cables, and used printed wiring boards, were considered waste materials and sent to waste treatment facilities. The transportation of these waste materials for further treatment was out of the scope of this study.

2.1. Spatial Modeling

Building on reverse logistics studies [15,21,23], a spatial modeling using GIS network analysis software was developed. This model enabled the development of an efficient transportation network of spent batteries along their supply chain including collection sites through dismantling hubs and recycling processing facilities. Transportation costs and GHG emissions can be reduced by identifying the shortest path between several origins and destinations. Network optimization in this study was performed using ArcGIS Pro© 3.0.1, software developed by Esri Inc., Toronto, ON, Canada.

Initially, a modeling framework combining GIS, MFA, and LCA was developed, and it is schematically summarized in Figure 2. The GIS optimization of dismantling hubs and recycling locations comprised four steps. First, network datasets were created; second, origin and destination sites were identified; third, the mapping of different sites was established; and finally, location-allocation was optimized [19,27]. Once the four steps were completed, optimized truck transportation distances from the collection sites to the recycling processing facilities were applied to the life-cycle GHG emission intensity of trucks on road transportation networks. This was done to obtain life-cycle environmental emissions related to the transportation of spent EV batteries, which complemented the life-cycle emission values of the battery recycling processing. Additionally, route heat maps were generated to highlight the hot spots of route transportation of each reverse logistics segment.

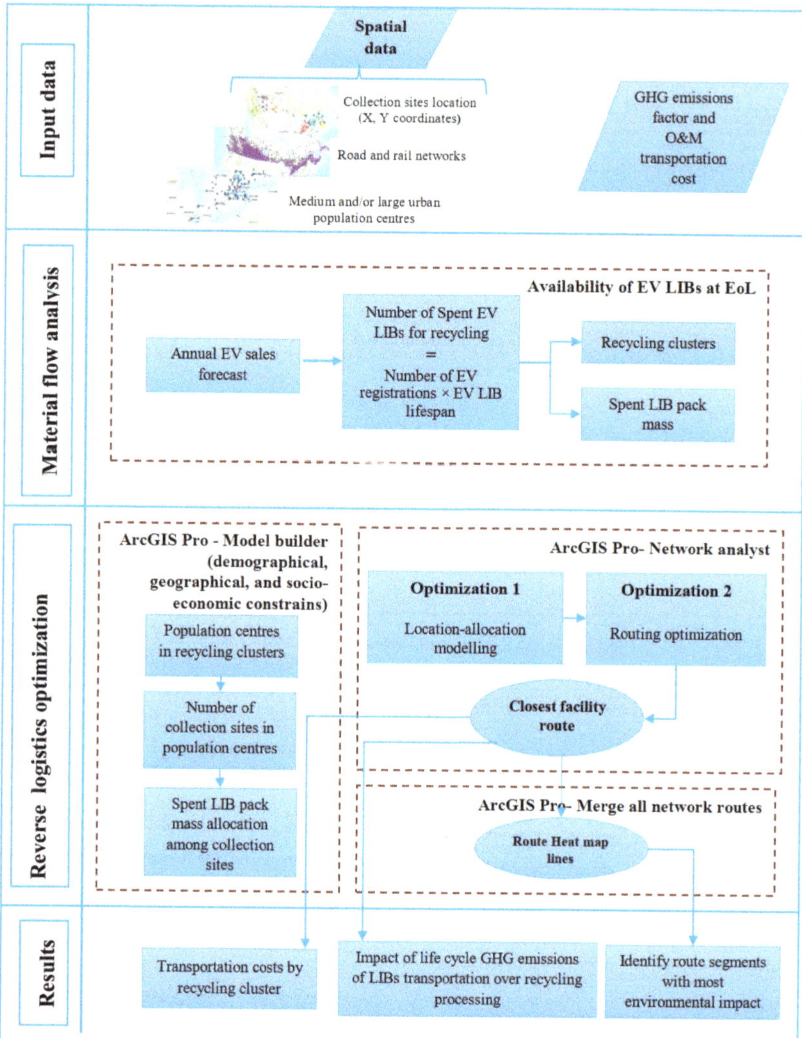

Figure 2. Proposed assessment framework combining GIS, MFA, and LCA tools to model transportation of spent EV batteries in a recycling cluster. This framework comprises three modules starting with input data (geospatial data, parameters and assumptions, and environmental and economic factors) follow by an MFA module to estimate recycling demand. Input data and MFA modules feed the reverse logistics optimization module to optimize EoL facility locations. Results: Module analyzed the economic and environmental impacts on recycling viability.

For this GIS-based spatial modeling approach, some assumptions were made to optimize locations and capacities of facilities, which directly affected truck transportation distances. The following GIS model criteria and constraints were listed for this study: (i) Availability of spent EV LIBs for EoL management by Canadian provinces, which depends on the annual demand forecast of EV LIBs packs; (ii) Allocation of available spent battery mass among collection sites in urban population centers. Distribution of the spent battery pack supplies among different collection sites may depend on demographic, geographic, and socioeconomic constraints; (iii) Black mass available for the recycling processing is based on average battery mass components contribution, which depends on battery chemistry; (iv) Potential number of dismantling and recycling facilities considers a flexible plant capacity with an estimated minimum and maximum plant capacity throughput based on its economic feasibility; (v) Placement of dismantling and recycling facilities to minimize the environmental impacts from on-site emissions should consider economic impacts, such as (a) access to a qualified labor force and (b) end point utilization with access to critical infrastructure, such as rail transportation systems, that may facilitate transportation to battery production facilities; (vi) GIS network analysis depends on the transportation travel type and mode (single/intermodal) and the location–allocation problem type under the network analysis tool.

Changes in criteria of facility placement are expected to have a significant impact on travel distances [19]. Consequently, this will affect GHG emissions and transportation costs.

2.1.1. Availability of Electric Vehicle LIBs at EoL

The spatial transportation model criteria were integrated to a material flow analysis (MFA) to estimate the number of spent EV LIBs that were available for collection. The MFA methodology used in the Moore, Russell, Babbitt, Tomaszewski and Clark [27] study, which used a market lifespan approach [27,28], was selected to be applied in this study. LIBs reaching their EoL were estimated based on past registrations of new EVs and average values of EV LIB life span (Equation (1)).

$$S_{B,t} = B_{(t-l)} \times P_l \tag{1}$$

$S_{B,t}$ represents the number of spent EV LIBs that are reaching EoL and are available for collection in year t. $B_{(t-l)}$ reflects the number of LIB packs starting useful life in EVs registered in the past year $(t - l)$ based on the expected LIB lifespan in years (l), and P_l reflects the percentage of EV LIBs that will reach their EoL after l year of lifespan. It was assumed that each EV includes a single LIB pack, and EoL options are recycling and repurposing (reuse) with an estimated share of 70% and 30%, respectively, over the total waste stream in year t. [14,29]. EV average lifespan distribution data (l and P_l) were adapted from Richa, et al. [30]. This study considered four lifespans of 6, 8, 10 and 15 years.

In this study, the collection of spent EV Li-ion batteries was limited to zero-emission electric vehicles (ZEVs) ($S_{B,t}$), including full battery electric vehicles (BEVs) and plug-in hybrid electric vehicles (PHEVs) as defined by Transport Canada [31]. To estimate the quantity of batteries placed on the market each year, the historical number of new ZEVs registered in Canada from 2011 to 2021 by Statistics Canada [32] and the ZEV forecast from 2022 to 2040 [33] were considered. On 1 May 2019, the Government of Canada launched the national rebate program as part of the Incentive for Zero-Emission Vehicles, offering a rebate of up to CAD 5,000 with the purchase of a zero-emission vehicle. ZEV reached a 5.2% share of total light-duty vehicle registrations in 2021 [34]. In addition, BEVs and PHEVs represented 68% and 32%, respectively, of the total new ZEV registrations in 2021. Table S1 in the Supplementary Information shows the annual new ZEV registrations from 2011 to 2021.

In Canada, the new 2030 Emissions Reduction Plan presents the goal to reach a new climate target of cutting emissions by 40% below 2005 levels by 2030 in order to achieve net-zero emissions by 2050 [24]. Since transportation accounted for 25% of total GHG emissions in 2019, and its carbon footprint has risen 16% in the last two decades, a 100% ZEV sales target by 2035 for light-duty vehicle (passenger cars and light commercial vehicles) was set [24].

The ZEV registrations forecast for 2040 can consider two scenarios, a baseline scenario and a net-zero target scenario. Sections S2 and S3 in the Supplementary Information describe the assumptions for both forecast scenarios, and Tables S2 and S3 summarize the EV LIB inflow and waste streams for recycling for the baseline and net-zero MFA scenarios, respectively.

Annual new electric vehicle registrations in Canada are not the same across all provinces [34]. In 2021, Quebec, Ontario, and British Columbia contributed the most registrations at 43%, 23%, and 28%, respectively, of total ZEV sales. Note that the ZEV incentive program, which is a key driver of EV adoption, is a mandate in British Columbia and Quebec as there is an Internal Combustion Engine (ICE) ban by 2030 and 2040, respectively. However, Ontario, the largest province by population, has no ZEV mandate, and there is no ICE ban [35].

Battery pack availability across Canada at the provincial level by 2040 was estimated by assuming the market share of the number of ZEVs by province in 2040 would be the same as the market share in 2021 [34]. Based on this estimated spent battery distribution by province, this study assumed three geospatial centralized recycling clusters to place dismantling hubs and recycling processing facilities in Canada with the capacity to satisfy the annual disposed battery flows in Canada. One was located in the West (British Columbia) covering spent batteries from four provinces (BC and the Prairies provinces AB, MB, and SK), one was located in Ontario (covering Ontario), and one was located in Quebec (covering QC and the Maritimes provinces NB and NS). The centralized recycling cluster scenarios assumed that batteries are transported to processing facilities from different provinces to allow for larger centralized facilities that would benefit from economies of scale and easier access to batteries and markets for recovered materials beyond a specific province. Tables S4 and S5 indicate the battery mass allocation among provinces for a baseline scenario and a net-zero target scenario, respectively. For this study, the spent battery mass allocation baseline scenario was used for the reverse logistics optimization case study developed in the following sections.

2.1.2. Reverse Logistics Optimization

To model the reverse logistics, an optimization analysis was used to estimate the optimal number of dismantling hubs and recycling processing facilities and their locations in the road transportation networks of the West and East recycling clusters. The objective function of location-allocation optimization was to minimize the transportation distances, expressed as total ton–kilometers transported, between the collection sites and the intermediate and final EoL processing destinations.

Allocation of Available Spent Battery Mass among Collection Sites in Population Centers

Some spent batteries may come through official original equipment manufacturer (OEM) channels and other car dealers, given the need for trained mechanics to service EVs. However, car dealerships are not required to take back spent EV batteries for disposal. Rather, most EV batteries at EoL will come through the auto dismantling or auto recycling supply chain [36]. In Canada, the Automotive Recyclers of Canada Association (ARC) represents approximately 400 end-of-life vehicle recycling and dismantling facilities throughout Canada. Among them, some typical vehicle scrapyard facilities have started to collect an increasing number of spent EV batteries. At that point, scrapyard staff at these facilities have been receiving training on EV battery EoL management, including diagnostics and safety discharging of EV batteries [36]. This study assumed that spent EV battery collection facilities would be confined to pre-existing EV scrapyards across Canada, which can provide the infrastructure to store, dismantle from EVs, and handle EV batteries properly. A registry of automotive scrapyards exists online through the website of the Automotive Recyclers of Canada Association [37]. The geospatial data layer of EV scrapyards is illustrated in Figure S1 in the Supplementary Information and shows a map of the geolocations of spent EV battery collection sites across Canada.

The disposed EV battery pack mass per provincial cluster for the baseline scenario was estimated in Section 2.1.1 and is shown in Table S4 in the Supplementary Information and is allocated among the different collection sites located in population centers (PCs) across Canada, to which geospatial data are provided by Statistics Canada [38,39] and illustrated in Figures S2 and S3 in the Supplementary Information. In order to identify strategic geolocations for collection sites to decrease payload distances, demographic, geographic, and socioeconomic constraints were applied by using geoprocessing tools. Collection sites would be located within 30 km outside the borders of medium and large urban PCs, whose population size is more than 30,000 based on population counts from the 2016 census [40], and applied a weighted allocation of the spent battery pack mass based on household income data. Using ModelBuilder of ArcGIS Pro© 3.0.1, software developed by Esri Inc., Canada, a geoprocessing model was built to automate workflows that string together sequences of geoprocessing tools, feeding the output of one tool into another tool as input [41]. A description of the workflow for the allocation of battery mass among collection sites is presented in Section S7 in the Supplementary Information.

Location-Allocation of Dismantling Hubs and Recycling Processing Facilities and Routing Optimization

To optimize the closest facility siting, transportation network datasets were set up using ArcGIS Pro© software environment. Building geospatial road transportation networks and other required geospatial data, such as locations of major cities, rail stations, and borders of cities, were sourced from transport networks as digital cartographic reference products produced by Natural Resources Canada [42,43] and Statistics Canada [44].

The locations of the dismantling and recycling facility candidates were assumed to be industrial zones located within the medium and large urban population centers, serving as a useful approximation for the purposes of this study [45]. Table S6 in the Supplementary Information shows the potential dismantling and recycling facility location candidates. An important feature of the model is that it allows the user to limit the number of facility candidates to those that fulfill certain siting criteria, such as proximity to large PCs and rail infrastructure. At this stage, the entire spent EV battery packs are transported to dismantling hub facilities where the black mass, i.e., battery cell components (cathode/anode materials), is produced and assumed to be 41% of the total battery pack mass allocated to each dismantling facility chosen. The black mass is delivered to either traditional recycling processing facilities to recover Co, Ni, and Cu (pyrometallurgical plants) or Li, Ni, Co, and Mn (hydrometallurgical plants). Plastics and electrolytes from battery cells and solvents and electronic scrap from BOS represent 20% of the battery pack mass and are delivered to a waste treatment facility. The remaining 39% of the battery pack mass from the battery cell and balance of system containing nonhazardous materials (Cu, Al, and steel scraps) are sent to recycling processing using smelters, where copper, aluminum, and steel scraps represent 18%, 19% and 2%, respectively, of the total battery pack mass allocated to each dismantling facility chosen. NMC811 was assumed to be the only EV LIB chemistry in the spent battery stream because it is one of the predominant battery chemistries in the current EV market [46]. Mass contribution of the EV LIB pack was based on Dai, Kelly, Gaines and Wang [22]'s study, which presented detailed material compositions at the cell, module, and pack levels.

We assumed that the actual smelters used as recycling processing facilities to process aluminum, copper, and steel scraps were located in the West and East clusters. Table S7 in the Supplementary Information shows the aluminum, steel, and copper smelting facility location candidates.

To identify the best locations for dismantling and recycling processing facilities, an algorithm was developed for the geospatial optimization of the dismantling and recycling facility sites using a location-allocation methodology that integrated the economic and environmental metrics into the segments of a GIS network [19,47]. The location-allocation optimization process was applied by using the ArcGIS Network analysis tool in ArcGIS Pro© software, in which locations of the collection sites represented where the demand point locations and potential dismantling sites were considered as facilities. For the other GIS segments, the dismantling facilities chosen were the demand points, and the recycling and smelter locations were the facilities. This optimization problem that is to minimize weighted impedance (P-Median) was solved by the location-allocation solver. Given N candidate facilities and M demand points with a weight, we chose a subset of the facilities, P, such that the sum of the weighted distances from each M to the closest P was minimized based on the Dijkstra's algorithm [48]. The values of the fields' demand weight and impedance corresponded to the allocated battery mass to each demand point and travel distance between the demand point and facility chosen, respectively. The allocated battery mass for the recycling and smelting facilities was calculated using geoprocessing tools in ArcGIS Pro©.

Routing optimization was used to generate route stops from location-allocation outputs. This network analysis type in ArcGIS Pro© optimized and merged the routes of the overlapping three GIS segments into one and summed their weight, i.e., collection sites to dismantling hubs chosen, dismantling sites chosen to recycling processing facilities chosen, and dismantling sites chosen to smelters, by using the ModelBuilder and the Join, Merge, Buffer, Intersect, and Dissolve geoprocessing tools [49].

2.1.3. Life-Cycle GHG Emissions and Transportation Costs

Utilizing a life cycle perspective in evaluating the overall spent EV battery transportation from collection sites through dismantling hubs to a centralized recycling facility is essential in fully understanding the environmental consequences of this infrastructure expansion. An LCA approach was used in this study to estimate the life-cycle GHG emissions of the reverse logistics of EV LIBs. A transportation system boundary comprised collection sites to recycling processing facilities (battery cell and other battery pack metal recycling processing). In order to integrate transportation LCA results with the life cycle emissions from spent EV battery pack recycling processing, an average life-cycle GHG emissions for recycling processing was assumed as proxy for each recycling cluster with a value of 0.678 kgCO_{2e}/kg of the spent battery pack [50–52]. The assumptions to estimate the GHG emission intensities and the truck transportation unit costs used in this study are explained in Section S10 in the Supplementary Information.

3. Results and Discussion

3.1. Location-Allocation of EoL Processing Facilities in Recycling Clusters and Route Optimization for Transportation of Spent EV LIBs to Recycling Facilities

The first step on the spatial analysis model in this study was to identify the suitable collection sites in population centers for each recycling cluster in Canada based on siting criteria, which included close distance to population centers with a population of more than 50,000. In addition to the criteria to select the recycling clusters explained in Section 2.1.1, these three recycling cluster scenarios were selected based on the following: (i) Three major cluster regions including Ontario or ON covering the center of Canada, Quebec (QC)–Maritimes covering Eastern Canada, and British Columbia (BC)—Prairies covering Western Canada; (ii) Current recycling plants are located in the provinces of BC, ON, and QC as the base of operations [7,53]; and (iii) ON, QC–Maritimes, and BC and Prairies are geoeconomic clusters based on employment (number of jobs) in the electrical equipment manufacturing sector [54]. In the ON, QC–Maritimes, and BC–Prairies recycling clusters, 81, 63 and 107 collection sites, respectively, were filtered over a total of 97, 100 and 140 potential collection sites, respectively. The number of collection sites chosen in the QC–Maritimes cluster was the smallest in comparison with the other two clusters due to the low population density in the PCs located in the Atlantic Provinces (seven collection sites chosen over 25 candidates). The spent battery mass from each Canadian province is allocated to collection sites located in selected PCs using an additional criterion based on the number of PCs with a household annual income over CAD 100,000. In the ON cluster, the three major PCs with the higher battery mass allocated are Toronto, Hamilton, and Ottawa with 16, 16, and 5 collection sites, respectively. Figure 3 illustrates a sample of this analysis for the ON cluster.

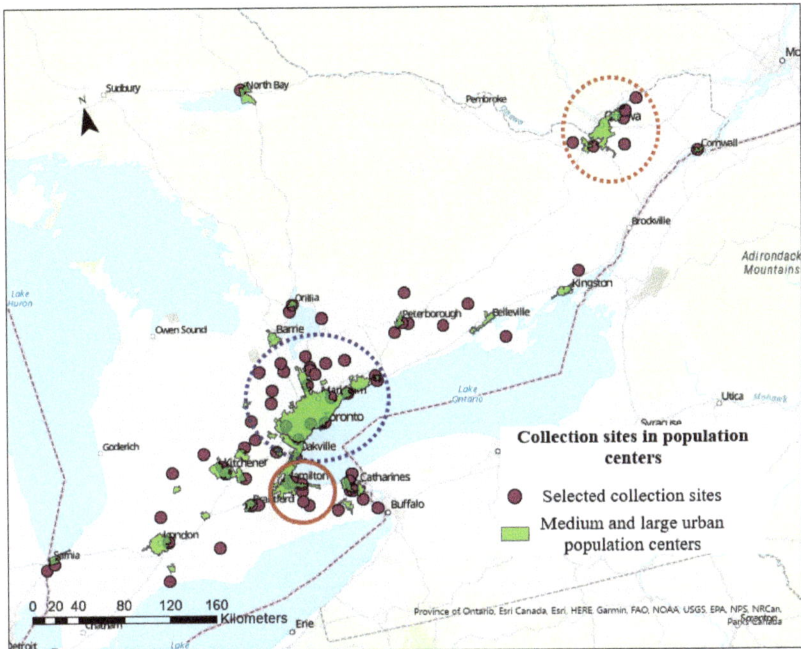

Figure 3. Analysis of a sample spatial analysis of suitable collection sites based on the siting criteria of near distance to large population areas, Ontario recycling cluster. The three major population centers with more collection sites and higher battery mas allocated are Toronto inside the dotted blue circle, Hamilton in the brown circle, and Ottawa inside the dotted red circle.

Moreover, the algorithms considered the battery mass allocation at the collection sites, which is why the optimal dismantling hub is located closer to the zone with the greatest number of scrapyards and the recycling facility is not halfway between the dismantling facilities chosen considering the criterion to be located closest to rail stations. In the end, recycling processing facilities can be located within or near high population centers and with high accessibility to a large quantity of spent LIBs from electric vehicles. The results from the location-allocation optimization of the dismantling hubs, recycling processing plants, and aluminum, copper, and steel smelting facilities are illustrated in Figure 4. The number of facilities (dismantling hubs and recycling processing plants) were iterative and estimated in order to minimize the truck transport distance between the demand points and the facilities chosen.

In the ON recycling cluster, the best locations for dismantling hubs are Toronto, Hamilton, Barrie, London, Oshawa, and Ottawa. The Toronto, Hamilton, and Barrie dismantling hubs represent 26%, 23% and 19%, respectively, of the total battery mass allocated in the ON cluster. The dismantling hub located in Toronto met the criteria to be selected as the centralized recycling processing facility in the Ontario recycling cluster. Because all available aluminum smelters are in Quebec, all the dismantling hubs located in Ontario feed the Alcoa Lte aluminum smelter located in Bécancour, QC. The dismantling hubs located in Oshawa, Barrie, and Toronto feed the steel smelter of Gerdau Whitby Steel Mill in Whitby; meanwhile, the dismantling hubs located in London and Hamilton feed the steel smelter of Stelco located in Hamilton, and the Ivaco Rolling Mills Ltd. steel smelter located in L'Original, ON is fed by the dismantling hub located in Ottawa. All six dismantling hubs feed the Glencore Canada Ltd. copper smelter, Horne Foundry, located in Rouyn-Noranda, QC.

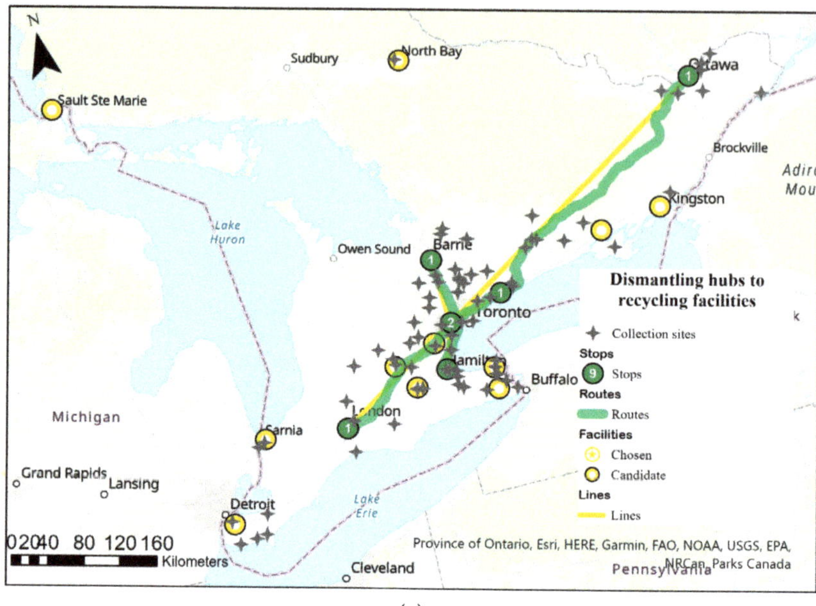

(**a**)

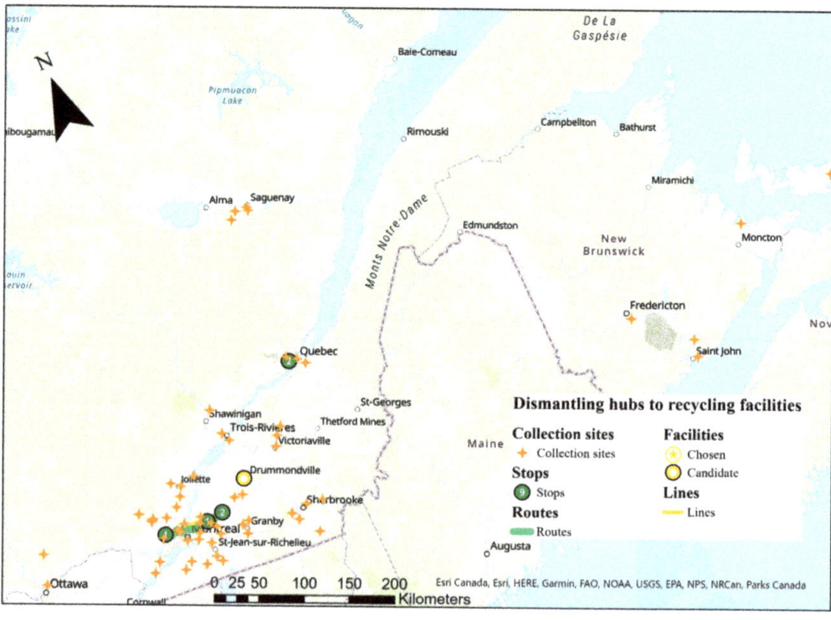

(**b**)

Figure 4. *Cont.*

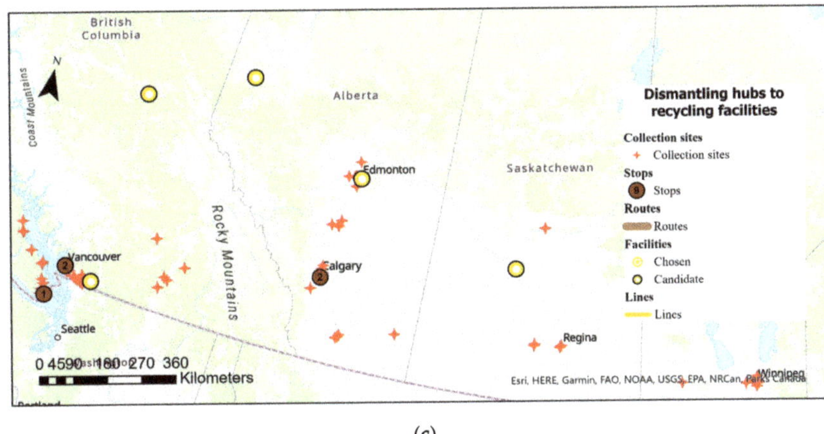

(c)

Figure 4. Geographical location of collection sites and optimal locations of dismantling hubs and recycling processing facilities. Transit routes from EV LIB dismantling hub to the closest recycling site using road transportation network. Dismantling hubs are identified as stop #1 (origin) and recycling processing facilities as stop #2 (destination), where the routes follow the direction toward recycling facilities. (**a**) Ontario recycling cluster; (**b**) Quebec–Maritimes recycling cluster; and (**c**) British Columbia–Prairies recycling cluster.

In the QC–Maritimes recycling cluster, two dismantling hubs located in Montreal and Beloeil feed the recycling facility in St. Hyacinthe, QC, and the dismantling hub located in Quebec, QC is also the same location to operate a recycling facility. The Montreal and Beloeil dismantling hubs represent 41% and 40%, respectively, of the total battery mass allocated in the QC–Maritimes cluster. These two dismantling hubs feed the Alcoa Lte aluminum smelter in Bécancour, QC, and the dismantling hub in Quebec feeds the aluminum Alcoa Lte aluminum smelter in Deschambault, QC. There were two steel smelters chosen; the ArcelorMittal Contrecœur, which is fed by the dismantling hub in Quebec, and the ArcelorMittal Montreal in St. Patrick, which is fed by the two dismantling hubs located in Montreal and Beloeil. All three dismantling hubs feed the Horne Foundry copper smelter located in Rouyn-Noranda, QC.

In the BC–Prairies recycling cluster, three dismantling hubs were chosen: Victoria and Vancouver in BC and Calgary in Alberta, where the dismantling hub in Vancouver represents 76% of the total allocation of battery mass in the BC–Prairies cluster. These dismantling hubs located in BC and AB also met the criteria to be chosen as recycling processing facilities, where the stops 1 (origin) and 2 (destination) have the same geolocation. There is only one aluminum smelter located in Kitimat, BC fed by all the dismantling hubs; similarly there is only one steel smelter in Edmonton, AB. The aluminum smelter belongs to Rio Tinto Alcan and the steel smelter to Alta Steel.

Transportation route optimization was performed for each road transportation network of the three recycling clusters to obtain accurate geodesic travel distances between origin (stop 1) and destination (stop 2) by finding the shortest path between stops. Figure 4 also illustrates the routes from the dismantling hubs to the recycling processing facilities.

3.2. Transportation Payload Distance and Environmental and Economic Impacts

The GIS optimization process is based on the minimization of the total ton–kilometers (t·km) transported between the collection sites and the EoL processing destination. The number of ton–kilometers is the weight in tons of battery pack/component transported multiplied by the number of kilometers driven. This is the appropriate objective function to minimize the economic and environmental costs corresponding to the total distance traveled by applying specific metrics and emission factors.

Currently, truck transportation is assumed as the primary mode for transporting spent batteries. Table S8 in the Supplementary Information presents the truck transportation payload distance of spent EV LIBs from demand points as either collection sites or dismantling hubs to EoL processing facilities (dismantling hubs and recycling and smelting facilities). The total ton–kilometers in the ON, QC–Maritimes, and BC–Prairies recycling clusters represents 19%, 38%, and 43%, respectively, of the total payload distance of the reverse logistics network. BC–Prairies has the longest distance travelled because of the distance from the dismantling facilities to the only aluminum smelter in the BC–Prairies cluster; meanwhile, QC–Maritimes has the second longest distance travelled because of the longest distance from all dismantling facilities to the only copper smelter in Canada.

The payload distance from the collection sites to the dismantling facilities varies among the three clusters. The ON cluster presents the shortest distance (12% of the total distance travelled to the dismantling facilities in all three clusters); meanwhile, QC–Maritimes and BC–Prairies represent 33% and 56%, respectively. In the Ontario scenario, there is a small volume and short distance because the dismantling hubs are located closer to scrapyard areas with more density; meanwhile, the QC–Maritimes cluster presents a long distance, and there are small-volume, out-of-province collection sites in the Maritimes. In the BC–Prairies cluster scenario, there is a long distance from the Prairies and a large volume in BC. This can be impacted by the number of dismantling facilities and differences in demand.

In the Ontario scenario, the truck transportation payload distance from the collection sites to the dismantling hubs and then to a recycling facility is 27% of total ton–kilometers transported within the ON cluster. The travelled distance from the dismantling facilities to the aluminum and copper smelters represents 73% of the total ton–kilometers transported within the ON cluster because there is out-of-province transportation from the dismantling hubs in ON to the smelter facilities in QC.

In the Quebec–Maritimes scenario, transportation from collection sites to the dismantling facilities and from dismantling hubs to the copper smelter facility represents 21% and 62%, respectively, of the total ton–kilometers transported within the QC–Maritimes cluster. Similarly, transportation from the collection sites to the dismantling facilities in the BC–Prairies cluster represents 31% of the total ton–kilometers in the out-of-province scenario in the West cluster, considering that the travelled distance from the dismantling facilities to the aluminum smelting facility represents 63% of the total ton–kilometers transported within the BC-Prairies cluster.

Aluminum, steel, and copper scrap smelter locations have a fixed effect in the model. This is reflected by the fact that most aluminum smelters are in QC. Therefore, the QC–Maritimes travel distance to aluminum smelters is shorter than the ON travel distance. At this point, increasing or decreasing the number of dismantling facilities depends on the demand of spent batteries, which can be controlled for the optimization process. For example, the longest ON travel distance to aluminum smelters could potentially be even further if there were more dismantling facilities in ON as consequence of an increase in available spent batteries. Currently, ON is not leading the EV sales in Canada despite being the most populated province. As we move toward 2035 (or 2050) where ICEs are to be removed and EV adoption is tied to population, LIB recycling demand will increase in ON.

The transportation payload distance has a direct influence on the environmental and economic impacts of implementing a recycling end-of-life EV LIB infrastructure.

Life-cycle GHG emission results for transporting 1 kg of the spent battery packs from the EV collection sites to the spent battery processing facilities are indicated in Table S9 in the Supplementary Information. This outcome shows the importance of understanding the environmental impacts of spent battery logistics at different points in the reverse logistics network when assessing the environmental sustainability of spent battery recycling for all three geospatial scenarios. The life-cycle GHG emissions for battery transportation along all the reverse logistics networks in the QC–Maritimes East cluster present the lowest environmental impact regardless of the long distance from the collection centers in the Maritimes because of its low tonnage. Meanwhile, the BC–Prairies West cluster ended up with the highest GHG emissions due to a lack of infrastructure, and the ON East cluster ended up in the middle.

In Figure 5, the life-cycle environmental impact for transporting 1 kg of spent EV LIB in the ON, QC–Maritimes and BC–Prairies recycling clusters is 0.063, 0.045 and 0.081 kg CO_{2e}, respectively. The transportation emissions from the collection sites to the dismantling and recycling facilities represents 27%, 28% and 28%of the total emissions in ON, QC–Maritimes, and BC–Prairies, respectively; meanwhile, the emissions for the transportation of aluminum scrap from the dismantling hubs to a smelting facility represents 41% and 66%, respectively, of total emissions in ON and BC–Prairies.

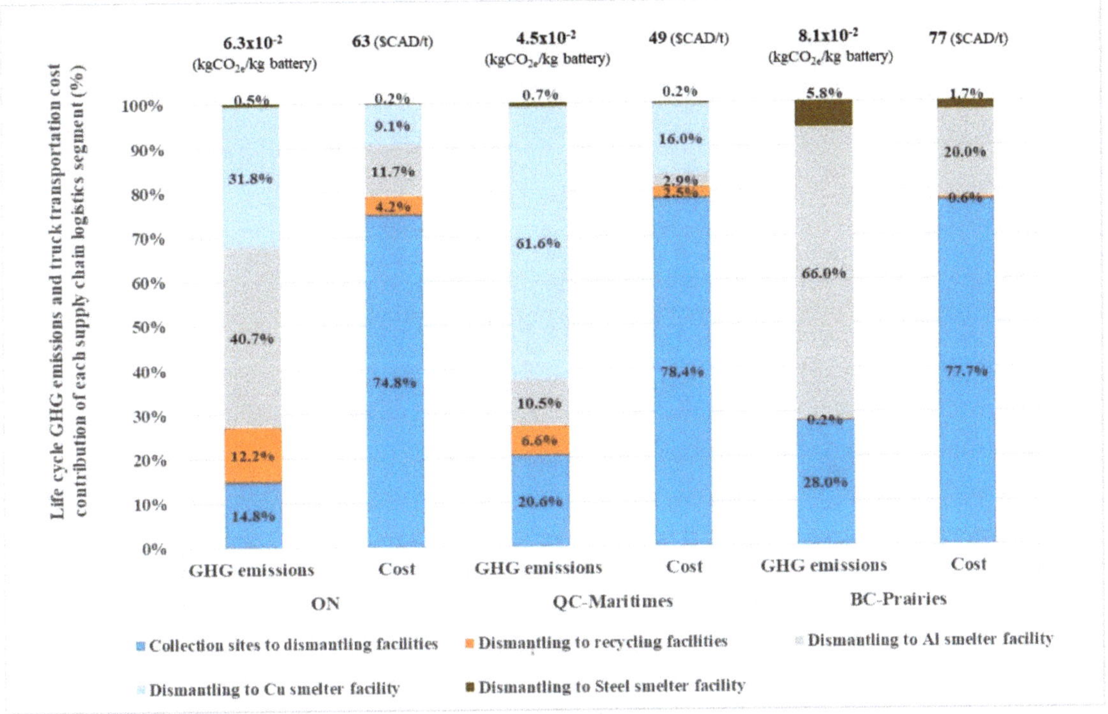

Figure 5. Overall GHG emissions (1 kg of spent EV LIB pack) and cost (CAD/t of spent EV LIB pack) impact of the overall logistics for three regional clusters. Contribution breakdown (%) is provided for each reverse logistics segment (Bottom).

In regards to the integrated life-cycle environmental impact of the spent battery pack at EoL, including transportation and recycling processing in the LCA system boundary, the total life-cycle GHG emissions for 1 kg of the spent battery pack recycling in the ON, QC–Maritimes, and BC–Prairies recycling clusters is estimated to be 1.17, 1.15 and 1.19 kg CO_{2e}/kg battery pack, respectively, where the emissions for transportation represents 5%, 4% and 7%, respectively, of the total emissions by battery pack recycling processes. The relative share of the environmental impact of only battery cathode material recovery, i.e., the total life-cycle GHG emissions of battery cathode materials recovered from recycling processing and emissions of transportation from collection sites to dismantling hubs and then to recycling facilities, accounts for 24% and 7%, respectively, of the total emissions of battery-grade cathode and battery pack production from virgin materials. The share of the life-cycle GHG emissions of the battery cathode of the total emissions of battery pack production from virgin materials accounts for 28%. The average estimated values of life-cycle GHG emissions of recycled cathode raw materials, battery cathode production from virgin materials, and battery pack production from virgin materials (0.7, 2.93 and 10.4 kg CO_{2e}/kg of the spent battery pack, respectively), including transportation are indicated in Section S11 and Table S10 in the Supplementary Information.

The economic costs of the transportation network of the spent battery supply chain for recycling processes is calculated based on the estimated transportation unit cost for each of the segments of the reverse logistics network, i.e., collection sites to dismantling facilities and from the later locations to recycling/smelting facilities, considering Table S11 and the correspondents optimized payload distances in Table S8 in the Supplementary Information.

As a result, the truck transportation costs of 1 t of the spent battery packs from the EV collection sites to the battery processing facilities are indicated in Table S12 in the Supplementary Information. Taking into account only the transportation costs of the spent battery cells for recycling, i.e., truck transportation from collection sites to dismantling hubs and then to the centralized recycling facility for each geospatial scenario, these costs are CAD 50/t, CAD 40/t, and CAD 60/t for the ON, QC–Maritimes, and BC–Prairies clusters, respectively, which include the regular loading/unloading and distance-dependent travel costs, as well as the handling fee and other costs for transporting class 9 hazardous goods, such as LIBs. The average estimated transportation costs of the spent batteries to recycling facilities in Canada is CAD 50/t of the spent battery pack, which represents 4% of the total operating costs of battery recycling. Based on the study by Baxter [55], the operating costs of spent battery recycling processing is estimated at CAD 1244/t of the spent battery pack for a hydrometallurgical recycling facility in Canada. Hence, the average estimated cost of recycled cathode raw materials is CAD 1.29/kg of the spent battery pack, including transportation.

In Figure 5 showing the ON recycling cluster, truck transportation costs to the dismantling facilities represent 74.8% of the total truck transportation; meanwhile in QC–Maritimes and BC–Prairies, these costs contribute 78.4% and 77.7% of the total truck transportation, respectively. By adding the transportation costs of scrap metal recycling, i.e., truck transportation from the dismantling facilities to aluminum, copper, and steel smelting facilities for each geospatial scenario, the total truck transportation costs of the spent battery pack recycling increase by CAD 13.3/t, CAD 9.4/t, and CAD 16.6/t for the ON, QC–Maritimes, and BC–Prairies clusters, respectively. Note that all the aluminum smelting candidates in the East cluster are located in Quebec and the aluminum scrap from the dismantling facilities in Ontario is transported to the chosen aluminum smelter in QC, thereby increasing the payload distance. In the West cluster, however, there is only one aluminum smelter candidate in northern BC, and it is assumed, one steel smelter candidate in Alberta that increased the total payload distance. Importing copper to a closer smelter facility in the US was not evaluated. The QC–Maritimes cluster presents 29% and 44% less life-cycle carbon footprint than the ON and BC–Prairies clusters, respectively. The total truck transportation costs are 22% and 36% lower than ON and BC–Prairies clusters, respectively.

Overall, the truck transportation from the collection to dismantling facilities has a significant impact on the overall costs but much less on the GHG emissions. However, this conclusion may change if fuel prices increase due to the introduction of the carbon tax and higher crude oil prices.

3.3. Carbon Intensity of Transportation Routes

The optimized transportation routes from collection sites to dismantling hubs, dismantling hubs to recycling processing facilities, and dismantling hubs to smelters, were merged as a single route network using different geoprocessing tools (Merge, Buffer, Intersect, and Dissolve) of ArcGIS Pro© software. This helped with the generation of a linear heat map [56] as $kgCO_{2e}$/km, linking hot spots of GHG emissions with the most frequently traveled trucking routes. These linear heat maps for each recycling cluster are shown in Figure 6. Linear heat maps show high-intensity lines (hot spot) in red and low-intensity lines in yellow.

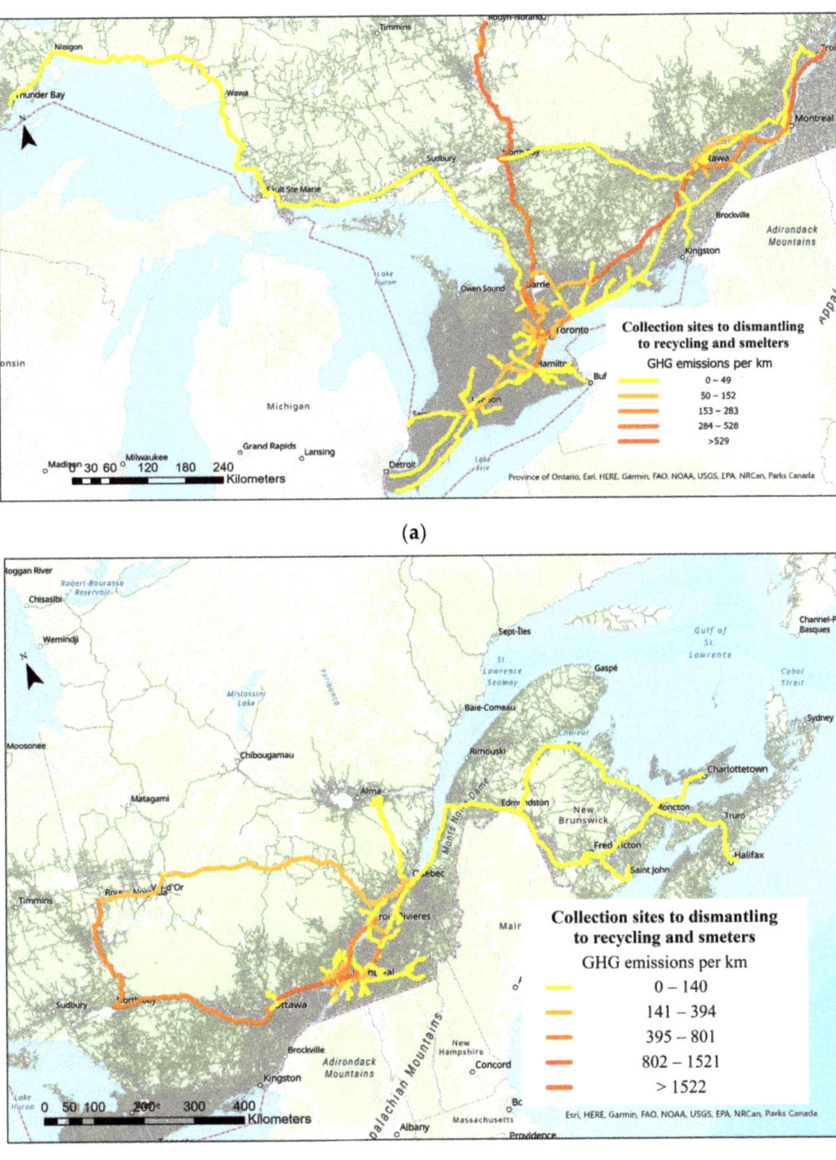

Figure 6. *Cont.*

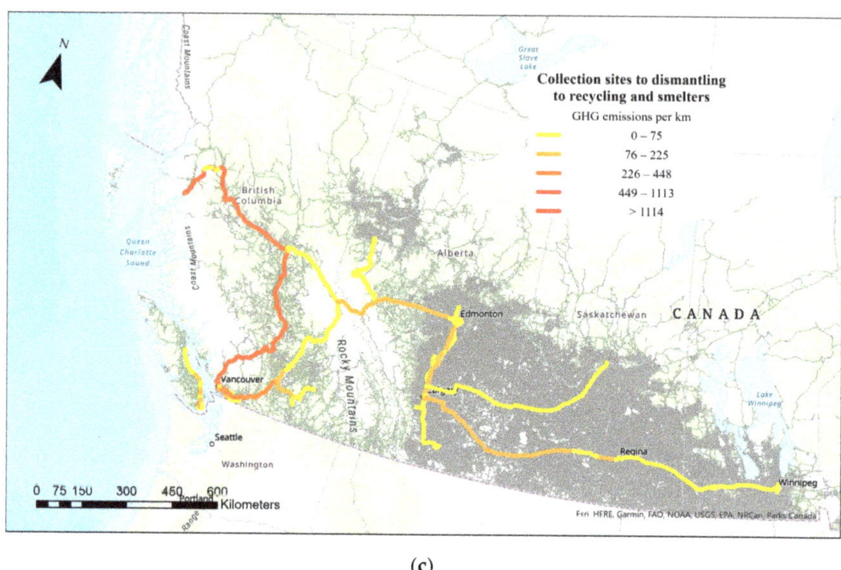

(**c**)

Figure 6. Linear heat map of carbon intensity of merged transportation routes of collection sites to dismantling hubs to recycling processing and smelting facilities for the Ontario, Quebec–Maritimes, and the BC–Prairies recycling clusters. (**a**) Ontario recycling cluster; (**b**) QC–Maritimes recycling cluster; and (**c**) BC–Prairies recycling cluster.

Overall, the low-intensity lines (in yellow) correspond to the routes of collection sites to dismantling hubs for all three recycling clusters. However, the emission intensity ranges vary among clusters where the QC–Maritimes cluster has the highest range up to 140 kg CO_{2e}/km and Ontario presents the lowest, with up to 49 kg CO_{2e}/km due to the scattered location of the collection sites in the QC–Maritimes cluster, a similar scenario to the BC–Prairies cluster. There are mild-intensity lines (in brown) that reach the location of the centralized recycling processing facility, i.e., Toronto (ON cluster) and St. Hyacinthe, (QC–Maritimes cluster) because of the concurrent routes from collection sites to dismantling hubs and from dismantling hubs to recycling processing facilities. In the QC–Maritimes cluster, there are mild-intensity lines from the dismantling hubs to the aluminum, copper, and steel smelters. In the BC–Prairies cluster, there are route segments with mild-intensity lines from the dismantling hubs in Vancouver, BC and Calgary, AB to the steel smelter in Edmonton, AB. The emissions intensities per km travelled are in the range of 153–283, 395–801, and 226–448 kg CO_{2e}/km. The high-intensity lines (in red) correspond to the routes of dismantling hubs to aluminum smelters and dismantling hubs to copper smelters for the ON cluster; the segment routes from the collection sites to the dismantling hub in Montreal integrated with the segment routes from the Montreal dismantling hub to the copper smelter; and the route segment from the dismantling hub in Vancouver, BC to the aluminum smelter in Kitimat in the BC–Prairies cluster.

3.4. Forecasted Recycling Processing Capacity

A baseline MFA scenario for spent EV LIBs was used in this study to forecast the annual number of LIBs available for recycling during the time period of 2022 to 2040. In this scenario, the number of EVs in Canada, specifically ZEVs, is expected to increase at a CAGR of 10%, and the number of LIBs for recycling in 2040 would be 130 times the potential current level (Table S2). Likewise, under a net-zero MFA scenario for the same period of time, which is described in Section 2.1.1, the quantity of EV registrations would grow at a CAGR of 25% to meet Canada's 2035 goal of reaching a 100% share of net-zero EVs over total passenger vehicle sales (Table S3). Under the analysis of these two scenarios and regardless of the uncertainty inherent to the assumptions made to determine long-term forecasts, the results suggest a substantial increase in spent EV LIBs to be diverted to recycling processing facilities in Canada by 2040. The challenges of building an EV battery EoL infrastructure are even greater in terms of facility capacity, GHG emissions, and economic costs per kilometer. The annual recycling processing capacity of the ON, QC–Maritimes, and BC–Prairies recycling clusters estimated in the net-zero MFA scenario would increase by a factor of 4.5, 4.6 and 4.3, respectively, in comparison with the annual capacity estimated in the baseline MFA scenario. The GHG emissions and transportation costs per kilometer present the same number of increments per recycling cluster. Keeping the location of dismantling hubs and recycling processing facility candidates and the optimal locations of the collection sites simulated for the baseline MFA scenario, the allocation of the battery mass of the net-zero MFA scenario was used to calculate the optimal recycling processing capacity and payload transportation distances. The estimated annual recycling processing capacity under these two scenarios by 2040 is illustrated in Figure 7.

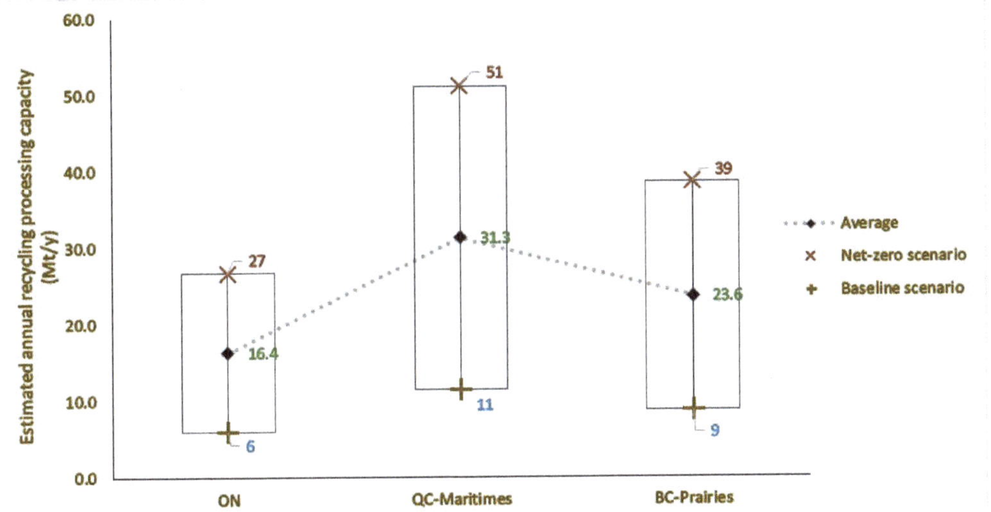

Figure 7. Estimated annual recycling processing facility capacity for the Ontario, Quebec–Maritimes, British Columbia–Prairies recycling clusters assumed to be viable by 2040 under the baseline and net-zero MFA scenarios. The lowest and highest potential recycling processing capacities presented in Ontario and QC–Maritimes clusters reflects the level of viability of local supply of spent EV LIBs in these provinces.

Recycling profitability can be achieved through economies of scale; for example, in a UK recycling facility, the profitability threshold could increase to >50,000 tons per year for pyrometallurgical and circa 17,000 tons per year for hydrometallurgical recycling [16]. In Figure 7, the baseline scenario presents low levels of economic viability for all the recycling clusters; hence, complementing this with an out-of-country recycling model would be an alternative to compare with the levels of viability of the in-country recycling model presented in the net-zero MFA scenario.

3.5. Further Work

The reverse logistics model of EV LIBs described in this work is flexible and can be adapted to integrate other jurisdictions in North America. The integration of US processing facilities closer to Canadian facilities could generate different optimal EoL management locations and optimal transportation routes. The network routing optimization in this model could be enhanced by integrating an LIB tracing and tracking system to ensure safe and environmentally friendly EoL battery management. The development of block chain technology solutions is an opportunity to increase efficiency and transparency in the spent battery supply chain through securely monitoring and answering the questions where, when, and how a spent battery can be collected to be transported to an EoL management facility [57]. Addressing battery traceability issues avoids the increased risk of diverting spent batteries to landfills and recycles them instead. Nevertheless, policy frameworks and regulatory mechanisms need to be addressed in terms of the final destination of the batteries [29].

In the context of the circular economy, the recovery of critical and strategic metals, such as cobalt and nickel, to produce cathode LIBs in Canada could generate a more reliable and complete set of environmental and economic data to improve hotspot analyses for the recycling process of spent EV batteries [58]. A closed-loop approach can be applied by integrating this spatial model, and its findings relate to the environmental implications of transporting spent batteries to recycling facilities with a life cycle assessment of spent EV LIB recycling processing options and its environmental credits as a result of recovering battery cathode material to be reused in the production phase of LIBs. A regionalized LCA of specific geographical locations, such as the provinces of Ontario, Quebec, and British Columbia, would be performed to be consistent with the battery transportation GHG emissions calculated in this study.

4. Conclusions

This study presents a spatial modeling framework to quantify the environmental and economic effects of the expansion of the supporting infrastructure network for EV LIB end-of-life management in Canada, based on the integration of geographic information system, material flow analysis, life cycle assessment (truck transportation emission intensity), and lithium-ion battery truck transportation costs. Although battery recycling processing options will be critical to diverting EV batteries from EV waste streams, the overall impacts of the reverse logistics (collecting, dismantling, and recycling) of these batteries must be considered. Because the collection and transportation of electric vehicle batteries have an important contribution on the environmental and economic overloads of the end-of-life infrastructure, there is a need for an optimal management and responsive recovery at end-of-life logistics.

The main conclusions of this case study are the following:

The allocation of battery mass at the collection sites, the road network travel distance minimization, and the facility placement criteria are some of the most important parameters that define the optimal locations of the dismantling hubs and recycling processing facilities and enable the analysis of the environmental and economic impacts on the end-of-life infrastructure network for EV LIBs.

The assumptions considered in the model design criteria have impacted the results of this study. Spent battery mass availability assumptions would change with EV market share projections, evolving battery chemistries, and used battery imports from other provinces or countries outside Canada. Using different ZEV adoption policies and government incentives could also adjust the total battery mass generated and the spatial distribution, thus affecting the optimal facility locations, transportation distances, and overall environmental and economic impacts.

In the three regional clusters for the collection and transportation of spent EV batteries in Canada, the transportation network optimization provided the best solution to allocate spent batteries to dismantling facilities and to locate the closest recycling/smelting facilities based on the payload distance parameter.

Life-cycle carbon footprints and transportation costs were calculated for all the optimized routes along the spent battery supply chain for recycling processes in Canada. The recycling cluster in Quebec presents the lowest life-cycle GHG emissions and transportation unit costs in comparison to the two recycling clusters located in Ontario and British Columbia due to Quebec having the largest share in EV registrations, which is largely influenced by government incentives.

The overall life-cycle GHG emissions of the spent battery pack recycling was obtained by adding life-cycle GHG emissions from transportation to emissions resulting from battery cell recycling processing and other metal recovery. In terms of the relative share of the environmental impacts of only battery cathode recycling, GHG emissions for transportation account for 2% to 3% of total life-cycle GHG emissions of battery cathode recycling. Furthermore, the latter represents 7% of the total life-cycle GHG emission of battery pack production from virgin materials that is lower than the 28% of the environmental impact share of the battery cathode production from virgin materials.

Truck transportation from the collection to the dismantling facilities had a significant impact on the overall costs but much less on the GHG emissions. There was a differentiation of transportation costs between the collection sites to the dismantling facility segments in comparison with the other segments in the transportation network due to the added costs for the complexity of hazardous material transportation of spent LIB packs. The same GHG emission factor was applied to each segment Thus, the proposed optimal facility siting that minimizes transport distance to dismantling hubs was suggested to reduce the overall costs of transportation, ensure safety compliance, and facilitate the feasibility of building a recycling facility. However, it is also important to take into account the regional regulatory framework related to the operational transportation costs that may lead to set up place-specific parameters.

Supplementary Materials: The following supporting information can be downloaded at: https://www.mdpi.com/article/10.3390/su142215321/s1. Supplementary data: EV demand forecast tables, geospatial data inputs, estimated truck transportation distances, GHG emissions and costs tables. References [59–62] are cited in the supplementary materials.

Author Contributions: Conceptualization, G.G.-C., B.Y. and F.B.; Data curation, G.G.-C.; Formal analysis, G.G.-C.; Investigation, G.G.-C.; Methodology, G.G.-C.; Project administration, B.Y.; Supervision, B.Y. and F.B.; Validation, G.G.-C.; Writing—original draft, G.G.-C.; Writing—review and editing, G.G.-C., B.Y. and F.B. All authors have read and agreed to the published version of the manuscript.

Funding: This research received no external funding.

Institutional Review Board Statement: Not applicable.

Informed Consent Statement: Not applicable.

Data Availability Statement: Data supporting reported results in Supplementary Information File.

Acknowledgments: The authors gratefully acknowledge financial support from the Office of Energy Research and Development (OERD) of the Natural Resources Canada (project number NRC-19-109) and the Advanced Clean Energy (ACE) program of the National Research Council of Canada.

Conflicts of Interest: The authors declare no conflict of interest.

References

1. IEA. *Net Zero by 2050-A Roadmap for the Global Energy Sector*; International Energy Agency: Paris, France, 2021.
2. Ziemann, S.; Müller, D.B.; Schebek, L.; Weil, M. Modeling the potential impact of lithium recycling from EV batteries on lithium demand: A dynamic MFA approach. *Resour. Conserv. Recycl.* **2018**, *133*, 76–85. [CrossRef]
3. Bobba, S.; Bianco, I.; Eynard, U.; Carrara, S.; Mathieux, F.; Blengini, G.A. Bridging Tools to Better Understand Environmental Performances and Raw Materials Supply of Traction Batteries in the Future EU Fleet. *Energies* **2020**, *13*, 2513. [CrossRef]
4. Mayyas, A.; Steward, D.; Mann, M. The case for recycling: Overview and challenges in the material supply chain for automotive li-ion batteries. *Sustain. Mater. Technol.* **2019**, *19*, e00087. [CrossRef]
5. Alipanah, M.; Saha, A.K.; Vahidi, E.; Jin, H. Value recovery from spent lithium-ion batteries: A review on technologies, environmental impacts, economics, and supply chain. *Clean. Technol. Recycl.* **2021**, *1*, 152–184. [CrossRef]
6. Rajaeifar, M.A.; Ghadimi, P.; Raugei, M.; Wu, Y.; Heidrich, O. Challenges and recent developments in supply and value chains of electric vehicle batteries: A sustainability perspective. *Resour. Conserv. Recycl.* **2022**, *180*, 106144. [CrossRef]

7. Slattery, M.; Dunn, J.; Kendall, A. Transportation of electric vehicle lithium-ion batteries at end-of-life: A literature review. *Resour. Conserv. Recycl.* **2021**, *174*, 105755. [CrossRef]
8. Wilson, M.; Goffnett, S. Reverse logistics: Understanding end-of-life product management. *Bus. Horiz.* **2022**, *65*, 643–655. [CrossRef]
9. Gaines, L.; Richa, K.; Spangenberger, J. Key issues for Li-ion battery recycling. *MRS Energy Sustain.* **2018**, *5*, 12. [CrossRef]
10. Kellner, R.; Goosey, E. Chapter 9 The Recycling of Lithium-ion Batteries: Current and Potential Approaches. In *Electronic Waste Management (2)*; The Royal Society of Chemistry: Cambridge, UK, 2020; pp. 246–277.
11. Li, L.; Dababneh, F.; Zhao, J. Cost-effective supply chain for electric vehicle battery remanufacturing. *Appl. Energy* **2018**, *226*, 277–286. [CrossRef]
12. Thompson, D.L.; Hartley, J.M.; Lambert, S.M.; Shiref, M.; Harper, G.D.J.; Kendrick, E.; Anderson, P.; Ryder, K.S.; Gaines, L.; Abbott, A.P. The importance of design in lithium ion battery recycling—A critical review. *Green Chem.* **2020**, *22*, 7585–7603. [CrossRef]
13. Wang, L.; Wang, X.; Yang, W. Optimal design of electric vehicle battery recycling network—From the perspective of electric vehicle manufacturers. *Appl. Energy* **2020**, *275*, 115328. [CrossRef]
14. Neubauer, J.; Smith, K.; Wood, E.; Pesaran, A. *Identifying and Overcoming Critical Barriers to Widespread Second Use of PEV Batteries*; NREL- National laboratory of the U.S. Department of Energy: Washington, DC, USA, 2015; p. 93.
15. Richa, K.; Babbitt, C.W.; Nenadic, N.G.; Gaustad, G. Environmental trade-offs across cascading lithium-ion battery life cycles. *Int. J. Life Cycle Assess.* **2017**, *22*, 66–81. [CrossRef]
16. Lander, L.; Cleaver, T.; Rajaeifar, M.A.; Nguyen-Tien, V.; Elliott, R.J.R.; Heidrich, O.; Kendrick, E.; Edge, J.S.; Offer, G. Financial viability of electric vehicle lithium-ion battery recycling. *iScience* **2021**, *24*, 102787. [CrossRef]
17. Hoyer, C.; Kieckhäfer, K.; Spengler, T.S. Technology and capacity planning for the recycling of lithium-ion electric vehicle batteries in Germany. *J. Bus. Econ.* **2015**, *85*, 505–544. [CrossRef]
18. Tadaros, M.; Migdalas, A.; Samuelsson, B.; Segerstedt, A. Location of facilities and network design for reverse logistics of lithium-ion batteries in Sweden. *Oper. Res.* **2022**, *22*, 895–915. [CrossRef]
19. Hendrickson, T.P.; Kavvada, O.; Shah, N.; Sathre, R.; Scown, C.D. Life-cycle implications and supply chain logistics of electric vehicle battery recycling in California. *Environ. Res. Lett.* **2015**, *10*, 014011. [CrossRef]
20. Rallo, H.; Sánchez, A.; Canals, L.; Amante, B. Battery dismantling centre in Europe: A centralized vs decentralized analysis. *Resour. Conserv. Recycl. Adv.* **2022**, *15*, 200087. [CrossRef]
21. Nguyen-Tien, V.; Dai, Q.; Harper, G.D.J.; Anderson, P.A.; Elliott, R.J.R. Optimising the geospatial configuration of a future lithium ion battery recycling industry in the transition to electric vehicles and a circular economy. *Appl. Energy* **2022**, *321*, 119230. [CrossRef]
22. Dai, Q.; Kelly, J.C.; Gaines, L.; Wang, M. Life Cycle Analysis of Lithium-Ion Batteries for Automotive Applications. *Batteries* **2019**, *5*, 48. [CrossRef]
23. Dai, Q.; Spangenberger, J.; Gaines, L.; Kelly, J.C.; Wang, M. *EverBatt: A Closed-Loop Battery Recycling Cost and Environmental Impacts Model*; Argonne National Laboratory: Lemont, IL, USA, 2019; p. 88.
24. ECCC. *Canada 2030 Emissions Reduction Plan: Canada's Next Steps for Clean Air and a Strong Economy*; Environment and Climate Change Canada: Gatineau, QC, Canada, 2022.
25. Sommerville, R.; Zhu, P.; Rajaeifar, M.A.; Heidrich, O.; Goodship, V.; Kendrick, E. A qualitative assessment of lithium ion battery recycling processes. *Resour. Conserv. Recycl.* **2021**, *165*, 105219. [CrossRef]
26. Yeh, A.G.-O.; Chow, M.H. An integrated GIS and location-allocation approach to public facilities planning—An example of open space planning. *Comput. Environ. Urban Syst.* **1996**, *20*, 339–350. [CrossRef]
27. Moore, E.A.; Russell, J.D.; Babbitt, C.W.; Tomaszewski, B.; Clark, S.S. Spatial modeling of a second-use strategy for electric vehicle batteries to improve disaster resilience and circular economy. *Resour. Conserv. Recycl.* **2020**, *160*, 104889. [CrossRef]
28. Kalmykova, Y.; Berg, P.E.O.; Patrício, J.; Lisovskaja, V. Portable battery lifespans and new estimation method for battery collection rate based on a lifespan modeling approach. *Resour. Conserv. Recycl.* **2017**, *120*, 65–74. [CrossRef]
29. Propulsion. *Study of Extended Producer Responsibility for Electric Vehicle Lithium-Ion Batteries in Quebec*; Propulsion Quebec: Montreal, QC, Canada, 2020; p. 98.
30. Richa, K.; Babbitt, C.W.; Gaustad, G.; Wang, X. A future perspective on lithium-ion battery waste flows from electric vehicles. *Resour. Conserv. Recycl.* **2014**, *83*, 63–76. [CrossRef]
31. Transport Canada. Zero Emission Vehicles. 2022. Available online: https://tc.canada.ca/en/road-transportation/innovative-technologies/zero-emission-vehicles (accessed on 10 March 2022).
32. Statistics Canada. Table 20-10-0021-01 New motor vehicle registrations. 2022. Available online: https://www150.statcan.gc.ca/t1/tbl1/en/tv.action?pid=2010002101 (accessed on 2 May 2022).
33. Balyk, J.; Livingston, B.; Hastings-Simon, S.; Bishop, G. *Driving Ambitions: The Implications of Decarbonizing the Transportation Sector by 2030. Commentary 604*; C.D. Howe Institute: Toronto, ON, Canada, 2021; p. 32.
34. Statistics Canada. New Motor Vehicle Registrations: Quarterly Data Visualization Tool. 2022. Available online: https://www150.statcan.gc.ca/n1/pub/71-607-x/71-607-x2021019-eng.htm (accessed on 10 March 2022).
35. Frost&Sullivan. *Strategic Analysis of the Canadian EV Market, 2019*; Frost&Sullivan: San Antonio, TX, USA, 2020; p. 66.

36. Kelleher Environmental. *Research Study on Reuse and Recycling of Batteries Employed in Electric Vehicles: The Technical, Environmental, Economic, Energy and Cost Implications of Reusing and Recycling EV Batteries*; Kelleher Environmental: Toronto, ON, Canada, 2019.
37. ARC. Registry of Automotive Scrapyards. Available online: https://autorecyclers.ca/ (accessed on 5 September 2021).
38. Statistics Canada. 2016 CENSUS—Population Centers—Cartographic Boundary Files. 2019. Available online: https://www12.statcan.gc.ca/census-recensement/2011/geo/bound-limit/bound-limit-2016-eng.cfm (accessed on 2 October 2021).
39. Statistics Canada. Population and Dwelling Counts, for Canada, Provinces and Territories, and Population Centres, 2016 and 2011 Censuses—100% data. 2018. Available online: https://www12.statcan.gc.ca/census-recensement/2016/dp-pd/hlt-fst/pd-pl/Table.cfm?Lang=Eng&T=801&S=47&O=A (accessed on 4 October 2021).
40. Statistics Canada; Population Centre (POPCTR). Dictionary, Census of Population, 2016. 2017. Available online: https://www12.statcan.gc.ca/census-recensement/2016/ref/dict/geo049a-eng.cfm (accessed on 2 October 2021).
41. ESRI. Use ModelBuilder. Available online: https://pro.arcgis.com/en/pro-app/2.7/help/analysis/geoprocessing/modelbuilder/modelbuilder-quick-tour.htm (accessed on 14 March 2022).
42. CANVEC. Transport networks in Canada n.d. Available online: https://open.canada.ca/data/en/dataset/8ba2aa2a-7bb9-4448-b4d7-f164409fe056 (accessed on 1 August 2021).
43. NRCan. National Railway Network—NRWN—GeoBase Series. 2021. Available online: https://open.canada.ca/data/en/dataset/ac26807e-a1e8-49fa-87bf-451175a859b8 (accessed on 2 February 2022).
44. Statistics Canada. Road Network File. 2020; Catalogue no. 2-500-X. Available online: https://www150.statcan.gc.ca/n1/en/catalogue/92-500-X (accessed on 2 October 2021).
45. Vidovic, M.; Dimitrijevic, B.; Ratkovic, B.; Simic, V. A novel covering approach to positioning ELV collection points. *Resour. Conserv. Recycl.* **2011**, *57*, 1–9. [CrossRef]
46. Crenna, E.; Gauch, M.; Widmer, R.; Wäger, P.; Hischier, R. Towards more flexibility and transparency in life cycle inventories for Lithium-ion batteries. *Resour. Conserv. Recycl.* **2021**, *170*, 105619. [CrossRef]
47. Corbett, J.J.; Hawker, J.S.; Winebrake, J.J. *Development of a California Geospatial Intermodal Freight Transport Model with Cargo Flow Analysis*; California Air Resources Board; the California Environmental Protection Agency: Sacramento, CA, USA, 2010.
48. ESRI. Location-allocation Analysis Layer. Available online: https://pro.arcgis.com/en/pro-app/2.7/help/analysis/networks/location-allocation-analysis-layer.htm (accessed on 14 March 2022).
49. ESRI. Route Analysis Layer. Available online: https://pro.arcgis.com/en/pro-app/2.7/help/analysis/networks/route-analysis-layer.htm (accessed on 14 March 2022).
50. Aichberger, C.; Jungmeier, G. Environmental Life Cycle Impacts of Automotive Batteries Based on a Literature Review. *Energies* **2020**, *13*, 6345. [CrossRef]
51. Wernet, G.; Bauer, C.; Steubing, B.; Reinhard, J.; Moreno-Ruiz, E.; Weidema, B. The ecoinvent database version 3 (part I): Overview and methodology. *Int. J. Life Cycle Assess.* **2016**, *21*, 1218–1230. [CrossRef]
52. Cusenza, M.A.; Bobba, S.; Ardente, F.; Cellura, M.; Di Persio, F. Energy and environmental assessment of a traction lithium-ion battery pack for plug-in hybrid electric vehicles. *J. Clean. Prod.* **2019**, *215*, 634–649. [CrossRef]
53. Baum, Z.J.; Bird, R.E.; Yu, X.; Ma, J. Lithium-Ion Battery Recycling–Overview of Techniques and Trends. *ACS Energy Lett.* **2022**, *7*, 712–719. [CrossRef]
54. Statistics Canada. Canadian Cluster Mapping Portal. 2016. Available online: https://www.clustermap.ca/app/scr/is/ccmp/web#!/en/map/CMA/2020/2024 (accessed on 15 August 2022).
55. Baxter, C.; Plugatyr, A.; Yu, B. Processing Black Mass into a Mixed Nickel/Cobalt Hydroxide Precipitate and Industrial Grade Lithium Carbonate: A Techno-Economic Analysis Perspective. In Proceedings of the 61st Annual Conference of Metallurgists, Montreal, QC, Canada, 21–24 August 2022; p. 11.
56. ESRI. Creating a Linear Heatmap with Track Data. Available online: https://www.esri.com/arcgis-blog/products/tracker/field-mobility/creating-a-linear-heatmap-with-track-data/ (accessed on 14 May 2022).
57. Reneos. Reneos -Tailor-Made Solutions for Collecting and Recycling Li-Ion Batteries. 2022, p. 16. Available online: https://www.reneos.eu (accessed on 15 August 2022).
58. Larouche, F.; Tedjar, F.; Amouzegar, K.; Houlachi, G.; Bouchard, P.; Demopoulos, G.P.; Zaghib, K. Progress and Status of Hydrometallurgical and Direct Recycling of Li-Ion Batteries and Beyond. *Materials* **2020**, *13*, 801. [CrossRef] [PubMed]
59. NRCan. Aluminum Facts. Available online: https://www.nrcan.gc.ca/our-natural-resources/minerals-mining/minerals-metals-facts/aluminum-facts/20510 (accessed on 5 May 2021).
60. Kelleher Environmental. *Preliminary Resource Recovery Report Card and Gaps Assessment for Canada*; Kelleher Environmental: Toronto, ON, Canada, 2020.
61. NREL. Battery Second Use for Plug-in Electric Vehicles. Available online: https://www.nrel.gov/transportation/battery-second-use.html (accessed on 4 September 2021).
62. ATRI. *An Analysis of the Operational Costs of Trucking: 2020 Update*; American Transportation Research Institute: Austin, TX, USA, 2020.

sustainability

Article

An Assessment of Drivers and Barriers to Implementation of Circular Economy in the End-of-Life Vehicle Recycling Sector in India

Altaf Hossain Molla , Hilal Shams, Zambri Harun * , Mohd Nizam Ab Rahman and Hawa Hishamuddin

Department of Mechanical and Manufacturing Engineering, Faculty of Engineering and Built Environment, Universiti Kebangsaan Malaysia (UKM), Bangi 43600, Malaysia
* Correspondence: zambri@ukm.edu.my; Tel.: +60-3-8921-6518

Abstract: The circular economy (CE) has been frequently in the news recently, as it offers a regenerative system that substitutes the end-of-life concept with restoration. Despite several benefits yielded by the CE from a triple-bottom-line perspective, India's end-of-life vehicle (ELV) recycling sector is striving against numerous impediments to implementing the CE approach. Therefore, this paper attempts to shine a spotlight on India's ELV recycling sector, to identify the potential drivers and barriers to CE implementation. This study has employed an explorative approach to determine the impediments and drivers regarding implementing CE in India's ELV recycling sector. This research reveals that economic viability (25 percent), environmental degradation (17 percent), and global agenda (15 percent) are the three leading primary drivers. In contrast, limited technology (18 percent), financial constraints (15 percent), and a lack of knowledge and expertise (12 percent) are significant barriers that thwart CE implementation in India's ELV recycling sector. This paper has made the first attempt to explore the drivers and barriers to implementing CE in the ELV recycling sector in India. Therefore, besides advancing our understanding of opportunities for and threats to implementing CE, this investigation may assist the Indian authorities in devising appropriate policies and strategies and developing a regulatory and legal framework that is conducive to CE and sustainability.

Keywords: end-of-life vehicle (ELV); recycling; circular economy (CE); drivers; barriers

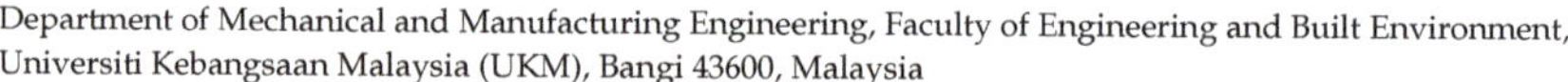

Citation: Molla, A.H.; Shams, H.; Harun, Z.; Ab Rahman, M.N.; Hishamuddin, H. An Assessment of Drivers and Barriers to Implementation of Circular Economy in the End-of-Life Vehicle Recycling Sector in India. *Sustainability* **2022**, *14*, 13084. https://doi.org/10.3390/su142013084

Academic Editors: Subrata Hait, Anshu Priya, Varsha Bohra and Nallapaneni Manoj Kumar

Received: 26 August 2022
Accepted: 18 September 2022
Published: 12 October 2022

Publisher's Note: MDPI stays neutral with regard to jurisdictional claims in published maps and institutional affiliations.

1. Introduction

Globalization, the industrialized economy, continuous technological advancements, and surging demand for products expedite the rate at which vehicles are replaced, consequently engendering an overwhelming increase in abandoned vehicles [1]. Manufacturing vehicles requires a considerable quantity of non-renewable resources and substances. As natural resources are declining exponentially, vehicle manufacturing industries strive to keep raw material supplies steady [2]. Vehicles that are no longer operating, stripped, or wrecked due to mechanical failure, which are known as end-of-life vehicles (ELVs), contain myriad precious, rare, and industrially valuable materials and components. Besides holding significant economic value, ELVs also contain certain pollutants. Hence, sustainably handling ELVs is imperative to avoid their inevitable threat to society. Sustainability, which protects against the dramatic depletion of natural resources and preserves them for long-term application, relies heavily on the products and materials loop [3], which is inevitable in the case of a traditional linear economic system, where products and materials are dumped as waste after usage. For this purpose, several policies are implemented at the production and consumption stages, emphasizing the importance of waste minimization [4]. The annual production of vehicles was around 78 million new cars around the globe in 2020, which indicates a future deluge of ELV needs to handle [5]. ELVs present a risk and menace from an economic, social, and environmental perspective, as nearly 5% of industrial waste is generated by automobiles [4]. They represent hazardous waste with immense potential to pollute the environment if not properly disposed of [6,7]. One such practice is CE, which tends to work for "recycling, reusing, and recovering" materials from the manufacturing to consumption stages, instead of discarding them to pollute the environment, enhancing sustainability through restorative design [8,9] wherein products are reused, recycled, and reduced as an effective strategy for eco-friendly processing of ELVs [10]. The purpose of recovery is to avoid using virgin material resources [11]. CE, considering its triple bottom line, is a challenging field and is still in development with regard to ELVs. Using tactical and strategic principles, optimization, and interconnection among the different stakeholders in the supply chain are crucial and imperative [12] as much waste is generated in terms of personal cars and private entities [13]. This has led researchers to work on all aspects of ELVs, which are considered a growing research area. Several methods are adopted to achieve CE. One such method is closed loop supply chains (CLSCs), where products and materials are reused, recycled, and reduced ("3Rs") to protect the environment and minimize the economic impact [14]. However, adopting such practices in the case of ELVs poses significant challenges while working on materials flow in a closed-loop system [15]. This includes tactical, operational, and strategic stages such as production volume [16], along with government legislation and policies [17], human perception [18], and environmental issues [19]. Table 1 summarizes previous relevant research works in the context of this study.

Table 1. A summary of the previous works.

Author and Year	Objective	Methodology	Focus Area	Reference
Chakraborty et al. (2019)	To find out enablers and barriers in automotive product remanufacturing	Fuzzy interpretive structural modeling (FISM)	Indian market	[20]
Mohan and Amit (2018)	Recycling of ELVs in an unregulated market	System dynamics model	Indian market	[21]
Numfor et al. (2021)	Challenges and opportunities for ELV recycling	Review + SWOT analysis	Developing countries = 8; Nigeria; Cameroon; South Africa; Kenya, Mexico; Malaysia; India; Egypt	[22]
Sharma and Pandey (2020)	(1) To study the Indian ELV sector by focusing on hatchback automobiles (2) Estimating recycling units capability for recycling and recovering materials	SWOT analysis + Primary data collection through Interviews, On-site observations	Mayapuri market, India	[23]
Arora et al. (2018)	How to improve the sustainability of ELV management in the Indian market through a business model by engaging stakeholders	Framework based on shared responsibility including all stakeholders + SWOT analysis	Indian market	[24]
Ravi and Shankar (2017)	To analyze interactions among variables (enablers, indicators, inhibitors) affecting reverse logistics in automobile industries	Interpretive structural modeling (ISM)	Automobile industries	[25]
Luthra et al. (2011)	Identifying barriers to implementing GSCM in the Indian automobile industry	Interpretive structural modeling (ISM)	Indian automobile industry	[26]
Govindan et al. (2016)	Barriers evaluation for Indian automotive parts remanufacturing	The fuzzy analytic network process	Indian automobile industry	[27]
Nag et al. (2021)	Multi-theoretical framework to identify and evaluate drivers and sub-drivers of the circular supply chain in automotive industries in case of emerging economies	Grey DEMATEL	Emerging economies	[28]
Sinha et al. (2017)	(1) Empirical investigation of the Indian automobile industry (2) To identify the reasons behind the slow growth of remanufacturing businesses	Empirical study	India	[29]
Zhou, Fuli et al. (2019)	Identifying drivers in the Chinese ELV recycling business from three perspectives (1) government (2) consumers, and (3) recycling organizations	(1) Literature review (2) Interpretive structural modeling (ISM)	China ELV recycling business	[30]

Table 1. *Cont.*

Author and Year	Objective	Methodology	Focus Area	Reference
Hu, Shuhan and Zongguo Wen (2017)	(1) To evaluate the impact of treatment scenarios on society by analyzing three sectors: (a) the advanced formal sector, (b) the informal sector, and (c) the common formal sector (2) Modeling the cost of ELV treatment on social values in these three scenarios	Data collection through: (1) Literature surveys (2) questionnaires (3) site visits	China	[31]
Mamat et al. (2016)	(1) Developing a framework for establishing an ELV management system	(1) Survey of 300 respondents was factor analyzed using SPSS (2) Structural equation modeling for confirming linear relationship among those factors	Malaysia	[32]
Zhang and Chen (2017)	To prioritize four dismantling strategies for vehicle manufacturers, battery manufacturers, and ELV dismantlers, based on the survey	(1) Survey on ELV stakeholders (2) Analytical hierarchy process (AHP)	China	[33]
Sitinjak et al. (2022)	Identifying social factors for ELV acceptance	(1) Survey (2) SEM was used for the model estimation	Indonesia	[34]
Edun and Vermette (2021)	Investigating the effect of ELV tires by using them as building envelope material for residential building design	(1) Energy performance analysis (2) Life cycle analysis	Canada	[35]

Research suggests that products can be reused after their end-of-life stage by adopting CLSC principles [36]. Countries and organizations are focusing on the concept of the CE from an adoption and implementation point of view for efficient resource utilization. However, CE policies are primarily implemented in developed countries. The developing countries face several barriers in policymaking and when implementing CE principles [28], which are discussed in detail in Section 4. The Indian ELVs market is not regulated, since most firms have not as yet adopted recycling practices [24,28]. Being a vast market, implementing recycling processes and policies is imperative for achieving resource efficiency in the case of automobile components. Despite several benefits offered by the CE model from a triple-bottom-line perspective, India's end-of-life vehicle (ELV) recycling sector is striving against numerous impediments in implementing the CE approach holistically. The literature reveals that limited research has been performed to explore the potential drivers of and barriers to CE implementation in developing countries, especially in India, and little of the research has endeavored to address the issues that arise when implementing the CE in India from other industries' perspectives, based on a particular industrial zone in India. The characteristics of the drivers and barriers to CE implementation vary significantly, depending on the sector and regions; hence, the significance and implications of outcomes of the previously reported research are limited to that particular industry and zone. The lack of a comprehensive study that sheds light on the drivers and barriers of change from the ELV recycling industry perspective is prominent, although the ELV recycling industry in India is burgeoning at an exponential rate as India has one of the world's largest automobile markets.

The absence of an inclusive and general study that embraces prominent ELV recycling sectors across India and that can provide crucial outcomes in a broader sense is forcing the Indian authorities to devise and enact policy, legal, and technical frameworks uniformly across India to counter the issues thwarting the implementation of the CE approach and expedite sustainability in ELV recycling in India. Furthermore, the complete lack of explorative studies encompassing face-to-face interviews and field investigation to record the stakeholders' attitudes and aspirations toward CE and to reflect ground-level practices that yield more precise and reliable data and information about implementing CE in India's ELV recycling sector is conspicuous. Therefore, this paper attempts to shine a spotlight on India's ELV recycling sector holistically by including five prominent automobile hubs across India, to identify the potential drivers and barriers to CE implementation.

An explorative approach has been employed for this study, as it offers more accurate screening by capturing attitudes and recording the aspirations of stakeholders, which are instrumental for decision-making. Besides exploring the enablers and impediments, this research offers specific practical recommendations for implementing and enhancing the CE initiatives in the ELV recycling sector in India. Moreover, these research findings reflect the stakeholders' perceptions and aspirations regarding the ELV recycling system. Therefore, besides advancing our understanding of opportunities and threats for implementing CE, this investigation may assist the Indian authorities in devising appropriate policies and strategies and developing a regulatory and legal framework that is conducive to CE and sustainability. The findings of this study may encourage ELV enterprises and other sectors to embrace the CE framework since it provides a robust financial system that produces significant economic benefits while protecting the environment. The study is based on the following steps:

Step 1: Specific automobile hubs (five selected industrial zones) were focused on as the sample, which is discussed in Section 3.

Step 2: A mixed method research mode was adopted for this study based on primary and secondary research approaches, as shown in Table 2.

Table 2. The approach used to obtain information and data.

Indian City	ELV Sector	Stakeholder	Method and Tool	Sample Size
Kolkata (West Bengal)	• Mallick Bazar • Phoolbagan • Panagarh	▪ Key stakeholder ▪ Academician ▪ Vehicle owner ▪ Government-authorized agency ▪ Expert ▪ Individuals in general	✓ Structured questionnaire ✓ Focused group interviews ✓ Individual in-depth interview ✓ Field investigation	144
Chennai (Tamil Nadu)	• Boarder Thottam • Pudupet	▪ Key stakeholder ▪ Academician ▪ Vehicle owner ▪ Government-authorized agency ▪ Expert ▪ Individuals in general	✓ Structured questionnaire ✓ Focused group interviews ✓ Individual in-depth interview ✓ Field investigation	123
Mumbai (Maharashtra)	• Kurla West • Mumbai	▪ Key stakeholder ▪ Academician ▪ Vehicle owner ▪ Government-authorized agency ▪ Expert ▪ Individuals in general	✓ Structured questionnaire ✓ Focused group interviews ✓ Individual in-depth interview ✓ Field investigation	112
Delhi (Delhi)	• Mayapuri • Jama Masjid	▪ Key stakeholder ▪ Academician ▪ Vehicle owner ▪ Government-authorized agency ▪ Expert ▪ Individuals in general	✓ Structured questionnaire ✓ Focused group interviews ✓ Individual in-depth interview ✓ Field investigation	98
Jamshedpur (Jharkhand)	• Jugsalai • Jamshedpur	▪ Key stakeholder ▪ Academician ▪ Vehicle owner ▪ Government-authorized agency ▪ Expert ▪ Individuals in general	✓ Structured questionnaire ✓ Focused group interviews ✓ Individual in-depth interview ✓ Field investigation	83

Step 3: A structured questionnaire was developed, and data was gathered through surveys based on specific research questions, which is discussed in Section 3.

Step 4: The data is analyzed in detail in Section 4.

Step 5: The managerial implications of the findings and a few practical recommendations are discussed in Sections 5 and 6, respectively.

In order to interlink the literature and the survey results, this study aims to answer the following research questions, based on step 3.

RQ1: What drivers enable CE implementation for the ELV recycling sector in India?

RQ2: What barriers hinder CE implementation for the ELV recycling sector in India?

The article is based on the following structure: Section 2 provides the background to this study, Section 3 gives the methodology, Section 4 is related to the critical findings and analysis, Section 5 provides a discussion of the managerial implications of this study, and Section 6 offers practical recommendations. Section 7 presents the research limitations of this study, Section 8 offers future prospects and suggestions, and Section 9 is the final section, presenting our conclusions.

2. Literature Review

Sustainability is not an easy task to achieve. Waste is being generated at lightning speed around the globe. This has brought the attention and focus of academia and professionals who are looking at sustainability and technology within such a context [37,38]. Consumer priorities and the demand for new vehicles have resulted in a need for the proper disposal and treatment of ELVs. Automobiles are made of several parts and different materials, such as ferrous metals, plastic, glass, rubber, fluids, and other materials [39,40]. In addition, most vehicles are composed of iron, which can be traded and utilized as a valuable secondary resource [41]. However, Tarrar et al. (2021) [40] suggest that a state-of-the-art disassembly unit is imperative for implementing sustainable strategies. This is mainly due to the threats posed to the environment by toxic materials and chemicals during the recycling and recovery process [5]. For such purposes, the treatment process of ELVs comprises depollution, shredding, and dismantling [42]. On the other hand, several strategies, including recycling, reusing, and reducing, commonly known as the 3Rs, are considered the primary way to achieve sustainability in the ELVs industry [30]. The 3Rs have different implementation stages. Reuse is where products and materials can be reused elsewhere, while recycling comes later in the waste hierarchy [5]. As a result, this will help the vehicle industry in terms of efficient resource utilization and saving energy [43]. Therefore, several countries are imposing laws for vehicle recovery by creating legal requirements [6,37,44], as there is a high probability of illegal transportation and processing of ELVs, considering the diverging interests of stakeholders and actors in the end-of-life (EOL) system. The generation of waste at the end of a product's life cycle is a considerable problem exacerbated by increased consumption as the number of vehicles increases around the globe, especially around Asia, considering its immense population [45]. ELVs are generating waste in terms of fuel, volume, resource consumption [46], and environmental threats [19], due to this increased consumption [47]. The ELV's life cycle can be divided into five stages: pre-production, production, distribution, usage, and disposal [48].

Zhang and Chen (2017) [33] conducted a survey based on key stakeholders for ELV by prioritizing dismantling modes, such as direct shredding, manual dismantling, dismantling machines, and disassembly lines for the Chinese market. Their paper was based on a case study approach by selecting particular 17 automobile enterprises involved in manufacturing, dismantling, and recycling in Shanghai, China. Kayikci et al. (2021) proposed a sustainable CE policy based on four aspects: "technology", "consumers", "policy", and "producers", working on a macro level that focused on the automotive industry [49]. Sitinjak et al. (2022) [34] performed a study based on a survey concerning the societal factors affecting the public's acceptance of ELVs in Indonesia. They identified that people have a lack of knowledge and poor attitudes, societal trust, and perceptions about ELV management. Nitish et al. (2019) found that most of the Indian ELV market is not regulated;

the ELV sector is handled informally by scrapping and dismantling parts for recovery, refurbishment, and recycling after being sold to the second-hand market [24].

Tarrar et al. (2021) [40] performed a literature review to study ELV treatment and dismantling systems. They identified four key areas for improvement in ELV management. These areas comprise: (1) battery recycling, (2) workforce for each unit, (3) the structure of ownership and investment, and (4) plastics recycling. Both plastics and battery recycling are related to the economic and environmental aspects of sustainability. The ELV share in total plastic waste is around 5% [6]. D'Adamo et al. (2020) [37] performed a linear regression model identifying the correlation between generated and recycled volumes of ELVs. Their estimated result for generated and recycled ELVs is around 9.3 and 8.3 million tons, respectively, as projected to 2030. Jang et al. (2022) [5] employed statistical data analysis to conduct a study based on sample collection from five ELV shredding facilities in South Korea. They identified two main challenges in the recycling process: (1) a lack of economic incentives, and (2) no proper support systems for low-value materials. Most recently, Nag et al. (2021) [28] investigated the drivers of circular supply chains with a focus on the emerging economy, based on the grey-DEMATEL approach. They recommended several drivers, such as "circular value marketing", "circular services", "circular product design", and "reverse flow drivers". Agyemang et al. (2019) [50] identified several drivers and barriers in the context of Pakistan's automobile manufacturing industry. According to their findings, "profitability share", "cost reduction", and "business concerns" are the top three drivers, while the top three barriers to the implementation of CE principles are "unawareness", "financial and cost constraints", and a "lack of expertise".

Gardas et al. (2018) [51] analyzed the barriers to reverse logistics (RL) in the case of oils obtained from automobile service stations. Their results showed that "inadequate government policies", "a lack of top management commitment", and "organizational policies" were the most significant barriers while adopting the interpretive structural modeling technique. These results were based on developing an interrelationship between obstacles and drivers. J. Li et al. (2014) [38] performed a study on recycling and pollution control in the case of ELVs in China. They studied the current laws and regulations for ELV management, identified several potential problems, and proposed various solutions. Such identified barriers can hinder the transition process of ELV recycling. It is evident that customers will buy refurbished products rather than expensive new ones. As Macarthur (2013) [52] suggested that new product sales may well decrease due to increased sales of refurbished products. Conversely, Mohan and Amit (2018) [21] proposed that the combination of automobiles with enhanced technology, government-enforcing environmental regulations, and high incomes led to shortened life-cycle automobiles, ultimately leading to increasing numbers of ELVs. Another issue is the role of RL in recycling ELVs, as significant barriers can slow or hinder the effectiveness of the whole process [51].

3. Methodology

This study has employed an explorative approach to determine impediments and drivers to implementing CE in the ELV recycling system in India. It incorporates research design and sampling, data collection, quantitative, qualitative, secondary research, and data analysis.

3.1. Research Design and Sampling

This study aims to determine the potential general drivers and barriers to implementing CE in the ELV recycling system in India. This study delves into the five fastest-growing automotive sectors in India's major cities, namely, Kolkata, Chennai, Delhi, Mumbai, and Jamshedpur, to identify the barriers and drivers to India adopting the circular economy in the ELV recycling system in the broader sense. This study has adopted a mixed method that entails quantitative and qualitative research to meet the objectives of this study. For the purposes of determining the sample of individuals concerned, this study uses a stratified and systematic sampling technique. For this investigation, we interviewed a total of

560 selected respondents, incorporating key stakeholders, government agencies, experts, academicians, vehicle owners, and general individuals. The authors developed a well-structured questionnaire that has been reviewed and approved by eminent professionals in the CE and sustainability fields (see Appendices A–C for more details), then sent this questionnaire to the chosen respondents to glean data and relevant information regarding the present research aims. The data were collected and stored in a well-organized way for ease of analysis. For the data analysis, descriptive statistics that employed the weighted-average method, together with inferential statistics, were performed to investigate and develop a profound understanding of the drivers and barriers to implementing CE in the ELV recycling system in India. A thematic analysis has also been carried out to interpret the interviews and group discussions. Figure 1 provides specific demographic information for the survey sample, while Table 2 illustrates the approach used to gather information and data.

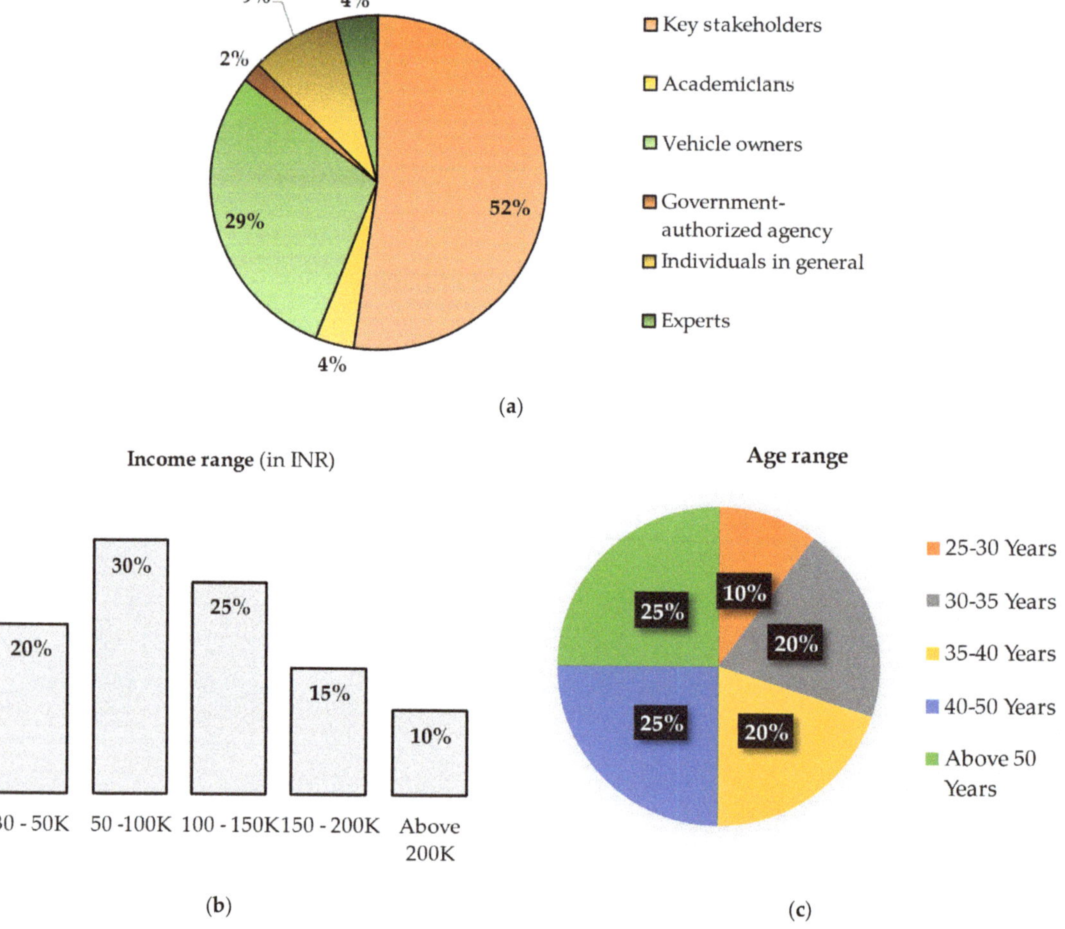

Figure 1. *Cont.*

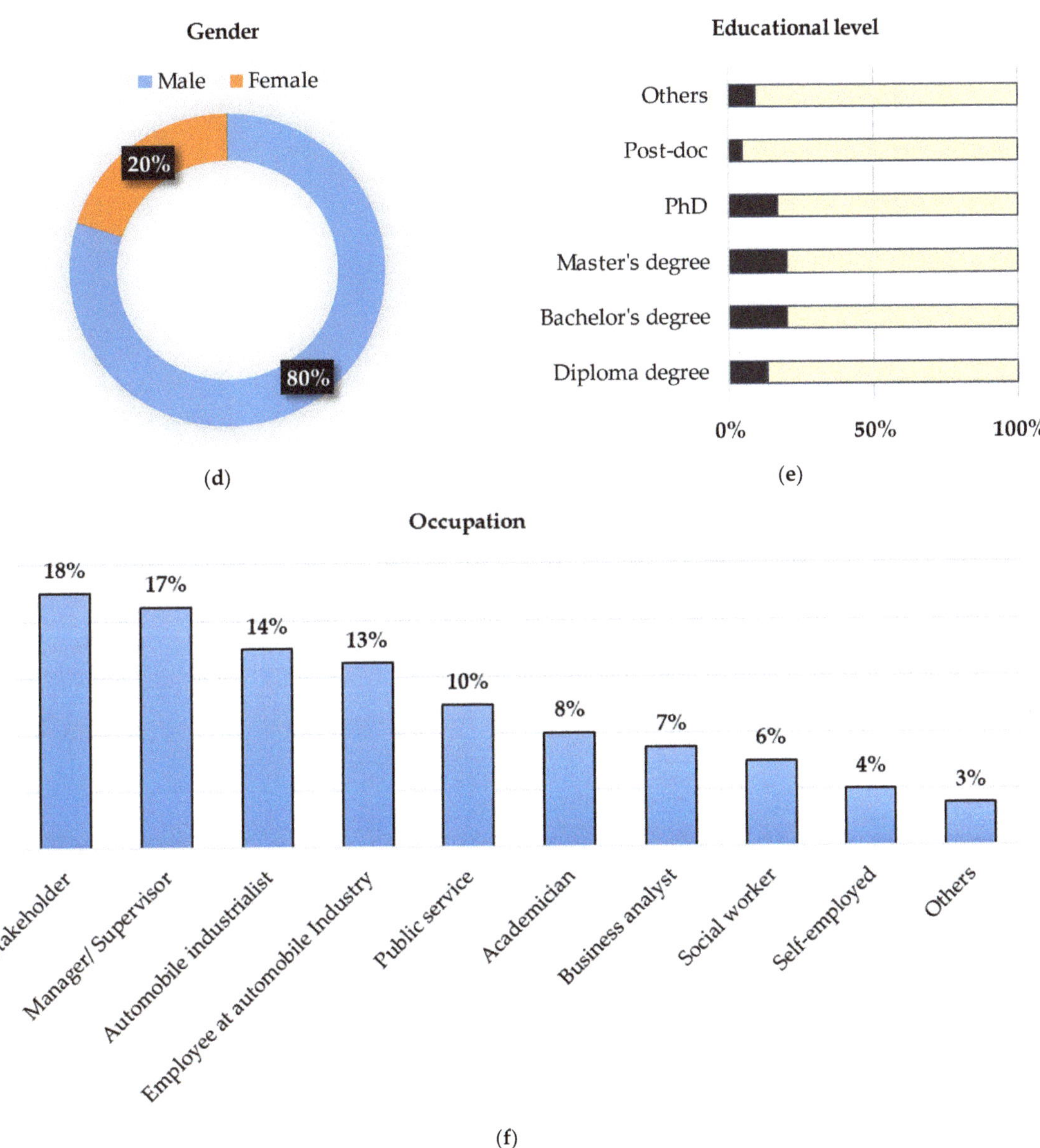

Figure 1. *Cont.*

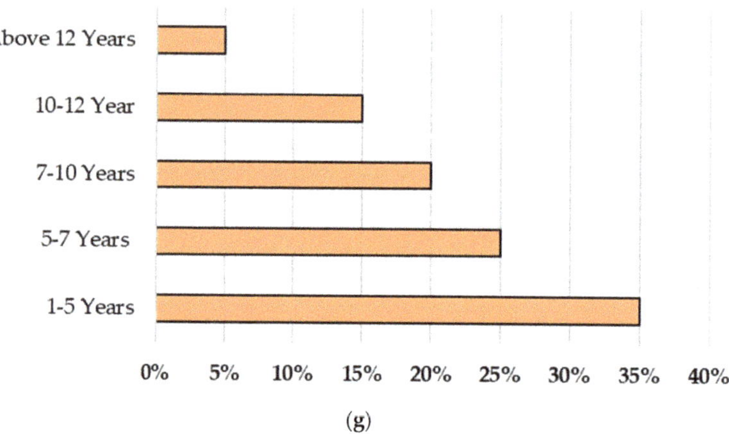

Figure 1. Demographic data for the survey sample: (**a**) interview composition; (**b**) income range; (**c**) age range; (**d**) gender; (**e**) educational level; (**f**) occupation; (**g**) stakeholders' number of years of experience in ELV recycling.

3.2. Data Collection

This study has employed certain methods for data collection. The data collection method for this research entailed interviews with structured questionnaires, focus group interviews, individual in-depth interviews, and field investigations. Table 2 provides details of the procedures for data collection. Figure 2 illustrates the overall methodology used in this research. Structured interviews have been performed by deploying questionnaires to obtain data and information from participants. A pilot test was performed before the actual data collection to assure the accuracy and precision of the data. Focus group interviews have been conducted to elicit general information about the practices, opinions, and aspirations of the stakeholders. Individual in-depth interviews and a two-way organized conversation have been carried out to gather information regarding the objective assessments of this study. Field investigation is instrumental, as it reveals the actual, precise, and authentic information and practice of the organization. This research has conducted field investigations to obtain information about current practices of enterprises, the new strategies the organization will implement, present market trends, raw-material availability, marketing methodologies, the application of existing technology, frameworks, infrastructure, guidelines, and direction.

3.3. Quantitative Research

This study adopts quantitative research methods to conduct objective measurements by sending out a carefully structured questionnaire to the selected survey participants. The survey questionnaire comprises three sections: impediments that thwart the implementation of the circular economy in India's ELV recycling sector, drivers toward the circular economy, and supplementary data for enhancing our in-depth understanding of the subject matter. The collected data were stored systematically. Descriptive statistics were employed to analyze the data. This research has applied the weighted-average method to quantify and analyze the data obtained from the sample. The qualitative data is instrumental in evaluating and identifying the critical factors regarding implementing the circular economy in India's ELV recycling sector and yields critical real-life data that is imperative for ensuring appropriate and effective decision-making and strategies [50].

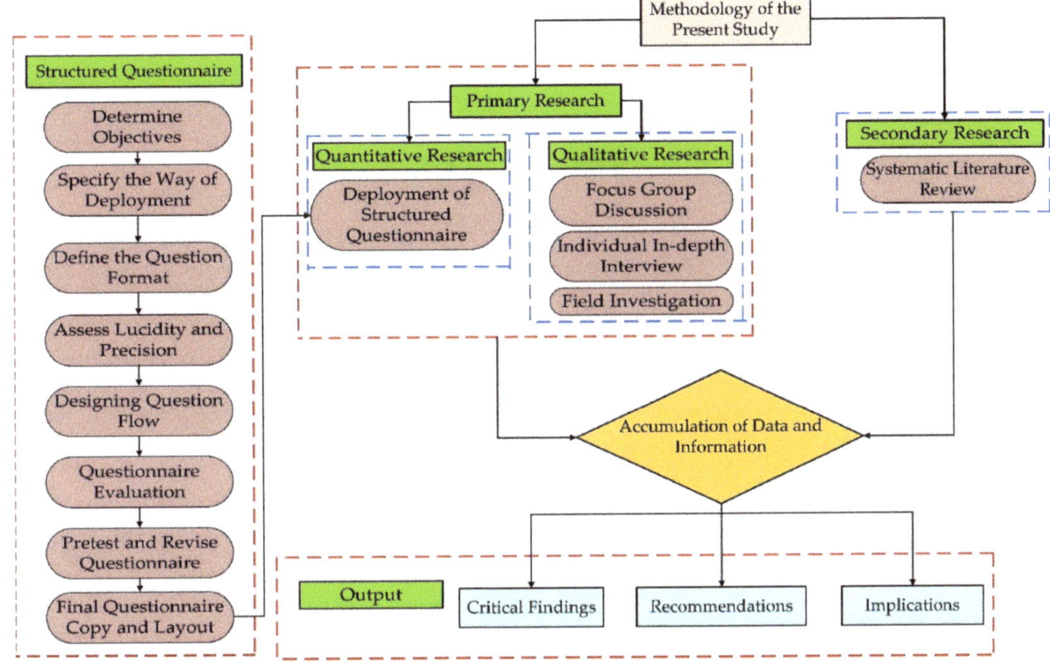

Figure 2. The overall methodology of this study.

3.4. Qualitative Research

This research employs the qualitative approach to investigate ELV recycling comprehensively. The qualitative analysis included focus-group interviews, individual in-depth interviews, and field investigations with survey participants to glean information and data regarding India's ELV recycling system. The authors performed a thematic analysis to interpret and analyze the interviews and group discussions. Focus-group interviews, together with personal interviews, revealed actual practices, whereas the field investigations provided crucial insights into the potential barriers and drivers to adopting the circular economy nationwide in the ELV recycling system in India [53,54].

3.5. Secondary Research

Secondary research is essential to this study's methodology and is a prerequisite step to collecting primary data [55]. This research has extensively explored the relevant literature to examine and evaluate the current landscape of available information on ELV recycling practices in India and gain insights into general industry trends. Secondary research yielded a robust foundation for this investigation and enabled the development of a comprehensive understanding of India's ELV recycling system from the circular economy perspective, and aided the research and sample design in this investigation. Secondary research assists in the interpretation of data collected from quantitative as well as qualitative research.

3.6. Data Analysis

Analyzing data to extract insights for rational decision-making is a crucial step of this investigation. Analyzing collected primary qualitative and quantitative data is a convoluted and arduous process. To examine and gain a thorough understanding of the factors that facilitate and impede the implementation of CE in India's ELV recycling system, descriptive statistics, which use the weighted-average method, along with inferential statistics have been employed to interpret and analyze the quantitative data obtained from the sample, while the interviews and group discussions have been interpreted using thematic analysis. This investigation employs sophisticated software to perform high-quality analysis. This study used Microsoft Excel and SPSS to analyze the quantitative data, while it employed ATLAS and Nvivo software to interpret the qualitative data.

4. Critical Findings and Analysis

This section reveals the critical findings and analyzes the results from the perspective of the research objectives.

4.1. Drivers That Enable CE implementation in India's ELV Recycling System

Figure 3 demonstrates the potential drivers for implementing CE in the ELV recycling system in India; among the identified potential drivers that facilitate the CE initiatives in the ELV recycling system in India, economic viability is the leading enabler for primary stakeholders involved in ELV recycling that is necessary for them to embrace and adopt the CE initiatives. In contrast, productivity and stability are the least significant drivers in India for adopting CE in the ELV recycling system. Drivers are classified into internal and external drivers, according to their characteristics [56].

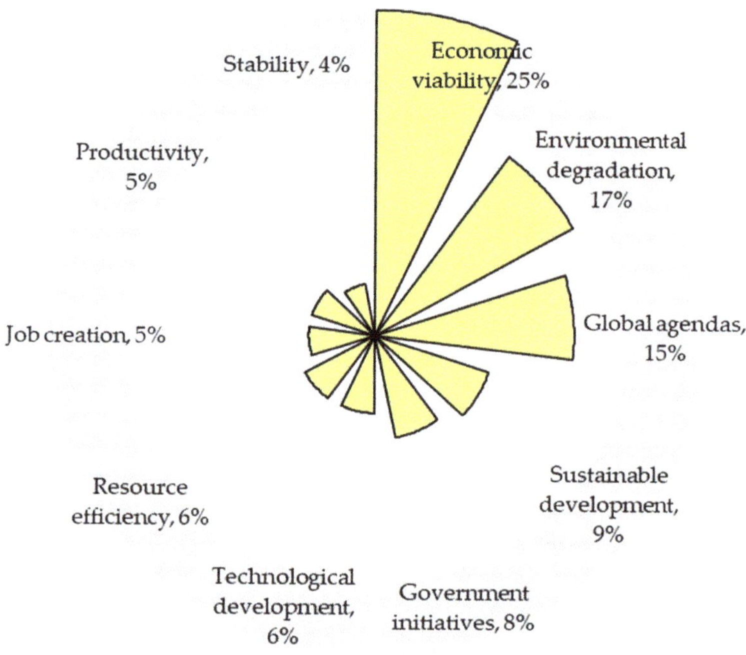

Figure 3. Drivers that enable CE implementation in India's ELV recycling system.

4.1.1. Internal Drivers

The internal potential drivers in implementing CE in the ELV recycling sector encompass various factors, from economic viability to productivity. These internal factors are elucidated in detail below.

Economic Viability

Figure 3 reveals the finding that several stakeholders favored embracing and implementing the CE in the pursuit of economic viability, gains in profit, and an approach to boosting the market share. This perspective is consistent and congruous with the observations reported in the current literature, as the central concept of CE is to recycle components in an environmentally friendly way to maximize economic gain and conserve natural resources [57]. Exactly one-quarter of the total responses from survey participants revealed that financial benefit is the more significant driver of implementing CE than improving the quality of the environment. Interacted stakeholders perceived CE as an approach to enhancing recycling and recovery rates from ELVs while preserving natural resources and safeguarding the environment. Respondents viewed the CE as an approach to developing closed-loop business systems that facilitate more significant economic gain while minimizing resource consumption. Hence, it is conspicuous that stakeholders have underscored the economic viability of adopting and implementing CE in the ELV recycling system in India.

Environmental Degradation

The interviewed respondents considered that concern for mitigating environmental degradation caused by the complex interaction of socio-economic, industrial, and technical activities is a crucial factor for adopting CE in India's ELV recycling system. The mitigation of environmental deterioration is the top priority of the United Nations' sustainable development goals and the primary objective of the CE concept [58,59]. Figure 3 clearly shows that the significant majority of survey participants (17%) perceive that implementing the CE in India's ELV recycling system is a pragmatic approach to handling ELVs and making the harmful waste generated from ELV dismantling operations more environmentally friendly. Therefore, it is significant that survey respondents have underlined the environmental degradation issue in terms of adopting and implementing CE in the ELV recycling system in India, which complies appropriately with the present literature and the country's sustainable development objectives [60].

Sustainable Development

Sustainable development is a roadmap for a better, more sustainable future for everyone on this planet, intertwined and inextricably linked with CE [61,62]. A significant number of survey respondents (9%) saw sustainable development as a potential driver in adopting and implementing CE in the ELV recycling system in India. The inappropriate handling of ELVs causes environmental degradation and generates a significant amount of harmful waste; hence, the sustainable handling of ELVs is of paramount importance [63]. Many stakeholders involved in ELV recycling have expressed a propensity toward adopting a sustainable development model to safeguard the environment, optimize the use of resources, and maximize materials recovery and recycling rates.

Technological Development

Technological development significantly impacts materials extraction rates from ELVs and reduces landfill waste [64]. India has witnessed phenomenal technological advancement, consequently becoming a technological hub worldwide. In total, 6% of survey respondents considered that technological progress is also a significant driver for adopting and implementing the CE in ELV recycling systems in India, as it enables higher materials recovery and recycling rates and, consequently, increases economic gain.

Resource Efficiency

Natural resources are declining dramatically; hence, optimum resource use is paramount. The core concept of the CE is the optimum use of resources [65]. The participants' responses revealed that a total of 6% of total survey respondents perceived that efficiently utilizing resources is a substantial driver for adopting and implementing the CE in India's ELV recycling sector, as it enables the sustainable use of the world's limited resources while respecting the environment.

Job Creation

Implementing the CE sustainably creates employment as it proposes a robust economic model by creating new markets and products. Of the total survey respondents, 5% considered that creating jobs is a crucial driver for implementing the CE in India's ELV recycling system, as the Indian authority has been endeavoring to create jobs for its youth.

Productivity

The traditional linear "take-make-dispose" strategy is prevalent in the ELV recycling sector in India, as it is in other developing countries. The CE provides a paradigm framework by operating via a closed-loop system to enhance productivity while respecting the environment [66,67]. In total, 5% of all respondents believed that the CE has the significant potential to improve the productivity of businesses and customers as it entails swiftly moving away from a reliance on natural resources toward recycling and repurposing resources after a product's life cycle is complete, through the existence of a practical framework and innovation. CE can also substantially mitigate the emissions from manufacturing and industrial hubs, which reduces the negative impact on the environment and enhances productivity.

Stability

CE involves sustainable production and consumption, which eventually cherishes and nurtures stability; the primary goal of CE is to create stability in the business arena. Of all the surveyed participants, 4% deemed that CE has tremendous potential to create stability by employing new business paradigms and technology and by mitigating pollution in the ELV recycling sector in India.

4.1.2. External Drivers

The ELV recycling industry has external drivers for implementing the CE, which extend from global agenda to government initiatives. Below, each of these external elements is addressed in greater depth.

Global Agenda

As the CE promotes the sustainable handling of waste and the safe use and conservation of natural resources, much more interest is being poured into the CE worldwide. International authorities are encouraging nations to adopt and implement the CE approach to enhance our planet's sustainable development, offering various incentives for promoting the CE. Overall, 15% of the survey respondents reported that global agenda for promoting sustainable development for the welfare of our future world play an imperative role in adopting and implementing the CE in ELV recycling in India.

Government Initiatives

A significant majority of the interviewed respondents (8%) considered that government initiatives inspired stakeholders involved in the ELV recycling system and also believed that initiatives from the authorities that encouraged adopting and implementing a CE in ELV recycling are significant and crucial. The Indian authorities have taken initial steps to promote and adopt the CE in different areas by enacting legislation and providing guidelines [68]. Even though many stakeholders perceive that the Indian government has taken a few rudimentary steps, these steps can lead to the swift adoption of the CE.

4.2. Barriers That Impede CE Implementation in India's ELV Recycling System

Figure 4 reveals the significant barriers that thwart CE implementation in India's ELV recycling system, as perceived by respondents. Like the potential drivers, the perceived barriers are split between internal and external impediments [69]. Among the identified significant barriers that hinder the CE's implementation in India's ELV recycling sector, limited technology is the leading impediment to embracing and adopting CE initiatives. In contrast, the lack of industrial collaboration and support is the least significant barrier to adopting a CE in India's ELV recycling sector.

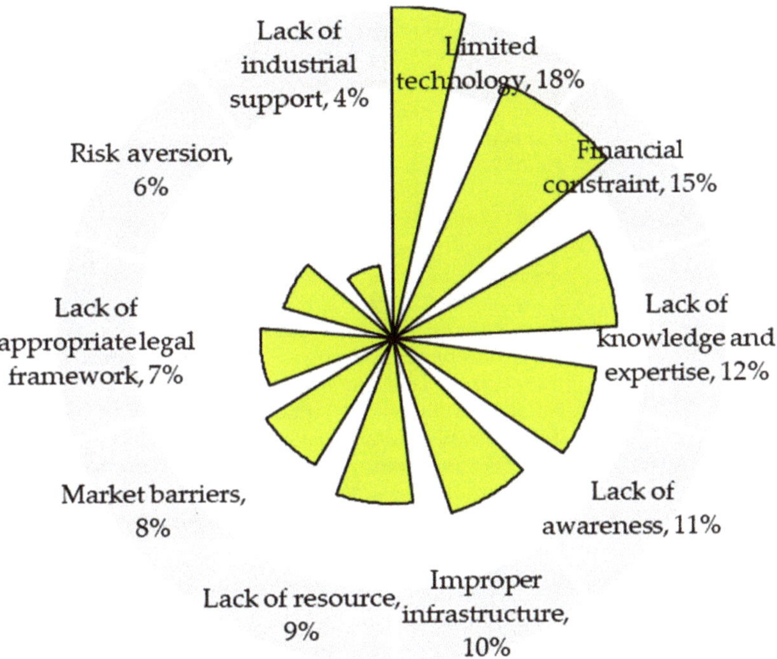

Figure 4. Barriers that impede CE implementation in India's ELV recycling system.

4.2.1. Internal Barriers

Internal potential barriers to implementing CE in the ELV recycling sector encompass a wide range of factors, from technological constraints to a lack of awareness. These internal factors are elucidated in detail below.

Limited Technology

Figure 4 reveals that significant respondents perceive limited technology as the primary hurdle for transitioning from linear to circular economy initiatives in India's ELV recycling sector. This prerequisite factor for adopting CE in India's ELV recycling sector is yet to meet. Overall, 18% of participants expressed concern regarding inadequate and poor technology, which consequently became the most significant obstacle to CE implementation in the ELV recycling sector in India. Our detailed interviews recorded stakeholders' aspirations in terms of the ELV recycling system, revealing that stakeholders expressed discontent with the lack of advanced technology, which prompts limited materials recovery and recycling rates. This perspective is consistent and harmonious with the observation reported in the present literature that poor technology is a prevalent cause for adopting a CE in developing and developed countries [70]. Stakeholders need and seek technological assistance from the government to enhance the recycling rate and sustain the business.

Financial Constraints

Figure 4 reveals that a substantial number of survey participants considered that at the initial stages of CE implementation, stakeholders involved in ELV recycling considered that the expenditure would be significantly increased as they would need to invest in making operations more sustainable. Overall, 15% of the respondents voiced concerns regarding economic constraints that can hinder the adoption of CE approaches in India's ELV recycling sector, which consequently became the second most significant obstacle to CE implementation. Our investigation revealed that stakeholders recognized that reconstructing and re-designing ordinary plants for sustainable operations requires substantial investment; most stakeholders are currently going through financial difficulties, especially after the COVID-19 pandemic. Hence, it is conspicuous that stakeholders have underscored the financial constraints for adopting and implementing a CE in the ELV recycling system in India. Recent research has emphasized how costs and economic conditions play a crucial role in impeding the adoption of CE projects as this is a common hindrance and prerequisite factor when implementing a CE [71,72].

Lack of Knowledge and Expertise

Figure 4 demonstrates that the lack of comprehensive knowledge and the relevant expertise for the transition from a linear to a circular economy is a significant setback for adopting the CE in the ELV recycling sector in India. The ELV recycling industry in India has yet to overcome this setback in adopting a CE. Overall, 12% of participants voiced their concern over the absence of thorough knowledge and the appropriate expertise to transition from a linear to a circular economy; consequently, this has emerged as one of the primary barriers to CE adoption in India's ELV recycling industry. Our thorough interviews documented the ambitions of stakeholders in the ELV recycling system, revealing that stakeholders expressed resentment at the lack of advanced expertise to enhance materials extraction rates, resulting in low rates of materials recovery and recycling. This viewpoint is in line with the observations reported in the current literature [53].

Lack of Awareness

A lack of awareness regarding the CE and its benefits is an obtrusive impediment to adopting the circular economy approach. The discussion about the CE is as yet confined to research works, and few attempts have been made for its implementation in real-world enterprises [73,74]. Figure 4 reveals that a significant number of respondents (11%) perceived that a lack of awareness and promotion regarding the CE is one of the primary obstacles that hinder the adoption of a CE in the ELV recycling sector in India. Our detailed interviews documented the aspirations and attitudes of stakeholders in the ELV recycling system, highlighting that significant stakeholders still have little knowledge and vague concepts about the CE approach and its benefits, especially regarding long-term viability. The Indian authorities have yet to promote and increase awareness at the ground level.

Improper Infrastructure

Proper and adequate infrastructure is paramount for adopting a CE approach, but the absence of an appropriate and proper infrastructure is prevalent in India's ELV recycling sector [75]. Overall, 10% of the interviewed respondents perceive inappropriate infrastructure as a primary impediment to embracing the CE in the ELV recycling sector and ranks fifth among all potential obstacles. The interviewed stakeholders stated that they were carrying out operations involved in ELV recycling without proper infrastructure, and no appropriate guidelines from authorities have been provided to them. Appropriate infrastructure is the foundation of modern civilization; hence, inappropriate infrastructure in the ELV recycling sector causes significant pollution, wastes resources, and endangers our planet [76].

Lack of Resource

Lack of resources is a perennial issue in the ELV recycling sector in India; it thwarts sustainability in ELV recycling and affects materials extraction rates. Contrary to the traditional materials and process flows in a linear economy, the CE separates economic output from the dwindling and limited resources of this planet and creates a resilient system that requires a closed chain of product flow. This requires a significant initial investment from stakeholders; such investments encompass human resources development, economic assets, sophisticated technologies, proper infrastructure, and industrial symbiosis and collaboration. Figure 4 reveals that overall, 9% of respondents believe that the lack of different resources is one of the primary impediments to transitioning from linear to circular economy initiatives in India's ELV recycling sector. During the detailed interviews, a substantial number of respondents stated, "We are striving to have viable resources to sustain our business in the long term; hence, effectively implementing the CE is an arduous task", whereas developed countries such as the US, Japan, and the EU have significant viable resources for adopting the CE in practice [4,77,78].

Market Barriers

The market demand and economic gains play an imperative role in adopting the CE in the ELV recycling sector in India. The demand for recycled components in the automobile market attracts substantial investment in ELV recycling industries [79]. A significant number of respondents perceived that the market for recycled components and remanufacturing parts for the automobile is a fledgling one, but this is still one of the primary hurdles for transitioning from linear to circular economy initiatives in India's ELV recycling sector. This prerequisite factor for adopting a CE in India's ELV recycling sector is yet to be met. Figure 4 shows that overall, 8% of interviewed participants expressed concern regarding the market for recycled products. During the detailed interviews with respondents, a substantial number of respondents said that "there is no market for recycled and remanufactured components near to us; we must go far when selling recovered and recycled products".

Risk Aversion

Many researchers are skeptical about the implementation of the CE in real-world business; they consider that the implementation of CE would be associated with risks and uncertainties as it is still in an embryonic stage [77,80]. Overall, a total of 6% of respondents believed that the CE is in an embryonic stage and requires significant time to fledge for real-world business applications. According to the interviewees, transitioning from the linear economy to the CE involves inherent uncertainties and great risks. In interviews, one of the stakeholders said, "We have to change all set-ups and business policies for adopting the CE approach, it involves extra investment and needs assistance and guidelines from authorities; hence, immediately, we would not adopt CE in our business until we get well-documented pieces of evidence."

4.2.2. External Barriers

External potential barriers to implementing CE in the ELV recycling sector are the lack of appropriate legal framework, industrial support, and collaboration. These external factors are elucidated in detail below.

Lack of Appropriate Legal Framework

The lack of an appropriate legal framework is a perennial issue in the ELV recycling sector in India, one that thwarts sustainability in ELV recycling and affects materials extraction rates. Informal ELV recycling centers predominate in India's ELV recycling sector; their operation does not comply with any standard guidelines, and economic gains drive only their business; a lack of framework is prevalent in ELV recycling industries [23]. CE creates a viable and robust pioneering financial system, which may require a closed loop of product flow that necessitates an appropriate legal framework [81–83]. Figure 4 reveals that overall, 7% of respondents perceived that the lack of an appropriate legal framework impeded transitioning from linear to CE initiatives in India's ELV recycling sector. During the interview, respondents stated, "The lack of an appropriate legal framework is strangling sustainability and making it more challenging to implement CE in the ELV recycling sector in India."

Lack of Industrial Support

Industrial cooperation is imperative for adopting the CE in the ELV recycling sector in India. The lack of collaboration between ELV recycling and the parent automobile manufacturing industries is prominent. Figure 4 demonstrates that, overall, 4% of respondents believed that the lack of industrial support and collaboration hindered transitioning from linear to CE initiatives in India's ELV recycling sector. The interviews revealed that respondents expressed resentment for not receiving any assistance from the parent automobile manufacturing industry. The lack of collaboration between the ELV recycling industry and the parent automobile manufacturing industry made the implementation of the CE more strenuous and cumbersome.

5. Discussions and Managerial Implications

As the circular economy offers numerous benefits to society according to different aspects that yield opportunities to ameliorate the environmental quality, secure the raw materials supply, bolster economic growth, and reduce the application of non-renewable resources; hence, in developed countries, such as the European Union (EU), Japan, China, and the USA, the authorities have taken several initiatives to adopt and advance the CE approach, and the enterprises and stakeholders are also firmly committed to implementing and enhancing CE initiatives [5,7,41,58,63]. Nonetheless, in India, many stakeholders have expressed a keen inclination toward adopting the CE, but they are not well-equipped to do so as yet [23,68].

This research determines and yields a critical, insightful interpretation of the potential drivers and barriers to implementing the CE in the ELV recycling sector in India, which may be instrumental in developing a subsequent legislative policy and framework to expedite the transition from the linear economy to the CE. This paper has determined ten potential drivers; out of all the potential drivers identified, economic viability is the prime factor in persuading and encouraging the principal stakeholders in the ELV recycling sector in India to embrace and implement CE initiatives. The primary stakeholders and enterprises have expressed their concerns regarding environmental degradation, which is a crucial aspect of their corporate value; environmental degradation, along with other significant factors, including global agenda, sustainable development, government initiatives, technological development, and resource efficiency, urges the top management of several industries to adopt and implement the CE approach. Conversely, productivity and stability are the least important factors that influence the implementation of CE initiatives in the ELV recycling sector in India. These findings are consistent with contemporary research [50,62,65,72].

However, this research also sheds light on the significant barriers that thwart CE implementation in India in the form of an ELV recycling system. Several ELV recyclers are withdrawing from adopting and implementing the CE in their business because of limited technology [84]. ELV recyclers should upgrade their technology to enhance material recycling and recovery rates. Financial constraints have emerged as a prominent hindrance; the CE operates on a closed-loop system that requires significant investment to develop the system for implementing a CE. The Indian authorities can offer financial aid to the primary stakeholders to adopt and implement the CE approach. The lack of knowledge and expertise of key stakeholders regarding the CE approach impedes the implementation of CE in the business, although the authorities have been showing great interest. Numerous ELV recyclers are unaware of and oblivious to the CE approach; this lack of awareness is thwarting the implementation of the CE in the ELV recycling sector in India [49,85]. The Indian authorities need to initiate a campaign to enhance awareness and knowledge regarding CE. This program may be conducive to implementing the CE initiative in the ELV recycling sector in India.

The present study's findings reveal that the majority of obstacles as well as enablers for implementing CE initiatives in the ELV recycling sector in India comprise internal rather than external factors. As these internal factors are instrumental, the ELV recyclers should take into account all the primary internal factors that thwart the transition from the linear economy to the CE. The enterprise can control and resolve internal obstacles by modifying its mission, vision, operations, and performance, whereas the government should control the requisite external factors to facilitate the implementation of the CE in India's ELV recycling sector.

6. Recommendations

This research has made a few practical recommendations for implementing and enhancing the CE initiatives in the ELV recycling sector in India; the recommendations can be organized into five themes, as below.

6.1. Development of an Appropriate Framework for the CE

Implementing the CE in India's ELV recycling sector requires a practical framework to meet the CE's goals and enhance the CE initiatives. With the application of information and communication technologies, an innovative framework needs to be developed to enhance sustainability and facilitate the implementation of a CE [86,87]. That framework appropriately encompasses the development of concepts and key terminology, setting targets, identifying indicators to assess enhancements toward the goals, designing a proper methodology, data collection and analysis, enhancing materials recovery, and protecting the environment. The lack of an appropriate framework can thwart the enhancement of CE initiatives and sustainability in ELV recycling. Therefore, the government of India should develop a "common reference framework" for all stakeholders to achieve set goals, evaluate progress, monitor the materials recycling and recovery rate, identify the waste, and eventually monitor and assess the environmental impact.

6.2. Waste Prevention and Using Waste as a Resource

One of the primary aspects of the CE initiative is to emphasize resource recovery from waste through recycling, but the authority should underscore waste prevention. The administration may enact the policy to reduce waste through actions that entail monitoring and data collection, promoting public awareness, and enhancing technology, while respecting the environment. The authorities may revise the policy for the age of obsolescence of vehicles and develop proper guidelines regarding ELVs. The management should perform a cost-benefit analysis of waste to understand the better values that can be extracted from waste. Switching focus from waste to resource management is imperative for extracting higher values from waste or ELVs. This transition can be facilitated by enacting policies, such as prohibiting dumping sites and landfills, the mandatory deregistration of vehicles, nurturing extended producer responsibility (EPR), enhancing the recycling rate, developing a market for recycled and remanufactured components, and promoting sustainability in ELV recycling.

6.3. Foster Excellent Governance

The absence of appropriate regulations that can adequately govern India's ELV recycling sector is prevalent, and India's ELV recycling sector is dominated by the informal sector, which hinders the sustainability of the ELV recycling sector, whereas developed countries, such as Japan, the EU, China, and the USA, have already introduced legislation regarding waste and ELVs and have issued proper guidelines to stakeholders. The Indian authorities should urgently introduce the relevant legislation to govern the ELV recycling sector toward sustainability. Implementing the policy will enhance the CE initiatives by promoting and sharing best practices, enhancing responsible consumption, and extracting higher values from ELVs.

6.4. Taxation and Subsidies

Taxation and subsidies are instrumental in achieving higher resource recovery efficiency and in sustainably disposing of hazardous substances. The tax will prompt the vehicle owner to embrace more sustainable practices, whereas incentives will encourage the vehicle owner to dispose of the vehicle in a sustainable and environmentally friendly way and urge stakeholders to adopt the CE approach. Overall, tax and subsidies are conducive to implementing the CE in the ELV recycling sector in India.

6.5. Assist the Business Sector

The CE operates on a closed-loop system; hence, transforming the business from a linear economy to a CE necessitated a significant change in the operating system; the ELV recycler needs financial aid from the authorities to upgrade the instrument. Besides financial assistance, ELV stakeholders require information about best practices and guidelines from the authorities. Collaboration between the vehicle manufacturer and the ELV recycler significantly impacts materials extraction and recovery. Therefore, the Indian government should take the initiative of collaboration between vehicle manufacturers and ELV recyclers as a crucial step to meeting resource recovery objectives in support of the CE.

7. Research Limitations

This research has certain potential limitations.

- This study has focused exclusively on the ELV recycling sector in India; hence, the implications of the present research findings may be limited to the Indian subcontinent.
- This research has selected five prominent automobile hubs, namely, Kolkata, Chennai, Mumbai, Delhi, and Jamshedpur, to determine the potential drivers and barriers to CE implementation in India's ELV recycling sector. By including more automobile sectors throughout India, this research could have provided more accurate outcomes.
- This study has interviewed a total of 560 selected respondents to perform objective measurements; a larger sample size might have yielded more precise findings.
- A literature review is essential for developing the foundation of any research; very little research has been performed about the drivers and barriers to implementing the CE in ELV recycling in India. The lack of contemporary literature in this research area might have made the selection of drivers and barriers a little harder and more limited.

8. Future Prospects and Suggestions

Based on the current literature review and the potential limitations of this research, this study suggests certain proposals for future study to advance the CE initiatives and facilitate the adoption and implementation of the CE.

- This study has identified the drivers and barriers to implementing the CE in the ELV recycling sector in India. However, the co-relationship between drivers and obstacles, as well as the impact of these drivers and barriers to sustainability from Indian market perspectives, require further investigation.
- The drivers and barriers to implementing the CE alter significantly depending on the country in question. The development of a better understanding of the nature of the drivers and barriers to implementing the CE and their impacts on sustainability in different countries, especially between developed and developing countries, prompts further exploration.
- The lack of an appropriate framework thwarts the achievement of sustainability in the ELV recycling sector. A sustainable, multi-faceted framework blending the CE and the Internet of Things (IoT) for the ELV recycling sector needs to be developed to facilitate sustainable development.
- Limited technology is a perennial issue that constrains the material recycling and recovery rates from ELVs. Advanced technology is required to maximize material recycling and recovery rates and meet sustainability goals. Hence, further study is necessary to address technological issues to expedite sustainable development.
- Designing an appropriate value chain for ELV recycling is imperative for adopting and implementing CE initiatives. The vehicle manufacturers and ELV recyclers are operating on different value chains, which is a primary hindrance to implementing CE initiatives in the ELV recycling sector and needs further exploration.

9. Conclusions

The CE yields a regenerative mechanism that replaces the end-of-life concept with restoration. India's end-of-life vehicle (ELV) recycling sector is encountering several challenges in implementing the CE initiatives, despite numerous benefits offered by the CE from a triple-bottom-line perspective. The literature reveals that limited research has been performed to explore the potential drivers and barriers to CE implementation in developing countries. Therefore, this study has made attempts to investigate the potential drivers and barriers to implementing the circular economy in the end-of-life vehicle recycling industry, to promote the sustainability of the ELV recycling sector in India by employing an explorative approach that encompasses qualitative, quantitative, and secondary research. The findings of this research reveal that economic viability, environmental degradation, global agenda, and sustainable development are the prime drivers that encourage enterprises to adopt and implement the CE initiative, while limited technology, financial constraints, a lack of knowledge and expertise, and a lack of awareness are the primary impediments hindering enterprises from adopting and implementing the CE initiative. This study provides a critical and insightful interpretation of the potential drivers and obstacles for the stakeholders and top management of the enterprises, which can be crucial in developing a subsequent policy, strategy, and framework to expedite the transition from the linear economy to a CE. Moreover, based on these observations, this research has offered certain insightful recommendations for implementing and enhancing CE initiatives in the ELV recycling sector in India.

These research findings reflect stakeholders' perceptions and aspirations regarding the ELV recycling system. Therefore, besides advancing the understanding of opportunities and threats for implementing the CE and raising awareness about the CE, this investigation may assist the Indian authorities in devising appropriate policies and strategies and developing a regulatory and legal framework conducive to the CE and sustainability. The outcomes of this study may inspire ELV enterprises, as well as other industries, to adopt the CE paradigm as it offers a resilient system of economy that generates substantial economic benefits while respecting the environment. India's ELV recycling sector is dominated by informality, driven by financial gain only, regardless of environmental quality; this research may lead to a change in the attitudes of informal enterprises as a CE offers sustainability as well as profitability.

This research has delved into India's ELV recycling sector exclusively; hence, further investigation may emphasize the findings' relevance to other large-scale as well as small and medium enterprises (SMEs). This study focuses solely on India's ELV recycling industry, and the implications of the findings may be limited to the Indian subcontinent.

Supplementary Materials: The following supporting information can be downloaded at: https://www.mdpi.com/article/10.3390/su142013084/s1. Supplementary data: Survey and demographic data of this study.

Author Contributions: Each author's contributions are listed as follows: conceptualization, A.H.M. and Z.H.; methodology, A.H.M.; software, A.H.M. and H.S.; validation, A.H.M. and H.S.; formal analysis, A.H.M.; investigation, A.H.M.; resources, Z.H. and A.H.M.; data curation, A.H.M.; writing—original draft preparation, A.H.M. and H.S.; writing—review and editing, Z.H., M.N.A.R. and H.H.; visualization, A.H.M. and H.S.; supervision, Z.H., M.N.A.R. and H.H.; project administration, Z.H.; funding acquisition, Z.H. All authors have read and agreed to the published version of the manuscript.

Funding: This research was funded by the Transdisciplinary Research Grant Scheme TRGS/1/2020/UKM/02/1/1 and UKM internal grant GUP-2018-012.

Institutional Review Board Statement: Not applicable.

Informed Consent Statement: Not applicable.

Data Availability Statement: The data presented in this study are available in this published article and Supplementary Material.

Conflicts of Interest: The authors declare that they have no known competing financial interests or personal relationships that could have appeared to influence the work reported in this paper.

Appendix A. Primary Survey Questions

1. What are the drivers that could enable circular economy implementation for the end-of-life vehicle recycling sector in India?

    (a)   Economic viability
    (b)   Environmental degradation
    (c)   Global agenda
    (d)   Sustainable development
    (e)   Government initiatives
    (f)   Technological development
    (g)   Resource efficiency
    (h)   Job creation
    (i)   Productivity
    (j)   Stability.

2. What are the barriers that could hinder circular economy implementation for the end-of-life vehicle recycling sector in India?

    (a)   Limited technology
    (b)   Financial constraint
    (c)   Lack of knowledge and expertise
    (d)   Lack of awareness
    (e)   Improper infrastructure
    (f)   Lack of resource
    (g)   Market barriers
    (h)   Lack of appropriate legal framework
    (i)   Risk aversion
    (j)   Lack of industrial support.

Appendix B. Interview Questions

1. Do you have any idea about sustainability?
2. Do you have any idea about the circular economy?
3. Do you know about end-of-life vehicle (ELV) recycling?
4. Has the ELV recycling business expanded or shrunk in the last ten years?
5. Have you upgraded the types of equipment to enhance the recycling rate from ELV?
6. Is there any supporting industry for raw materials nearby you?
7. Do you know that ELV dismantling causes environmental problems?
8. Do you follow proper guidelines for disposing of hazardous waste?
9. Have you been getting any government subsidies?
10. What types of assistance do you want from the government?

Appendix C. Demographic Questions

1. What is your gender?
2. What is your age?
3. To which Indian state do you belong?
4. What is your education level?
5. What is your occupation?
6. What is your income range?
7. Do you own a vehicle?
8. How long have you (stakeholders) been involved in the ELV recycling sector?

References

1. Zhang, L.; Lu, Q.; Yuan, W.; Jiang, S.; Wu, H. Characterizing end-of-life household vehicles' generations in China: Spatial-temporal patterns and resource potentials. *Resour. Conserv. Recycl.* **2022**, *177*, 105979. [CrossRef]
2. Liu, B.; Chen, D.; Zhou, W.; Nasr, N.; Wang, T.; Hu, S.; Zhu, B. The effect of remanufacturing and direct reuse on resource productivity of China's automotive production. *J. Clean. Prod.* **2018**, *194*, 309–317. [CrossRef]
3. Zimon, D.; Madzík, P. Standardized management systems and risk management in the supply chain. *Int. J. Qual. Reliab. Manag.* **2020**, *37*, 305–327. [CrossRef]
4. Karagoz, S.; Aydin, N.; Simic, V. End-of-life vehicle management: A comprehensive review. *J. Mater. Cycles Waste Manag.* **2020**, *22*, 416–442. [CrossRef]
5. Jang, Y.-C.; Choi, K.; Jeong, J.-H.; Kim, H.; Kim, J.-G. Recycling and material-flow analysis of end-of-life vehicles towards resource circulation in South Korea. *Sustainability* **2022**, *14*, 1270. [CrossRef]
6. Petronijević, V.; Đorđević, A.; Stefanović, M.; Arsovski, S.; Krivokapić, Z.; Mišić, M. Energy recovery through end-of-life vehicles recycling in developing countries. *Sustainability* **2020**, *12*, 8764. [CrossRef]
7. Zhao, Q.; Chen, M. A comparison of ELV recycling system in China and Japan and China's strategies. *Resour. Conserv. Recycl.* **2011**, *57*, 15–21. [CrossRef]
8. Pigosso, D.C.A.; De, M.; Pieroni, P.; Kravchenko, M.; Awan, U.; Sroufe, R.; Bozan, K. Designing value chains for Industry 4.0 and a circular economy: A Review of the literature. *Sustainability* **2022**, *14*, 7084. [CrossRef]
9. Kirchherr, J.; Reike, D.; Hekkert, M. Conceptualizing the circular economy: An analysis of 114 definitions. *Resour. Conserv. Recycl.* **2017**, *127*, 221–232. [CrossRef]
10. Lindahl, P.; Robèrt, K.-H.; Ny, H.; Broman, G. Strategic sustainability considerations in materials management. *J. Clean. Prod.* **2014**, *64*, 98–103. [CrossRef]
11. Smol, M.; Kulczycka, J.; Henclik, A.; Gorazda, K.; Wzorek, Z. The possible use of sewage sludge ash (SSA) in the construction industry as a way towards a circular economy. *J. Clean. Prod.* **2015**, *95*, 45–54. [CrossRef]
12. Modoi, O.-C.; Mihai, F.-C. E-Waste and end-of-life vehicles management and circular economy initiatives in Romania. *Energies* **2022**, *15*, 1120. [CrossRef]
13. Li, Y.; Liu, Y.; Chen, Y.; Huang, S.; Ju, Y. Projection of end-of-life vehicle population and recyclable metal resources: Provincial-level gaps in China. *Sustain. Prod. Consum.* **2022**, *31*, 818–827. [CrossRef]
14. Kaviani, M.A.; Tavana, M.; Kumar, A.; Michnik, J.; Niknam, R.; de Campos, E.A.R. An integrated framework for evaluating the barriers to successful implementation of reverse logistics in the automotive industry. *J. Clean. Prod.* **2020**, *272*, 122714. [CrossRef]
15. Geng, Y.; Doberstein, B. Developing the circular economy in China: Challenges and opportunities for achieving 'leapfrog development'. *Int. J. Sustain. Dev. World Ecol.* **2008**, *15*, 231–239. [CrossRef]
16. Al-Quradaghi, S.; Zheng, Q.P.; Betancourt-Torcat, A.; Elkamel, A. Optimization model for sustainable end-of-life vehicle processing and recycling. *Sustainability* **2022**, *14*, 3551. [CrossRef]
17. Yuan, X.; Liu, M.; Yuan, Q.; Fan, X.; Teng, Y.; Fu, J.; Ma, Q.; Wang, Q.; Zuo, J. Transitioning China to a circular economy through remanufacturing: A comprehensive review of the management institutions and policy system. *Resour. Conserv. Recycl.* **2020**, *161*, 104920. [CrossRef]
18. Farahani, S.; Otieno, W.; Barah, M. Environmentally friendly disposition decisions for end-of-life electrical and electronic products: The case of computer remanufacture. *J. Clean. Prod.* **2019**, *224*, 25–39. [CrossRef]
19. Onat, N.C.; Kucukvar, M.; Afshar, S. Eco-efficiency of electric vehicles in the United States: A life cycle assessment based principal component analysis. *J. Clean. Prod.* **2019**, *212*, 515–526. [CrossRef]
20. Chakraborty, K.; Mondal, S.; Mukherjee, K. Critical analysis of enablers and barriers in extension of useful life of automotive products through remanufacturing. *J. Clean. Prod.* **2019**, *227*, 1117–1135. [CrossRef]
21. Mohan, T.V.K.; Amit, R.K. Dismantlers' dilemma in end-of-life vehicle recycling markets: A system dynamics model. *Ann. Oper. Res.* **2018**, *290*, 591–619. [CrossRef]
22. Numfor, S.A.; Omosa, G.B.; Zhang, Z.; Matsubae, K. A review of challenges and opportunities for end-of-life vehicle recycling in developing countries and emerging economies: A SWOT analysis. *Sustainability* **2021**, *13*, 4918. [CrossRef]
23. Sharma, L.; Pandey, S. Recovery of resources from end-of-life passenger cars in the informal sector in India. *Sustain. Prod. Consum.* **2020**, *24*, 1–11. [CrossRef]
24. Arora, N.; Bakshi, S.K.; Bhattacharjya, S. Framework for sustainable management of end-of-life vehicles management in India. *J. Mater. Cycles Waste Manag.* **2018**, *21*, 79–97. [CrossRef]
25. Ravi, V.; Shankar, R. An ISM-based approach analyzing interactions among variables of reverse logistics in automobile industries. *J. Model. Manag.* **2017**, *12*, 36–52. [CrossRef]
26. Luthra, S.; Kumar, V.; Kumar, S.; Haleem, A. Barriers to implement green supply chain management in automobile industry using interpretive structural modeling technique: An Indian perspective. *J. Ind. Eng. Manag.* **2011**, *4*, 231–257. [CrossRef]
27. Govindan, K.; Madan Shankar, K.; Kannan, D. Application of fuzzy analytic network process for barrier evaluation in automotive parts remanufacturing towards cleaner production—A study in an Indian scenario. *J. Clean. Prod.* **2016**, *114*, 199–213. [CrossRef]
28. Nag, U.; Sharma, S.K.; Govindan, K. Investigating drivers of circular supply chain with product-service system in automotive firms of an emerging economy. *J. Clean. Prod.* **2021**, *319*, 128629. [CrossRef]

29. Sinha, A.; Mondal, S.; Boone, T.; Ganeshan, R. Analysis of issues controlling the feasibility of automobile remanufacturing business in India. *Int. J. Serv. Oper. Manag.* **2017**, *26*, 459–475. [CrossRef]

30. Zhou, F.; Lim, M.K.; He, Y.; Lin, Y.; Chen, S. End-of-life vehicle (ELV) recycling management: Improving performance using an ISM approach. *J. Clean. Prod.* **2019**, *228*, 231–243. [CrossRef]

31. Hu, S.; Wen, Z. Monetary evaluation of end-of-life vehicle treatment from a social perspective for different scenarios in China. *J. Clean. Prod.* **2017**, *159*, 257–270. [CrossRef]

32. Raja Mamat, T.N.A.; Saman, M.Z.M.; Sharif, S.; Simic, V. Key success factors in establishing end-of-life vehicle management system: A primer for Malaysia. *J. Clean. Prod.* **2016**, *135*, 1289–1297. [CrossRef]

33. Zhang, C.; Chen, M. Prioritising alternatives for sustainable end-of-life vehicle disassembly in China using AHP methodology. *Technol. Anal. Strat. Manag.* **2017**, *30*, 556–568. [CrossRef]

34. Sitinjak, C.; Ismail, R.; Bantu, E.; Fajar, R.; Samuel, K. The understanding of the social determinants factors of public acceptance towards the end of life vehicles. *Cogent Eng.* **2022**, *9*, 2088640. [CrossRef]

35. Edun, A.; Hachem-Vermette, C. Energy and environmental impact of recycled end of life tires applied in building envelopes. *J. Build. Eng.* **2021**, *39*, 102242. [CrossRef]

36. MahmoumGonbadi, A.; Genovese, A.; Sgalambro, A. Closed-loop supply chain design for the transition towards a circular economy: A systematic literature review of methods, applications and current gaps. *J. Clean. Prod.* **2021**, *323*, 129101. [CrossRef]

37. D'Adamo, I.; Gastaldi, M.; Rosa, P. Recycling of end-of-life vehicles: Assessing trends and performances in Europe. *Technol. Forecast. Soc. Chang.* **2020**, *152*, 119887. [CrossRef]

38. Li, J.; Yu, K.; Gao, P. Recycling and pollution control of the end of life vehicles in China. *J. Mater. Cycles Waste Manag.* **2014**, *16*, 31–38. [CrossRef]

39. Nakano, K.; Shibahara, N. Comparative assessment on greenhouse gas emissions of end-of-life vehicles recycling methods. *J. Mater. Cycles Waste Manag.* **2015**, *19*, 505–515. [CrossRef]

40. Tarrar, M.; Despeisse, M.; Johansson, B. Driving vehicle dismantling forward—A combined literature and empirical study. *J. Clean. Prod.* **2021**, *295*, 126410. [CrossRef]

41. Sakai, S.-I.; Yoshida, H.; Hiratsuka, J.; Vandecasteele, C.; Kohlmeyer, R.; Rotter, V.S.; Passarini, F.; Santini, A.; Peeler, M.V.; Li, J.; et al. An international comparative study of end-of-life vehicle (ELV) recycling systems. *J. Mater. Cycles Waste Manag.* **2014**, *16*, 1–20. [CrossRef]

42. Cossu, R.; Lai, T. Automotive shredder residue (ASR) management: An overview. *Waste Manag.* **2015**, *45*, 143–151. [CrossRef]

43. Chen, K.-C.; Huang, S.-H.; Lian, I.-W. The development and prospects of the end-of-life vehicle recycling system in Taiwan. *Waste Manag.* **2010**, *30*, 1661–1669. [CrossRef]

44. Li, Y.; Fujikawa, K.; Wang, J.; Li, X.; Ju, Y.; Chen, C. The Potential and trend of end-of-life passenger vehicles recycling in China. *Sustainability* **2020**, *12*, 1455. [CrossRef]

45. Chen, Y.; Ding, Z.; Liu, J.; Ma, J. Life cycle assessment of end-of-life vehicle recycling in China: A comparative study of environmental burden and benefit. *Int. J. Environ. Stud.* **2019**, *76*, 1019–1040. [CrossRef]

46. Rovinaru, F.I.; Rovinaru, M.D.; Rus, A.V. the economic and ecological impacts of dismantling end-of-life vehicles in Romania. *Sustainability* **2019**, *11*, 6446. [CrossRef]

47. Melo, A.C.S.; Braga, A.E.; Leite, C.D.P.; Bastos, L.D.S.L.; Nunes, D.R.D.L. Frameworks for reverse logistics and sustainable design integration under a sustainability perspective: A systematic literature review. *Res. Eng. Des.* **2020**, *32*, 225–243. [CrossRef]

48. Ramani, K.; Ramanujan, D.; Bernstein, W.Z.; Zhao, F.; Sutherland, J.; Handwerker, C.; Choi, J.-K.; Kim, H.; Thurston, D. Integrated sustainable life cycle design: A review. *J. Mech. Des. Trans. ASME* **2010**, *132*, 0910041–09100415. [CrossRef]

49. Kayikci, Y.; Kazancoglu, Y.; Lafci, C.; Gozacan, N. Exploring barriers to smart and sustainable circular economy: The case of an automotive eco-cluster. *J. Clean. Prod.* **2021**, *314*, 127920. [CrossRef]

50. Agyemang, M.; Kusi-Sarpong, S.; Khan, S.A.; Mani, V.; Rehman, S.T.; Kusi-Sarpong, H. Drivers and barriers to circular economy implementation: An explorative study in Pakistan's automobile industry. *Manag. Decis.* **2019**, *57*, 971–994. [CrossRef]

51. Gardas, B.B.; Raut, R.D.; Narkhede, B. Reducing the exploration and production of oil: Reverse logistics in the automobile service sector. *Sustain. Prod. Consum.* **2018**, *16*, 141–153. [CrossRef]

52. Macarthur, E. Founding Partners of the Towards the Circular Economy, Vol. 1, Economic and Business Rationale for an Accelerated Transition. Available online: https://ellenmacarthurfoundation.org/towards-the-circular-economy-vol-1-an-economic-and-business-rationale-for-an (accessed on 25 August 2022).

53. Grafström, J.; Aasma, S. Breaking circular economy barriers. *J. Clean. Prod.* **2021**, *292*, 126002. [CrossRef]

54. Hina, M.; Chauhan, C.; Kaur, P.; Kraus, S.; Dhir, A. Drivers and barriers of circular economy business models: Where we are now, and where we are heading. *J. Clean. Prod.* **2022**, *333*, 130049. [CrossRef]

55. Tan, J.; Tan, F.J.; Ramakrishna, S. Transitioning to a circular economy: A systematic review of its drivers and barriers. *Sustainability* **2022**, *14*, 1757. [CrossRef]

56. Mont, O.; Plepys, A.; Whalen, K.; Nußholz, J.L.K. *Business Model Innovation for a Circular Economy: Drivers and Barriers for the Swedish Industry—The Voice of REES Companies*; Uniwersytet śLąski: Lund, Sweden, 2017; pp. 343–354.

57. Ariztia, T.; Araneda, F. A "win-win formula:" environment and profit in circular economy narratives of value. *Consum. Mark. Cult.* **2022**, *25*, 124–138. [CrossRef]

58. Liu, Z.; Adams, M.; Cote, R.P.; Chen, Q.; Wu, R.; Wen, Z.; Liu, W.; Dong, L. How does circular economy respond to greenhouse gas emissions reduction: An analysis of Chinese plastic recycling industries. *Renew. Sustain. Energy Rev.* **2018**, *91*, 1162–1169. [CrossRef]
59. Kumar, R.; Verma, A.; Shome, A.; Sinha, R.; Sinha, S.; Jha, P.K.; Kumar, R.; Kumar, P.; Shubham; Das, S.; et al. Impacts of plastic pollution on ecosystem services, sustainable development goals, and need to focus on circular economy and policy interventions. *Sustainability* **2021**, *13*, 9963. [CrossRef]
60. Zhang, Z.; Malik, M.Z.; Khan, A.; Ali, N.; Malik, S.; Bilal, M. Environmental impacts of hazardous waste, and management strategies to reconcile circular economy and eco-sustainability. *Sci. Total Environ.* **2021**, *807*, 150856. [CrossRef]
61. Barros, M.V.; Salvador, R.; do Prado, G.F.; de Francisco, A.C.; Piekarski, C.M. Circular economy as a driver to sustainable businesses. *Clean. Environ. Syst.* **2021**, *2*, 100006. [CrossRef]
62. Skvarciany, V.; Lapinskaite, I.; Volskyte, G. Circular economy as assistance for sustainable development in OECD countries. *Oeconomia Copernic.* **2021**, *12*, 11–34. [CrossRef]
63. Wang, J.; Sun, L.; Fujii, M.; Li, Y.; Huang, Y.; Murakami, S.; Daigo, I.; Pan, W.; Li, Z. Institutional, technology, and policies of end-of-life vehicle recycling industry and its indication on the circular economy—Comparative analysis between China and Japan. *Front. Sustain.* **2021**, *2*, 13. [CrossRef]
64. Mallampati, S.R.; Lee, B.H.; Mitoma, Y.; Simion, C. Sustainable recovery of precious metals from end-of-life vehicles shredder residue by a novel hybrid ball-milling and nanoparticles enabled froth flotation process. *J. Clean. Prod.* **2018**, *171*, 66–75. [CrossRef]
65. López Ruiz, L.A.; Roca Ramón, X.; Gassó Domingo, S. The circular economy in the construction and demolition waste sector—A review and an integrative model approach. *J. Clean. Prod.* **2020**, *248*, 119238. [CrossRef]
66. Aguilar Esteva, L.C.; Kasliwal, A.; Kinzler, M.S.; Kim, H.C.; Keoleian, G.A. Circular economy framework for automobiles: Closing energy and material loops. *J. Ind. Ecol.* **2020**, *25*, 877–889. [CrossRef]
67. Kristoffersen, E.; Blomsma, F.; Mikalef, P.; Li, J. The smart circular economy: A digital-enabled circular strategies framework for manufacturing companies. *J. Bus. Res.* **2020**, *120*, 241–261. [CrossRef]
68. Venkatesan, M.; Annamalai, V.E. An institutional framework to address end-of-life vehicle recycling problem in India. *SAE Tech. Pap.* **2017**, *10*, 11. [CrossRef]
69. Cantú, A.; Aguiñaga, E.; Scheel, C. Learning from failure and success: The challenges for circular economy implementation in SMEs in an emerging economy. *Sustainability* **2021**, *13*, 1529. [CrossRef]
70. Zhou, X.; Song, M.; Cui, L. Driving force for China's economic development under Industry 4.0 and circular economy: Technological innovation or structural change? *J. Clean. Prod.* **2020**, *271*, 122680. [CrossRef]
71. Bockholt, M.T.; Hemdrup Kristensen, J.; Colli, M.; Meulengracht Jensen, P.; Vejrum Wæhrens, B. Exploring factors affecting the financial performance of end-of-life take-back program in a discrete manufacturing context. *J. Clean. Prod.* **2020**, *258*, 120916. [CrossRef]
72. Kaya, D.I.; Pintossi, N.; Dane, G. An empirical analysis of driving factors and policy enablers of heritage adaptive reuse within the circular economy framework. *Sustainability* **2021**, *13*, 2479. [CrossRef]
73. Jaeger, B.; Upadhyay, A. Understanding barriers to circular economy: Cases from the manufacturing industry. *J. Enterp. Inf. Manag.* **2020**, *33*, 729–745. [CrossRef]
74. Prieto-Sandoval, V.; Torres-Guevara, L.E.; Ormazabal, M.; Jaca, C. Beyond the circular economy theory: Implementation methodology for industrial SMEs. *J. Ind. Eng. Manag.* **2021**, *14*, 425–438. [CrossRef]
75. Ellsworth-Krebs, K.; Rampen, C.; Rogers, E.; Dudley, L.; Wishart, L. Circular economy infrastructure: Why we need track and trace for reusable packaging. *Sustain. Prod. Consum.* **2022**, *29*, 249–258. [CrossRef]
76. Coenen, T.B.J.; Haanstra, W.; Jan Braaksma, A.J.J.; Santos, J. CEIMA: A framework for identifying critical interfaces between the Circular Economy and stakeholders in the lifecycle of infrastructure assets. *Resour. Conserv. Recycl.* **2020**, *155*, 104552. [CrossRef]
77. Soo, V.K.; Doolan, M.; Compston, P.; Duflou, J.R.; Peeters, J.; Umeda, Y. The influence of end-of-life regulation on vehicle material circularity: A comparison of Europe, Japan, Australia and the US. *Resour. Conserv. Recycl.* **2021**, *168*, 105294. [CrossRef]
78. Baldassarre, B.; Maury, T.; Mathieux, F.; Garbarino, E.; Antonopoulos, I.; Sala, S. Drivers and barriers to the circular economy transition: The case of recycled plastics in the automotive sector in the European Union. *Procedia CIRP* **2022**, *105*, 37–42. [CrossRef]
79. Yi, S.; Lee, H. Economic analysis to promote the resource circulation of end-of-life vehicles in Korea. *Waste Manag.* **2021**, *120*, 659–666. [CrossRef]
80. Yu, Z.; Tianshan, M.; Rehman, S.A.; Sharif, A.; Janjua, L. Evolutionary game of end-of-life vehicle recycling groups under government regulation. *Clean Technol. Environ. Policy* **2020**, *2020*, 1–12. [CrossRef]
81. Ghosh, S.K.; Ghosh, S.K.; Baidya, R. Circular economy in India: Reduce, reuse, and recycle through policy framework. In *Circular Economy: Recent Trends in Global Perspective*; Springer: Singapore, 2021; pp. 183–217. [CrossRef]
82. Khan, S.A.R.; Godil, D.I.; Thomas, G.; Tanveer, M.; Zia-Ul-Haq, H.M.; Mahmood, H. The decision-making analysis on end-of-life vehicle recycling and remanufacturing under extended producer responsibility policy. *Sustainability* **2021**, *13*, 11215. [CrossRef]
83. Yu, J.; Wang, S.; Roy, K.; Serrona, B. Comparative analysis of ELV Recycling policies in the European Union, Japan and China. *Investig. Linguisticae* **2019**, *XLIII*, 34–56. [CrossRef]
84. Karuppiah, K.; Sankaranarayanan, B.; Ali, S.M.; Jabbour, C.J.C.; Bhalaji, R.K.A. Inhibitors to circular economy practices in the leather industry using an integrated approach: Implications for sustainable development goals in emerging economies. *Sustain. Prod. Consum.* **2021**, *27*, 1554–1568. [CrossRef]

85. Dulia, E.F.; Ali, S.M.; Garshasbi, M.; Kabir, G. Admitting risks towards circular economy practices and strategies: An empirical test from supply chain perspective. *J. Clean. Prod.* **2021**, *317*, 128420. [CrossRef]
86. Kumar, N.M.; Chopra, S.S. Leveraging Blockchain and Smart Contract Technologies to Overcome Circular Economy Implementation Challenges. *Sustainability* **2022**, *14*, 9492. [CrossRef]
87. Tools, M.; Abed Mohammed, M.; Jamal Abdulhasan, M.; Manoj Kumar, N.; Hameed Abdulkareem, K.; Mostafa, S.A.; Maashi, M.S.; Salman Khalid, L.; Saadoon Abdulaali, H.; Chopra, S.S. Automated waste-sorting and recycling classification using artificial neural network and features fusion: A digital-enabled circular economy vision for smart cities. *Multimedia Tools Appl.* **2022**, 1–16. [CrossRef]

sustainability

MDPI

Review

Recent Studies and Technologies in the Separation of Polyvinyl Chloride for Resources Recycling: A Systematic Review

Theerayut Phengsaart [1,2,*], Pongsiri Julapong [1,3], Chaiwat Manositchaikul [1], Palot Srichonphaisarn [1], Monthicha Rawangphai [1], Onchanok Juntarasakul [1], Kosei Aikawa [4], Sanghee Jeon [5], Ilhwan Park [4], Carlito Baltazar Tabelin [6,7] and Mayumi Ito [4]

1   Department of Mining and Petroleum Engineering, Faculty of Engineering, Chulalongkorn University, Bangkok 10330, Thailand
2   Applied Mineral and Petrology Research Unit (AMP RU), Department of Geology, Faculty of Science, Chulalongkorn University, Bangkok 10330, Thailand
3   Department of Mining and Materials Engineering, Faculty of Engineering, Prince of Songkla University, Songkhla 90110, Thailand
4   Division of Sustainable Resources Engineering, Faculty of Engineering, Hokkaido University, Sapporo 060-8628, Japan
5   Department of Earth Resource Engineering and Environmental Science, Faculty of International Resources Science, Akita University, Akita 010-0865, Japan
6   Department of Materials and Resources Engineering Technology, College of Engineering, Mindanao State University—Iligan Institute of Technology, Iligan City 9200, Philippines
7   Resource Processing and Technology Center, Research Institute of Engineering and Innovative Technology (RIEIT), Mindanao State University—Iligan Institute of Technology, Iligan City 9200, Philippines
*   Correspondence: theerayut.p@chula.ac.th

Citation: Phengsaart, T.; Julapong, P.; Manositchaikul, C.; Srichonphaisarn, P.; Rawangphai, M.; Juntarasakul, O.; Aikawa, K.; Jeon, S.; Park, I.; Tabelin, C.B.; et al. Recent Studies and Technologies in the Separation of Polyvinyl Chloride for Resources Recycling: A Systematic Review. *Sustainability* **2023**, *15*, 13842. https://doi.org/10.3390/su151813842

Academic Editors: Manoj Kumar Nallapaneni, Subrata Hait, Anshu Priya and Varsha Bohra

Received: 30 June 2023
Revised: 28 August 2023
Accepted: 11 September 2023
Published: 18 September 2023

Abstract: Material recycling and thermal treatment are the two most common recycling methods employed for plastic waste management. Thermal treatment for energy recovery is more widely applied compared with material recycling because the latter requires a high efficiency of separation and a high purity of products. Unfortunately, certain plastics like polyvinyl chloride (PVC) are unsuitable for thermal treatment because they contain additives like chloride (Cl^-) that have adverse effects on refractory materials used in boilers. As a result of this, mixed plastic wastes containing PVC generally end up in landfills. PVC-bearing mixed plastics, however, remain valuable resources as championed by the United Nation Sustainable Development Goals (UN-SDGs): Goal 12 "Responsible production and consumption", and their recycling after the removal of PVC is important. In this paper, recent studies (2012–2021) related to the separation of PVC from other types of plastics were systematically reviewed using the Preferred Reporting Items for Systematic Reviews and Meta-Analyses (PRISMA) guidelines. A total of 66 articles were selected, reviewed, and summarized. The results showed that various separation technologies conventionally applied to mineral processing—selective comminution, gravity separation, magnetic separation, electrical separation, and flotation—have been studied for PVC separation, and the majority of these works (>60%) focused on flotation. In addition, more advanced technologies including sorting and density-surface-based separation were introduced between 2019 and 2021.

Keywords: plastic; recycling; separation; flotation; polyvinyl chloride

1. Introduction

Plastics offer numerous advantages as flexible materials in various applications and critical industries, such as packaging, electrical/electronic equipment, and automotive components, due to their lightweight nature, chemical and moisture resistance, and excellent insulation properties [1]. According to a report published by Plastics Europe, global plastic production reached 391 million metric tons in 2021 [2,3]. Among the wide array of plastics, PVC stands out as a commonly used material with diverse applications, ranging from construction and automotive manufacturing to healthcare and consumer products [1,4]. Recent reports indicate that PVC production has reached approximately 32 million metric tons annually, accounting for 9.6% of total plastics production in 2021 [2]. Consequently, the escalating production and consumption of PVC-based products have led to a surge in PVC waste generation, necessitating the development of efficient and sustainable approaches for its removal and recovery. The disposal of PVC waste presents significant environmental challenges due to its non-biodegradable nature and the potential release of hazardous substances during incineration or landfilling [5,6]. Additionally, plastic debris or microplastics (MPs) derived from plastic waste accumulation can interact with organic pollutants or heavy metals, leading to concentrated toxicity in marine and freshwater systems [7].

Despite their many advantages, plastics are environmentally undesirable because they persist for a long time in nature. Plastic degradation and decomposition take hundreds of years under natural conditions, resulting in their accumulation in landfills, oceans, and other ecosystems [8]. The disposal of PVC waste presents significant environmental challenges due to its non-biodegradable nature and the potential release of hazardous substances during incineration or landfilling [5,6]. Incinerating PVC can release toxic pollutants, including dioxins and furans, which have detrimental effects on both human health and the environment [8]. Furthermore, PVC waste deposited in landfills contributes to long-term environmental contamination, as it persists for extended periods without significant degradation [8]. Additionally, plastic debris or microplastics (MPs) derived from plastic waste accumulation can interact with organic pollutants or heavy metals, leading to concentrated toxicity in marine and freshwater systems [7]. Moreover, landfills contribute to the contamination of land and aquifers through the release of leachable components trapped in plastic and washed away by rainwater. One example of such leaching is the release of bisphenol A and lead compounds from PVC waste, posing risks to human health and the environment [9,10]. Improper disposal and the littering of PVC waste pose significant threats to wildlife, marine life, and overall ecosystem health, highlighting the need for effective waste management practices.

In recent years, there has been a growing emphasis on shifting towards sustainable waste-management strategies, particularly within the concept of the circular economy, which prioritizes material recycling and resource recovery in line with the Sustainable Development Goals (SDGs): Goal 12 "Responsible consumption and production" [11]. This focus is crucial in reducing the environmental impact of plastics, including PVC, by decreasing the amount of waste for disposal and minimizing the industrial demand for virgin plastics derived from fossil fuels. As a result, the recycling of PVC and other plastics has become a central area of research and technological advancements. Effective strategies for PVC removal and recovery have garnered considerable attention to mitigate the environmental impacts associated with PVC waste and promote sustainable resource management. PVC recycling not only reduces dependence on virgin materials but also minimizes energy consumption and greenhouse gas emissions linked to the production of new PVC [12]. Furthermore, the recovery of valuable components from PVC waste presents an opportunity for resource conservation and aligns with the principles of the circular economy.

In general, conventional plastic waste management practices can be classified into two categories: (i) incineration and (ii) landfilling. Incineration, particularly the field burning of waste, is widely used but notorious due to its negative impacts on air quality, the release of toxic components, and the wastage of energy content in the waste. It is notoriously widespread in low-to-medium income countries with inadequate waste treatment infrastructure [13]. Despite the various options available, landfilling remains the dominant form of waste disposal for mixed plastic waste, including PVC-containing plastics like electronic wastes [14]. Although landfills occupy a relatively small portion of land, they have other serious environmental consequences. The decomposition of organic matter, such as food waste, in landfills leads to the release of odorous compounds and greenhouse gases, including biogas, which is a mixture of carbon dioxide (CO_2) and methane (CH_4).

Moreover, landfills contribute to the contamination of land and aquifers through the release of leachable components trapped in plastic and washed away by rainwater. One example of a leachable toxic component in PVC is bisphenol A and lead compounds [9,10]. Bisphenol A is an endocrine-disrupting chemical (EDC) because it can mimic the functions of estrogen in the body, while lead is a heavy metal notorious for causing neurological damage especially in the developing brain of babies and children [15–18].

PVC can undergo two main recycling processes: mechanical recycling and feedstock recycling [18]. On the one hand, mechanical recycling is the recommended method for PVC recycling, which involves repurposing and reprocessing the material directly within the production plant where the waste was generated. This type of waste arises during various manufacturing stages, such as the start-up and end of production, mechanical processing of finished products, or waste resulting from production errors. On the other hand, feedstock recycling is another approach for managing PVC waste and should be employed for those that cannot be mechanically recycled due to economic or environmental reasons [19]. Energy recovery, including the gasification of fuels or direct combustion in specialized thermal utilization plants, is one relatively straightforward method of feedstock recycling [13,19,20]. These processes require appropriately designed thermal decomposition facilities and often involve significant investment costs for constructing specialized plants. While this type of recycling may sometimes be considered uneconomical, it may be necessary to facilitate the closed-loop circulation of materials in the global economy. It is important to note that advancements in science and technology continue to offer new possibilities for processing PVC into alternative raw materials, as well as prospects for further enhancing existing recycling technologies to utilize recycled PVC and reduce the consumption of virgin PVC.

In general, plastic recycling encompasses various stages, including classification, separation, and production, which involve both chemical and mechanical recycling methods [21]. Mechanical recycling is suitable for various types of plastic waste but yields low-value materials [20]. Chemical recycling aims to convert plastic waste into plastic monomers, chemicals, fuels, feedstock, and value-added polymers while recovering energy through chemical reactions [22]. Consequently, separation plays a crucial role in plastic recycling, as impurities and non-targeted plastic within waste mixtures can disrupt the recycling process and diminish the expected benefits. Despite the relatively straightforward separation between plastic and other materials like steel, copper, aluminum, and metallic alloys, efficient and clean separation techniques hold significant importance in plastic–plastic separation and recycling. This is because separating homogeneous, high-purity plastics from plastic mixtures with similar densities and surface properties remains extremely challenging. As a result, various separation methods for plastic recycling, including manual separation, gravity separation, tribo-electrostatic separation, magnetic separation, spectroscopy sorting, and plastic flotation have been developed.

The objective of this systematic review is to provide a comprehensive overview of recent technological advancements in PVC removal and recovery, with a specific focus on their potential for resource conservation and recycling. By examining the state-of-the-art approaches and evaluating their advantages and limitations, this study aims to contribute to the development of sustainable practices for PVC waste management. The subsequent sections of this paper are organized as follows: Section 2 explains the review methods, Section 3 presents the comminution techniques for PVC separation from other materials, Section 4 discusses gravity separation techniques for PVC removal, Section 5 explores the application of magnetic separation in PVC recycling, Section 6 examines various approaches for electrical separation, Section 7 introduces sorting technologies for PVC recovery and recycling, Section 8 addresses the challenges and future prospects of flotation technology for PVC removal and recycling, Section 9 explores other relevant PVC recycling technologies, and Section 10 presents the conclusions of this work.

2. Review Methods

A systematic review was conducted to examine the current state of technologies related to the separation of PVC from other materials between 2012 and 2021. The review followed the Preferred Reporting Items for Systematic Reviews and Meta-Analyses (PRISMA) guidelines [23], as well as the guidelines recommended by Andrews [24]. Peer-reviewed journal publications were identified using specific keywords, including "PVC", "polyvinyl chloride", "separation", "recycling", "sorting", "recovery", and "removal". The databases of Web of Science and Scopus were utilized, and the publication dates were restricted to between 2012 and 2021. Initially, a total of 4720 articles were obtained based on these criteria; that is, 2407 papers from Web of Science and 2313 papers from Scopus. After removing duplicates, 3123 articles underwent screening, as shown in Figure 1. The screening involved examining titles, highlights, abstracts, and keywords to eliminate articles that did not focus on "PVC". Among the screened articles, 2821 papers were excluded, leaving 302 papers for the next step. During the eligibility evaluation, full-text articles were assessed [25]. The results revealed that 26 papers were inaccessible, 30 papers were not peer-reviewed, 13 papers were not written in English, 22 papers were either review papers, technical papers, features, focuses, letters, or case studies, and 137 papers were unrelated to PVC separation. Following the systematic selection process, a total of 66 papers remained and were included in this review. These papers were categorized based on their contents into the following: selective comminution (1 paper), gravity separation (6 papers), magnetic separation (1 paper), electrical separation (13 papers), sorting (2 papers), flotation (40 papers), and others (3 papers).

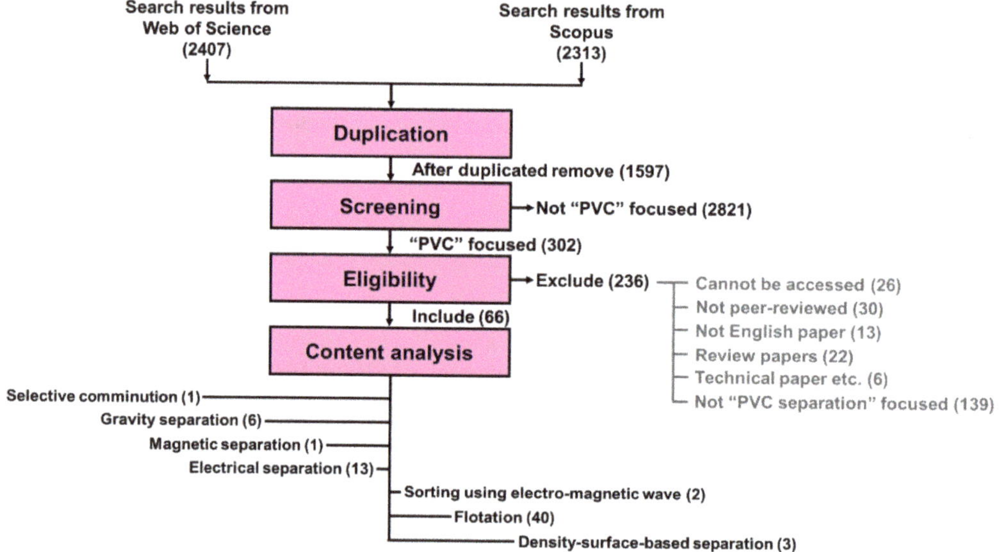

Figure 1. A schematic diagram of the study selection criteria and methodology to identify related research for this systematic review.

3. Selective Comminution

The comminution technique has been applied to isolate plastics from metals, as well as to separate PVC from wire harnesses which usually consist of copper (Cu) strands coated with thin PVC cables. Kumar et al. [26] reported the separation of PVC from Cu-wire harness using n-butyl acetate to swell the PVC coatings and remove them by comminution in a rod mill with an inner diameter of 160 mm and a length of 160 mm including seven stainless steel rods with 125 mm in length and 15 mm in diameter (Figure 2). The results show that PVC and Cu-wires (20 cm long) were completely separated within 60 min at a rotation speed of 15 rpm. In addition, it was observed that n-butyl acetate extracted approximately 90 wt% of the phthalate plasticizer from the PVC coating during the swelling treatment. This technique is promising because it efficiently reduced the environmental impacts of PVC by chemically separating its phthalate plasticizer component.

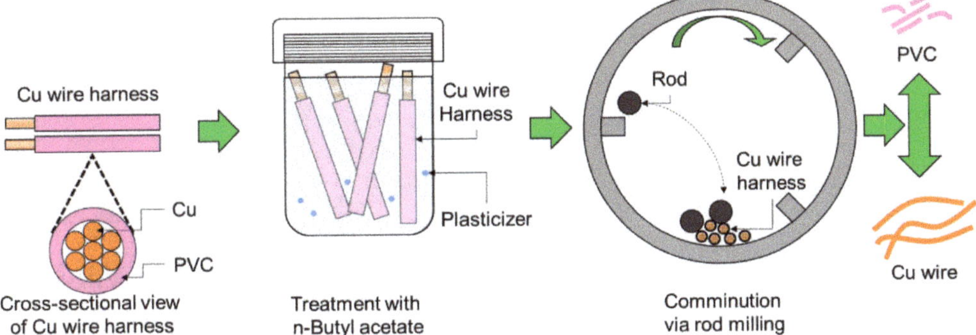

Figure 2. A schematic illustration of PVC separation from Cu wire harnesses via comminution.

4. Gravity Separation

Gravity separation is a well-established technique in various industries like mineral processing and coal cleaning because of its simplicity, low cost, low energy consumption, and high efficiency. This technique relies on the principle that different materials settle at different rates under the influence of gravity in a fluid medium like air, water, and dense media. It involves creating controlled conditions where the heavier plastics (i.e., high density) sink while the lighter plastics (i.e., low density) float. According to relevant research compiled in this systematic review, many gravity separation methods have been extensively investigated and employed for PVC recycling. These techniques include sink–float separation [27], hydrocyclone with suspension media [28,29], hydraulic separator [30], and jig separation [31,32], and they vary in terms of equipment requirements, operational parameters, and applications as summarized in Table 1.

The sink float separation technique was developed and has been employed as a method of separation based on varying the density of the aqueous media utilized during the density-based process. According to Quelal et al. [27], multiple separation steps were needed with different media density (i.e., 1.000, 1.090, 1.100, and 1.175 g/cm^3) to separate post-consumer plastics including PVC, polypropylene (PP), polyethylene (PE), polyethylene terephthalate (PET), polystyrene (PS), and acrylonitrile butadiene styrene (ABS). These authors reported an approximately 84.6% recovery of each plastic using sodium hydroxide (NaOH) and tap water as an aqueous media for separation.

Hydrocyclones are widely utilized in various industries for their exceptional capability to separate solid particles from liquid suspensions based on the principles of centrifugal force and fluid dynamics. As reported by Yuan et al. [28], a hydrocyclone was used to separate binary mixtures of PVC/PET with a 94.6% PVC recovery and a purity of 87.5%. In addition, calcium chloride (CaCl$_2$) was used by these authors as a medium with a density of 1.3 g/cm^3. Moreover, this work noted that to improve the separation performance, a series of cyclone separations known as LARCODEMS is required [29] to recover PVC from residual plastic products containing PVC, PET, PS, polycarbonate (PC), ABS, and polymethyl methacrylate (PMMA). They also highlighted that a 100% PVC recovery and purity was obtained when the media suspension (i.e., suspension of ground calcite) was adjusted to 1.27 g/cm^3.

An apparatus called Multidune, proposed by Lupo et al. [30], was developed as a hydraulic separator for the purpose of separating plastics based on variations in water flow rate. This device was demonstrated to be capable of achieving high levels of purity and recovery when separating PVC from PVC/PC mixtures, with a purity reaching up to 99.9% and recovery rates reaching up to 99.7%.

Table 1. A summary of recent gravity separation techniques for PVC.

Technique	Density [g/cm^3]							Size [mm]	Conditioning Details	Purity [%]	Recovery [%]	Reference
	PVC	PP	PE	PET	PS	PC	ABS					
Sink–float	1.35	0.91	0.95	1.35	1.04	1.20	1.07	3.0	Tap water (Density = 1.00 g/cm^3)	N/A	84.6	[27]
Cyclone with suspension media	1.44	–	–	1.34	–	–	–	0.75	NaOH (Density = 1.09; 1.10; 1.18 g/cm^3)	94.6	87.5	[28]
Cyclone with suspension media (LARCODEMS)	1.44	–	–	1.36	1.05	1.20	1.05	2.0–4.0	Suspension prepared from CaCl$_2$ (Density = 1.30 g/cm^3)	100.0	100.0	[29]
Hydraulic separator	1.61	–	–	–	–	1.21	–	2.0–4.8	Suspension prepared from ground calcite (Density = 1.09; 1.18; 1.27 g/cm^3)	99.9	99.7	[30]
Jig	1.28	–	–	–	1.05	–	–	1.0–5.6	Frequency of diaphragm movement 30 cycles/min; water displacement 30 mm	99.3	82.2	[31]
	1.38	–	–	1.31	–	–	–	2.0–8.0	Vary the water flowrate in multidune (700–1400 cm^3/s)	94.3	N/A	[32]

Note: "N/A" means "not available", and "–" means "not included in the experiments".

Jigging is another gravity concentration technique that operates within a pulsated bed where a combination of solid and water is contained in a perforated vessel. Vertical currents of water are introduced to generate water pulsation and induce the stratification of particles based on density differences among the constituent particles within the mixture. Consequently, particles with a higher density migrate downward and remain confined within the jigging chamber, while particles with lower density ascend and overflow from the system. As highlighted by Pita and Castilho [31] and Phengsaart et al. [32], jig separation has a high performance and it recovered PVC from binary mixtures of PVC/PS and PVC/PET with a purity between 94.3% and 99.3%.

5. Magnetic Separation

Magnetic separation is another physical separation technique that has been applied to recover PVC. Based on the literature, a novel magnetic separator was recently developed to separate PVC from various mixed plastics called magnetic projection [33]. This device consists of a feed unit, magnetic ring unit, and a separation area as illustrated in Figure 3. The feed unit is comprised of an entrance, a feeding channel, and a feeding pendulum driven by an actuator. Meanwhile, the separation area is composed of a paramagnetic solution (MnCl$_2$) and baffles to separate and collect particles of different densities. The separation process begins when the feeding pendulum pushes samples through the anti-magnetic force, which is generated by a ring magnet placed between the feeding part and the separation area. The samples are then projected and thrown into the separation area. These authors reported that the method can reach a 96.4% PVC recovery from mixtures of PP, ABS, PC, polylactic acid (PLA), and PET [33].

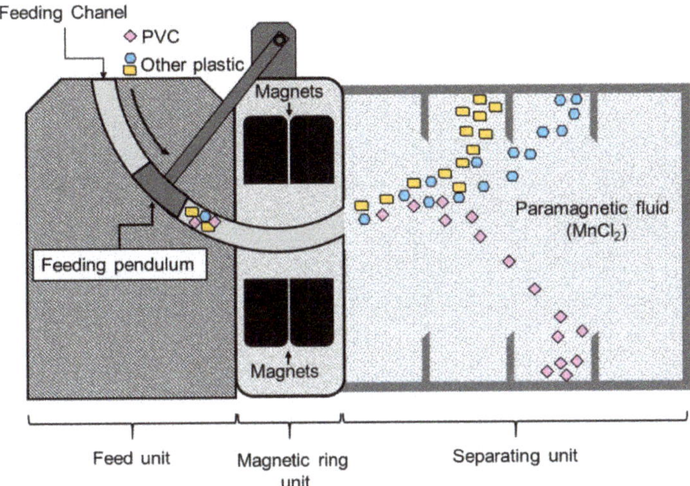

Figure 3. A schematic diagram illustrating the principles of magnetic projection [33].

6. Electrical Separation

Electrical separation is a technique that separates plastics from other materials using differences in electrical conductivity. In addition, this technology is beneficial due to its low cost, low environmental impact, high recovery rate, and high purity of products.

The tribo-electrostatic separator is the most frequently used electrical-based technique to separate PVC from other plastics reported in recent related research (Table 2). This separation method was found to be suitable for separating mm-sized particles as shown in Table 2, and its optimal size range is between 1 and 10 mm. Generally, the separator compartment consists of a feeding device, charging device, high-voltage electrical field, and collection bin. The separation step can be divided into two parts as displayed in Figure 4: (i) electrical charging on particles (i.e., tribo-charger), (ii) deflection process (i.e., electrostatic separator).

In the tribo-charger, particles are charged by an electric source or contact-friction between the particle–particle and particle–wall resulting in the creation of positive and negative electric charges that accumulate on plastic surfaces. In general, it can be classified into two types: solid single-phase and gas–solid two-phase. Solid single-phase includes vibration, corona, and friction rotating drum charging [34–37] while fluidized bed, propeller-type, and cyclone charging are categorized as gas–solid two-phase [34,35,38–44].

In the deflection step, charged particles from the first step are separated and deflected by an electric field depending on their polarity and amount of charge. There are typically three types of deflection process reported in the literature: free fall, drum-type, and belt separator. The free fall type allows particles to fall freely into the electrostatic field generated by applying a high voltage on the positive and negative electrodes of the separation system. Within the electric field, charged particles move towards the oppositely charged electrodes and are collected in the corresponding boxes [34,39,41]. For drum-type and belt separators, charged particles are placed on a drum or belt that is electrically grounded. Highly conducting particles lose their charge, are pulled away from the drum surface, and are placed in collection boxes in the zone farthest from the drum. In contrast, poorly conducting particles retain their charge, are held on the drum surface, and brushed off into the boxes closest to the drum [36,37,40,42–44].

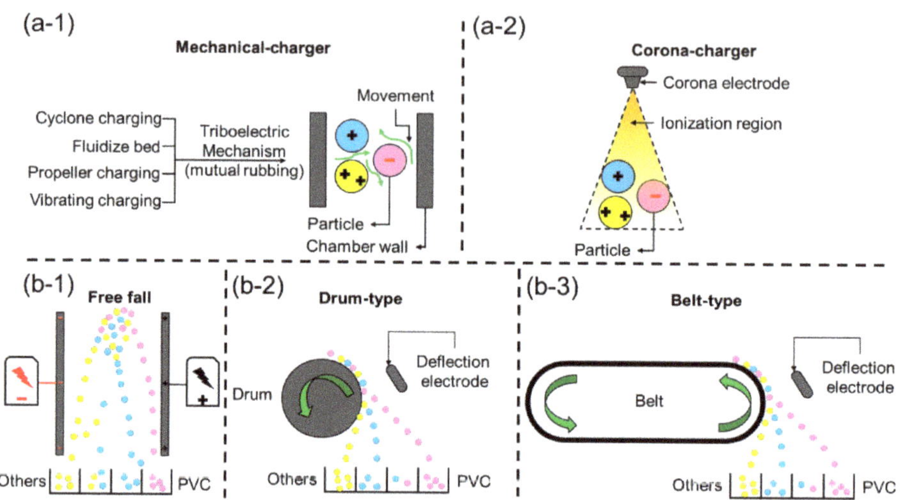

Figure 4. Schematic diagrams illustrating the principles of electrical separation; (**a**) electrical charging on particles ((**a-1**) mechanical-charger and (**a-2**) corona-charger) and (**b**) deflection process ((**b-1**) free fall, (**b-2**) drum-type, and (**b-3**) belt-type).

Table 2. Relate research on the separation of PVC from other materials using electrical separator.

Separators	Charging Mechanism	Moving Mechanism	Samples									Size [mm]	Purity [%]	Recovery [%]	Reference
			PVC	PP	PE	PET	PA	PC	ABS	Others	Metals				
Tribo-electric separator	Fluidized bed	Free fall	√				√					3.0–5.0	N/A	98	[34]
			√									0.01–0.1	N/A	90	[38]
			√				√					2.0–2.5	N/A	90	[39]
			√					√				2.0–2.5	N/A	92	[39]
			√			√						6.5–10	98	66	[41]
			√		√							3.5–6.5	91	77	[41]
			√		√	√						3.5–10	N/A	77	[41]
		Drum-type	√			√						2.0–2.8	N/A	96	[42]
			√						√			0.02	N/A	95	[43]
			√			√				√		2.7–4.0	N/A	<50	[44]
		Belt-type	√				√	√		√		1.0–2.0	93	N/A	[40]
	Propeller charging	Free fall	√				√					3.0–5.0	N/A	97	[34]
	Cyclone charging	Free fall	√	√	√				√	√	√	<5.0	N/A	N/A	[35]
	Vibrating charging	Free fall	√	√	√				√	√	√	<5.0	N/A	N/A	[35]
			√				√					3.0–5.0	N/A	85	[34]
	Corona charging	Belt-type	√	√						√		1.6–3.2	N/A	50–100	[36]
			√	√						√	√	1.6–3.2	96	99	[36]
	Friction rotating drum charging	Drum-type	√	√						√		4.0–8.0	97	41	[37]
Electrostatic adhesion	No charging	Vibrating inclined plane-type	√		√						√	0.5–5.0	99	95–100	[45]
Electrostatic separator	No charging	Drum-type	√								√	2.0–5.0	N/A	100	[46]

Note: "√" means "included in the experiments", "blank" means "not included in the experiments", "N/A" means "not available", "PA" means "polyamide", "Other" means "other plastics including PS, polyurethane (PU), polylactide (PLA), polycaprolactone (PCL), and poly(3-hydroxybutyrate-co-4-hydroxybutyrate) (P(3,4HB))", and "Metals" means "aluminum and copper".

Electrostatic separators like the drum-type electrostatic separator and electrostatic adhesion are suitable for separating electrically conducting from non-conducting materials and could be applied to separate PVC when some metals are present in the feed sample [45,46]. These separation techniques have been shown to achieve high separation performances with 95–100% PVC recovery.

7. Sorting Using Electromagnet Wave Sorting

In past decades, researchers have diligently strived to advance increasingly accessible and efficient methodologies for the purpose of plastic sorting. These techniques are widely employed in the non-destructive assessment of plastic samples through analysis of the unique molecular and elemental signatures emitted or absorbed by plastics. These techniques include laser-induced breakdown spectroscopy (LIBS), near-infrared spectroscopy (NIR), mid-infrared spectroscopy (MIR), Raman spectroscopy, and X-ray fluorescence spectroscopy (XRF) [47]. The underlying principle of these sorting technologies entails directing a light beam, such as a neodymium-doped yttrium aluminum garnet laser or halogen light, onto the surface of plastics. The atoms, ions, and molecules within the plastic sample become activated and transition to excited states, subsequently emitting photons of specific frequencies upon returning to a steady state. Light collectors are employed to gather the emitted light, which is then processed by computer software to generate a distinct spectrum. Subsequently, an air gun is employed to eject the specifically selected plastic, guided by signals from the operating software. Finally, each plastic is sorted into the appropriate container, as depicted in Figure 5. More recently, machine learning algorithms have emerged as powerful tools in plastic sorting technology. These algorithms leverage vast datasets of plastic samples to learn and recognize patterns, enabling the automated identification and sorting of plastics. Notably, Peng et al. [48] reported achieving 100% accuracy in plastic sorting through LIBS techniques by compressing the training image data of each plastic type, such as PVC, ABS, polyamide (PA), and PMMA, into a singular image for use by the machine-learning algorithms. Additionally, Duan and Li [47] explored the classification of distinct plastic categories using NIR spectroscopy, demonstrating that sorting plastics via NIR techniques could accurately distinguish between PVC, PP, PS, PET, high-density polyethylene (HDPE), low-density polyethylene (LDPE), and PC with 100% accuracy when incorporating multiple regions of the spectrum to enhance the machine-learning model.

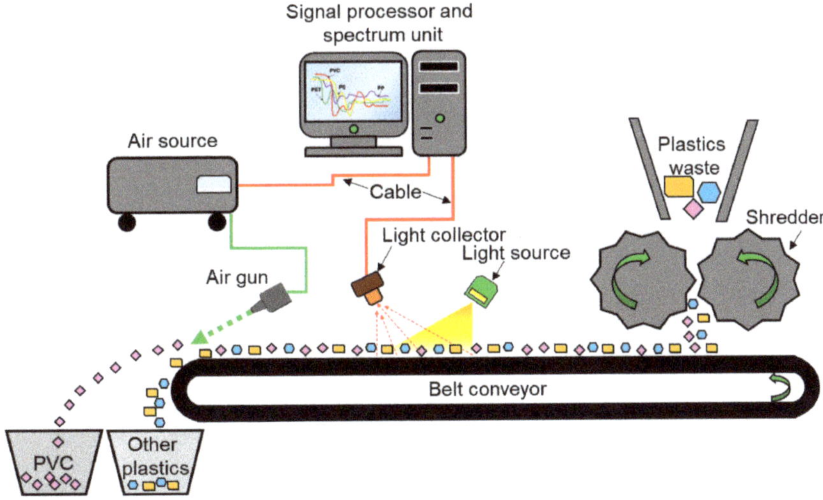

Figure 5. A schematic diagram of sorting technology for the separation of plastic waste.

Overall, the principle of plastic sorting technology lies in the systematic analysis and categorization of plastics using a combination of spectroscopic analysis, automated sorting systems, and machine-learning algorithms. By effectively segregating plastic materials, this technology facilitates the recycling and reutilization of plastics, contributing to sustainable waste management and resource-conservation efforts.

8. Flotation

Early research has demonstrated that flotation technology can be used as a separation approach for plastic wastes, due to its previous successful industry application in mineral processing. This technique is based on the difference in the surface wettability of materials utilizing bubbles as a particle carrier. Hydrophobic particles (i.e., water-hating) are brought to the water surface via bubble attachment, while hydrophilic particles (i.e., water-loving) remain in the flotation cell and sink to the bottom. In mineral flotation, collectors are commonly employed to enhance the hydrophobicity of mineral particles. However, the approach taken in plastics flotation differs significantly, as it primarily centers on enhancing the hydrophilicity of plastic particles to augment the disparity in their wettability. Nevertheless, the proximity in hydrophobicity among different plastic types presents challenges in segregating individual plastic types through flotation alone. Consequently, the field of plastics flotation heavily relies on modifying the hydrophobicity of plastics to improve flotation efficiency. Extensive investigations have been conducted on various reagents known as wetting agents to address this requirement. Figure 6a illustrates the fundamental concept of mineral flotation in comparison to plastic flotation (Figure 6b). Additionally, prior surface treatments of plastics are necessary before flotation can be carried out. These pretreatment steps are particularly effective for separating high-density plastics, notably PVC, owing to PVC's unique surface properties, characterized by a higher dielectric-loss coefficient compared with other plastics. The application of surfactants facilitates the selective separation of PVC through selective surface reactions on PVC; the abundance of hydrophilic functional groups, such as ether, hydroxyl, and carboxyl moieties, can be increased or decreased relative to other plastics [49]. Based on our assessment of the literature, this technique can be classified into two categories: flotation with surfactants and flotation without surfactants.

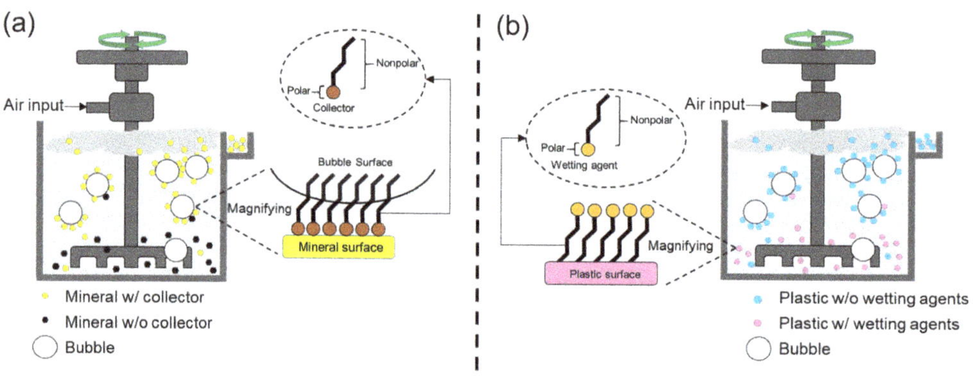

Figure 6. Schematic diagrams illustrating the flotation of (**a**) minerals and (**b**) plastics.

8.1. Flotation with Surfactants

The separation of polymeric materials poses challenges, as gravity separation and hand sorting are not effective due to the similar densities of different polymers. Therefore, froth flotation is utilized to separate materials that cannot be separated by gravity. Naturally, all types of plastics exhibit hydrophobic properties. In separation techniques, it is crucial to modify the surface properties of plastics, making one surface more hydrophilic while keeping the other hydrophobic. Plasticizer reagents or wetting agents are employed to alter the behavior of plastic surfaces [50–52] and various reagents have been applied to modify the surface characteristics of plastics (Table 3).

Table 3. A summary of flotation studies on PVC separation using flotation and various surfactants.

Flotation Type	Reagents	Concentration	Condition Time [min]	PVC	PET	PS	PC	ABS	Others	Size [mm]	Purity [%]	Recovery [%]	Ref.
Direct	LA Gelatin	250 g/t 1250 g/t	3	√	√					2.0–3.4	98.9	57.0	[52]
	PVA PEG MC TA	800 mg/L 2000 mg/L 2000 mg/L 1200 mg/L	N/A	√	√	√				N/A	N/A	N/A	[53]
Reverse	Diisooctyl Azelate	600 g/t	10	√	√					2.0–3.4	52.0	99.0	[50]
	LA DIB	0–500 g/t 0–1500 g/t	2	√	√					2.0–3.4	–	–	[51]
	LA	25–55 g/t	5	√	√					2.0–3.4	99.4	90.1	[54]
	Triton XL-100N DIB	1000 g/t 1000 g/t	3	√	√					2.0–3.4	86.4	100.0	[55]
	LS TA MC Triton X-100	N/A	10	√	√	√	√	√	√	<5	98.7	98.9	[56]
	TA	0–10 mg/L	5	√	√	√			√	1.0–5.6	95.8	94.4	[57]
	Saponin	10–30 mg/L	5	√	√	√			√	N/A	95.7	72.8	[58]
	SL CaCl$_2$	100–300 mg/L 30%	4	√			√	√	√	2.0–5.0	N/A	99.8	[59]

Note: "√" means "included in the experiments", "blank" means "not included in the experiments", "N/A" means "not available", and "Others" means "Other plastics including PP, PE, PA, PMMA, and PTFE".

In direct flotation in which PVC is recovered as a froth product, many reagents such as tannic acid (TA), polyethylene glycol (PEG), polyvinyl alcohol (PVA), methyl cellulose (MC), lignin alkali (LA), and gelatin have been used as wetting agents. Without the use of these reagents, PVC, PET, and PS simply sink to the bottom due to their higher density compared with the surrounding liquid media. The concentration of TA up to 1200 mg/L, for example, significantly affected the wettability of PVC, achieving 100% recovery by flotation [53]. Due to the presence of chlorine in the PVC structure, it facilitates the molecular absorption of TA on the PVC surface because chlorine possesses the highest electron affinity and the third-highest electronegativity among chemical elements. Other reagents such as PEG, PVA, and MC are also adsorbed on PVCs, but they exhibit less intensification of hydrophobicity compared to TA. Binary mixtures of chemical agents including PVA-PEG, PVA-MC, TA-MC, TA-PVA, PEG-MC, and TA-PEG have been investigated, resulting in a PVC floatability of 0%, 35%, 40%, 50%, 75%, and 85%, respectively [44]. The highest recovery of PVC from plastic mixtures (PET, PS, and PVC) is achieved using TA-PEG due to the more hydrophobic surface of PVCs after wetting-agent adsorption [53]. Additional research conducted by Yenial et al. [52] examined two other chemical reagents—lignin alkali and gelatin—and reported that both wetting agents affected PVC separation from PET/PVC mixtures. Both of these compounds are anionic substances, causing the surface of PET to become more hydrophilic, depressing PET particles and allowing PVC to float naturally.

In reverse flotation, PVC is recovered in the tail product by depressing PVC plastics using chemical reagents. Altering the wettability and contact angle of the plastic surface influences its floatability behavior. Guney et al. [51] investigated PET/PVC separation using LA and diethyelene glycol dibenzoate (DIB) as chemical agents, both of which increased the hydrophobicity of PET and reduced the hydrophobicity of PVC. These findings are consistent with the study of Yuce et al. [55], who also examined PET/PVC separation using LA. The concentration of LA was highlighted by these previous works as a critical factor in depressing PVC particles; that is, increasing the concentration of wetting reagent leads to a higher PVC content in the tail/sink product [55].

Contact angle is another important factor influencing the separation process during flotation. The addition of LA reduces the contact angle of PVC as the concentration increases. Experimental results have shown a reduction in the contact angle from 79° to 73°, indicating the increased wettability of PVC [51]. A high dosage of TA also resulted in contact angle reduction as depicted in Figure 7. However, it is unnecessary to reduce the contact angle of PVC to zero, as the contact angle of PVC is 49°, which leads to a nearly 100% depression rate [57]. Saponin and sodium lignosulfonate (SL) also influenced PVC flotation by reducing the contact angle [58].

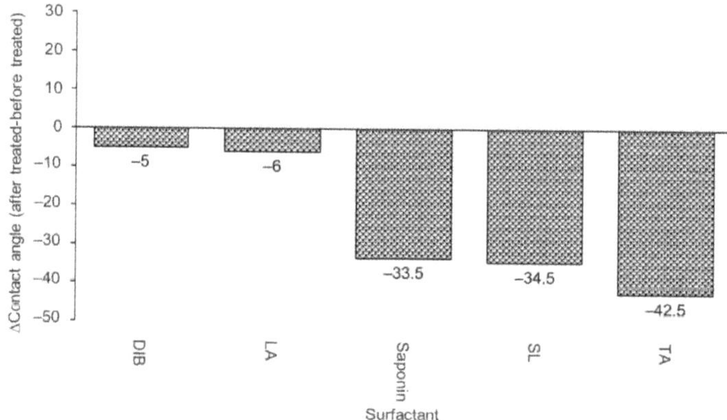

Figure 7. Changes in the PVC contact angle in the presence of various surfactants.

Furthermore, PVC can be depressed due to its higher density compared to other plastics, allowing the other plastics to float. $CaCl_2$ solution is utilized as a wetting agent for PVC separation from ABS, PC, and PA. Increasing the concentration of $CaCl_2$ enhanced the purity of separated PVC. However, a 30% $CaCl_2$ solution reduced the recovery potential of PVC, likely due to the similar densities of the solution and PVC [59].

8.2. Flotation without Surfactants

8.2.1. Pretreatment Using Reagents

The similar surface hydrophobicity of PVC with other plastics makes it impossible to freely isolate from plastics mixtures by flotation without surface modifiers. Hence, surface treatment with chemical reagents is required to increase the difference in surface wettability of PVC with other plastics. According to recent studies, many reagents have been investigated as summarized in Table 4. Direct flotation, whereby PVC is reported as a froth product, has been applied using various reagents for surface oxidation such as chlorine dioxide (ClO_2) [60], potassium ferrate (K_2FeO_4) [61], potassium permanganate ($KMnO_4$) [62,63], sodium persulfate ($Na_2S_2O_8$) [64,65], ammonium persulfate (($NH_4)_2S_2O_8$) [66], and combinations of potassium hydroxide and ethylene glycol (KOH and $(CH_2OH)_2$) [67], as well as sodium hydroxide (NaOH) [68–72] for surface hydrolysis. These reagents enhanced the hydrophilicity of other plastics, while the hydrophobicity of PVC was unaffected. The contact angle of PVC after treatment was stable when these reagents were used, but those of other plastics such as PET, PC, PS, ABS, and PMMA significantly decreased, resulting in the decrease in the floatability of these plastics and their separation from PVC (Figure 8).

Table 4. A summary of flotation studies on PVC separation using flotation without surfactants.

Flotation Type	Mechanism	Reagents	Concentration	Treatment Temperature	Treatment Time	PVC	PET	PS	PC	ABS	PMMA	Size [mm]	Purity [%]	Recovery [%]	Reference
Direct	Surface oxidation	ClO_2	0.5 g/L	70.0	70	√			√			0.8–5.0	N/A	100.0	[60]
		K_2FeO_4	0.18 M/L	75.0	11.5	√				√		3.0–4.0	98.4	98.4	[61]
		$KMnO_4$	5 mM/L	60.0	10.0	√			√		√	2.0–3.2	98.4	98.7	[62]
		$KMnO_4$	2 mM/L	N/A	1.0	√			√		√	3.2–4.0	95.0	98.6	[63]
		$Na_2S_2O_8$	0.1 M/L	70.0	30.0	√		√	√			0.9–4.0	99.8	100.0	[64]
		$Na_2S_2O_8$	0.1 M	20–70	10	√				√		2.0–4.0	99.7	100	[65]
		$(NH_4)_2S_2O_8$	0.2 M	70.0	30.0	√		√	√	√		3.0–4.0	100.0	99.7	[66]
		KOH and $(CH_2OH)_2$	2 g (KOH) and 10 mL $(CH_2OH)_2$	25.0	5.0	√	√		√	√		3.0–4.0	N/A	N/A	[67]
	Surface hydrolysis	NaOH	10%	70.0	20.0	√	√					2.0–4.0	94.6	94.9	[68]
		NaOH	40 g/l	90	20	√	√	√				2.0–4.0	N/A	95.9	[69]
		NaOH	1 M	N/A	10	√	√					2.0–3.4	N/A	N/A	[70]
		NaOH	N/A	60	20	√	√					5.0	N/A	100	[71]
		NaOH	10%	70	20	√	√					0.9–3.2	98.22	93.98	[72]
Reverse	Surface oxidation	$KMnO_4$	1.25 mM/L	60.0	50.0	√	√					0.9–4.0	99	99.7	[73]
		$KMnO_4$	4.6 mM/L	66.5	38.0	√	√					2.5–3.2	96.6	97.9	[74]
		$CaCO_3$	0.11 g	50.6	20.0	√	√					2.0–4.0	100.0	99.0	[75]
		H_2O_2	3%	30.0	30.0	√	√					5.0	99.8	100.0	[76]
	Surface coating	H_2O_2	3%	30.0	30.0	√	√		√	√		5.0	99.5	100.0	[77]
		$AlCl_3$	0.2 M	25.0	1.7	√			√			1.0–2.0	100.0	99.7	[78]
		Ca/CaO	5%	N/A	30	√	√		√	√	√	5.0	96.4	100	[79]
		Fe/Ca/CaO	0.5%	25.0	N/A	√	√		√	√	√	5.0	99	100	[80]
Direct and reverse	Surface oxidation	$Ca(ClO)_2$	0.2–0.5 g/L	70.0	30–50	√	√	√	√	√	√	2.0–2.5	73.0	99.6	[81]

Note: "√" means "included in the experiments", "blank" means "not included in the experiments", and "N/A" means "not available".

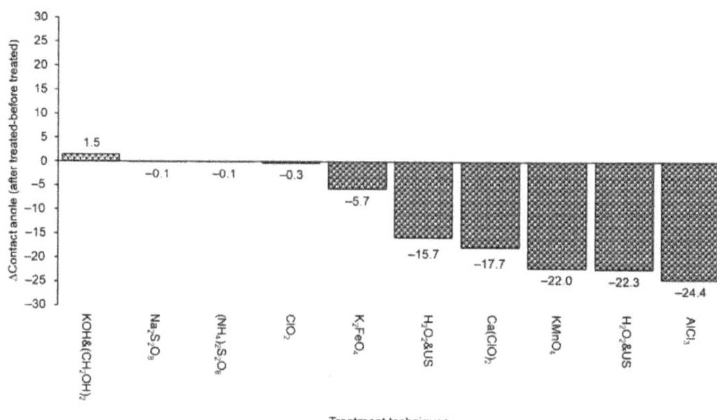

Figure 8. Changes in the PVC contact angle after treatment with various reagents.

Aside from increasing the surface wettability of other plastics, reverse flotation has also been utilized when the PVC surface is more sensitive during surface treatment than other plastics. During flotation, PVC particles remain in the flotation cell and are reported in the sink product. Table 4 summarizes the recent studies on the reverse flotation of PVC and the use of wetting agents and surface modifiers, including $KMnO_4$ [73,74], $CaCO_3$ [75], H_2O_2 [76,77], $AlCl_3$ [78], and $Ca(ClO)_2$ [79]. These previous works reported that these reagents render the PVC surface more hydrophilic compared with other plastic surfaces. As illustrated in Figure 8, these wetting agents directly decreased the contact angle of PVC by approximately 15.7–24.4°, while those of other plastics like PC, ABS, and PET only had minor changes. It is also interesting to note that the dechlorination of PVC surfaces induced by $KMnO_4$, $CaCO_3$, H_2O_2, and $AlCl_3$ was the primary mechanism responsible for the decrease in the hydrophobicity of PVC.

8.2.2. Pretreatment with Fenton Reaction

According to recent studies on plastics flotation, the utilization of the Fenton reaction is an emerging and promising pretreatment method for enhancing the efficiency of plastics flotation. Through an extensive analysis of the available literature, this section reviews the contributions made by various researchers, with particular emphasis on the works of Wang and Wang [82], Wang et al. [83], and Zhang et al. [84]. The Fenton reaction, which involves the generation of hydroxyl radicals (–OH) through the reaction between hydrogen peroxide (H_2O_2) and ferrous ions (Fe^{2+}), has been employed to improve the flotation efficiency of PVC in the presence of other plastics, such as PS, PC, and ABS. Experimental observations indicate that the Fenton treatment significantly reduced the hydrophobicity of PS, PC, and ABS, while no discernible change was observed on the surface of PVC. Wang and Wang [82] conducted experiments wherein PVC, PC, and PS were subjected to the Fenton reaction prior to flotation and found that the optimal treatment conditions consisted of (i) a molar ratio of H_2O_2/Fe^{2+} of 7500:1, (ii) a H_2O_2 concentration of 0.2 M, and (iii) a treatment time and temperature of 2 min and 25 °C, respectively. Moreover, Zhang et al. [84] expanded the scope of the study of Wang and Wang [82] by incorporating the use of green-synthesized nanoscale zero valent iron (GnZVI) for the Fenton reaction. These authors revealed that a lower molar ratio of H_2O_2/Fe^0, specifically 40:1, could be employed with a high performance and speculated that a reduced reagent dosage may be utilized in the Fenton treatment to render the PS surface hydrophilic. Meanwhile, Wang et al. [83] explored the separation of PVC and ABS through Fenton pretreatment and flotation, and demonstrated that a H_2O_2/Fe^{2+} molar ratio of 10,000:1 improved the selectivity and separation efficiency of PVC and ABS, resulting in a high recovery (100%) and purity (>99%) of PVC.

In terms of understanding the underlying mechanism of this treatment, previous works noted that the application of the Fenton reaction introduced hydrophilic functional groups (such as C–O–H and O=C–O) to the surfaces of PC, PS, and ABS, thereby rendering these plastics more hydrophilic and reducing their ability to float. In contrast, the hydrophobic nature of PVC remained largely unaffected following the Fenton treatment, a selective surface reaction that specifically enhanced the flotation efficiency of PVC.

8.2.3. Pretreatment with Thermal heat Treatment (Mild Heat and Microwave)

Thermal heat treatment has emerged as an environmentally friendly pretreatment technique for PVC recycling by enhancing the surface hydrophilicity of PVC. Previous studies have reported that mild heat or microwave pretreatment can generate reactive species, including free radicals, by disrupting chemical bonds within the polymer structure [49,85]. These reactive species have the potential to modify the surface chemistry of plastics [85]. Several investigations conducted by Truc and Lee [49], as well as Mallampati et al. [86], explored the impacts of microwave treatment on the contact angle of various plastics, such as PVC, PC, PS, ABS, and PMMA. The results of these previous works revealed a significant reduction in the contact angle of these plastics following pretreatment, particularly for PVC. Moreover, a combination of powder-activated carbon (PAC) coating and microwave treatment further decreased the contact angle of PVC [49,86]. Specifically, the contact angle of PVC decreased by approximately 17°, while the contact angle of other plastics remained relatively unchanged. This combined treatment involving PAC coating and microwave treatment holds promise in selectively depressing PVC particles while allowing other plastics to float during froth flotation. The optimal treatment conditions for this technique involved microwave treatment for 0.5–1 min at a power of 1120 Watts and a frequency of 2450 MHz. The optimized froth flotation process, based on these treatment conditions, achieved PVC recovery rates of 90–100% with a purity ranging from 82% to 100% when separating PVC from plastic mixtures. These findings offer valuable insights for the development of efficient strategies in the separation and recycling of PVC from mixed plastics. Furthermore, the proposed approach exhibited a great potential in advancing the sustainable management of plastic waste and fostering a circular economy within the plastics industry. Future research should focus on scaling up the process and optimizing treatment conditions for practical implementation in industrial settings.

8.2.4. Pretreatment with Corona Discharge

Corona discharge is one of the techniques for localized hydrophilization, which utilizes a high-voltage electric field to generate a localized plasma on the plastic surface. The corona discharge treatment leads to the formation of highly reactive species, such as oxygen radicals, which induce chemical reactions on the plastic surface. These reactions modify the plastic surface, leading to increased hydrophilicity and surface energy. The enhanced surface properties enable better wetting and the adhesion of flotation agents, improving the efficiency of plastics flotation. Numerous studies have investigated the application of corona discharge pretreatment in separating PVC from other plastics [87]. For instance, Zhao et al. [87] examined the impact of corona discharge pretreatment on the surface properties of PVC and HDPE. They found that a corona discharge activating energy of 12.0 kJ/m^2 significantly increased the hydrophilicity of HDPE, while causing a slight decrease in the hydrophobicity of PVC. This modification resulted in an improved difference in contact angle between the two plastics, ultimately enhancing the flotation efficiency of PVC. The study achieved an impressive 88.1% recovery with 94.2% purity for PVC, with the contact angle of HDPE decreasing to 63.5° (a decrease of 32° from the untreated surface), while the contact angle of PVC was maintained at 87.5° (a decrease of 2.7° from the untreated surface). Furthermore, when the activating energy was increased to 15.0 kJ/m^2, the hydrophobicity of PVC was significantly reduced [87]. Similarly, Zhao et al. [88] investigated the use of corona discharge pretreatment for PVC and PET before flotation. The results demonstrated that corona discharge treatment effectively modified the surface properties of PVC, leading to a substantial decrease in its contact angle from 90.2° to 70.1°. However, the impact on PET was found to be insignificant and only a slight decrease in its contact angle was observed. An additional advantage of corona discharge pretreatment is its dry operation and environmentally friendly nature. Unlike other methods that may involve the use of chemicals or solvents, corona discharge pretreatment relies solely on electricity to induce surface changes. This aspect makes corona discharge an appealing option for large-scale applications in plastic-recycling facilities.

9. Density-Surface-Based Separation

Recently, the techniques of gravity separation and flotation were combined as density-surface-based separation, which was applied for plastic separation including PVC.

One approach for the separation of PVC from other plastics involves the combination of elutriation and flotation, exploiting settling characteristics and differential hydrophobicity, respectively [89]. Researchers have successfully employed the elutriation principle in a teeter bed separator to achieve the density-based concentration of plastics [89]. The method utilizes surface active reagents, such as TA and $KMnO_4$, to modify the floatability of plastics and enhance selectivity during separation [89]. By integrating crossflow separation with froth flotation, a novel process flowsheet was developed, demonstrating a high efficiency and selectivity in the recovery of different plastics [89]. Notably, HDPE and PVC were recovered at high rates, showcasing the effectiveness of the approach [89].

Another promising technique in the separation of PVC from other plastics is the hybrid jig, which combines the principles of jig separation and flotation [90,91]. A two-step approach involving a pre-wetting step with a solution containing the wetting agent (AOT) and subsequent hybrid jig separation in water was proposed to separate PVC from PA [90]. To further optimize the hybrid jig separation process, the estimation of critical parameters such as the apparent specific gravity ($SG_{apparent}$) and attached-bubble volume on plastic particles during water pulsation have been investigated [91]. By utilizing this measurement method, researchers conducted hybrid jig separation experiments on various plastic mixtures including PET and PVC with similar specific gravities [91]. The results demonstrated that $SG_{apparent}$ and a newly proposed index called the apparent concentration criterion ($CC_{apparent}$) could be used to estimate the separation efficiency of the hybrid jig [91].

10. Conclusions

In this paper, research works between 2012 and 2021 related to the separation of PVC from other materials were systematically reviewed using the PRISMA guidelines. The findings revealed that most studies employed flotation techniques and the surface modification of plastics to facilitate the separation of PVC from other plastics. These techniques demonstrated impressive results, achieving PVC recovery rates of 82% to 100%, purity levels ranging from 94% to 100%, and efficacy within a specific size range of 1.0–5.6 mm. In comparison, electrical separation presented a broader operational window, catering to a size spectrum of 0.01–5.0 mm. Notably, this method excelled particularly when PVC was intermixed with other waste materials, especially those containing metals. The electrical separation approach has shown remarkable success in achieving high recovery and purity levels in such complex waste streams.

Flotation has emerged as a widely investigated method for PVC separation, driven by the inherently hydrophobic nature of plastics and its compatibility with plastic separation from various materials. Recent studies highlighted that flotation could achieve significant PVC recovery rates ranging from 57% to 100%, achieving purity levels of 52% to 99%, all without necessitating the use of surfactants. Previous works have also explored the modification of plastic surfaces using an array of surfactants, such as ClO_2, K_2FeO_4, NaOH, $KMnO_4$, $Na_2S_2O_8$, $(NH_4)_2S_2O_8$, KOH, $(CH_2OH)_2$, $CaCO_3$, H_2O_2, $AlCl_3$, and $Ca(ClO)_2$, and found that this approach was effective in enhancing PVC separation by amplifying the differences in hydrophobicity among plastics during flotation. Notably, both direct and reverse flotation methodologies have shown promise in recent investigations, achieving impressive PVC recovery rates of 94% to 100% alongside purity levels ranging from 73% to 100%.

A cutting-edge magnetic separation technique known as magnetic projection has emerged as a recent innovation for isolating PVC from a range of mixed plastics. This method achieved an impressive PVC recovery rate of 96% when dealing with a mixture comprising of PP, ABS, PC, PLA, and PET. Furthermore, the refinement of plastic sorting methodologies was evident in the widespread development of electromagnetic wave sorting. This technique significantly enhanced the precision of plastic separation. Within the scope of this systematic review, it is worth noting that both LIBS and NIR techniques have demonstrated remarkable accuracy in plastic sorting. Specifically, these techniques achieved a 100% accuracy rate when applied to mixed plastics containing ABS, PA, PMMA, PP, PS, PET, and PE. In addition, recent advancements in the field with more sophisticated technologies such as density-surface-based separation have been introduced.

The emergence of the UN-SDGs is expected to catalyze further research endeavors aimed at mitigating environmental impacts and promoting resource conservation across the entire life cycle of materials and products. Recycling, recognized as a pivotal approach towards achieving sustainability, offers the dual benefit of reducing environmental burdens and decreasing reliance on finite natural resources. By continually improving the separation processes and exploring innovative recycling techniques, we can make significant strides in addressing the challenges associated with PVC separation and contribute to a more sustainable future.

Author Contributions: Conceptualization, T.P.; methodology, T.P., C.M., P.S. and O.J.; software, C.M. and P.S.; validation, T.P., P.J., C.M. and P.S.; formal analysis, T.P., P.J., C.M. and P.S.; data curation, P.J., C.M. and P.S.; writing—original draft preparation, T.P., P.J., C.M., P.S., M.R., O.J., K.A., S.J., I.P., C.B.T. and M.I.; writing—review and editing, T.P., P.J., K.A., S.J., I.P., C.B.T. and M.I.; visualization, P.J.; supervision, T.P., O.J., C.B.T. and M.I.; project administration, T.P.; funding acquisition, T.P. All authors have read and agreed to the published version of the manuscript.

Funding: This research was funded by the Asahi Glass Foundation, and Chulalongkorn University.

Institutional Review Board Statement: Not applicable.

Informed Consent Statement: Not applicable.

Data Availability Statement: Not applicable.

Acknowledgments: The authors wish to thank the members of the Faculty of Engineering, Chulalongkorn University, for their support to this project and gratefully acknowledge the editors and anonymous reviewers for their valuable input to this paper.

Conflicts of Interest: The authors declare no conflict of interest.

References

1. Geyer, R.; Jambeck, J.R.; Law, K.L. Production, use, and fate of all plastics ever made. *Sci. Adv.* **2017**, *3*, e1700782. [CrossRef] [PubMed]
2. Key Figures of the European Plastics Industry. In *Plastics—The Facts 2022*; Plastics Europe: Brussels, Belguim, 2022. Available online: https://plasticseurope.org/wp-content/uploads/2021/09/Plastics_the_facts-WEB-2020_versionJun21_final.pdf (accessed on 25 June 2023).
3. Rafey, A.; Pal, K.; Bohre, A.; Modak, A.; Pant, K.K. A State-of-the-Art Review on the Technological Advancements for the Sustainable Management of Plastic Waste in Consort with the Generation of Energy and Value-Added Chemicals. *Catalysts* **2023**, *13*, 420. [CrossRef]
4. Miliute-Plepiene, J.; Fråne, A.; Almasi, A.M. Overview of polyvinyl chloride (PVC) waste management practices in the Nordic countries. *Clean. Eng. Technol.* **2021**, *4*, 100246. [CrossRef]
5. Onwudili, J.A.; Insura, N.; Williams, P.T. Composition of products from the pyrolysis of polyethylene and polystyrene in a closed batch reactor: Effects of temperature and residence time. *J. Anal. Appl. Pyrol.* **2009**, *86*, 293–303. [CrossRef]
6. Miandad, R.; Barakat, M.A.; Aburiazaiza, A.S.; Rehan, M.; Nizami, A.S. Catalytic pyrolysis of plastic waste: A review. *Process Saf. Environ.* **2016**, *102*, 822–838. [CrossRef]

7. Xia, Y.; Niu, S.; Yu, J. Microplastics as vectors of organic pollutants in aquatic environment: A review on mechanisms, numerical models, and influencing factors. *Sci. Total Environ.* **2023**, *887*, 164008. [CrossRef]

8. Hopewell, J.; Dvorak, R.; Kosior, E. Plastics recycling: Challenges and opportunities. *Philos. Trans. R. Soc. B* **2009**, *364*, 2115–2126. [CrossRef]

9. Xu, S.-Y.; Zhang, H.; He, P.-J.; Shao, L.-M. Leaching behavior of bisphenol A from municipal solid waste under landfill environment. *Environ. Technol.* **2011**, *32*, 1269–1277. [CrossRef]

10. Tsunekawa, M.; Ito, M.; Yuta, S.; Tomoo, S.; Hiroyoshi, N. Removal of lead compounds from polyvinylchloride in electric wires and cables using cation-exchange resin. *J. Hazard. Mater.* **2011**, *191*, 388–392. [CrossRef]

11. United Nations. Transforming Our World: The 2030 Agenda for Sustainable Development. 2015. Available online: https://sustainabledevelopment.un.org/post2015/summit (accessed on 29 June 2023).

12. Andrady, A.L.; Neal, M.A.; Ji, R. Applications and societal benefits of plastics. *Philos. Trans. R. Soc. B* **2019**, *374*, 20180198. [CrossRef]

13. Lange, J.-P. Managing Plastic Waste—Sorting, Recycling, Disposal, and Product Redesign. *ACS Sustain. Chem. Eng.* **2021**, *9*, 15722–15738. [CrossRef]

14. Tabelin, C.B.; Park, I.; Phengsaart, T.; Jeon, S.; Villacorte-Tabelin, M.; Alonzo, D.; Yoo, K.; Ito, M.; Hiroyoshi, N. Copper and critical metals production from porphyry ores and E-wastes: A review of resources availability, processing/recycling challenges, socio-environmental aspects, and sustainability issues. *Resour. Conserv. Recy.* **2021**, *170*, 105610. [CrossRef]

15. Silwamba, M.; Ito, M.; Hiroyoshi, N.; Tabelin, C.B.; Fukushima, T.; Park, I.; Jeon, S.; Igarashi, T.; Sato, T.; Nyambe, I.; et al. Detoxification of lead-bearing zinc plant leach residues from Kabwe, Zambia by coupled extraction-cementation method. *J. Environ. Chem. Eng.* **2020**, *8*, 104197. [CrossRef]

16. Wu, P.; Tang, Y.; Jin, H.; Song, Y.; Liu, Y. Consequential fate of bisphenol-attached PVC microplastics in water and simulated intestinal fluids. *Environ. Sci. Ecotechnol.* **2020**, *2*, 100027. [CrossRef] [PubMed]

17. Jordan, V.C. The New Biology of Estrogen-induced Apoptosis Applied to Treat and Prevent Breast Cancer. *Endocr. Relat. Cancer* **2023**, *22*, R1–R31. [CrossRef]

18. Chen, Y.; Gao, Y.; Huang, X.; Li, S.; Zhang, Z.; Zhan, A. Incorporating adaptive genomic variation into predictive models for invasion risk assessment. *Environ. Sci. Ecotechnol.* **2024**, *18*, 100299. [CrossRef]

19. Lewandowski, K.; Skórczewska, K.A. Brief Review of Poly(Vinyl Chloride) (PVC) Recycling. *Polymers* **2022**, *14*, 3035. [CrossRef]

20. Lu, L.; Li, W.; Cheng, Y.; Liu, M. Chemical recycling technologies for PVC waste and PVC-containing plastic waste: A review. *Waste Manag.* **2023**, *166*, 245–258. [CrossRef]

21. Jiang, H.; Zhang, Y.; Wang, H. Surface Reactions in Selective Modification: The Prerequisite for Plastic Flotation. *Environ. Sci. Technol.* **2020**, *54*, 9742–9756. [CrossRef]

22. Jehanno, C.; Alty, J.W.; Roosen, M.; Meester, S.D.; Dove, A.P.; Chen, E.Y.; Leibfarth, F.A.; Sardon, H. Critical advances and future opportunities in upcycling commodity polymers. *Nature* **2022**, *603*, 803–814. [CrossRef]

23. Moher, D.; Liberati, A.; Tetzlaff, J.; Altman, D.G. Preferred reporting items for systematic reviews and meta-analyses: The PRISMA statement. *BMJ* **2009**, *339*, b2535. [CrossRef] [PubMed]

24. Andrews, R. The place of systematic reviews in education research. *Br. J. Educ. Stud.* **2005**, *53*, 399–416. [CrossRef]

25. Jindal-Snape, D.; Hannah, E.F.S.; Cantali, D.; Barlow, W.; MacGillivray, S. Systematic literature review of primary–secondary transitions: International research. *Rev. Educ.* **2020**, *8*, 526–566. [CrossRef]

26. Kumar, H.; Kumagai, S.; Kameda, T.; Saito, Y.; Yoshioka, T. Highly efficient recovery of high-purity Cu, PVC, and phthalate plasticizer from waste wire harnesses through PVC swelling and rod milling. *React. Chem. Eng.* **2020**, *5*, 1805–1813. [CrossRef]

27. Quelal, W.O.M.; Velázquez-Martí, B.; Gisbert, A.F. Separation of virgin plastic polymers and post-consumer mixed plastic waste by sinking-flotation technique. *Environ. Sci. Pollut. Res.* **2022**, *29*, 1364–1374. [CrossRef]

28. Yuan, H.; Fu, S.; Tan, W.; He, J.; Wu, K. Study on the hydrocyclonic separation of waste plastics with different density. *Waste Manag.* **2015**, *45*, 108–111. [CrossRef]

29. Gent, M.; Sierra, H.M.; Menéndez, M.; de Cos Juez, F.J. Evaluation of ground calcite/water heavy media cyclone suspensions for production of residual plastic concentrates. *Waste Manag.* **2018**, *71*, 42–51. [CrossRef]

30. Lupo, E.; Moroni, M.; Marca, F.L.; Fulco, S.; Pinzi, V. Investigation on an innovative technology for wet separation of plastic wastes. *Waste Manag.* **2016**, *51*, 3–12. [CrossRef]

31. Pita, F.; Castilho, A. Influence of shape and size of the particles on jigging separation of plastics mixture. *Waste Manag.* **2016**, *48*, 89–94. [CrossRef]

32. Phengsaart, T.; Ito, M.; Kimura, S.; Azuma, A.; Hori, K.; Tanno, H.; Jeon, S.; Park, I.; Tabelin, C.B.; Hiroyoshi, N. Development of a restraining wall and screw-extractor discharge system for continuous jig separation of mixed plastics. *Miner. Eng.* **2021**, *168*, 106918. [CrossRef]

33. Zhang, X.; Zhang, W.; Xie, J.; Zhang, C.; Fu, J.; Fu, J.; Zhao, P. Automatic magnetic projection for one-step separation of mixed plastics using ring magnets. *Sci. Total Environ.* **2021**, *786*, 147217. [CrossRef] [PubMed]

34. Boukhoulda, M.F.; Rezoug, M.; Aksa, W.; Miloudi, M.; Medles, K.; Dascalescu, L. Triboelectrostatic separation of granular plastics mixtures from waste electric and electronic equipment. *Part. Sci. Technol.* **2017**, *35*, 621–626. [CrossRef]

35. Li, J.; Wu, G.; Xu, Z. Tribo-charging properties of waste plastic granules in process of tribo-electrostatic separation. *Waste Manag.* **2015**, *35*, 36–41. [CrossRef]

36. Li, J.; Gao, K.; Xu, Z. Charge-decay electrostatic separation for removing Polyvinyl chloride from mixed plastic wastes. *J. Clean. Prod.* **2017**, *157*, 148–154. [CrossRef]
37. Li, T.; Yu, D.; Zhang, H. Triboelectrostatic separation of polypropylene, polyurethane, and polyvinylchloride used in passenger vehicles. *Waste Manag.* **2018**, *73*, 54–61. [CrossRef] [PubMed]
38. Brahami, Y.; Tilmatine, A.; Bendimerad, S.-E.; Miloudi, M.; Zelmat, M.E.-M.; Dascalescu, L. Tribo-aero-electrostatic Separation of Micronized Mixtures of Insulating Materials using "Back-and-Forth" Moving Vertical Electrodes. *IEEE Trans. Dielectr. Electr. Insul.* **2016**, *23*, 2. [CrossRef]
39. Fekir, D.E.; Miloudi, M.; Kimi, I.; Miloua, F.; Tilmatine, A. Experimental and numerical analysis of a new tribo-electrostatic separator with coaxial cylindrical electrodes for plastic binary granular mixtures. *J. Electrostat.* **2020**, *108*, 103523. [CrossRef]
40. Messafeur, R.; Mahi, I.; Ouiddir, R.; Medles, K.; Dascalescu, L.; Tilmatine, A. Tribo-electrostatic separation of a quaternary granular mixture of plastics. *Part. Sci. Technol.* **2019**, *37*, 764–769. [CrossRef]
41. Rodrigues, B.M.; Saron, C. Electrostatic separation of polymer waste by tribocharging system based on friction with PVC. *Int. J. Environ. Sci. Technol.* **2022**, *19*, 1293–1300. [CrossRef]
42. Silveira, A.V.M.; Cella, M.; Tanabe, E.H.; Bertuol, D.A. Application of tribo-electrostatic separation in the recycling of plastic wastes. *Process Saf. Environ. Prot.* **2018**, *114*, 219–228. [CrossRef]
43. Zelmat, M.E.-M.; Tilmatine, A.; Touhami, S.; Bendaoud, A.; Medles, K.; Ouiddir, R.; Dascalescu, L. Experimental investigation of a new Tribo-Aeroelectrostatic separation process for micronized plastics from WEEE. *IEEE Trans. Ind. Appl.* **2017**, *53*, 4950–4956. [CrossRef]
44. Żenkiewicz, M.; Żuk, T.; Pietraszek, J.; Rytlewski, P.; Moraczewski, K.; Stepczyńska, M. Electrostatic separation of binary mixtures of some biodegradable polymers and poly(vinyl chloride) or poly(ethylene terephthalate). *Polimery* **2016**, *61*, 835–843. [CrossRef]
45. Louati, H.; Zouzou, N.; Tilmatine, A.; Zouaghi, A.; Ouiddir, R. Experimental investigation of an electrostatic adhesion device used for metal/polymer granular mixture sorting. *Powder Technol.* **2021**, *391*, 301–310. [CrossRef]
46. Richard, G.; Salama, A.; Medles, K.; Zeghloul, T.; Dascalescu, L. Comparative study of three high-voltage electrode configurations for the electrostatic separation of aluminum, copper and PVC from granular WEEE. *J. Electrostat.* **2017**, *88*, 29–34. [CrossRef]
47. Duan, Q.; Li, J. Classification of common household plastic wastes combining multiple methods based on near-infrared spectroscopy. *ACS EST Eng.* **2021**, *1*, 1065–1073. [CrossRef]
48. Peng, X.; Xu, B.; Xu, Z.; Yan, X.; Zhang, N.; Qin, Y.; Ma, Q.; Li, J.; Zhao, N.; Zhang, Q. Accuracy improvement in plastics classification by laser-induced breakdown spectroscopy based on a residual network. *Opt. Express* **2021**, *29*, 33269–33280. [CrossRef]
49. Truc, N.T.T.; Lee, B.-K. Sustainable and selective separation of PVC and ABS from a WEEE plastic mixture using microwave and/or mild-heat treatment with froth flotation. *Environ. Sci. Technol.* **2016**, *50*, 10580–10587. [CrossRef]
50. Guney, A.; Poyraz, M.I.; Kangal, O.; Burat, F. Investigation of thermal treatment on selective separation of postconsumer plastics prior to froth flotation. *Waste Manag.* **2013**, *33*, 1795–1799. [CrossRef]
51. Guney, A.; Ozdilek, C.; Kangal, M.O.; Burat, F. Flotation characterization of PET and PVC in the presence of different plasticizers. *Sep. Purif. Technol.* **2015**, *151*, 47–56. [CrossRef]
52. Yenial, U.; Kangal, O.; Guney, A. Selective flotation of PVC using gelatin and lignin alkali. *Waste Manag. Res.* **2013**, *31*, 613–617. [CrossRef]
53. Negari, M.S.; Movahed, S.O.; Ahmadpour, A. Separation of polyvinylchloride (PVC), polystyrene (PS) and polyethylene terephthalate (PET) granules using various chemical agents by flotation technique. *Sep. Purif. Technol.* **2018**, *194*, 368–376. [CrossRef]
54. Yuce, A.E.; Killic, M. Separation of PVC/PET mixture from plastic waste using column flotation technique. *J. Environ. Prot. Ecol.* **2015**, *16*, 705–715.
55. Yenial, U.; Burat, F.; Yuce, A.E.; Guney, A.; Kangal, M.O. Separation of PET and PVC by flotation technique without using alkaline treatment. *Miner. Process. Extr. Metall. Rev.* **2013**, *34*, 412–421. [CrossRef]
56. Wang, H.; Wang, C.Q.; Fu, J.G.; Gu, G.H. Flotability and flotation separation of polymer materials modulated by wetting agents. *Waste Manag.* **2014**, *34*, 309–315. [CrossRef] [PubMed]
57. Pita, F.; Castilho, A. Separation of plastics by froth flotation. The role of size, shape and density of the particles. *Waste Manag.* **2017**, *60*, 91–99. [CrossRef]
58. Pita, F.; Castiho, A. Plastics floatability: Effect of saponin and sodium lignosulfonate as wetting agents. *Polimeros* **2019**, *29*, e2019035. [CrossRef]
59. Ghinea, C.; Paduret, S. Separation of waste plastic resulting from electrical products by forced aeration flotation. *Ovidius Univ. Ann. Chem.* **2021**, *32*, 104–109. [CrossRef]
60. Zhang, Y.; Li, C.; Wang, L.; Wang, H. Application of froth flotation in the separation of polyvinyl chloride and polycarbonate for recycling of waste plastic based on a novel surface modification. *Waste Manag.* **2020**, *110*, 43–52. [CrossRef]
61. Wang, J.; Wang, H.; Yue, D. Optimizing green ferrate (VI) modification towards flotation separation of waste polyvinylchloride and acrylonitrile-butadiene-styrene mixtures. *Waste Manag.* **2020**, *101*, 83–93. [CrossRef]
62. Wang, C.-Q.; Wang, H.; Huang, L.-L. A novel process for separation of polycarbonate, polyvinyl chloride and polymethyl methacrylate waste plastics by froth flotation. *Waste Manag.* **2017**, *65*, 3–10. [CrossRef]

63. Huang, L.; Wang, H.; Wang, C.; Zhao, J.; Zhang, B. Microwave-assisted surface modification for the separation of polycarbonate from polymethylmethacrylate and polyvinyl chloride waste plastics by flotation. *Waste Manag. Res.* **2017**, *35*, 294–300. [CrossRef] [PubMed]
64. Chen, S.; Zhang, Y.; Guo, C.; Zhong, Y.; Wang, K.; Wang, H. Separation of polyvinyl chloride from waste plastic mixtures by froth flotation after surface modification with sodium persulfate. *J. Clean. Prod.* **2019**, *21*, 167–172. [CrossRef]
65. Zhang, Y.; Chen, S.; Wang, H.; Luo, M. Separation of polyvinylchloride and acrylonitrile-butadiene-styrene combining advanced oxidation by $S_2O_8^{2-}/Fe^{2+}$ system and flotation. *Waste Manag.* **2019**, *91*, 80–88. [CrossRef] [PubMed]
66. Zhang, Y.; Jiang, H.; Wang, H.; Wang, C. Separation of hazardous polyvinyl chloride from waste plastics by flotation assisted with surface modification of ammonium persulfate: Process and mechanism. *J. Hazard. Mater.* **2020**, *389*, 121918. [CrossRef] [PubMed]
67. Wang, K.; Zhang, Y.; Zhong, Y.; Luo, M.; Du, Y.; Wang, L.; Wang, H. Flotation separation of polyethylene terephthalate from waste packaging plastics through ethylene glycol pretreatment assisted by sonication. *Waste Manag.* **2020**, *105*, 309–316. [CrossRef]
68. Pita, F.; Castilho, A. Separation of PET from other plastics by flotation combined with alkaline pretreatment. *Polímeros* **2020**, *30*, e2020026. [CrossRef]
69. Carvalho, M.T.; Ferreira, C.; Santos, L.R.; Paiva, M.C. Optimization of Froth Flotation Procedure for Poly(ethylene terephthalate) Recycling Industry. *Polym. Eng. Sci.* **2012**, *52*, 157–164. [CrossRef]
70. Shafeeq, M.; Aslam, R.; Shahid, S.A.; Muhammad, A.; Ijaz, A.; Ali, A.M.; Ullah, I.; Mushtaq, M.; Rashid, U. Poly(vinyl alcohol): An Economical Substitute to Calcium Lignosulphonate for PVC/PET Separation by Froth Flotation. *Asian J. Chem.* **2013**, *25*, 7509–7512. [CrossRef]
71. Saisinchai, S. Separation of PVC from PET/PVC Mixtures Using Flotation by Calcium Lignosulfonate Depressant. *Eng. J.* **2014**, *18*, 45–53. [CrossRef]
72. Wang, C.Q.; Wang, H.; Liu, Y.N. Separation of polyethylene terephthalate from municipal waste plastics by froth flotation for recycling industry. *Waste Manag.* **2015**, *35*, 42–47. [CrossRef]
73. Wang, C.; Wang, H.; Fu, J.; Zhang, L.; Luo, C.; Liu, Y. Flotation separation of polyvinyl chloride and polyethylene terephthalate plastics combined with surface modification for recycling. *Waste Manag.* **2015**, *45*, 112–117. [CrossRef] [PubMed]
74. Wang, C.; Wang, H.; Lui, Y.; Huang, L. Optimization of surface treatment for flotation separation of polyvinyl chloride and polyethylene terephthalate waste plastics using response surface methodology. *J. Clean. Prod.* **2016**, *139*, 866–872. [CrossRef]
75. Zhang, Y.; Jiang, H.; Wang, K.; Wang, H.; Wang, C. Green flotation of polyethylene terephthalate and polyvinyl chloride assisted by surface modification of selective $CaCO_3$ coating. *J. Clean. Prod.* **2020**, *242*, 118441. [CrossRef]
76. Truc, N.T.T.; Le, H.A.; Nguyen, D.T.; Pham, T.-D. Novel method for sustainable and selective separation of PVC and PET by the homogeneous dissociation of H_2O_2 using ultrasonication. *J. Mater. Cycles Waste Manag.* **2019**, *21*, 1085–1094. [CrossRef]
77. Truc, N.T.T.; Lee, B.-K. Sustainable hydrophilization to separate hazardous chlorine PVC from plastic wastes using H_2O_2/ultrasonic irrigation. *Waste Manag.* **2019**, *88*, 28–38. [CrossRef]
78. Jiang, H.; Zhang, Y.; Wang, C.; Wang, H. A clean and efficient flotation towards recovery of hazardous polyvinyl chloride and polycarbonate microplastics through selective aluminum coating: Process, mechanism, and optimization. *J. Environ. Manag.* **2021**, *299*, 113626. [CrossRef]
79. Mallampati, S.R.; Heo, J.H.; Park, M.H. Hybrid selective surface hydrophilization and froth flotation separation of hazardous chlorinated plastics from E-waste with novel nanoscale metallic calcium composite. *J. Hazard. Mater.* **2016**, *306*, 13–23. [CrossRef]
80. Mallampati, S.R.; Lee, B.H.; Mitoma, Y.; Simom, C. Selective sequential separation of ABS/HIPS and PVC from automobile and electronic waste shredder residue by hybrid nano-Fe/Ca/CaO assisted ozonisation process. *Waste Manag.* **2017**, *60*, 428–438. [CrossRef]
81. Wang, J.; Wang, H.; Wang, C.; Zhang, L.; Wang, T.; Zheng, L. A novel process for separation of hazardous poly(vinyl chloride) from mixed plastic wastes by froth flotation. *Waste Manag.* **2017**, *69*, 59–65. [CrossRef]
82. Wang, J.-C.; Wang, H. Fenton treatment for flotation separation of polyvinyl chloride from plastic mixtures. *Sep. Purif. Technol.* **2017**, *187*, 415–425. [CrossRef]
83. Wang, J.-C.; Wang, H.; Huang, L.-L.; Wang, C.-Q. Surface treatment with Fenton for separation of acrylonitrile-butadienestyrene and polyvinylchloride waste plastics by flotation. *Waste Manag.* **2017**, *67*, 20–26. [CrossRef] [PubMed]
84. Zhang, Y.; Jiang, H.; Wang, H.; Wang, C.; Du, Y.; Wang, L. Flotation separation of polystyrene and polyvinyl chloride based on heterogeneous catalytic Fenton and green synthesis of nanoscale zero valent iron (GnZVI). *J. Clean. Prod.* **2020**, *267*, 122116. [CrossRef]
85. Truc, N.T.T.; Lee, C.-H.; Lee, B.-K.; Mallampati, S.R. Development of hydrophobicity and selective separation of hazardous chlorinated plastics by mild heat treatment after PAC coating and froth flotation. *J. Hazard. Mater.* **2017**, *321*, 193–202. [CrossRef] [PubMed]
86. Mallampati, S.R.; Lee, C.-H.; Park, M.H.; Lee, B.-K. Processing plastics from ASR/ESR waste: Separation of poly vinyl chloride (PVC) by froth flotation after microwave-assisted surface modification. *J. Mater. Cycles Waste Manag.* **2018**, *20*, 91–99. [CrossRef]
87. Zhao, Y.; Han, F.; Guo, L.; Singh, S.; Zhang, H.; Zhang, J. Flotation separation of hazardous polyvinyl chloride from waste plastics based on green plasma modification. *J. Clean. Prod.* **2021**, *318*, 128569. [CrossRef]
88. Zhao, Y.; Han, F.; Guo, L.; Zhang, J.; Zhang, H.; Abdelaziz, I.I.M.; Ghazali, K.H. Flotation separation of poly (ethylene terephthalate and vinyl chloride) mixtures based on clean corona modification: Optimization using response surface methodology. *Waste Manag.* **2021**, *136*, 184–194. [CrossRef]

89. Baigabelov, A.; Das, A.; Young, C. Integrated density concentration and surface treatment for selective separation of plastics from a mixture. *J. Mater. Cycles Waste Manag.* **2021**, *23*, 2162–2178. [CrossRef]
90. Ito, M.; Takeuchi, M.; Saito, A.; Murase, N.; Phengsaart, T.; Tabelin, C.B.; Hiroyoshi, N.; Tsunekawa, M. Improvement of hybrid jig separation efficiency using wetting agents for the recycling of mixed-plastic wastes. *J. Mater. Cycles Waste Manag.* **2019**, *21*, 1376–1383. [CrossRef]
91. Ito, M.; Saito, A.; Murase, N.; Phengsaart, T.; Kimura, S.; Kitajima, N.; Takeuchi, M.; Tabelin, C.B.; Hiroyoshi, N. Estimation of hybrid jig separation efficiency using a modified concentration criterion based on apparent densities of plastic particles with attached bubbles. *J. Mater. Cycles Waste Manag.* **2020**, *22*, 2071–2080. [CrossRef]

 *sustainability*

Article

Adsorption of Fatty Acid on Beta-Cyclodextrin Functionalized Cellulose Nanofiber

Nor Hasmaliana Abdul Manas [1,2,*], Nurhidayah Kumar Muhammad Firdaus Kumar [1], Nurul Aqilah Mohd Shah [1], Guang Yik Ling [1], Nur Izyan Wan Azelee [1,2], Siti Fatimah Zaharah Mohd Fuzi [3], Nasratun Masngut [4], Muhammad Abd Hadi Bunyamin [5], Rosli Md. Illias [1,2] and Hesham Ali El Enshasy [1,2,6]

[1] Faculty of Chemical and Energy Engineering, Universiti Teknologi Malaysia, Skudai 81310, Johor, Malaysia
[2] Institute of Bioproduct Development, Universiti Teknologi Malaysia, Skudai 81310, Johor, Malaysia
[3] Department of Technology and Natural Resources, Faculty of Applied Sciences and Technology, Universiti Tun Hussein Onn Malaysia, Pagoh Campus, Panchor 84600, Johor, Malaysia
[4] Faculty of Chemical and Process Engineering Technology, Universiti Malaysia Pahang, Lebuhraya Tun Khalil Yaakob, Kuantan 26300, Pahang, Malaysia
[5] Faculty of Social Sciences and Humanities, Universiti Teknologi Malaysia, Skudai 81310, Johor, Malaysia
[6] City of Scientific Research and Technology Applications (SRTA), New Burg Al Arab, Alexandria 21934, Egypt
* Correspondence: hasmaliana@utm.my; Tel.: +60-7553-5691

Abstract: Fatty acids in wastewater contribute to high chemical oxygen demand. The use of cellulose nanofiber (CNF) to adsorb the fatty acids is limited by its strong internal hydrogen bonding. This study aims to functionalize CNF with β-cyclodextrin (β-CD) and elucidate the adsorption behaviour which is yet to be explored. β-CD functionalized CNF (CNF/β-CD) was achieved by crosslinking of β-CD and citric acid. Functionalization using 7% (w/v) β-CD and 8% (w/v) citric acid enhanced mechanical properties by increasing its thermal decomposition. CNF/β-CD was more efficient in removing palmitic acid, showcased by double adsorption capacity of CNF/β-CD (33.14% removal) compared to CNF (15.62% removal). CNF/β-CD maintained its adsorption performance after five cycles compared to CNF, which reduced significantly after two cycles. At 25 °C, the adsorption reached equilibrium after 60 min, following a pseudo-second-order kinetic model. The intraparticle diffusion model suggested chemical adsorption and intraparticle interaction as the controlling steps in the adsorption process. The maximum adsorption capacity was 8349.23 mg g^{-1} and 10485.38 mg g^{-1} according to the Sips and Langmuir isotherm model, respectively. The adsorption was described as monolayer and endothermic, and it involved both a physisorption and chemisorption process. This is the first study to describe the adsorption behaviour of palmitic acid onto CNF/β-CD.

Keywords: electrospinning; adsorption kinetics; adsorption isotherms; β-cyclodextrin; oily wastewater

Citation: Abdul Manas, N.H.; Kumar, N.K.M.F.; Mohd Shah, N.A.; Ling, G.Y.; Azelee, N.I.W.; Fuzi, S.F.Z.M.; Masngut, N.; Bunyamin, M.A.H.; Illias, R.M.; El Enshasy, H.A. Adsorption of Fatty Acid on Beta-Cyclodextrin Functionalized Cellulose Nanofiber. *Sustainability* **2023**, *15*, 1559. https://doi.org/10.3390/su15021559

Academic Editors: Nallapaneni Manoj Kumar, Subrata Hait, Anshu Priya and Varsha Bohra

Received: 5 December 2022
Revised: 4 January 2023
Accepted: 9 January 2023
Published: 13 January 2023

1. Introduction

Oily wastewater is not a new issue when it comes to water pollution. A large amount of oily wastewater effluents was discharged into rivers by industries such as oil and gas, textile, food, and petrochemical processing. The discharge of oily effluent into the sea or river will have a negative influence on the ecosystem and living beings. A presence of oil in emulsified form in the oily wastewater is extremely challenging to remove from the aqueous media [1]. A droplet size smaller than 20 μm is a stable emulsion in the emulsified oily wastewater. The effective separation of the emulsified oily wastewater is one of the most challenging issues in the field of wastewater treatment because of its stable dynamic structure and nonuniform small droplet size [2]. The difficulty of treating the oily wastewater lies in the existence of the stable oil-in-water emulsions. The traditional treatment methods were rendered by the presence of the stable oil-in-water emulsions [3]. Nanofibrous materials hold excellent potential for various environmental applications, including in the treatment of wastewater, due to its high porosity, bigger surface area, and better connectivity [4]. Various types of materials can be used, like polymer, ceramic and carbon, thus making the scalable synthesis of it much easier. Cellulose is inexpensive and comes as the most abundant natural bio polymer to be considered in developing a cost-effective wastewater treatment technology [5]. In addition to the expansion of nanotechnology for fabricating polymer-based materials, cellulose nanofiber (CNF) produced by an electrospinning technique seems to be a promising method for use in wastewater treatment. The key success of nanofibrous materials was due to its voids among fibers, which led to a better selectivity. Nanofibrous materials were said to have higher sorption capacity compared to non-nanofibrous or non-porous materials. However, the high hydrophilicity of this material [6] has become one of the main drawbacks; therefore, its surface needs to be modified in order to expand the application of CNF.

Cyclodextrins (CD) are a group of cyclic oligomers composed of α- (1,4) linked glucopyranose subunits having a 3D structure that looks like a cup or a shallow, truncated cone with a hydrophobic core and hydrophilic exterior. The capability of the CDs to act as hosts for guests in forming noncovalent host-guest inclusion complexes has made them an attractive compound [7]. α- cyclodextrin (α- CD), β- cyclodextrin (β-CD) and γ- cyclodextrin (γ- CD) are the most common native CDs and are differentiated by the number of glucose units. On the other hand, β-CD is a type of CD that has many advantages as it is cheap and accessible, and is thus the most studied CD [8]. Inclusion complexes, or host-guest complexes between CDs and other molecules or pollutants, are stabilized by weak forces, which implies an equilibrium between free and complex species [9]. However, CDs are soluble in water; therefore, it cannot be directly applied in water for adsorption of pollutants. In order to improve its performance, several works had been done on the immobilization of CDs on a good, insoluble support including nanopolymers. Surface functionalization of nanofibers with CD would be interesting for designing an efficient filtering material. β-CD functionalized nanofibers produced via electrospinning has drawn a positive result in improving the adsorption and separation of various types of pollutants from aqueous solution including dyes, polycyclic aromatic hydrocarbon (PAH), heavy metals, oil and others [10]. However, to the best of our knowledge, the study on β-CD functionalized CNF for removal of fatty acids has not been previously performed. Palmitic is one of the long chain fatty acids (LCFAs) found in high concentration in the palm oil mill effluent that is recalcitrant to biodegradation [11]. It was also a major fatty acid found in industrial dairy wastewater (65%) [12]. LCFAs at concentrations higher than 0.5 mM could potentially inhibit in anaerobic digester of wastewater treatment [13] and at concentrations above 16 mM was shown to cause a lag in methane production [14]. Therefore, there is an urgency in removing LCFAs from the wastewater.

In this study, the surface modification of CNF with β-CD (CNF/β-CD) was achieved by a polymerization reaction between β-CD and citric acid as crosslinking agent. Different concentrations of β-CD and citric acid were used in order to optimize the factors affecting functionalization process. The morphological, surface and thermal decomposition properties of CNF/β-CD were characterized. The adsorption performance of the CNF/β-CD was investigated by removal of a model fatty acid (palmitic acid) from aqueous solution. The adsorption kinetics and isotherms were elucidated to precisely describe the adsorption behaviour.

2. Materials and Methods

The materials and methodology for the study are described in the following subsection. All chemicals and materials were used without further purifcation.

2.1. Materials

Cellulose acetate powder (average M_n ~30,000 by GPC), β-cyclodextrin (≥97% Sigma-Aldrich, Missouri, United States), citric acid, sodium hydrophosphite hydrate (SHPI), acetone, dimethylacetamide (DMAc), palmitic acid powder, 1-propanol (99% of purity).

2.2. Synthesis of CNF by Electrospinning Process

First, cellulose acetate powder was used as the polymer to form nanofibers that was then functionalized with β-CD. The optimum parameters used were based on the method adapted from Liu and Hsieh [15]. Cellulose acetate solution with concentration of 15% (w/v) was prepared in acetone: dimethylacetamide mixtures (2:1). Next, the electrospinning was carried out in which 8 mL of cellulose acetate solution was placed into a syringe fitted with a metallic needle having an inner diameter of 0.8 mm, and the syringe was horizontally placed at the syringe pump. The flow rate of polymer solution during electrospinning process was set at 1 mL h^{-1} with 15 cm distance between needle tip and metal collector covered with a piece of aluminium foil. The voltage was set at 15 kV throughout the process, and it was carried out at room temperature in an enclosed Plexiglas box.

2.3. Functionalization of Cellulose Nanofiber with β-cyclodextrin

For functionalization of CNF with β-CD, different concentration of β-CD solutions (5%, 6%, 7%, 8%, 9% and 10% w/v) were prepared in 50 mL aqueous solution at 50 °C. Then, different concentrations of citric acid (5%, 6%, 7%, 8%, 9% and 10% w/v) and 1.25% w/v of sodium hydrophosphite hydrate (SHPI) that act as a catalyst were added to each β-CD solutions separately and stirred continuously using a magnetic stirrer with mixing speed of 150 rpm for about 30 min at 50 °C. After all reactants were completely dissolved in aqueous solution, a rectangular shaped 4 cm × 4 cm of CNF (weighed about 0.05 g) was immersed into each different mixture of β-CD solutions and kept at 50 °C for a duration of 3 h. Then, the β-CD functionalized cellulose nanofibers (CNF/β-CD) were dried at 70 °C for 3 h in order to ensure the crosslinking of β-CD with cellulose nanofibers. The dried CNF/β-CD mats were washed two times with 40 °C warm water for removal of unreacted β-CD and citric acid. Lastly, all of the mats were dried again at 70 °C for 3 h or until they achieved constant weight.

2.4. Morphological, Surface and Thermal Characterization of Nanofibers

Scanning electron microscope (SEM) was performed by using Hitachi TM3000 SEM to check the changes that occurred on the morphology and fiber diameter of CNF and CNF/β-CD mats. The CNF and CNF/β-CD mats were coated with 5 nm Au/Pd prior to SEM analysis, and each of the samples was captured at magnifications (×10,000) randomly at any spot to examine the average fiber diameter (AFD) of the nanofibers. It was very vital to ensure that the modification performed during crosslinking process did not deform the original structure of cellulose nanofibers.

Next, the surface chemical characterization was carried out by using Fourier transform infrared spectroscopy (FTIR) for both CNF and CNF/β-CD. Dried CNF and CNF/β-CD samples were pressed into thin transparent films. FTIR analysis was carried out by using a Shimadzu IRTracer- 100 (ATC) (Shimadzu Scientific Instruments, Maryland, United States) to observe changes of the functional groups on the surface of CNF/β-CD. Each spectrum was obtained with a wavelength in the range of 400 cm^{-1} and 4000 cm^{-1}.

The thermogravimetric analyzer (TGA) was used to investigate the thermal characterization. CNF and CNF/β-CD were analysed by using Shimadzu TGA-50 equipment under a itrogen (N$_2$) atmosphere with a purge rate of 100 mL min^{-1} with temperatures ranging from 50 °C to 900 °C at a heating rate of 10 °C min^{-1}. Samples of 6–8 mg were used for each test. This TGA analysis was applied for CNF and CNF/β-CD in order to study the effect of functionalization of CNF with β-CD on the thermal stability.

2.5. Adsorption of Palmitic Acid

2.5.1. Comparison of Palmitic Acid Adsorption onto CNF and CNF/β-CD

The performance of the CNF and CNF/β-CD mats as adsorbent for palmitic acid was tested. Identical square-shaped CNF and CNF/β-CD mats weighing 0.1 g were immersed individually into 30 mL of palmitic acid solution with a concentration of 70,000 ppm in order to study the adsorption performance. The reduction of palmitic acid concentration was checked at different contact times (15 min, 30 min, 60 min, 90 min and 120 min), and the percentage of removal was calculated.

Reusability of the CNF and CNF/β-CD were also investigated. After immersing the CNFs in the palmitic acid solution for the first time, it was then washed with 50 mL 1-propanol at 50 °C for 3 h in a thermostatic water bath shaking incubator and dried in a vacuum oven at 70 °C for 3 h or until it was completely dried. These washing steps were repeated for two times in order to ensure the previous palmitic acid captured was totally detached from the CNF and CNF/β-CD. Next, the CNFs were immersed again in the 30 mL, 70,000 ppm palmitic acid solution, and the reduction of the palmitic acid concentration was monitored for 2 h. The changes on the adsorption performance for each cycle was recorded. The reusability study was continuously carried out until both CNFs show a significant drop in adsorption performance.

2.5.2. Adsorption Kinetics

The 5% (*w/v*) palmitic acid solution was prepared by dissolving 5 g of palmitic acid powder into 100 mL 1-propanol (99% of purity). The mixture was then continuously stirred by using a magnetic stirrer with 150 rpm at room temperature for about 30 min to have a well-mixed solution. Identical square-shaped CNF and CNF/β-CD mats weighing 0.1 g were immersed individually into 30 mL of palmitic acid solution with a concentration of 70,000 ppm at room temperature (25 °C). The reduction of palmitic acid concentration was checked at different contact time (15 min, 30 min, 60 min, 90 min and 120 min). The percentage removal efficiency and adsorption capacity (q_e) of pollutant by the adsorbent were calculated as described in Equation (1) and (2).

$$\text{Removal efficiency } (\%) = \frac{C_0 - C_t}{C_0} \times 100 \tag{1}$$

$$q_e \, (\text{mg/g}) = \frac{(C_0 - C_t) \times V}{W} \tag{2}$$

where C_0 (mg L^{-1}) is the concentration of pollutants at initial, C_t (mg L^{-1}) is the concentration of pollutants at certain time, W (g) is the weight of the adsorbent and V (L) is volume of the testing solution.

The kinetic behavior of the adsorption process was investigated by nonlinear data fitting into the pseudo-first-order model [16], pseudo-second-order model [17] and Elovich model [18] as in Equation (3)–(5) respectively.

$$q_t = q_e \left(1 - e^{-k_1 t}\right) \tag{3}$$

$$q_t = \frac{t}{\frac{t}{q_e} + \frac{1}{k_2 q_e^2}} \tag{4}$$

$$q_t = \frac{1}{\beta \, \ln(\alpha \, \beta \, t + 1)} \tag{5}$$

where q_t and q_e (mg g^{-1}) represent the adsorption capacity at certain time and equilibrium time respectively, while k_1 (min^{-1}) and k_2 (g mg^{-1} min^{-1}) are the pseudo-first-order model rate constant and pseudo-second-order model rate constant respectively. For the Elovich model, α is the initial adsorption rate (mg g^{-1} min^{-1}), and β is a desorption constant. Adsorption diffusion was investigated by using Weber–Morris intraparticle diffusion model [19] as shown in Equation (6).

$$q_t = k_{id} t^{1/2} + C \tag{6}$$

where k_{id} is the intraparticle diffusion rate constant and C is the boundary effect.

2.5.3. Adsorption Isotherms and Thermodynamic

CNF/β-CD weighing 0.1 g was immersed in various concentrations of palmitic acid solutions (10,000, 30,000, 50,000, 70,000 and 90,000 mg L^{-1})for 60 min. The final concentration of palmitic acid in the solutions was measured. Adsorption isotherm was investigated using non-linear data fitting into Langmuir [20], Freundlich [21], Sips [22] and Temkin [23] isotherm models, where the equation for the models are shown in Equations (7)–(10) respectively.

$$q_e = q_{max}\frac{K_L C_e}{1 + K_L C_e} \tag{7}$$

$$q_e = K_F C_e^{1/n_F} \tag{8}$$

$$q_e = \frac{q_m (K_s C_e)^{n_s}}{1 + (K_s C_e)^{n_s}} \tag{9}$$

$$q_e = b_T \ln(A K_T C_e) \tag{10}$$

where q_e (mg g^{-1}) is the amount of fatty acids adsorbed, q_m (mg g^{-1}) is the maximum adsorption, K_L (L mg^{-1}) is the Langmuir coefficient, C_e (mg L^{-1}) is the equilibrium concentration of fatty acids, K_F and n_F are the Freundlich constants, K_s (mg^{-1}) and n_s is the Sips equilibrium constants, b_T (J mol^{-1}) is heat of adsorption and K_T (L mg^{-1}) is the equilibrium binding constant. Dimensionless constant separation factor, R_L for Langmuir isotherm model was calculated using Equation (11) [24].

$$R_L = \frac{1}{1 + K_L C_0} \tag{11}$$

where K_L (L mg^{-1}) is the Langmuir coefficient and C_0 (mg L^{-1}) is the initial concentration. The Gibbs free energy of the adsorption was calculated using Equation (12) [25] to measure the spontaneity of the process.

$$\Delta G^o = -RT\ln K_L \tag{12}$$

where R is the gas constant (8.314 J mol^{-1} K^{-1}), T is the absolute temperature (K) and K_L is the Langmuir equilibrium constant (L mg^{-1}).

2.6. Palmitic Acid Quantification Using High Performance Liquid Chromatography

The concentration of palmitic acid in the solution during the time course of the adsorption experiment was measured by using Agilent 1200 Infinity Series high performance liquid chromatography (HPLC) with C18 column. Acetonitrile and isopropanol (80:20 *v/v*) at a flow rate of 1 mL min^{-1} were used as the mobile phase, and the concentration of palmitic acid taken at different contact times was monitored by the UV detector at the wavelength of 210 nm [17]. Each of the samples of palmitic acid solution taken at different contact times was filtered by using 0.45 μm disposable syringe filter and injected into 1.5 mL autosampler HPLC vials. As a result, the amount of remaining palmitic acid in each sample was measured from the area of palmitic acid peak observed in HPLC chromatograms. Palmitic acid solutions with different concentrations ranging from 0.5% to 5% (*w/v*) were prepared to obtain the calibration curve and R^2 was calculated as 0.9995.

3. Results and Discussion

3.1. The Synthesis of Beta-Cyclodextrin Functionalized Cellulose Nanofiber

The optimum parameters used to synthesize cellulose nanofiber (CNF) via electrospinning process in this study was found to be sufficient to form smooth and continuous CNF with an average weight of 0.05 g for each 4 cm × 4 cm dimension of the CNF mats. Acetone and dimethylacetamide (DMAc) with a ratio of 2:1 were used as solvents for producing 15% (*w/v*) cellulose acetate polymer. Acetone and DMAc that have solubility of (9.76 cal cm^{-3})$^{1/2}$ and 11.1 (cal cm^{-3})$^{1/2}$, respectively, were used as solvents, as they fit the solubility required to be solvent for cellulose acetate where the Hildebrand solubility parameter (δ) should be in a range between 9.5 and 12.5 (cal cm^{-3})$^{1/2}$ [15]. Moreover, the low surface tension offered by acetone (23.7 dyne cm^{-1}) and DMAc (32.4 dyne cm^{-1}) generated a mixture of these solvents that is suitable and efficient for electrospinning. Cellulose acetate with concentration range between 12.5 and 20% in 2:1 acetone: DMAc could form a good fibrous membrane due to high viscosity of the polymer liquids [15]. The jet produced by high viscosity liquid does not break up but travels as a jet to the grounded target and thus forms continuous fibers.

3.2. Functionalization of Cellulose Nanofiber with β-Cyclodextrin

Surface modification of the resulted CNF from electrospinning was achieved through crosslinking reaction between β-CD and citric acid, and no leaching of β-CD was found when the washing process was carried out. Surface modification of CNF is very essential in order to synthesize a better nanomaterial adsorbent. Different concentrations of β-CD and citric acid as crosslinking agents were the parameters studied in this case to synthesize the best adsorbent for removal of palmitic acid that acts as the model of fatty acids pollutants. The impregnation of CNF with different concentrations of β-CD (5, 6, 7, 8, 9 and 10% (w/v)) (Figure 1a) while the concentration of citric acid was maintained at 5% (w/v) during the crosslinking process, influenced the texture, structure and performance of the CNF that were used as adsorbent for palmitic acid removal. The amount of palmitic acid being removed increases with increasing concentration of β-CD. This is because larger amount of β-CD attached on the surface of CNF after the functionalization process will eventually lead to more binding between β-CD and palmitic acid due to more vacant adsorption sites [26]. The percentage rate of removal of palmitic acid for CNF and CNF/β-CD functionalized with 7% (w/v) of β-CD concentration were 0.73% min^{-1} and 1.87% min^{-1}, respectively.

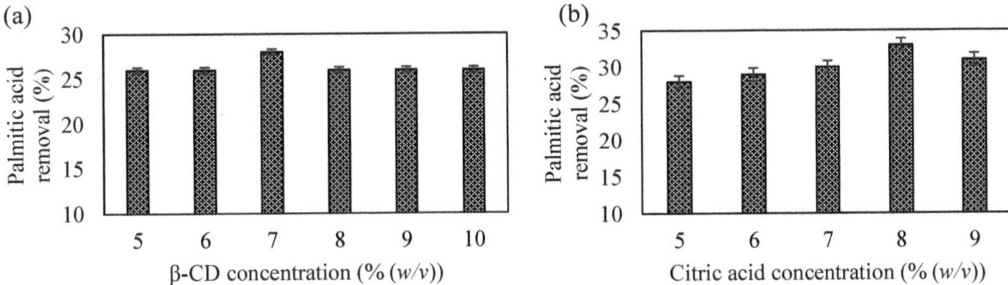

Figure 1. The effect of (**a**) β-CD concentration and (**b**) citric acid concentration on palmitic acid removal. The error bars represent the standard deviation of triplicate experiments.

Next, an increase in concentration of citric acid (5, 6, 7, 8, 9 and 10% (w/v)) (Figure 1b) during the crosslinking process was assumed to increase the carboxyl content of the CNF. Therefore, with a higher concentration of citric acid introduced during the crosslinking process, more β-CD will be attached to the CNF which acts as molecules that captured the palmitic acid and hence increase the reduction of palmitic acid concentration from the aqueous solution. It was found that 8% (w/v) was the optimum citric acid concentration to crosslink between CNF and 7% (w/v) of β-CD. The highest removal percentage of palmitic acid concentrations performed by CNF and optimum CNF/β-CD were 16% and 33%, respectively. Moreover, the concentration of palmitic acid became constant or achieved equilibrium after 60 min of contact time. Lastly, it could be concluded that a 7% of β-CD functionalized CNF with a dimension size of 4 cm × 4 cm needs 8% concentration of citric acid as a crosslinker to achieve the highest removal efficiency of palmitic acid at 60 min contact time. It clearly shows that the removal of palmitic acid amount from its aqueous solution was better when CNF/β-CD mats were used.

3.3. Morphological Characterization of the Nanofibers

Scanning electron microscope (SEM) analysis was performed to investigate the changes of the morphology of the CNF after the modification was made on its surface with β-CD during the functionalization process. The images of the morphology and its fiber diameter are represented as in Figure 2. The surface morphology for CNF and CNF/β-CD were obviously different, as clearly seen from the SEM images. For the CNF, it has a smooth and uniform surface structure compared to the CNF/β-CD that appeared rough due to the functionalized β-CD. Irregularities at certain point of CNF/β-CD were also observed. Resulting rough surface and irregularities of the nanofibers were also reported for modified electrospun polyester with CDs [27], cotton fabrics grafted with glycidyl methacrylate/β-CD [28] and woven PET vascular prosthesis grafted with CD [29]. Therefore, the roughness and irregularities on the surface of CNF indicated the successful attachment of β-CD onto CNF.

It is vital to ensure the functionalization process made on CNF did not deform the fibrous structure of CNF. This study has proven that functionalized CNF with β-CD maintained the fibrous structure as clearly seen from the SEM images. In addition, CNF was recorded to have a range of the average fiber diameter from 133 nm to 241 nm, and this result was supported by a finding from the study of preparation of cellulose- based nanofibers using electrospinning [30], where the fiber diameter generated from 15kV of electrospinning condition was between 100 nm to 200 nm. On the other hand, CNF/β-CD was observed to have larger average fiber diameter ranges from 262 nm to 378 nm. The increases of fiber diameter for after modification made on CNF was also demonstrated by a study for modified electrospun polyester with cyclodextrin polymer [27].

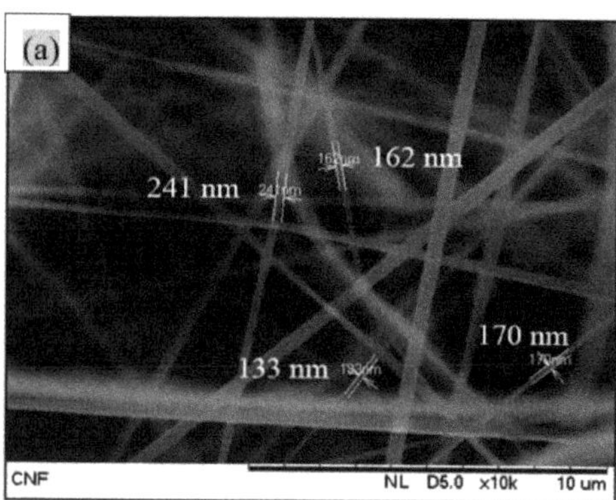

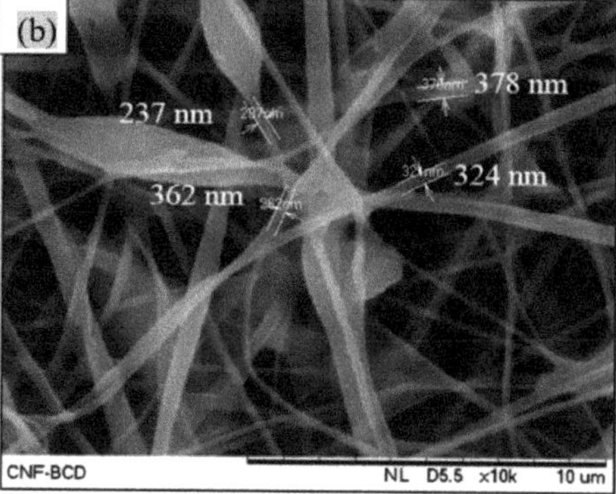

Figure 2. Representative images analyzed by scanning electron microscope in original magnification of 10,000 (**a**) cellulose nanofiber with average fiber diameter of 177 nm and (**b**) β-CD functionalized cellulose nanofiber with average fiber diameter of 312 nm.

3.4. Surface Chemical Characterization of Nanofiber

The surface chemical characterization for both types of CNF was performed by using Fourier-Transform Infrared Spectroscopy (FTIR) to further demonstrate the effect of the functionalization of CNF with β-CD as depicted in Figure 3. FTIR spectra represented in Figure 3 proves that changes occurred on the functional groups of the surface of CNF/β-CD. For the CNF, the absorption band between 3600 and 3000 cm^{-1} was observed, which attributed to hydroxyl groups of cellulose [31]. Next, the peak at 1100 cm^{-1} shown by the FTIR spectrum of CNF corresponds to C-C or C=C bonds. In comparison, CNF/β-CD was observed to have high intensity peak at 1740 cm^{-1} which may be due to the formation of carbonyl group (C=O) stretching vibration mode of an ester bond between citric acid with CNF and β-CD. The resulting adsorption band of carbonyl group confirmed the chemical linkages between CNF and citric acid via ester bonds [32]. On the other hand, the intensity of O-H peak at 3100–3550 cm^{-1} was slightly decreased for CNF/β-CD due to the consumption of cellulose hydroxyl groups in the crosslinking reaction. The result based on this FTIR spectrum clearly indicates the effective crosslinking of citric acid with β-CD and CNF. Similar results were also reported by other studies including cyclodextrin functionalized cellulose nanofiber composites [33] as well as cellulose fiber synthesis via electrospinning and crosslinking with β-cyclodextrin [34].

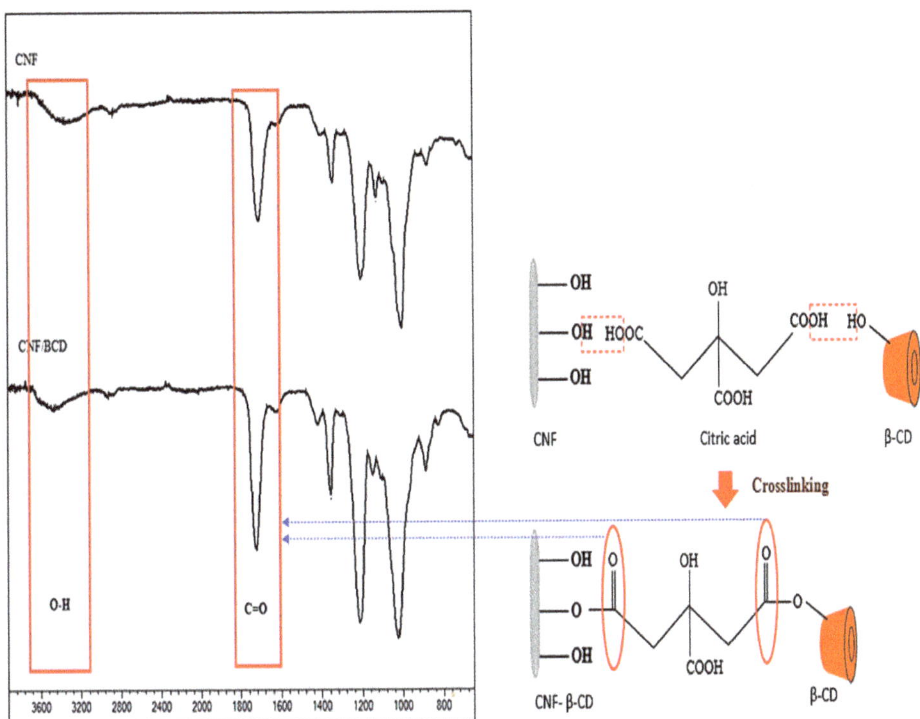

Figure 3. Representative Fourier-transform infrared (FTIR) spectroscopy spectra image of CNF and CNF/β-CD and the schematic diagram of the functionalization of CNF with β-CD by using citric acid as crosslinker.

3.5. Thermal Characterization of the Nanofibers

Thermogravimetric analyser (TGA) is used to study the thermal decomposition characteristics of the CNF and CNF/β-CD. The TGA thermograms and derivative TGA thermograms of CNF and CNF/β-CD are shown in Figure 4. It clearly shows that the main degradation of CNF occurred between 275 °C and 375 °C. For the CNF/β-CD, two major weight losses were drawn between 50–250 °C and 300–375 °C, which corresponded to main thermal degradation of β-CD and CNF. The first weight loss was attributed by the decomposition of citric acid and β-CD, which have melting points of 153 °C and 290 °C, respectively. By analysing the derivative weight percentage loss, it was recorded that the peak point of CNF at 370 °C shifted slightly to a higher temperature of 390 °C for CNF/β-CD. This slightly increased temperature of the peak points for CNF/β-CD was due to higher energy required for decomposition of the modified cellulose nanofiber that has a crosslinked structure. Increased 20 °C onset degradation temperature of CNF/β-CD compared to unmodified CNF suggested the successful modifications of the CNF. The result on increased thermal stability was also supported by cyclodextrin inclusion complexes grafted onto polyamide-6 fabric [35] as well as modification of electrospun polyester nanofibers with cyclodextrin [27]. Modification by crosslinking methods involved either cleaving or attaching the chemical group to alter the features of the original molecule. Citric acid has been acknowledged as a good crosslinker for cellulose materials for many years and had drawn several advantages, as it is an inexpensive organic acid [36] and could react with hydroxyl groups of starch molecules through the formation of esters due to the presence of carboxylic groups in its structure [32]. The formation of crosslinks between citric acid and CNF was due to the esterification reaction. When citric acid is heated at 50 °C during functionalization process, it allows the formation the cyclic anhydride intermediate, as shown in Figure 5, that play a role as the base mechanism responsible for the development of crosslinks with β-CD and CNF. Esterification of -OH functional groups of the CNF with the cyclic anhydride intermediate lead to new carboxylic acid units, and there might also be involvement of the primary –OH groups of the polysaccharides, as they are more reactive than the secondary –OH groups in the esterification process [37,38]. The summary of comparison of CNFs characteristics is shown in Table 1.

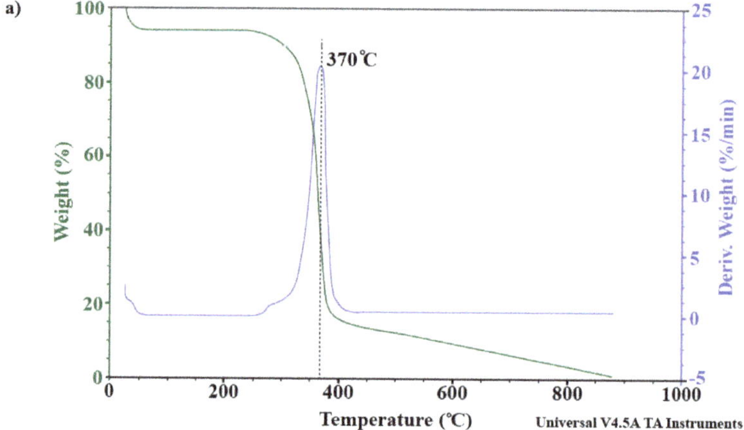

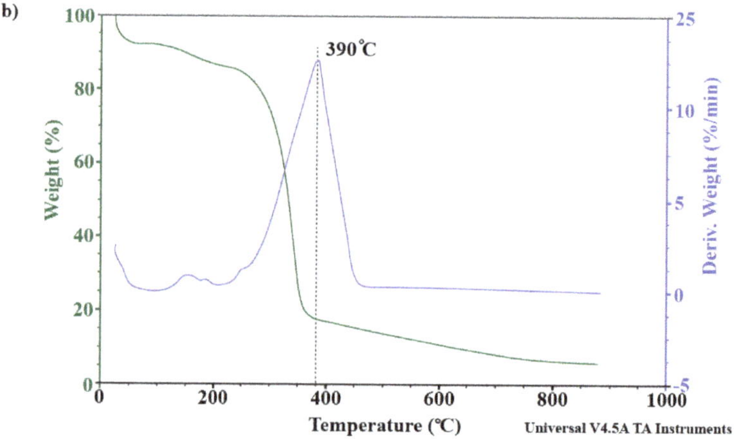

Figure 4. Thermogravimetric analyzer (TGA) thermograms of the (**a**) CNF and (**b**) CNF/β-CD. Green line and blue line represent the weight percentage and derivative weight percentage loss, respectively.

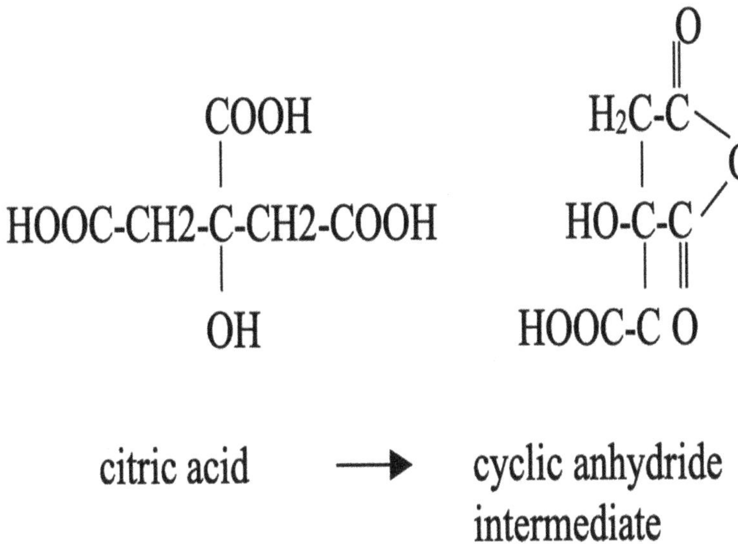

Figure 5. Esterification reaction of citric acid and β-CD on the CNF.

Table 1. Comparison of material characteristics of cellulose nanofiber (CNF) before and after functionalization with β-cyclodextrin (β-CD).

Material Characteristics	CNF	CNF/β-CD
Physical appearance	Smooth surface texture	Rough surface texture
Morphology (SEM analysis)	Regular and uniform fiber	Irregular fiber
Fiber diameter (SEM analysis)	133 nm to 241 nm	262 nm to 378 nm.
Functional groups (FTIR analysis)	-OH (3600 and 3000 cm^{-1}) -C-C or C=C (1100 cm^{-1})	-OH (3100–3550 cm^{-1}) -C=O (1740 cm^{-1})
Thermal decomposition (TGA analysis)	370 °C (cellulose nanofiber)	153 °C (citric acid) 290 °C (β-cyclodextrin) 390 °C (cellulose nanofiber)

3.6. Adsorption Behavior of Palmitic Acid onto CNF/β-CD

3.6.1. Comparison of Adsorption Performance of the CNF and CNF/β-CD

The adsorption capability of CNF and CNF/β-CD was tested using palmitic acid as a model fatty acid. Figure 6a illustrates the cumulative percentage decreases of palmitic acid concentration over time when CNF and CNF/β-CD mats have been kept in aqueous of palmitic acid at room temperature. It clearly shows that the removal of palmitic acid amount from its aqueous solution by CNF/β-CD mats was about two times better than that by CNF. The palmitic acid removal increased sharply for the first 15 min to 29.85% for CNF/β-CD and increased gradually as the time passed. However, the CNF was only able to remove approximately half the amount of those shown by the CNF/β-CD. The result obtained from the experiment indicated that the host-guest complexes between β-CD and palmitic acid was able to improve the adsorption capacity of the nanofiber by two times than those without the functionalization. The β-CD and palmitic acid complex formed almost immediately, as there were no significant differences observed on reduction of palmitic acid between 15 min and other contact times. From surface characterisation, it is noticeable that CNF have a larger average fiber diameter (AFD) and smaller surface area after modification. However, the adsorption efficiency was still further improved for modified CNF due to the β-CD structure onto nanofibers, which plays vital role in molecular capturing of palmitic acid through host-guest interaction. When the concentration of β-CD and citric acid were compared, it showed that 7% (*w*/*v*) of β-CD and 8% (*w*/*v*) concentration of citric acid as a crosslinker were the optimum parameters needed for synthesized CNF/β-CD achieving the highest removal efficiency of palmitic acid at 15 min contact time.

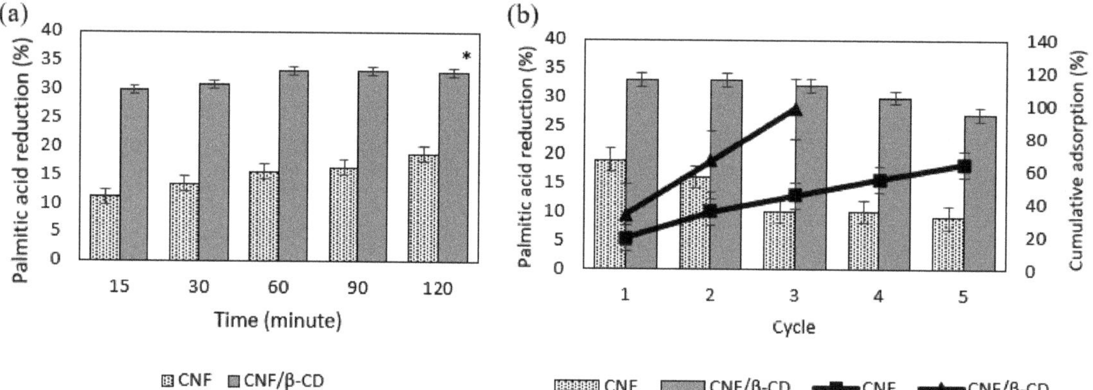

Figure 6. (**a**) Comparison between CNF and CNF/β-CD with 7% (*w*/*v*) of β-CD and 8% (*w*/*v*) of citric acid crosslinker on the reduction palmitic acid concentration, (**b**) Reusability analysis of cellulose nanofiber (CNF) and β-CD functionalized CNF (CNF/β-CD). The statistical significance was determined using regression data analysis and cellulose nanofibers (CNF) without functionalization as a control. $r(4) = 0.63$, * $p < 0.2$.

Reusability of CNF and CNF/β-CD were tested in a repetitive batch adsorption process. The results are illustrated in Figure 6b. Based on the presented results, significance decrease was not observed between the first and second cycle for both CNFs, as there was only a 5% drop in performance for CNF, while the performance for CNF/β-CD was maintained. Next, it was clearly seen that CNF recorded a significant drop in its performance to 47% from first cycle to third cycle. The performance of CNF/β-CD was maintained until being recycled four times, where there was significance decrease of 18% in the fifth cycle. Cumulatively, CNF/β-CD exhibited nearly 100% palmitic acid adsorption after three cycles. Meanwhile, CNF was only able to adsorb 64% of palmitic acid after five cycles. Therefore, CNF/β-CD showed remarkable adsorption capacity with very high reusability.

3.6.2. Adsorption Kinetics

The adsorption capacity as a function of contact time corresponding to a pseudo-first-order kinetic model and a pseudo-second-order kinetic model was studied (Figure 7a). The data were fitted into the non-linear model equations. According to the fitting results (Table 2), the R^2 value for the pseudo first order kinetic model is 0.6615, and this value indicated that this adsorption process does not favor this kinetic model. On the other hand, the pseudo-second-order kinetic model shows a larger R^2 value that is equal to 0.8983. Based on the R^2 values, the pseudo-second-order kinetic model is more valid and better describes time-dependent adsorption behavior for this process than the pseudo-first-order kinetic model. Thus, adsorption of palmitic acid onto CNF/β-CD possibly occurred via chemical adsorption, also known as chemisorption. The adsorption is carried out by the surface exchange reactions [39]. Palmitic acid molecules diffuse inside the CNF/β-CD where inclusion complexes, hydrogen bonds or hydrophobic interactions could take place. As a whole, this model could strongly describe the adsorption process that occurred in this case. The similar result on the pseudo-second-order kinetic behaviour was also reported before for β-CD-epichlorohydrin polymer used for removal of direct blue

(DB78) dye from wastewater, chemisorption of rhodamine B dye from aqueous solution through hydroxypropyl-β-CD cavity, as well as water-insoluble β-CD polymer for phenol uptake [40–42].

Table 2. Kinetics parameters for describing the adsorption of palmitic acid onto functionalized CNF at room temperature (298 K).

Adsorption Kinetics Models	Parameters	R^2
Pseudo-first-order	$K_1 = 0.1559 \text{ min}^{-1}$ $q_e = 6940.68 \text{ mg g}^{-1}$	0.6615
Pseudo-second-order	$K_2 = 6.42 \times 10^{-5} \text{ g mg}^{-1} \text{ min}^{-1}$ $q_e = 7181.89 \text{ mg g}^{-1}$	0.8983
Elovich	$\alpha = 1.0122 \times 10^{10}$ $\beta = 0.0033 \text{ mg g}^{-1} \text{ min}^{-1}$	0.8533
Intra-particle diffusion	$K_{id} = 107.04 \text{ mg g}^{-1} \text{ min}^{1/2}$ $C = 5990.80$	0.8365

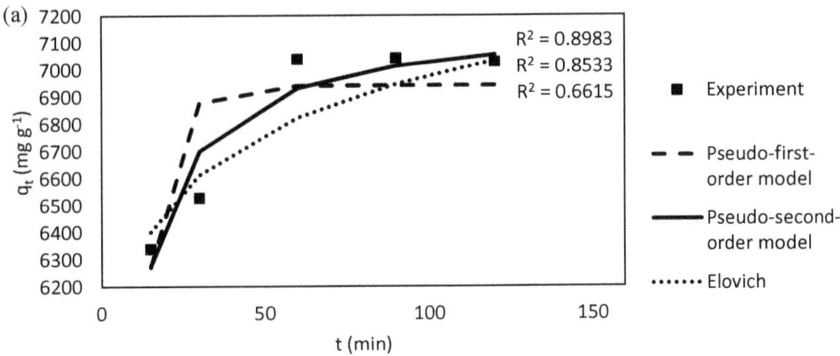

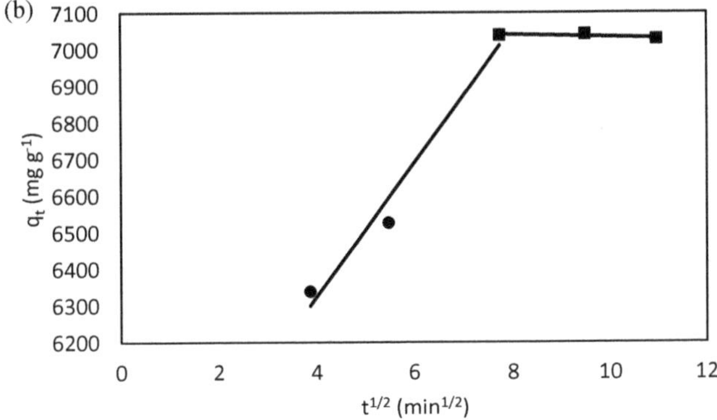

Figure 7. Non-linear fittings of palmitic acid adsorption onto CNF/β-CD with (**a**) pseudo-first-order model, pseudo-second-order model and Elovich model. (**b**) Linear fitting of intra-particle diffusion model.

The Elovich model (Figure 7a) also produced a good data fitting with R^2 value of 0.8533. Several research studies done on adsorption kinetics using the Elovich model suggested that it describes chemisorption between the adsorbent and adsorbate. For example, the adsorption between acetaminophen and chitosan/β-CD composite [43] and the adsorption between Fe(II) and activated carbon [44]. Hence, this result further supports the chemisorption adsorption behavior as suggested by the pseudo-second-order kinetic model, with an initial adsorption rate of 0.0033 mg g^{-1} min^{-1}. Adsorption of palmitic acid on the CNF/β-CD is chemisorption involving the hydrogen bond and host-guest interaction. This theory is strengthened by other research where the ester forms a hydrogen bond with hydroxypropyl-β-CD [45] and the removal of organic pollutants by hydroxypropyl-β-CD occurred through host-guest inclusion complexes [5].

The intra-particle diffusion model was also used to investigate the controlling step in the adsorption process. As depicted in Figure 7b, there is multilinearity indicating that there are more than one mass transfer mechanisms in the adsorption process. Furthermore, the constant C is not zero, indicating there are mass transfer mechanisms other than intraparticle diffusion. The first mechanism is chemical adsorption where the particle adsorbs from bulk solution to the β-CD active site surface of the adsorbent, forming the host-guest interaction. Fast adsorption rates can be observed in this first stage, indicating a rapid transportation of the adsorbate from the solution to the active site. The second mechanism indicates the intraparticle interaction where the mass transfer occurred on the interior sites within the pores of the adsorbent and reached equilibrium, as shown by the plateau. Pellicer et al. [46] also described the chemical and intraparticle interaction as controlling steps in the adsorption of Direct Red 83.1 onto CD-based adsorbents. Feng et al. [47] reported two main stages of liquid film diffusion and intraparticle diffusion in naproxen adsorption onto β-CD immobilized reduced graphene oxide composite.

3.6.3. Adsorption Isotherms

The isotherms data for palmitic acid adsorption at 25 °C were simulated with Langmuir, Freundlich, Sips and Temkin isotherm models using nonlinear fitting. The data q_e against C_e for experimental and simulated models are shown in Figure 8. The fitting data were tabulated in Table 3. Based on the R^2 value, Sips isotherm was the best to simulate the adsorption of palmitic acid onto CNF/β-CD, which proposed a monolayer adsorption process. Langmuir isotherms also gave high R^2 values that suggested the adsorption process underwent monolayer adsorption on homogeneous surface. The maximum adsorption capacity for palmitic acid obtained from Sips and Langmuir models were 8349.23 mg g^{-1} and 10485.38 mg g^{-1}, respectively. In the research done on adsorption of Cu (II) on β-CD by Lv et. al. [4], the surface suggested is homogeneous as the R^2 value of 0.999, indicating the experimental data fitted the isotherm best. Furthermore, in the study of adsorption isotherm between organic pollutant and β-CD done by Chen et. al. [48], the result obtained is similar to this study with a high R^2 value of 0.999 on the Langmuir isotherm.

The separation factor, R_L, for the Langmuir isotherm model was greater than zero and less than one, which indicated a favourable adsorption process. Furthermore, the data fitted into the Freundlich model with the value of n_f ranging between 0 to 10, which also indicates that the adsorption was favourable [46]. The value of n_f (1.9444) obtained in this study was higher than one which indicates a physical adsorption [48]. The Temkin isotherm produced heat of adsorption, b_T of 2.4 kJ mol^{-1} (equivalent to 0.57 kcal mol^{-1}). Heat of adsorption value less than 1 kcal mol^{-1} indicates a physical adsorption process. In contrast to the pseudo-second-order kinetic model that proposed a chemisorption process, the Freundlich and Temkin isotherm models suggest a physisorption process. The best explanation of this discrepancy probably lies on the adsorption of palmitic acid that involved the formation of palmitic acid-β-CD complex with host-guest interaction through noncovalent bonding. While most chemisorption processes were described to involve the formation of covalent bonds, the host-guest interaction between palmitic acid and β-CD molecules was associated with the formation of forces weaker than the covalent bond, such as ionic bonding, hydrogen bonding, van der Waals forces and hydrophobic interactions. Lv et al. [4] describes the plausible adsorption mechanism of bisphenol pollutants onto β-CD modified cellulose nanofiber through electrostatic interactions, π-π stacking interactions, hydrogen bonding, and hydrophobic interaction. Furthermore, the cellulose nanofiber itself could possibly adsorb the palmitic acid through physisorption process. Further analysis is required to reveal the underlying mechanism of interaction for this study. Therefore, it could be concluded that palmitic acid adsorbs onto the CNF/β-CD via both physisorption and chemisorption.

Gibbs free energy was calculated to reveal the spontaneity of the adsorption process. The Gibbs free energy value was 1.627 kJ mol^{-1} at 298 K (25°), suggesting a non-spontaneous adsorption process. The positive value of Gibbs free energy stipulated an endothermic process that requires energy input to take place. This is consistent with the positive value heat of adsorption from the Temkin isotherm model that designated an endothermic process. The adsorption process is said to be more favorable to occur at a higher temperature. The positive b_T values were also reported previously in the adsorption between organic pollutants and β-CD polymers [1,48]

Generally, the experimental data fitted all the adsorption isotherms used in this study as the correlation coefficient obtained were satisfactory. The Sips model best explained the adsorption behaviour of palmitic acid onto functionalized CNF/β-CD. The Sips isotherm can describe the adsorption process the best as a monolayer process. However, multilayer interaction might as well exist as suggested by the Freundlich and Temkin isotherm models.

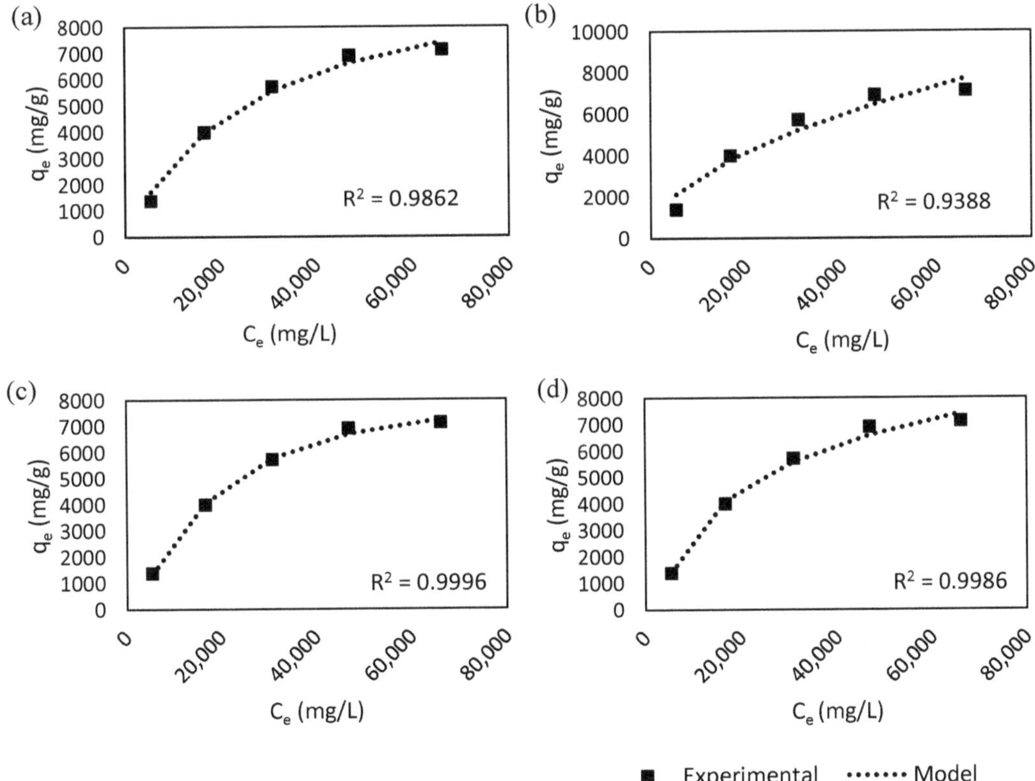

Figure 8. Non-linear fittings of palmitic acid adsorption onto CNF/β-CD with (**a**) Langmuir isotherm, (**b**) Freundlich isotherm, (**c**) Sips isotherm and (**d**) Temkin isotherm models.

Table 3. The coefficients and the constants of the Langmuir, Freundlich, Sips and Temkin isotherm models.

Isotherm Models	Parameters	Correlation Coefficient, R^2	Type of Surface/Adsorption Type
Langmuir	$K_L = 3.64 \times 10^{-5}$ L mg^{-1} $q_m = 10485.38$ mg g^{-1} $R_L = 0.7331$	0.9862	Homogeneous/monolayer
Freundlich	$n_f = 1.9444$ $K_f = 25.5563$ mg g^{-1} L mg$^{1/nf}$	0.9388	Heterogenous/multi-layer
Sips	$K_S = 5.7141 \times 10^{-5}$ L mg^{-1} $q_{ms} = 8349.23$ mg g^{-1} $n_S = 1.4177$	0.9996	Homogeneous or heterogenous/monolayer
Temkin	$K_T = 3.3 \times 10^{-4}$ L mg^{-1} $b_T = 2402.17$ J/mol (0.57 kcal/mol)	0.9986	Multi-layer

4. Conclusions

Functionalization of β-CD has improved the properties and adsorption capacity of cellulose nanofiber, and this serves as a promising material for uptake of various pollutant molecules from wastewater, including the hydrophobic molecules. Characterization study on CNFs by SEM analysis shows the maintenance of the nanofibrous structure but with increased fiber diameter after surface modification with β-CD. The presence of β-CD layer coating on the surface of CNF was supported by FTIR analysis that revealed changes on surface functional groups after modification was made. The resulted adsorption band of the carbonyl group confirmed the chemical linkages between CNF and citric acid via ester bonds. The thermal characterization analyzed by TGA observed that CNF/β-CD has higher thermal stability, which indicated more energy is required to decompose CNF that has a crosslinked structure.

Palmitic acid removal was recorded as two times higher when CNF/β-CD was used compared to CNF. CNF/β-CD reduced its adsorption performance after five cycles, while CNF adsorption capacity reduced after two cycles. The palmitic acid adsorption onto CNF/β-CD reached equilibrium after 60 min, and the pseudo-second-order kinetic is the best model to simulate the kinetic data of palmitic acid adsorption. According to the Sips and Langmuir isotherm models, the maximum adsorption capacity was 8349.23 mg g^{-1} and 10485.38 mg g^{-1}, respectively, at 25 °C. The isotherms indicated homogenous distribution of β-CD on the surface of CNF that leads to the uniform adsorption of palmitic acid to form a monolayer coverage.

The intraparticle diffusion model suggested chemical adsorption and intraparticle interaction as the controlling steps in the adsorption process. Thermodynamic analysis proposed the nonspontaneous endothermic process of adsorption which is favourable at higher temperatures. It can be concluded that the adsorption of palmitic acid onto CNF/β-CD was characterized as a combination of the chemisorption and physisorption processes, as the palmitic acid formed a complex with β-CD via host-guest interaction that involves noncovalent bonding. It is also speculated that the physical adsorption of palmitic acid onto CNF surfaces might also take place.

This is the first study to describe the adsorption behaviour of palmitic acid onto CNF/β-CD. This work provides an important fundamental starting point for further investigation of the adsorption mechanism of fatty acids onto CNF/β-CD and as a guideline for the practical application of oily wastewater treatment.

Author Contributions: Conceptualization, N.H.A.M.; Methodology, N.A.M.S. and G.Y.L.; Validation, N.I.W.A.; Formal Analysis, S.F.Z.M.F. and N.M.; Investigation, N.A.M.S. and G.Y.L.; Data Curation, M.A.H.B.; Writing—Original Draft Preparation, N.A.M.S. and G.Y.L.; Writing—Review & Editing, N.K.M.F.K. and N.H.A.M.; Visualization, N.K.M.F.K.; Supervision, R.M.I. and H.A.E.E.; Project Administration, N.H.A.M.; Funding Acquisition, N.H.A.M. All authors have read and agreed to the published version of the manuscript.

Funding: This study was funded by Malaysian Research University Network Young Researchers Grant Scheme, Ministry of Higher Education Malaysia (No. R.J130000.7851.4L904) and UTM Research University Grant (No. R.J130000.2651.17J89).

Institutional Review Board Statement: Not applicable.

Informed Consent Statement: Not applicable.

Acknowledgments: The authors would like to express their appreciation to Universiti Teknologi Malaysia and Ministry of Higher Education Malaysia for the support and sponsor involved.

Conflicts of Interest: The authors declare no conflict of interest.

References

1. Liu, D.; Huang, Z.; Li, M.; Sun, P.; Yu, T.; Zhou, L. Novel porous magnetic nanospheres functionalized by β-cyclodextrin polymer and its application in organic pollutants from aqueous solution. *Environ. Pollut.* **2019**, *250*, 639–649. [CrossRef] [PubMed]
2. Ying, T.; Su, J.; Jiang, Y.; Ke, Q.; Xu, H. A pre-wetting induced superhydrophilic/superlipophilic micro-patterned electrospun membrane with self-cleaning property for on-demand emulsified oily wastewater separation. *J. Hazard. Mater.* **2020**, *384*, 121475. [CrossRef] [PubMed]
3. Cai, Q.; Zhu, Z.; Chen, B.; Zhang, B. Oil-in-water emulsion breaking marine bacteria for demulsifying oily wastewater. *Water Res.* **2019**, *149*, 292–301. [CrossRef]
4. Lv, Y.; Ma, J.; Liu, K.; Jiang, Y.; Yang, G.; Liu, Y.; Lin, C.; Ye, X.; Shi, Y.; Liu, M.; et al. Rapid elimination of trace bisphenol pollutants with porous β-cyclodextrin modified cellulose nanofibrous membrane in water: Adsorption behavior and mechanism. *J. Hazard. Mater.* **2021**, *403*, 123666. [CrossRef]
5. Maniyazagan, M.; Chakraborty, S.; Pérez-Sánchez, H.; Stalin, T. Encapsulation of triclosan within 2-hydroxypropyl–β–cyclodextrin cavity and its application in the chemisorption of rhodamine B dye. *J. Mol. Liq.* **2019**, *282*, 235–243. [CrossRef]
6. Missoum, K.; Belgacem, M.N.; Bras, J. Nanofibrillated Cellulose Surface Modification: A Review. *Materials* **2013**, *6*, 1745–1766. [CrossRef]
7. Connors, K.A. The Stability of Cyclodextrin Complexes in Solution. *Chem. Rev.* **1997**, *97*, 1325–1358. [CrossRef]
8. Del Valle, E. Cyclodextrins and their uses: A review. *Process. Biochem.* **2004**, *39*, 1033–1046. [CrossRef]
9. Yin, C.; Cui, Z.; Jiang, Y.; van der Spoel, D.; Zhang, H. Role of Host–Guest Charge Transfer in Cyclodextrin Complexation: A Computational Study. *J. Phys. Chem. C* **2019**, *123*, 17745–17756. [CrossRef]
10. Dodero, A.; Schlatter, G.; Hébraud, A.; Vicini, S.; Castellano, M. Polymer-free cyclodextrin and natural polymer-cyclodextrin electrospun nanofibers: A comprehensive review on current applications and future perspectives. *Carbohydr. Polym.* **2021**, *264*, 118042. [CrossRef]
11. Wongfaed, N.; Kongjan, P.; Prasertsan, P.; O-Thong, S. Effect of oil and derivative in palm oil mill effluent on the process imbalance of biogas production. *J. Clean. Prod.* **2020**, *247*, 119110. [CrossRef]
12. Ekka, B.; Mierina, I.; Juhna, T.; Turks, M.; Kokina, K. Quantification of different fatty acids in raw dairy wastewater. *Clean. Eng. Technol.* **2022**, *7*, 100430. [CrossRef]
13. Sousa, D.Z.; Salvador, A.F.; Ramos, J.; Guedes, A.; Barbosa, S.L.; Stams, A.; Alves, M.; Pereira, M.A. Activity and Viability of Methanogens in Anaerobic Digestion of Unsaturated and Saturated Long-Chain Fatty Acids. *Appl. Environ. Microbiol.* **2013**, *79*, 4239–4245. [CrossRef] [PubMed]
14. Deaver, J.A.; Diviesti, K.I.; Soni, M.N.; Campbell, B.J.; Finneran, K.T.; Popat, S.C. Palmitic acid accumulation limits methane production in anaerobic co-digestion of fats, oils and grease with municipal wastewater sludge. *Chem. Eng. J.* **2020**, *396*, 125235. [CrossRef]
15. Liu, H.; Hsieh, Y.-L. Ultrafine fibrous cellulose membranes from electrospinning of cellulose acetate. *J. Polym. Sci. Part B Polym. Phys.* **2002**, *40*, 2119–2129. [CrossRef]
16. Lagergren, S. Zur Theorie Der Sogenannten Adsorption Gelöster Stoffe. *Z. Für Chem. Und Ind. Der Kolloide* **1907**, *2*, 15.
17. Ho, Y.S.; McKay, G. Pseudo-second order model for sorption processes. *Process Biochem.* **1999**, *34*, 451–465. [CrossRef]
18. Roginsky, S.; Zeldovich, Y.B. The Catalytic Oxidation of Carbon Monoxide on Manganese Dioxide. *Acta Phys. Chem. USSR* **1934**, *1*, 2019.
19. Weber, W.J., Jr.; Morris, J.C. Kinetics of Adsorption on Carbon from Solution. *J. Sanit. Eng. Div.* **1963**, *89*, 31–59. [CrossRef]
20. Langmuir, I. The adsorption of gases on plane surfaces of glass, mica and platinum. *J. Am. Chem. Soc.* **1918**, *40*, 1361–1403. [CrossRef]
21. Freundlich, H.M.F. Over the Adsorption in Solution. *J. Phys. Chem.* **1906**, *57*, 385–471.

22. Jeppu, G.P.; Clement, T.P. A modified Langmuir-Freundlich isotherm model for simulating pH-dependent adsorption effects. *J. Contam. Hydrol.* **2012**, *129–130*, 46–53. [CrossRef] [PubMed]
23. Tempkin, M.J.; Pyzhev, V. Recent Modifications to Langmuir Isotherms. *Acta Phys.-Chim. Sin.* **1940**, *12*, 217–225.
24. Hall, K.R.; Eagleton, L.C.; Acrivos, A.; Vermeulen, T. Pore- and Solid-Diffusion Kinetics in Fixed-Bed Adsorption under Constant-Pattern Conditions. *Ind. Eng. Chem. Fundam.* **1966**, *5*, 212–223. [CrossRef]
25. Smith, J.M.; Van Ness, H.C. *Introduction to Chemical Engineering Thermodynamics*, 3rd ed.; McGraw-Hill: New York, NY, USA, 1975.
26. Chen, G.; Shah, K.J.; Shi, L.; Chiang, P.-C. Removal of Cd(II) and Pb(II) ions from aqueous solutions by synthetic mineral adsorbent: Performance and mechanisms. *Appl. Surf. Sci.* **2017**, *409*, 296–305. [CrossRef]
27. Kayaci, F.; Aytac, Z.; Uyar, T. Surface modification of electrospun polyester nanofibers with cyclodextrin polymer for the removal of phenanthrene from aqueous solution. *J. Hazard. Mater.* **2013**, *261*, 286–294. [CrossRef]
28. Abdel-Halim, E.; Fouda, M.M.; Hamdy, I.; Abdel-Mohdy, F.; El-Sawy, S. Incorporation of chlorohexidin diacetate into cotton fabrics grafted with glycidyl methacrylate and cyclodextrin. *Carbohydr. Polym.* **2010**, *79*, 47–53. [CrossRef]
29. Uyar, T.; Havelund, R.; Hacaloglu, J.; Besenbacher, F.; Kingshott, P. Functional Electrospun Polystyrene Nanofibers Incorporating α-, β-, and γ-Cyclodextrins: Comparison of Molecular Filter Performance. *ACS Nano* **2010**, *4*, 5121–5130. [CrossRef]
30. Lim, Y.-M.; Gwon, H.-J.; Pyo, J.; Nho, Y.-C. Preparation of Cellulose-based Nanofibers Using Electrospinning. In *Nanofibers*; Kumar, A., Ed.; Intech: Rijeka, Croatia, 2010; pp. 179–188. ISBN 978-953-7619-86-2.
31. Korhonen, J.T.; Kettunen, M.; Ras, R.H.A.; Ikkala, O. Hydrophobic Nanocellulose Aerogels as Floating, Sustainable, Reusable, and Recyclable Oil Absorbents. *ACS Appl. Mater. Interfaces* **2011**, *3*, 1813–1816. [CrossRef]
32. Ortega-Toro, R.; Jiménez, A.; Talens, P.; Chiralt, A. Properties of starch–hydroxypropyl methylcellulose based films obtained by compression molding. *Carbohydr. Polym.* **2014**, *109*, 155–165. [CrossRef]
33. Yuan, G.; Prabakaran, M.; Qilong, S.; Lee, J.S.; Chung, I.-M.; Gopiraman, M.; Song, K.-H.; Kim, I.S. Cyclodextrin functionalized cellulose nanofiber composites for the faster adsorption of toluene from aqueous solution. *J. Taiwan Inst. Chem. Eng.* **2017**, *70*, 352–358. [CrossRef]
34. Kang, Y.; Choi, Y.K.; Kim, H.J.; Song, Y.; Kim, H. Preparation of anti-bacterial cellulose fiber via electrospinning and crosslinking with β-cyclodextrin. *Fash. Text.* **2015**, *2*, 11. [CrossRef]
35. Gawish, S.M.; Ramadan, A.M.; Mosleh, S.; Morcellet, M.; Martel, B. Synthesis and characterization of novel biocidal cyclodextrin inclusion complexes grafted onto polyamide-6 fabric by a redox method. *J. Appl. Polym. Sci.* **2006**, *99*, 2586–2593. [CrossRef]
36. Wilpiszewska, K.; Antosik, A.K.; Zdanowicz, M. The Effect of Citric Acid on Physicochemical Properties of Hydrophilic Carboxymethyl Starch-Based Films. *J. Polym. Environ.* **2019**, *27*, 1379–1387. [CrossRef]
37. Demitri, C.; Del Sole, R.; Scalera, F.; Sannino, A.; Vasapollo, G.; Maffezzoli, A.; Ambrosio, L.; Nicolais, L. Novel superabsorbent cellulose-based hydrogels crosslinked with citric acid. *J. Appl. Polym. Sci.* **2008**, *110*, 2453–2460. [CrossRef]
38. Xie, X.; Liu, Q. Development and Physicochemical Characterization of New Resistant Citrate Starch from Different Corn Starches. *Starch–Stärke* **2004**, *56*, 364–370. [CrossRef]
39. Crini, G.; Peindy, H.N.; Gimbert, F.; Robert, C. Removal of C.I. Basic Green 4 (Malachite Green) from aqueous solutions by adsorption using cyclodextrin-based adsorbent: Kinetic and equilibrium studies. *Sep. Purif. Technol.* **2007**, *53*, 97–110. [CrossRef]
40. Huang, H.; Fan, Y.; Wang, J.; Gao, H.; Tao, S. Adsorption kinetics and thermodynamics of water-insoluble crosslinked β-cyclodextrin polymer for phenol in aqueous solution. *Macromol. Res.* **2013**, *21*, 726–731. [CrossRef]
41. Murcia-Salvador, A.; Pellicer, J.A.; Fortea, M.I.; Gómez-López, V.M.; Rodríguez-López, M.I.; Núñez-Delicado, E.; Gabaldón, J.A. Adsorption of Direct Blue 78 Using Chitosan and Cyclodextrins as Adsorbents. *Polymers* **2019**, *11*, 1003. [CrossRef]
42. Pellicer, J.A.; Rodríguez-López, M.I.; Fortea, M.I.; Lucas-Abellán, C.; Mercader-Ros, M.T.; López-Miranda, S.; Gómez-López, V.M.; Semeraro, P.; Cosma, P.; Fini, P.; et al. Adsorption Properties of β- and Hydroxypropyl-β-Cyclodextrins Cross-Linked with Epichlorohydrin in Aqueous Solution. A Sustainable Recycling Strategy in Textile Dyeing Process. *Polymers* **2019**, *11*, 252. [CrossRef]
43. Rahman, N.; Nasir, M. Effective removal of acetaminophen from aqueous solution using Ca (II)-doped chitosan/β-cyclodextrin composite. *J. Mol. Liq.* **2020**, *301*, 112454. [CrossRef]
44. Kaveeshwar, A.R.; Ponnusamy, S.K.; Revellame, E.D.; Gang, D.D.; Zappi, M.E.; Subramaniam, R. Pecan shell based activated carbon for removal of iron(II) from fracking wastewater: Adsorption kinetics, isotherm and thermodynamic studies. *Process. Saf. Environ. Prot.* **2018**, *114*, 107–122. [CrossRef]
45. Qiao, X.; Yang, L.; Hu, X.; Cao, Y.; Li, Z.; Xu, J.; Xue, C. Characterization and evaluation of inclusion complexes between astaxanthin esters with different molecular structures and hydroxypropyl-β-cyclodextrin. *Food Hydrocoll.* **2021**, *110*, 106208. [CrossRef]
46. Pellicer, J.A.; Rodríguez-López, M.I.; Fortea, M.I.; Gómez-López, V.M.; Auñón, D.; Núñez-Delicado, E.; Gabaldón, J.A. Synthesis of New Cyclodextrin-Based Adsorbents to Remove Direct Red 83:1. *Polymers* **2020**, *12*, 1880. [CrossRef] [PubMed]
47. Feng, X.; Qiu, B.; Dang, Y.; Sun, D. Enhanced adsorption of naproxen from aquatic environments by β-cyclodextrin-immobilized reduced graphene oxide. *Chem. Eng. J.* **2021**, *412*, 128710. [CrossRef]
48. Chen, D.; Shen, Y.; Wang, S.; Chen, X.; Cao, X.; Wang, Z.; Li, Y. Efficient removal of various coexisting organic pollutants in water based on β-cyclodextrin polymer modified flower-like Fe_3O_4 particles. *J. Colloid Interface Sci.* **2021**, *589*, 217–228. [CrossRef] [PubMed]

 *sustainability*

Review

COVID-19 Biomedical Plastics Wastes—Challenges and Strategies for Curbing the Environmental Disaster

Siddharthan Selvaraj [1,†], Somasundaram Prasadh [2,†], Shivkanya Fuloria [3,*], Vetriselvan Subramaniyan [4], Mahendran Sekar [5], Abdelmoty M. Ahmed [6,7], Belgacem Bouallegue [6], Darnal Hari Kumar [8], Vipin Kumar Sharma [9], Mohammad Nazmul Hasan Maziz [4], Kathiresan V. Sathasivam [10], Dhanalekshmi U. Meenakshi [11] and Neeraj Kumar Fuloria [3,12,*]

1   Faculty of Dentistry, AIMST University, Bedong 08100, Malaysia; sidzcristiano@gmail.com
2   National Dental Center Singapore, 5 Second Hospital Avenue, Singapore 168938, Singapore; e0204949@u.nus.edu
3   Faculty of Pharmacy, AIMST University, Bedong 08100, Malaysia
4   Faculty of Medicine, Bioscience and Nursing, MAHSA University, Kuala Lumpur 42610, Malaysia; drvetriselvan@mahsa.edu.my (V.S.); mohammadnazmul@mahsa.edu.my (M.N.H.M.)
5   Department of Pharmaceutical Chemistry, Faculty of Pharmacy and Health Sciences, Royal College of Medicine Perak, Universiti Kuala Lumpur, Ipoh 30450, Malaysia; mahendransekar@unikl.edu.my
6   Department of Computer Engineering, College of Computer Science, King Khalid University Abha, Abha 61421, Saudi Arabia; amoate@kku.edu.sa (A.M.A.); belgacem.bouallegue2015@gmail.com (B.B.)
7   Department of Systems and Computer Engineering, Faculty of Engineering, Al Azhar University, Cairo 11651, Egypt
8   Jeffrey Cheah School of Medicine & Health Sciences, Monash University, No. 3 Jalan Masjid Abu Bakar, Johor Bahru 80100, Malaysia; hari.kumar@monash.edu
9   Department of Pharmaceutical Sciences, Gurukul Kangri (Deemed to Be University), Haridwar 249404, India; vipin@gkv.ac.in
10   Faculty of Applied Science, AIMST University, Bedong 08100, Malaysia; skathir@aimst.edu.my
11   College of Pharmacy, National University of Science and Technology, Muscat 130, Oman; dhanalekshmi@nu.edu.om
12   Center for Transdisciplinary Research, Department of Pharmacology, Saveetha Institute of Medical and Technical Sciences, Saveetha Dental College and Hospital, Saveetha University, Chennai 600077, India
*   Correspondence: shivkanya_fuloria@aimst.edu.my (S.F.); neerajkumar@aimst.edu.my (N.K.F.)
†   The authors contributed equally to this work.

Citation: Selvaraj, S.; Prasadh, S.; Fuloria, S.; Subramaniyan, V.; Sekar, M.; Ahmed, A.M.; Bouallegue, B.; Hari Kumar, D.; Sharma, V.K.; Maziz, M.N.H.; et al. COVID-19 Biomedical Plastics Wastes—Challenges and Strategies for Curbing the Environmental Disaster. *Sustainability* **2022**, *14*, 6466. https://doi.org/10.3390/su14116466

Academic Editors: Nallapaneni Manoj Kumar, Subrata Hait, Anshu Priya and Varsha Bohra

Received: 15 April 2022
Accepted: 22 May 2022
Published: 25 May 2022

Publisher's Note: MDPI stays neutral with regard to jurisdictional claims in published maps and institutional affiliations.

Abstract: The rise of the COVID-19 outbreak has made handling plastic waste much more difficult. Our superior, hyper-hygienic way of life has changed our behavioural patterns, such as the use of PPE (Personal Protective Equipment), the increased desire for plastic-packaged food and commodities, and the use of disposable utensils, as a result of the fear of transmission. The constraints and inefficiencies of our current waste management system, in dealing with our growing reliance on plastic, could worsen its mismanagement and leakage into the environment, causing a new environmental crisis. A sustainable, systemic, and hierarchical plastic management plan, which clearly outlines the respective responsibilities as well as the socioeconomic and environmental implications of these actions, is required to tackle the problem of plastic pollution. It will necessitate action strategies tailored to individual types of plastic waste and country demand, as well as increased support from policymakers and the general public. The situation of biomedical plastic wastes during the COVID-19 epidemic is alarming. In addition, treatment of plastic waste, sterilisation, incineration, and alternative technologies for transforming bio-plastic waste into value-added products were discussed, elaborately. Our review would help to promote sustainable technologies to manage plastic waste, which can only be achieved with a change in behaviour among individuals and society, which might help to safeguard against going from one disaster to another in the coming days.

Keywords: plastic pollution; waste management; COVID-19; personal protective equipment; biomedical waste

1. Introduction

Plastics in medical products, such as disposable syringes, tablets, capsule blister packaging, joint replacement prostheses, intravenous (IV) fluid tubes, blood bags, catheters, and heart valves, help in supporting human life [1]. The human body is implanted with medical devices made from plastics. While plastics' benefits are far-reaching, massive production and waste mismanagement have raised environmental concerns [2]. In 2018, plastic production totalled 359 million metric tonnes (Mt), with 6.9 Mt of waste generated (3.2 Mt for short-life products), of which approximately 22% was incinerated, 25% was recycled, and 42% was inefficiently treated (i.e., littered or improperly disposed of in dumps or open landfills) [3,4]. By 2050, an estimated 12 billion Mt of plastic litter will have accumulated in landfills and the natural environment [5], with greenhouse gas (GHG) emissions from the whole plastic lifetime accounting for 15% of the total global carbon budget [6]. Low biodegradation, combined with indiscriminate use, improper disposal, and mismanagement, has resulted in the accumulation of plastic debris in terrestrial and aquatic compartments around the world, affecting natural biota, agriculture, fisheries, and tourism, as well as posing a health and safety risk to humans [7].

The quick spike in demand, for the usage of plastic items to protect the general public, patients, and health and service employees, is one of the acute environmental repercussions of any pandemic crisis, such as COVID-19 [8]. The extensive usage of protective gear across the world, as a result of the epidemic, causes huge supply-chain interruptions and waste-disposal issues downstream. The demand trend for various plastic products, such as personal protective equipment (PPE) including gloves and masks for health workers, disposable plastic components for life support equipment, respirators, and general plastic supplies, such as syringes, is expected to follow the global pandemic curve. Used plastic goods are, commonly, infected with pathogens and should be treated as hazardous trash. Plastic waste management was, already, a serious environmental issue before the COVID-19 epidemic began, due to rising worries about contamination in terrestrial and marine ecosystems [1,9]. Garbage management systems throughout the world have, already, struggled to deal with current plastic waste, and the predicted rise in waste from the COVID-19 epidemic threatens to overload waste management systems as well as healthcare capacity. Medical waste from hospitals is particularly problematic, due to the need to destroy any residual pathogens [2,10].

Treatment facilities are, often, built to handle steady-state circumstances, in which medical waste is treated at a consistent flowrate and composition. Thermal procedures, such as cremation, steam treatment (autoclaving), plasma therapy, and microwave treatment, are used in a variety of treatment technologies. Multiple economic, technological, environmental, and social acceptability factors influence treatment selection [11]. Systems that are built for steady-state settings are likely to be disrupted by rapid waste volume scale-up. Experience in Wuhan has shown that optimization models may be utilised, to give decision assistance for the hospital waste management reverse-supply-chain problem [12]. A related issue is deciding where additional waste-handling facilities should be erected, to handle the rising volume of garbage. Economic considerations, pollution, safety, regulatory concerns, and public acceptance are all important considerations. Those sentiments, however, arrived too late, at the start of the epidemic. Since these systems were intended for trash amounts generated during regular operations, the projected volume of waste greatly exceeds the existing capability to handle hazardous medical waste. If suppressions alone are not enough, new facilities can be erected, or mobile units can be deployed to increase capacity [12,13].

A pinch point occurs, where this expanded capacity meets the suppressed curve's peak, and ensures that pathogen-contaminated waste hazards are managed. After the epidemic has passed, it is clear that there will be a substantial surplus of treatment capacity. These treatment facilities are unlikely to be incinerators with heat recovery capable of being reused for municipal solid trash. Given the variety of technological possibilities for medical waste treatment, life cycle assessment (LCA) and associated methodologies can help determine which solution is the most ecologically friendly. Medical waste incineration, combined with waste heat recovery, is one technique for recovering the chemical energy content of polymers for usable applications. An early LCA [14], with sensitivity analysis of heat-recovery efficiency, demonstrates that boosting energy recovery reduces environmental consequences. Even when non-thermal solutions, such as chemical disinfection, are considered, this has been corroborated by a more recent LCA [15]. The widespread use of incineration with heat recovery, however, has some challenges. Trace dioxin and furan emissions can cause public worry [16]. Contagion worries are expected to outweigh any concerns about environmental footprints, including GHG and pollutants, therefore, social-acceptability considerations may not play a significant role in the present epidemic. A mismatch between the supply of recovered heat and the demand for it is a major issue. Some scientists [17,18] predict that the epidemic would peak in the near future, when demand for heat in much of the Northern Hemisphere falls, owing to the colder weather. Since the safe disposal of hazardous waste takes precedence, waste-to-energy plants may not be conveniently positioned for energy recovery. Xu et al. [19] presented ideas for balancing heat supply and provided insights on imbalances towards management. However, it is unclear if systems that must be erected quickly or transportable units that must manage rapidly increasing medical waste quantities can be developed for maximum energy recovery. Even before the COVID-19 epidemic, the long-term viability of plastics had been questioned [16]. This review will give a prospective outlook on how the disruption caused by COVID-19 can act as a catalyst for short-term and long-term changes in plastic waste management practices throughout the world and measures to mitigate the plastic wastes.

2. The Impact of the Pandemic on Plastic Waste

Handling municipal solid waste (MSW) and hazardous medical waste has become extremely difficult because of the epidemic. China has the greatest information on this subject [20]. In Hubei Province, the production of medical waste surged dramatically, with a significant number of plastics. The total amount of medical waste in China was projected to be 207 kt from 20 January 2019 to 31 March 2019. Medical waste production in Wuhan grew from 40 t/d (tonnes per day) to almost 240 t/d, surpassing the maximum incineration capacity of 49 t/d [21–23]. Hazardous medical waste incineration costs in China are predicted to be 281.7–422.6 USD/t (US dollars per tonne), compared to 14.1 USD/t for MSW [21]. Treatment systems intended for normal waste quality and quantity must deal with extreme fluctuations, which compel abnormal operations. To guarantee that these systems can cope with the pandemic's dynamic and changeable character, engineering analysis is required. Another issue is that there is still a lot of unknown information about the virus, as well as what items and methods would be required to handle the pandemic. The COVID-19 problem has brought to light the importance of plastic in everyday life. Even while disposability is mostly viewed as an environmental liability, in most other uses, the virus demands single-use plastic [24]. Plastic products have significant environmental footprints, which may be summarised using a realistic evaluation technique. During the COVID-19 epidemic, demand for medical items and packaging has been skyrocketing. The amount and quality of plastic garbage fluctuates, as a result of the various mitigation or suppression strategies, undertaken in various nations. Single-use plastics are seen by consumers as a safe choice, for a variety of uses. Van Doremalen et al. [25] investigated and found that viral pathogens can be seen in various surfaces and have the ability to survive on plastics. These findings were confirmed by Kampf et al. [26]. Despite the fact that plastics are no better than other materials in terms of virus retention, customers who value hygiene see disposability as a critical benefit. Even for non-medical purposes, this has resulted in a rise in the usage and disposal of plastic items. In contrast, in the probable global economic crisis, plastic demand in other areas (such as aerospace and automotive applications) is declining. As a result of the lockdown, the amount of packaging utilised to convey food and consumables to residents has increased. These changes may intensify environmental concerns about plastics, which persisted even before the epidemic. Despite the fact that this rise is inevitable, environmental protection activities must continue. Metrics for system design and comparing alternatives, as well as footprints, should be created and properly utilised. [4]. The top 10 countries' daily global plastic waste generation, prior to management, were elucidated in Table 1.

Table 1. Estimated daily global plastic waste generation by country, prior to management [27].

Rank	Country	Population	Total Estimated Plastic Waste (Tonnes)
1	China	1,439,323,776	107,949,283.20
2	India	1,380,004,385	103,500,328.90
3	United States	331,002,651	24,825,198.80
4	Brazil	212,559,417	15,941,956.30
5	Indonesia	273,523,615	20,514,271.10
6	Japan	126,476,461	9,485,734.58
7	Russia	145,934,462	10,945,084.70
8	Mexico	128,932,753	9,669,956.48
9	Nigeria	206,139,589	15,460,469.20
10	Pakistan	220,892,340	16,566,925.50

3. Challenges in Waste Management

Many sorts of extra medical and hazardous waste are created during an epidemic, including contaminated masks, gloves, and other protective equipment, as well as a greater number of non-infected products of the same nature. Recent instances of airborne transmission [28] have prompted suggestions that people wear masks in public places. The proper handling of this waste can help to avoid negative consequences for human health or the environment. The identification, collection, separation, storage, transportation, treatment, and disposal of biomedical and healthcare waste, as well as crucial connected issues such as disinfection, staff protection, and training, are all required for effective management. Even the most modern healthcare institutions are struggling to keep up with the continually growing number of infected people. Patients who are self-isolated at home, due to minor symptoms, create polluted MSW. This necessitates a significant structural adjustment in trash management, ranging from sorting regulations, collection, and waste treatment to garbage collection personnel safety protocols, which has been followed by ACR, an international network that plays a major role in promoting the sustainable consumption of resources and management of waste among various nations around the globe [29,30], including a number of safety precautions regarding waste management.

They pose logistical obstacles for waste management systems, so other economic and environmental concerns have been pushed aside, in the wake of the coronavirus outbreak. The most prevalent methods for the thermal treatment of hazardous medical waste are incineration and steam sterilisation (90 min, 120 °C). After an effective decontamination cycle, in accordance with non-hazardous solid waste regulations, the residue from these operations can be properly handled [31,32]. In Germany, incineration temperatures must be kept at 1000 °C, to ensure safe annihilation. The WHO recommends a temperature range of 900 °C to 1200 °C for healthcare waste [33]. The fundamental issue is that COVID-19 is causing a waste spike, which might easily surpass treatment capacity. In this dire scenario, whether to use MSW incinerator capacity for medical waste remains an unanswered subject. Cement facilities in Spain are said to be able to co-incinerate waste on demand [34,35].

Due to flexibility in responding to changing demands, on-site and mobile therapy is preferred in China. There have always been benefits and costs, and they are subject to contextual limits. Plastics have similar calorific values as traditional fuels. The plastics fraction of MSW is predicted to have a calorific value of 25% [36]. The assumptions established during the development of waste management strategy (e.g., incentives, taxes, oversimplification on specific plastic composition, collection method) are, suddenly, no longer totally true. They were justified by the necessity to meet governmental goals for collection, recycling, and recovery, which resulted in the undersizing of retrieval and dumping facilities, promoting recycling, even when it was neither practicable nor sustainable, during the existing epidemic.

Pyrolysis and gasification are in development, stimulated by the request for more sustainable-waste-treatment options [37]. An economic assessment proposes that the present scenario is sustained by a tipping fee, which is continuously rising due to the high costs of transportation towards the treatment processes, both for recovery and for disposal [38]. Many countries have restricted the use of plastic bags. In the EU, even if the food packaging is plastic, the carrying bag is made of paper. However, the environmental footprint advantage of paper bags is questionable, since they, mostly, have limited potential for reuse. The typical paper bag (2.62 MJ/bag) has a higher energy footprint than a typical plastic bag (0.76 MJ/bag), which is much lighter. This reduced weight, also, incurs reduced footprints elsewhere in the supply chain [38].

4. Scientific Strategies for Mitigating Medical Waste Plastics

4.1. Recycled Polymers for 3D Printing

The worldwide manufacturing of components composed of plastics has risen significantly in the recent past, as it has soared as high as 359 million tons in 2019, per PEMRG (Plastics Europe Market Research Group); a little higher than half of the production is in Asia, and around 17% is in Europe. The production of virgin plastics requires close to 4% of the total oil that is manufactured, which is tantamount to 1.3 billion barrels annually [39]. These plastics are composed of polymers, which do not degrade and, instead, stay in the landscape for several hundred years; thus, the pollution associated with this plastic waste should be taken very seriously [40]. A maximum of 90% of the plastics can be reused, however, 80% of the wastes are dumped in the landfill, and a very minute percentage is being recycled. The primary issue is that plastics composed of PVC, PP LDPE, and HDPE are dumped, mainly, in landfills and release greenhouse gases [41]. The goods composed of polymers, such as PLA, have a weaker influence on environmental pollution, however, the usage is limited due to their low durability. These plastics' stability declines sharply upon reusing, causing harmful effects to human beings [42].

The technique of 3D printing has led to the production of complex structures on a smaller scale. This technique could be used to tackle the increasing amount of post-production waste [43,44]. Activities, such as selective material separation, decontamination, purification, grinding, re-melting, and extrusion, determine these polymers' recycling process. The logistics and economic aspects hinder this process, as no profits are gained upon recycling, and the original's market cost determines the cost of these recycled products [45]. However, given the increasing environmental restrictions and recycling of plastic waste, this could be a potential solution, despite the lack of exact economic profitability [45,46] (Figure 1).

Polypropylene (PP), polyvinyl chloride (PVC), high- and low-density polyethylene (HDPE, LDPE), polystyrene (PS), polyethylene terephthalate (PET), and the "other" category, mostly acrylonitrile-butadiene-styrene (ABS) and polycarbonate (PC), are now being recycled globally. All of the aforementioned categories have been studied for their possible application in 3D-printing filaments, according to the literature. PLA's natural origin was, primarily, investigated in comparison to other polymers. The impact of repeated material recycling [47,48], as well as the potential of adding an additional strengthening component [49–51], were investigated. The material is separated and cleansed in the primary step, before being ground. After that, the ground material is extruded at a hot concentration (subject to the polymer type). This filament material is inserted into the 3D printer, where various kinds of analysis are carried out, such as mechanical, rheological, and structural characteristics, to name a few. This tested sample is milled again [48]. If the material requires any modification, in the primary case, an additional component and a binder, such as silicone oil, are added to the mixed material, followed by extrusion [52].

Figure 1. Strategic scheme for recycled plastics wastes [45].

4.1.1. Impact of Recycling on the Material Properties

Mechanical stresses, such as shear stress, are combined with temperature and oxygen occurrence, while extruding leads to the degradation of the polymers, such as PLA and resistant P.E. [53]. This change in physical characteristics has a considerable effect on the production of high-quality extrusion. The multiple extrusion of polymers influences their change in viscosity, molecular weight, and breaking strength. Changes in properties are generated by temperature and the amount of extrusion of one material [48].

4.1.2. Mechanical Properties

Filaments composed of polylactic acid (PLA) and acrylonitrile butadiene styrene (ABS) are the most common thermoplastics available for 3D printing. Their costs are close to 200 times as much as the raw plastics [43], however, recycling thermo-mechanically leads to a lower printing cost. Despite its toxicity, ABS synthesised from oil is used for several applications. PLA, a biodegradable and biocompatible polymer, is sensitive to high temperatures, so degradation occurs upon exposure [54]. A reduction in PLA chains can be accomplished, by repeating the number of heating cycles, leading to shorter polymer chains [50]. The shorter polymer chains can, effectively, reorganise themselves into more ordered crystals, due to the increased melt flow rate. There was a small decrease observed in the tensile strength and strain, at break. The largest reduction in strain was recorded upon the first extrusion, which was 4%, and the largest decline in stress was 8%. The lower cohesion is responsible for reducing stress at break, whereas the decline of strain resulted from a reduction in the chain's length and a higher degree of crystallinity [48,50].

The decrease in the cold crystallization temperature and the decline in the melting point results in PLA's heating. A small reduction in the molecular weight is subjected to one reprocessing step [46,55]. The degradation rises to 30% after three cycles and 60% after seven cycles [50,56]. The loss of weight, repeatedly, upon extrusion was confirmed by T.G. analysis, which starts at 320 °C and continues until evaporation at 600 °C. The transesterification is encouraged in the presence of free radicals, upon thermo-mechanical recycling [48]. Furthermore, scientists have, also, discovered that the extrusion cycles increased water vapour transmission by 40% and oxygen transmission by 20%. Both activating substances are considered free radical reactions' antecedents. Hydrolysis and transesterification with residual catalysts are two further possibilities for degradation. There is a minor decrease in the intrinsic velocity, upon hydrolytic degradation without washing. On the contrary, a significant decline can be seen in the viscosity in washed PLA wastes, due to elevated temperatures and shear stress while reprocessing polymers. This degradation during accelerated aging can, also, contribute to this process [57]. The oxidative stabilisers (quinone) and residual catalyst stabilisers (tropolone) decrease rheological degradation. Upon recycling, the mechanical strength of the PLA decreases. The coating of the recycled polymer with polydopamine (PDA) can tackle this issue, where PDA is adsorbed on the water-repelling surface of PDA, thus, developing cohesive strength during self-polymerization. The thermal stability of this coated polymer exists up to 200 °C. It possesses higher tensile strength and strain at break, and its surface exhibits higher adhesion than uncoated PLA [58].

Another technique was proposed by Anderson et al. (2017) [53], where direct recycling of the utilised PLA filament was carried out by grounding and re-extrusion into 3D-printing filament. The material regains the same dimensions and surface finish, after two cycles of extrusion and one process of 3D printing, but the mechanical properties decline slightly. This process's primary limitation is reducing viscosity, due to chain scission, prohibiting the PLA filaments' utilization for further printing [51]. The lamellar structure rearranges the polymer chains' randomization, due to the reduction in the molecular weight. As a result of 3D printing, there is an increase in crystallinity as well as the number and average size of the pinholes, in twice-recycled PLA filament, contrary to the notion that repeated extrusion was the cause. Due to the thermal process, the shortening of the polymer chains led to easier crystallization, with a higher crystal population [59]. The recycled and shredded PLA filament can be enhanced by the virgin PLA addition, where the viscosity, mechanical, and thermal properties improve [51,58]. Thus, this contributes to the closed-looped recycling of PLA filament, which can be done in a benchtop machine at home [60].

The combination of a recycle bot and an open-source self-replicating 3D printer [59] has made recycling of PLA and ABS wastes very promising. Computer wastes were mechanically cleaned, to remove the impurities affecting the filament consistency. It may, also, lead to clogging in the nozzle of the 3D printer. The temperature while heating was regulated, at a temperature lower than the decomposition of structures and higher than the glass-transition temperature, to ensure that the printout does not degrade [61]. ABS must be dried and crushed, to ensure no bubble creation on the filament surface. Cruz Sanchez et al. (2017) [43] reported degradation of the PLA filament in five reprocessing cycles. The data obtained showed a considerable reduction in tensile strength and breaking strength as well as nominal deformation at the break. It was reported that the decomposition of the material correlates to the reduction in the crystallinity, viscosity, and molecular composition. The degradation mechanism majorly involves the five processes, namely: (1) formation of oligomers (hydroxyl and carboxyl); (2) esterification; (3) intermolecular transesterification, including interchanging of ester units between different chains; (4) thermo-oxidation; and (5) micro-compounding process. Moreover, 3D printing plays a role in the degradation of filaments, where uneven heating and cooling lead to stress accumulation, thereby affecting mesostructured and fibre-to-fibre bond strength [62]. Among other factors, one can mention the neck's growth between filaments and layers, randomization of the polymer chains on the contact surface, molecular diffusion, and internal defects (e.g., voids and the staircase effect) to the material during printing [63].

4.2. Methods to Recycle and Reuse Biomedical Plastics Waste

The 3Rs, namely Recycle, Reduce, and Reuse, have been the significant motto used for decreasing waste accumulation. The recycling of biomedical wastes is required to avoid accumulation, as in the case of conventional waste. However, before recycling, these wastes must be sterilised, to ensure the non-transmission of pathogens. Reports have suggested that the coronavirus survives on the PPEs for four days after contact [64]. Several methods of sterilization have been discussed below. The application of these methods on a large scale is a challenge.

4.2.1. Thermal Processes

Thermal-based systems, involving low heat (93 °C to 177 °C) generated by steam and microwaves, are being used. It is widely known that exposure to heat and microwaves can degrade plastics and lead to more inferior characteristics. Moreover, very high capital is essential, for such large quantities of waste to be treated by microwaves. Besides, the contaminant degassing and a release of toxic fumes are highly likely. The pyrolysis occurs at mid-range temperatures (177 °C to 540 °C); however, it consumes a lot of energy, and the products formed are not useful. The processes involving higher heat, such as plasma, lead to plastics's incineration into carbonaceous substances, leading to an increase in the carbon footprint, thus polluting the atmosphere [65].

4.2.2. Chemical Processes

The chemical processes, majorly, involve using compounds generating chlorine, such as sodium hypochlorite and chlorine dioxides [66], however, it is slower, as the PPEs must be dismantled, before exposure to chlorine–alcohol-based solutions. PPE's metallic components must be safeguarded from corrosive chlorine mists, which are applied along with alcohol.

Battelle CCDSTM (USA) has developed a concentrated-hydrogen-peroxide system in the vapor phase, for decontaminating PPEs. They offer complete pickup of contaminated PPEs and drop off at the healthcare facility after decontamination. They decontaminate 80,000 PPEs at a time [67,68]. This technology has been approved for use by the USFDA and has already been deployed for use in the US cities of New York, Seattle, Ohio, and Washington, D.C. [67]. Moreover, Lynntech has perfected the use of ozone for disinfection of plastics. Although this is a quick method that does not leave behind much residue, plastics are more susceptible to damage due to ozone use [68]. The mist's ability to penetrate through the layer of fabrics, as seen in respirators, has made chemical systems advantageous, to ensure a greater extent of decontamination. Unlike common plastics such as LDPE, HDPE, and PP, condensation polymers, such as polyesters, nylon, and a few others, degrade [68].

4.3. Use of Ionisation and Energetic Radiation

Electron beam radiations and other ionizing radiations can harm any living organism's DNA and neutralise it, but these radiations must not be applied in streams consisting of metals. A containment system, composed of a concrete bunker several feet thick, is required, making it the most expensive and challenging to construct in a short amount of time [67]. In the recent past, the application of UV rays in PPE's disinfection has been an area of focus. In specific, the pulsed xenon ultraviolet light is useful during the removal of worn PPE [69].

The ultraviolet-light spectrum consists of three sub-classifications: UV-A (320–340 nm), UV-B (280–320 nm), and UV-C (200–280 nm), where the UV-C rays have the highest germicidal properties. RNA and DNA absorb UV-C, leading to structural damage to the molecules through photodimersation, resulting in virus inactivation. As SARS-CoV-2 is a recent find, there is not sufficient data to analyse the survival of the virus under different conditions, however, scientists have treated this virus similarly to SARS-CoV-1 [70]. Although UV-C has been significant, factors such as the inoculum size, culture medium, geometry, and size of material used in PPEs play a significant role, which leads to incoherent findings. At 360 mJ/cm^2, SARS-CoV-1 had the highest UV D90 among nearly 130 different types of viruses [71]. It has been reported that the majority of aerosols are captured on the initial layers of respirators [72]. Their efficacy remains questionable, especially in complex geometries, despite the lower costs and rapid throughput of disinfection. The mechanical degradation upon exposure to UV-C deems it unsuitable for use.

5. Environmentally Sustainable Management of Used Personal Protective Equipment

Face masks, gloves, goggles, gowns, and aprons are examples of personal protective equipment (PPE) that can help protect people from infections and toxins. PPE used against pathogens has, traditionally, been mostly in the hospital setting. However, due to the worldwide COVID-19 epidemic, personal protective equipment (PPE) is, increasingly, commonly employed in residential circumstances, resulting in supply chain shortages and a fast accumulation of potentially infected PPE in household-solid-waste streams [73]. Manufacturing, building, the oil and gas industry, transportation, firefighting, and food production have all been impacted by the enormous domestic need for personal protective equipment, in reaction to the epidemic. Since the COVID-19 epidemic, there has been a tremendous surge in the development of plastic-based PPE equipment. For example, the worldwide market for PPE grew at a compound annual rate of 6.5%, between 2016 and 2020, from around USD 40 billion to USD 58 billion [23].

The World Health Organization, on the other hand, estimates that PPE supplies will need to grow by 40% monthly, to adequately combat the COVID-19 epidemic. In total, 89 million medical masks, 76 million pairs of medical gloves, and 1.6 million pairs of goggles are among the needed PPE [74]. PPE demand is not predicted to fall much in the post-pandemic period, with the supply of face and surgical masks expected to expand at a compound annual rate of 20%, from 2020 to 2025 [23,73]. The long-term management of PPE is a major problem.

The lack of a coordinated international strategy to manage the PPE production and waste lifecycle is threatening to stymie progress, towards achieving key components of the United Nations' Sustainable Development Goals (SDGs), such as SDG 3, good health and well-being; SDG 6, clean water and sanitation; SDG 8, decent work and economic growth; SDG 12, responsible consumption and production; and SDG 13, climate action [75]. We offer product lifecycle techniques that should be included into public–private-partnership-based solutions. Increases in PPE production and distribution result in an increase in waste, which is exacerbated by health and environmental dangers along the waste treatment chain, particularly in countries with poor infrastructure (Figure 2).

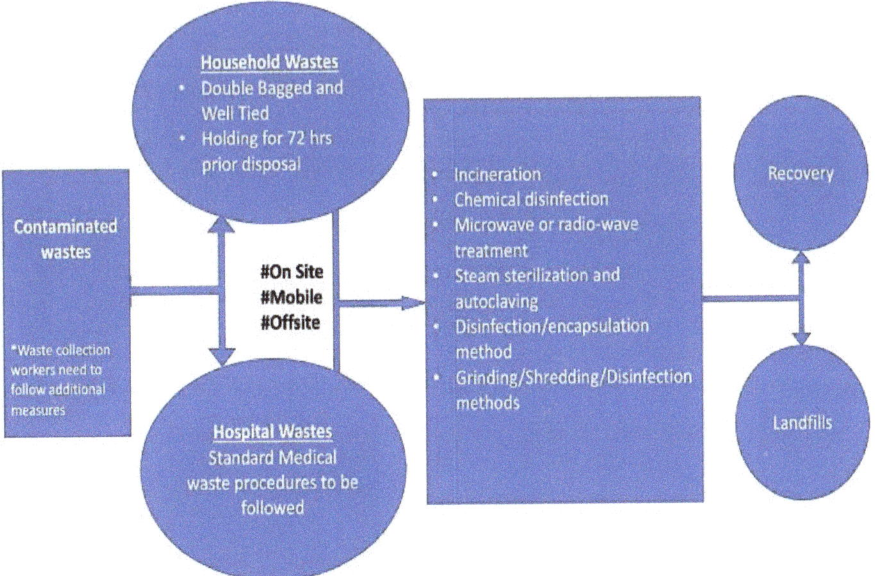

Figure 2. Methods for handling contaminated wastes in pandemic situations.

At the peak of the pandemic in Wuhan, China produced 240 tonnes of medical waste per day, which was six times more than before the epidemic. As a result, the city's waste management department set up mobile incinerators across the city to dispose of the massive amounts of abandoned face masks, gloves, and other contaminated single-use protective equipment. Across the globe, similar increases in the number of abandoned face masks, hand gloves, and safety eyewear have been reported. Moreover, over seven million Hong Kong residents, for example, use single-use masks on a regular basis. Discarded masks have been reported in the water and on Hong Kong's beaches and nature paths [76]. The outbreak has had an impact on how solid waste management is handled. Waste management and resource recycling were declared non-essential and placed on lockdown.

This interruption of normal waste management services has been observed across the world, and it has been compounded by China's previous limits on importing "recyclable" solid waste, which were imposed in 2019. As a result, several governments have implemented informal protocols for collecting and recycling discarded PPE, a practise that may pose a risk, owing to insufficient decontamination [77]. Viral infections can be transmitted to healthcare and recycling employees, if infected trash is improperly disposed of or handled. It has been estimated that up to 30% of hepatitis B rates, 13% of hepatitis C rates, and 0.3% of HIV rates have been transmitted from patients to healthcare workers, as a result of incorrect medical waste disposal. A higher-than-normal incidence of viral infection among solid waste collectors may be related, directly, to pathogens in contaminated wastes, according to studies undertaken in Pakistan, Greece, Brazil, Iran, and India [23]. The Basel Convention on the Transboundary Movement of Hazardous Wastes and their Disposal, a treaty by the United Nations, has recently urged member countries to treat waste management as an urgent and essential public service, in the wake of COVID-19, in order to minimise potential secondary effects on health and the environment. As a result, safe and long-term recovery as well as treatment of PPEs should be prioritised. It is essential to clarify the role of informal recyclers in developing countries, where medical waste has not been, adequately, regulated.

The PPE reaction to the COVID-19 epidemic has had an influence on plastic recovery and recyclability, as well as landfilling and pollution. PPE materials have polymers as a main ingredient, accounting for 20–25%of the total weight. If they are not recycled, they contribute, significantly, to dangerous environmental contaminants, such as dioxins and toxic metals [73]. Contrary to World Health Organization recommendations, which encourage safe practises that reduce the volume of waste generated and ensure proper waste segregation at the source, plastic-based PPEs discarded from households are mixed with other domestic plastic wastes, such as single-use plastic bags, the use of which has increased rapidly, since grocery stores prohibited customers from bringing their own bags for fear of introducing additional virus-transmission routes. PPEs, such as N-95 masks, Tyvek protective suits, gloves, and medical face shields, all include polypropylene. Polypropylene, also, accounts for a significant amount of the nearly 25 million tonnes of plastic materials disposed of in US landfills each year, with just 3% of the polypropylene plastic created being recovered and recycled [13,78–80]. It is difficult to recover polymers from mixed healthcare waste, which includes PPE. Individuals operating as recyclers in middle- and low-income countries are constrained in their ability to recycle, without risking infection, due to the low proportion (15–25%) of healthcare waste that is not contaminated. Furthermore, poor plastic waste recycling rates throughout the world, as well as a lack of coordinated government laws requiring minimum recycling content in new goods, will almost certainly result in an increase in virgin plastic manufacture in the post-pandemic period. The plastics manufacturing industry in the United States has asked more than USD 1 billion in emergency financing to deal with the increased demand caused by COVID-19 [79]. Restriction of emergency funding is necessary to promote investments in the research and development of used PPE collecting, sorting, and recycling, to ensure that growing plastic PPE manufacturing does not lead to greater pollution. Public–private collaborations will help to implement a long-term PPE waste management strategy. The importance of artisanal solid-waste collectors and recyclers in nations in transition is unquestionable. It is difficult to develop safe and sustainable PPE management, outside of healthcare settings (hospitals and clinics) in emergency situations, since it necessitates a thorough understanding of best practises, as well as the monitoring and enforcement of laws and regulations. Thermal, chemical, irradiative, and biological processes can be applied locally or scaled up in regional facilities, where waste collection and transportation are viable in healthcare settings. Single-use PPE is not a sustainable practise, so addressing the PPE pollution problem will need multidisciplinary technical skills, encompassing biological sciences, environmental science, public health, materials science, and engineering. PPE disinfection and reuse can be done on a large-scale, using methods such as infusion of hydrogen-peroxide vapour, ultraviolet or gamma irradiation, ethylene-oxide gasification, application of spray-on disinfectants, and infusion of base materials with antimicrobial nanoparticles, according to new research published since the start of the current pandemic [81–84]. Many disinfection technologies are, still, in the early stages of development, and they must be calibrated to guarantee that material deterioration during each disinfection cycle does not jeopardise the fundamental purpose of PPEs, which is to prevent pathogen penetration and human exposure. During and after the present epidemic, the circular economy idea of reducing, reusing, and recycling materials should govern PPE management policy development. Plastic producers should be required to include a minimum recycling content in new items under national rules, and product price should reflect environmental and health externalities. Policy implementation, monitoring, and enforcement should also include public education initiatives, to encourage proper PPE stewardship. In low-income nations, infrastructure development is critical to ensure the safety of informal garbage collection and recycling. PPE management policies must be linked into economic models that encourage the use of green technology and alternative evaluations, to find and implement safer methods based on thorough material-life-cycle assessments and customer preferences, if they are to be sustainable. In conclusion, the COVID-19 epidemic has put pressure on worldwide solid waste management, as well as emphasising supply chain bottlenecks in PPE manufacturing, demand–supply, usage, and disposal. PPEs will continue to be in high demand, thus, now is the time to invest in innovative PPE materials that decrease waste creation and enhance methods for the safe and sustainable storage of worn PPE, with worldwide policy guidelines [75].

6. Conclusions and Future Perspective

Every country in this globe is, currently, focusing on tackling the COVID-19 epidemic. This requires consideration of the economic and environmental factors as well. In this paper, the issues pertaining to PPE disposal and safe disposal strategies have been elaborated. The precipitous rise in plastic waste, especially for protection and healthcare purposes, has been one of the significant consequences of this coronavirus outbreak. The measurement of environmental issues, in terms of footprints as PF and PWF, along with the strategy for safe disposal, has been discussed in this paper.

Considering the current trend, it is necessary to, immediately, reassess goals and priorities, without harming the environment, society, or economy. Massive quantities of plastic waste are landfilled or incinerated. Only a small part is recycled, leading to the disposal of 4–12 million tonnes/year of plastics into the seas and oceans [85,86]. The health of our environment and human beings is interdependent. To ensure a future, sustainability is essential. Scientists must guide world leaders and corporations' management, to implement an efficient plastic waste management system for plastic waste recovery, governed by strict rules and regulations in the manufacture and consumption of plastic products. It is, also, vital to seek sustainable techniques to reduce plastic waste. One way is to use biobased plastics; however, more research has to be carried out, to scale up the economic and environmental aspects. Besides, the manufacture of sustainable products must be complemented by the producer's responsibility, and the cost of waste management must be adopted by the distributors and sellers. Therefore, in every country, it is essential that plastics are of primary political concern, to minimise plastic wastage as well as promote a circular economy and sustainability.

This circular economy principle should centre on reducing, reusing, and recycling principles for PPE management, throughout the pandemic and post-pandemic situations. National policies ought to be devised, to ensure that plastic production involves minimum recycling content in new products. However, the cost of the products must reflect environmental and health externalities. In lower-income countries, the advancement of infrastructure is necessary, to safeguard informal waste collection. To ensure sustainability, PPE management policies need be incorporated into fiscal policies, to encourage green technology as well as find and implement safer practices.

Author Contributions: Conceptualization, S.S. and S.P.; methodology, S.P.; validation, S.P.; data curation, S.P.; writing–original draft preparation, S.S. and S.P.; writing–review and editing, S.S. and S.P.; supervision, N.K.F., S.F., V.S., M.N.H.M., M.S., A.M.A., B.B., D.H.K., V.K.S., K.V.S. and D.U.M.; funding acquisition; N.K.F., S.F., V.S., M.N.H.M., M.S., A.M.A., B.B., D.H.K., V.K.S., K.V.S. and D.U.M. All authors have read and agreed to the published version of the manuscript.

Funding: No funding was received to carry out this research.

Institutional Review Board Statement: Not applicable.

Informed Consent Statement: Not applicable.

Data Availability Statement: Not applicable.

Conflicts of Interest: The authors declare no conflict of interest.

Abbreviations

IV	Intravenous
Mt	Metric tonne
GHG	Greenhouse gas
PPE	Personal protective equipment
LCA	Life cycle assessment
MSW	Municipal solid waste

References

1. Rajmohan, K.V.S.; Ramya, C.; Viswanathan, M.R.; Varjani, S. Plastic pollutants: Effective waste management for pollution control and abatement. *Curr. Opin. Environ. Sci. Health* **2019**, *12*, 72–84. [CrossRef]
2. Windfeld, E.S.; Brooks, M.S.-L. Medical waste management—A review. *J. Environ. Manag.* **2015**, *163*, 98–108. [CrossRef] [PubMed]
3. Plastics Europe—The Facts 2016. An Analysis of European Plastics Production, Demand and Waste Data. *Plast. Eur.* **2016**, 1–38. Available online: https://www.plasticseurope.org/application/files/4315/1310/4805/plastic-the-fact-2016.pdf (accessed on 26 April 2020).
4. Hahladakis, J.N.; Velis, C.A.; Weber, R.; Iacovidou, E.; Purnell, P. An overview of chemical additives present in plastics: Migration, release, fate and environmental impact during their use, disposal and recycling. *J. Hazard. Mater.* **2018**, *344*, 179–199. [CrossRef] [PubMed]
5. Geyer, R.; Jambeck, J.; LAW, K. Producción, uso y destino de todos los plásticos jamás fabricados. *Sci. Adv.* **2017**, *3*, 1207–1221.
6. Zheng, J.; Suh, S. Strategies to reduce the global carbon footprint of plastics. *Nat. Clim. Chang.* **2019**, *9*, 374–378. [CrossRef]
7. Jambeck, J.; Geyer, R.; Wilcox, C.; Siegler, T.; Perryman, M.; Andrady, A.; Naray, R. Plastic waste inputs from land into the ocean. *Science* **2015**, *347*, 768–771. [CrossRef]
8. MacKenzie, D. COVID-19 goes global. *New Sci.* **2020**, *245*, 7. [CrossRef]
9. Babaji, P.; Singh, A.; Lau, H.; Lamba, G.; Somasundaram, P. Deletion of short arm of chromosome 18, Del (18p) syndrome. *J. Indian Soc. Pedod. Prev. Dent.* **2014**, *32*, 68. [CrossRef]
10. Selvaraj, D.; Raja, J.; Prasath, S. Interdisciplinary approach for bilateral maxillary canine: First premolar transposition with complex problems in an adult patient. *J. Pharm. Bioallied Sci.* **2013**, *5* (Suppl. 2), S190. [CrossRef]
11. Liu, H.-C.; You, J.-X.; Lu, C.; Chen, Y.-Z. Evaluating health-care waste treatment technologies using a hybrid multi-criteria decision making model. *Renew. Sustain. Energy Rev.* **2015**, *41*, 932–942. [CrossRef]
12. Yu, H.; Sun, X.; Solvang, W.D.; Zhao, X. Reverse Logistics Network Design for Effective Management of Medical Waste in Epidemic Outbreaks: Insights from the Coronavirus Disease 2019 (COVID-19) Outbreak in Wuhan (China). *Int. J. Environ. Res. Public Health* **2020**, *17*, 1770. [CrossRef] [PubMed]

13. Kumar, N.M.; Mohammed, M.A.; Abdulkareem, K.H.; Damasevicius, R.; Mostafa, S.A.; Maashi, M.S.; Chopra, S.S. Artificial intelligence-based solution for sorting COVID related medical waste streams and supporting data-driven decisions for smart circular economy practice. *Process Saf. Environ. Prot.* **2021**, *152*, 482–494. [CrossRef]

14. Zhao, W.; Van Der Voet, E.; Huppes, G.; Zhang, Y. Comparative life cycle assessments of incineration and non-incineration treatments for medical waste. *Int. J. Life Cycle Assess.* **2008**, *14*, 114–121. [CrossRef]

15. Hong, J.; Zhan, S.; Yu, Z.; Hong, J.; Qi, C. Life-cycle environmental and economic assessment of medical waste treatment. *J. Clean. Prod.* **2018**, *174*, 65–73. [CrossRef]

16. Makarichi, L.; Jutidamrongphan, W.; Techato, K.-A. The evolution of waste-to-energy incineration: A review. *Renew. Sustain. Energy Rev.* **2018**, *91*, 812–821. [CrossRef]

17. Bertschinger, N. Visual explanation of country specific differences in COVID-19 dynamics. *arXiv* **2020**, arXiv:2004.07334.

18. Saha, S.; Samanta, G.P.; Nieto, J.J. Epidemic model of COVID-19 outbreak by inducing behavioural response in population. *Nonlinear Dyn.* **2020**, *102*, 455–487. [CrossRef]

19. Xu, Z.; Wang, R.; Yang, C. Perspectives for low-temperature waste heat recovery. *Energy* **2019**, *176*, 1037–1043. [CrossRef]

20. Dincer, B.; Inangil, D. The effect of Emotional Freedom Techniques on nurses' stress, anxiety, and burnout levels during the COVID-19 pandemic: A randomized controlled trial. *Explore* **2021**, *17*, 109–114. [CrossRef]

21. Huang, H.; Tang, L. Treatment of organic waste using thermal plasma pyrolysis technology. *Energy Convers. Manag.* **2007**, *48*, 1331–1337. [CrossRef]

22. Singh, N.; Tang, Y.; Ogunseitan, O.A. Environmentally Sustainable Management of Used Personal Protective Equipment. *Environ. Sci. Technol.* **2020**, *54*, 8500–8502. [CrossRef] [PubMed]

23. Singh, N.; Tang, Y.; Zhang, Z.; Zheng, C. COVID-19 waste management: Effective and successful measures in Wuhan, China. *Resour. Conserv. Recycl.* **2020**, *163*, 105071. [CrossRef] [PubMed]

24. Tikkinen, K.A.O.; Malekzadeh, R.; Schlegel, M.; Rutanen, J.; Glasziou, P. COVID-19 clinical trials: Learning from exceptions in the research chaos. *Nat. Med.* **2020**, *26*, 1671–1672. [CrossRef]

25. Van Doremalen, N.; Bushmaker, T.; Morris, D.H.; Holbrook, M.G.; Gamble, A.; Williamson, B.N.; Tamin, A.; Harcourt, J.L.; Thornburg, N.J.; Gerber, S.I. Aerosol and surface stability of SARS-CoV-2 as compared with SARS-CoV-1. *N. Engl. J. Med.* **2020**, *382*, 1564–1567. [CrossRef] [PubMed]

26. Kampf, G.; Todt, D.; Pfaender, S.; Steinmann, E. Persistence of coronaviruses on inanimate surfaces and their inactivation with biocidal agents. *J. Hosp. Infect.* **2020**, *104*, 246–251. [CrossRef]

27. Benson, N.U.; Bassey, D.E.; Palanisami, T. COVID pollution: Impact of COVID-19 pandemic on global plastic waste footprint. *Heliyon* **2021**, *7*, e06343. [CrossRef]

28. Bourouiba, L. Turbulent gas clouds and respiratory pathogen emissions: Potential implications for reducing transmission of COVID-19. *JAMA* **2020**, *323*, 1837–1838. [CrossRef]

29. Petrov, A.; Petrova, D. Sustainability of Transport System of Large Russian City in the Period of COVID-19: Methods and Results of Assessment. *Sustainability* **2020**, *12*, 7644. [CrossRef]

30. Prüss, A.; Giroult, E.; Rushbrook, P. Treatment and disposal technologies for health-care waste. In *Safe Management of Wastes from Healthcare Activities*; WHO: Geneva, Switzerland, 1999; pp. 77–111.

31. Moriarty, D.G.; Zack, M.M.; Kobau, R. The Centers for Disease Control and Prevention's Healthy Days Measures–Population tracking of perceived physical and mental health over time. *Health Qual. Life Outcomes* **2003**, *1*, 1–8. [CrossRef]

32. *Obtenido de Project Management Guide*; CDC: Centers for Disease Control and Prevention: Atlanta, GA, USA. Available online: https://www2a.cdc.gov/cdcup/library (accessed on 26 April 2020).

33. Klemeš, J.J.; Van Fan, Y.; Tan, R.R.; Jiang, P. Minimising the present and future plastic waste, energy and environmental footprints related to COVID-19. *Renew. Sustain. Energy Rev.* **2020**, *127*, 109883. [CrossRef] [PubMed]

34. Chapple, K. *Planning Sustainable Cities and Regions: Towards More Equitable Development*; Routledge: London, UK, 2014.

35. Ravetz, J. Integrated assessment for sustainability appraisal in cities and regions. *Environ. Impact Assess. Rev.* **2000**, *20*, 31–64. [CrossRef]

36. Al-Salem, S. Energy production from plastic solid waste (PSW). In *Plastics to Energy*; Elsevier: Cambridge, MA, USA, 2019; pp. 45–64. [CrossRef]

37. Vanapalli, K.R.; Samal, B.; Dubey, B.K.; Bhattacharya, J. Bhattacharya, Emissions and environmental burdens associated with plastic solid waste management. In *Plastics to Energy*; Elsevier: Cambridge, MA, USA, 2019; pp. 313–342. [CrossRef]

38. Mastellone, M.L. Technical description and performance evaluation of different packaging plastic waste management's systems in a circular economy perspective. *Sci. Total Environ.* **2020**, *718*, 137233. [CrossRef] [PubMed]

39. Singh, N.; Hui, D.; Singh, R.; Ahuja, I.P.S.; Feo, L.; Fraternali, F. Recycling of plastic solid waste: A state of art review and future applications. *Compos. Part B Eng.* **2017**, *115*, 409–422. [CrossRef]

40. Gu, L.; Ozbakkaloglu, T. Use of recycled plastics in concrete: A critical review. *Waste Manag.* **2016**, *51*, 19–42. [CrossRef]

41. Aboulkas, A.; El Harfi, K.; El Bouadili, A. Thermal degradation behaviors of polyethylene and polypropylene. Part I: Pyrolysis kinetics and mechanisms. *Energy Convers. Manag.* **2010**, *51*, 1363–1369. [CrossRef]

42. Lithner, D.; Larsson, Å.; Dave, G. Environmental and health hazard ranking and assessment of plastic polymers based on chemical composition. *Sci. Total Environ.* **2011**, *409*, 3309–3324. [CrossRef]

43. Cruz Sanchez, F.A.; Boudaoud, H.; Hoppe, S.; Camargo, M. Polymer recycling in an open-source additive manufacturing context: Mechanical issues. *Addit. Manuf.* **2017**, *17*, 87–105. [CrossRef]

44. Vatsal, A.; Prasadh, S.; Deepamala, S.; Patil, A.; Sulochana, K.; Shruthi, D. Comparative evaluation of dimensional changes of elastomeric impression materials after disinfection with glutaraldehyde and microwave irradiation. *J. Int. Oral Health* **2015**, *7*, 44–46.

45. Hopewell, J.; Dvorak, R.; Kosior, E. Plastics recycling: Challenges and opportunities. *Philos. Trans. R. Soc. B Biol. Sci.* **2009**, *364*, 2115–2126. [CrossRef]

46. Mikula, K.; Skrzypczak, D.; Izydorczyk, G.; Warchoł, J.; Moustakas, K.; Chojnacka, K.; Witek-Krowiak, A. 3D printing filament as a second life of waste plastics—A review. *Environ. Sci. Pollut. Res.* **2020**, *28*, 12321–12333. [CrossRef] [PubMed]

47. Saravanan, P.; Ettappan, M.; Kumar, N.M.; Elangkeeran, N. Exhaust Gas Recirculation on a Nano-Coated Combustion Chamber of a Diesel Engine Fueled with Waste Plastic Oil. *Sustainability* **2022**, *14*, 1148. [CrossRef]

48. Anderson, I. Mechanical properties of specimens 3D printed with virgin and recycled polylactic acid. *3D Print. Addit. Manuf.* **2017**, *4*, 110–115. [CrossRef]

49. Żenkiewicz, M.; Richert, J.; Rytlewski, P.; Moraczewski, K.; Stepczyńska, M.; Karasiewicz, T. Characterisation of multi-extruded poly (lactic acid). *Polym. Test.* **2009**, *28*, 412–418. [CrossRef]

50. Gkartzou, E.; Koumoulos, E.; Charitidis, C.A. Production and 3D printing processing of bio-based thermoplastic filament. *Manuf. Rev.* **2017**, *4*, 1. [CrossRef]

51. Pillin, I.; Montrelay, N.; Bourmaud, A.; Grohens, Y. Effect of thermo-mechanical cycles on the physico-chemical properties of poly (lactic acid). *Polym. Degrad. Stab.* **2008**, *93*, 321–328. [CrossRef]

52. Zhao, X.G.; Hwang, K.-J.; Lee, D.; Kim, T.; Kim, N. Enhanced mechanical properties of self-polymerized polydopamine-coated recycled PLA filament used in 3D printing. *Appl. Surf. Sci.* **2018**, *441*, 381–387. [CrossRef]

53. Pan, G.-T.; Chong, S.; Tsai, H.-J.; Lu, W.-H.; Yang, T.C.-K. The Effects of Iron, Silicon, Chromium, and Aluminum Additions on the Physical and Mechanical Properties of Recycled 3D Printing Filaments. *Adv. Polym. Technol.* **2018**, *37*, 1176–1184. [CrossRef]

54. Yuan, X.; Kumar, N.M.; Brigljević, B.; Li, S.; Deng, S.; Byun, M.; Lee, B.; Lin, C.S.; Tsang, D.C.; Lee, K.B.; et al. Sustainability-inspired upcycling of waste polyethylene terephthalate plastic into porous carbon for CO_2 capture. *Green Chem.* **2022**, *24*, 1494–1504. [CrossRef]

55. Badia, J.; Strömberg, E.; Karlsson, S.; Ribes-Greus, A. Material valorisation of amorphous polylactide. Influence of thermo-mechanical degradation on the morphology, segmental dynamics, thermal and mechanical performance. *Polym. Degrad. Stab.* **2012**, *97*, 670–678. [CrossRef]

56. Chariyachotilert, C.; Joshi, S.; Selke, S.; Auras, R. Assessment of the properties of poly (L-lactic acid) sheets produced with differing amounts of postconsumer recycled poly (L-lactic acid). *J. Plast. Film Sheeting* **2012**, *28*, 314–335. [CrossRef]

57. Brüster, B.; Addiego, F.; Hassouna, F.; Ruch, D.; Raquez, J.-M.; Dubois, P. Thermo-mechanical degradation of plasticized poly (lactide) after multiple reprocessing to simulate recycling: Multi-scale analysis and underlying mechanisms. *Polym. Degrad. Stab.* **2016**, *131*, 132–144. [CrossRef]

58. Beltrán, F.; Lorenzo, V.; Acosta, J.; de la Orden, M.; Urreaga, J.M. Effect of simulated mechanical recycling processes on the structure and properties of poly (lactic acid). *J. Environ. Manag.* **2018**, *216*, 25–31. [CrossRef]

59. Zhao, P.; Rao, C.; Gu, F.; Sharmin, N.; Fu, J. Close-looped recycling of polylactic acid used in 3D printing: An experimental investigation and life cycle assessment. *J. Clean. Prod.* **2018**, *197*, 1046–1055. [CrossRef]

60. Baechler, C.; DeVuono, M.; Pearce, J.M. Distributed recycling of waste polymer into RepRap feedstock. *Rapid Prototyp. J.* **2013**, *19*, 118–125. [CrossRef]

61. Di Maria, A.; Eyckmans, J.; Van Acker, K. Downcycling versus recycling of construction and demolition waste: Combining LCA and LCC to support sustainable policy making. *Waste Manag.* **2018**, *75*, 3–21. [CrossRef]

62. Zhong, S.; Pearce, J.M. Tightening the loop on the circular economy: Coupled distributed recycling and manufacturing with recyclebot and RepRap 3D printing. *Resour. Conserv. Recycl.* **2018**, *128*, 48–58. [CrossRef]

63. Tymrak, B.M.; Kreiger, M.; Pearce, J.M. Mechanical properties of components fabricated with open-source 3D printers under realistic environmental conditions. *Mater. Des.* **2014**, *58*, 242–246. [CrossRef]

64. Sun, Q.; Rizvi, G.; Bellehumeur, C.; Gu, P. Effect of processing conditions on the bonding quality of FDM polymer filaments. *Rapid Prototyp. J.* **2008**, *14*, 72–80. [CrossRef]

65. Derraik, J.G.; Anderson, W.A.; Connelly, E.A.; Anderson, Y.C. Rapid evidence summary on SARS-CoV-2 survivorship and disinfection, and a reusable PPE protocol using a double-hit process. *MedRxiv* **2020**, *17*, 6117.

66. Nema, S.; Ganeshprasad, K. Plasma pyrolysis of medical waste. *Curr. Sci.* **2002**, *83*, 271–278.

67. Jinadatha, C.; Simmons, S.; Dale, C.; Ganachari-Mallappa, N.; Villamaria, F.C.; Goulding, N.; Tanner, B.; Stachowiak, J.; Stibich, M. Disinfecting personal protective equipment with pulsed xenon ultraviolet as a risk mitigation strategy for health care workers. *Am. J. Infect. Control* **2015**, *43*, 412–414. [CrossRef] [PubMed]

68. Uloma, A.A.; Nkem Benjamin, I.; Kiss, I. Knowledge, Attitude and Practice of Healthcare Workers Towards Medical Waste Management: A Comparative Study of Two Geographical Areas. *J. Waste Manag. Dispos.* **2022**, *5*, 101.

69. Mahanwar, P.A.; Bhatnagar, M.P. Medical Plastics Waste. 2020. Available online: https://www.scirp.org/(S(351jmbntv-nsjt1aadkposzje))/reference/referencespapers.aspx?referenceid=2812826 (accessed on 26 April 2020).

70. Puro, V.; Pittalis, S.; Chinello, P.; Nicastri, E.; Petrosillo, N.; Antonini, M.; Ippolito, G. Disinfection of personal protective equipment for management of Ebola patients. *Am. J. Infect. Control* **2015**, *43*, 1375–1376. [CrossRef]

71. Coronaviridae Study Group of the International. The species Severe acute respiratory syndrome-related coronavirus: Classifying 2019-nCoV and naming it SARS-CoV-2. *Nat. Microbiol.* **2020**, *5*, 536. [CrossRef]

72. Kowalski, W. *Ultraviolet Germicidal Irradiation Handbook: UVGI for Air and Surface Disinfection*; Springer Science & Business Media: Berlin, Germany, 2010.

73. Ogunseitan, O.A. The Materials Genome and COVID-19 Pandemic. *Jom* **2020**, *72*, 1–3.

74. Tan, G.S.E.; Linn, K.Z.; Soon, M.M.L.; Vasoo, S.; Chan, M.; Poh, B.F.; Ng, O.-T.; Ang, B.S.-P.; Leo, Y.-S.; Marimuthu, K. Effect of extended use N95 respirators and eye protection on personal protective equipment (PPE) utilization during SARS-CoV-2 outbreak in Singapore. *Antimicrob. Resist. Infect. Control* **2020**, *9*, 1–3. [CrossRef]

75. Vijayakumar, V. Personal Protection Prior to Preoperative Assessment—Little more an anaesthesiologist can do to prevent SARS-CoV-2 transmission and COVID-19 infection. *Ain-Shams J. Anesthesiol.* **2020**, *12*, 1–2. [CrossRef]

76. Mahmood, S.U.; Crimbly, F.; Khan, S.; Choudry, E.; Mehwish, S. Strategies for Rational Use of Personal Protective Equipment (PPE) Among Healthcare Providers During the COVID-19 Crisis. *Cureus* **2020**, *12*, e8248. [CrossRef]

77. Mallapur, C. Sanitation Workers at Risk from Discarded Medical Waste Related to COVID-19, IndiaSpend. URL. Available online: https://www.indiaspend.com/sanitation-workersat-risk-from-discarded-medical-waste-related-tocovid-19/ (accessed on 26 April 2020).

78. Wu, S.; Montalvo, L. Repurposing waste plastics into cleaner asphalt pavement materials: A critical literature review. *J. Clean. Prod.* **2021**, *280*, 124355. [CrossRef]

79. Prasadh, S.; Raguraman, S.; Wong, R.; Gupta, M. Metallic Foams in Bone Tissue Engineering. In *Nanoscale Engineering of Biomaterials: Properties and Applications*; Springer: Singapore, 2022; pp. 181–205.

80. Prasadh, S.; Parande, G.; Gupta, M.; Wong, R. Compositional Tailoring of Mg–2Zn–1Ca Alloy Using Manganese to Enhance Compression Response and In-Vitro Degradation. *Materials* **2022**, *15*, 810. [CrossRef] [PubMed]

81. Saini, J.; Choudhary, S.; Kataria, P. Awareness of Biomedical Waste Management among Nursing Students: A Hospital Based Study in Haryana. *Indian J. Prev. Soc. Med.* **2020**, *51*, 8.

82. Kujur, M.S.; Manakari, V.; Parande, G.; Prasadh, S.; Wong, R.; Mallick, A.; Gupta, M. Effect of samarium oxide nanoparticles on degradation and invitro biocompatibility of magnesium. *Mater. Today Commun.* **2021**, *26*, 102171. [CrossRef]

83. Prasadh, S.; Suresh, S.; Hong, K.L.; Bhargav, A.; Rosa, V.; Wong, R.C. Biomechanics of alloplastic mandible reconstruction using biomaterials: The effect of implant design on stress concentration influences choice of material. *J. Mech. Behav. Biomed. Mater.* **2020**, *103*, 103548. [CrossRef]

84. Prasadh, S.; Krishnan, A.V.; Lim, C.Y.; Gupta, M.; Wong, R. Titanium versus magnesium plates for unilateral mandibular angle fracture fixation: Biomechanical evaluation using 3-dimensional finite element analysis. *J. Mater. Res. Technol.* **2022**, *18*, 2064–2076. [CrossRef]

85. Price, A.D.; Cui, Y.; Liao, L.; Xiao, W.; Yu, X.; Wang, H.; Zhao, M.; Wang, Q.; Chu, S.; Chu, L.F. Is the fit of N95 facial masks effected by disinfection? A study of heat and UV disinfection methods using the OSHA protocol fit test. *medRxiv* **2020**. [CrossRef]

86. Prasadh, S.; Gupta, M.; Wong, R. In vitro cytotoxicity and osteogenic potential of quaternary Mg-2Zn-1Ca/X-Mn alloys for craniofacial reconstruction. *Sci. Rep.* **2022**, *12*, 8259. [CrossRef]

Review

Heavy Metal, Waste, COVID-19, and Rapid Industrialization in This Modern Era—Fit for Sustainable Future

Muhammad Adnan [1,2] , Baohua Xiao [1,*], Peiwen Xiao [1,2], Peng Zhao [1,2] and Shaheen Bibi [3,4]

1   State Key Laboratory of Environmental Geochemistry, Institute of Geochemistry, Chinese Academy of Sciences, Guiyang 550081, China; adnan@mail.gyig.ac.cn (M.A.); xiaopeiwen@mail.gyig.ac.cn (P.X.); zhaopeng@mail.gyig.ac.cn (P.Z.)
2   University of Chinese Academy of Sciences, Beijing 100049, China
3   Lanzhou Veterinary Research Institute, Chinese Academy of Agricultural Sciences, Lanzhou 730046, China; shaheenwazir12@gmail.com
4   Graduate School, Chinese Academy of Agricultural Sciences, Beijing 100081, China
*   Correspondence: xiaobaohua@mail.gyig.ac.cn

Abstract: Heavy metal contamination, waste, and COVID-19 are hazardous to all living things in the environment. This review examined the effects of heavy metals, waste, and COVID-19 on the ecosystem. Scientists and researchers are currently working on ways to extract valuable metals from waste and wastewater. We prefer Tessier sequential extraction for future use for heavy metal pollution in soil. Results indicated that population growth is another source of pollution in the environment. Heavy metal pollution wreaks havoc on soil and groundwater, especially in China. COVID-19 has pros and cons. The COVID-19 epidemic has reduced air pollution in China and caused a significant reduction in CO_2 releases globally due to the lockdown but has a harmful effect on human health and the economy. Moreover, COVID-19 brings a huge amount of biomedical waste. COVID-19's biomedical waste appears to be causing different health issues. On the other hand, it was discovered that recycling has become a new source of pollution in south China. Furthermore, heavy metal contamination is the most severe ecological effect. Likewise, every problem has a remedy to create new waste management and pollution monitoring policy. The construction of a modern recycling refinery is an important aspect of national waste disposal.

Keywords: heavy metals; COVID-19; waste; biotoxicity; SARS-CoV-2; CO_2; circular economy

Citation: Adnan, M.; Xiao, B.; Xiao, P.; Zhao, P.; Bibi, S. Heavy Metal, Waste, COVID-19, and Rapid Industrialization in This Modern Era—Fit for Sustainable Future. *Sustainability* **2022**, *14*, 4746. https://doi.org/10.3390/su14084746

Academic Editors: Nallapaneni Manoj Kumar, Subrata Hait, Anshu Priya and Varsha Bohra

Received: 18 March 2022
Accepted: 12 April 2022
Published: 15 April 2022

Publisher's Note: MDPI stays neutral with regard to jurisdictional claims in published maps and institutional affiliations.

1. Introduction

One of humankind's greatest issues in the twenty-first century is heavy metal pollution. In recent decades, fast development, which has happened in the majority of places all over the world, has increased concern about and attention to soil quality [1]. In China, heavy metal poisoning of farming soil has been a major problem [2]. Heavy metal bioaccumulation can harm humans through various routes, such as food intake, particle inhalation, particle ingestion, and skin absorption [3]. More than 10 million polluted sites are known to exist worldwide, with heavy metal(loid) contamination found in >50% of sites [4]. By the end of 2000, China had 3.2 million hectares of wasteland, and this number is growing at a rate of 46,700 hectares each year [5]. In China, HMs pollute around 20 million acres of cropland and 12 million tons of grain every year [6,7]. Toxic heavy metals such as lead (Pb), cadmium (Cd), chromium (Cr), and arsenic (As) are found in about 82 percent of polluted agricultural soils in China [8]. Between 2005 to 2013, 1.50 percent of soil samples in China were polluted with Pb, according to the first National Soil Pollution Investigation [9]. Around 80 million hectares of soil are contaminated by heavy metals in China [10].

Heavy metals are mainly obtained from natural and anthropogenic origins. Volcanic emissions, continental dust movement, and the weathering of metal-enriched rocks are all examples of natural sources [11]. Heavy metals originating from mining operations are one of the most hazardous contaminants in surrounding areas [12]. This is especially true when soils are utilized for the discharge of inadequately treated liquid effluents, solid waste disposal, and deposition of exhaust gas from enterprises [13]. One of the biggest causes of heavy metals pollution is atmospheric pollution, notably, dust from zinc and lead industrial processes. Heavy metals from atmospheric deposition could be accumulated in topsoil by sedimentation, impaction, and interception [14]. Toxic metals penetrate the environment via nonferrous metal mining and smelting, through enduring and draining sewage sludge, discharge of contaminated water, or atmospheric particles from smelter piles [15].

Energy scarcity, pollution, and climate change, all linked to population expansion and the combustion of fossil fuels, have become key problems that humanity must address in the twenty-first century [16]. The rapid growth of the world's population, coupled with urban growth and technological advancement, has increased the production of complicated waste materials [17]. China has an important role in almost every aspect of the world economy. China is the most populous country on the planet. Urbanization and advancements in current technology, which contain the innovation of electrical and electronic tools, have a significant impact on a country's economy. It is common knowledge that all electronic tools include a variety of toxic metals, such as Pb, Cd, As, mercury (Hg), zinc (Zn), cadmium (Cd), copper (Cu), and aluminum (Al), which instantly affect public health and the environment [18].

Apart from heavy metals, landfilling modifications have a considerable impact on the biota of the ecosystem. China is a developing country, and while the industry is the main cause of hazardous waste in developing nations, the threats posed by industrial hazardous waste sources are greater. E-waste recycling utilizing rudimentary technologies is being removed quite vigorously in limited sites in south China, driven by profitability. It is rapidly becoming a significant novel source of contamination in these areas [19]. China has become the world's leading distributor and recycler of e-waste, accepting more than one million tons of e-waste each year from the United States and Europe [20]. Hazardous waste management is given priority because of its poisonous nature. This ensures that such wastes are controlled to avoid contaminating the environment, which could negatively affect human, plant, and animal health and biodiversity. In 2015, 191 million tons of municipal solid waste (MSW) were gathered, with almost 94.1% of it being preserved in sterile condition, 63.7% of MSW was disposed of in sterile landfills, 34.3% was handled in furnaces, and 2.0% was preserved through biological procedures [21]. Direct landfilling of uncooked food wastes has been prohibited in Korea since 1 January 2005, to address a lack of landfill space, preserve groundwater and soil from pollution, and encourage food waste recycling as a viable resource [22].

Heavy metal pollution's toxicity may lead to diseases [23]. The coronavirus disease, also known as COVID-19, was announced as a worldwide virus outbreak by the World Health Organization (WHO) on 11 March 2020 [24]. Due to poor respiratory functioning, chronic inflammation, and decreased resistance to diseases, environmental contamination has been deemed one of the threat issues for COVID-19 intensity and death rates, with indirect data from China and Northern Italy supporting this theory [25]. The collapse of numerous organelles, a weakened immune system, impairment to the central nervous system, kidney involvement, fracture, and a decrease in children's IQ are all effects of heavy metal exposure [26]. Soil risk factors are challenging to decompose and can emigrate to plants and humans via food chains and water supply systems, posing a direct or indirect threat to food security and human health [27,28].

This review aimed to highlight the impact of heavy metal and waste on environmental pollution. While providing a comparative valuation of the COVID-19 pandemic and its effect on the atmosphere, humans, and economy. Furthermore, a comprehensive review was conducted on how waste recycling, waste, and biomedical waste management will not only help in fighting against diseases but will also achieve a more circular economy.

2. Heavy Metal Extraction in Soil

There are a variety of forms used for soil analysis, and it depends on the goal of the investigation. In the geochemical investigation and environmental geochemistry, sequential extraction of components from soil and sediment is commonly used [29]. Sequential extraction procedures (SEPs) have grown in popularity rapidly since their inception in the late 1970s [30]. Identifying the primary binding locations, the potency of the bind among metal and soil mixtures, and the step connections of trace elements in soils can all be obtained using SEPs [31]. Several studies have employed the sequential extraction approach with selective chemical agents to partition solid-phase metals in river sediments [32,33]. SEPs, established on the sensible usage of a sequence of additional or particular small reagents selected to solubilize the various mineralogical particles for maintaining the more considerable part of metals sequentially, are the most popular and easiest method to determine the states in which metals are discovered in soils [34]. SEPs have been used to analyze the physicochemical states of metals and offer a more useful interpretation of the mechanisms determining their availability and assessing the efficacy of soil remediation systems and identifying underlying mechanisms [35]. Sequential extraction has recently become popular for evaluating the environmental impact of human activities such as mining [30] and smelting. It is critical to distinguish the accessible and inaccessible states of metals in soil contaminated by metals to assure that the soil is managed to avoid the inaccessible states becoming accessible [34]. The problem of the partial selectivity of chemicals used in sequential extraction schemes (SESs) to dissolve one stage emerges in the assault on other stages, and they may be ineffective in entirely dissolving the stage; modifications to the experimental conditions, including the extraction period, the extractant sample fraction, the chemical content, the extraction temperature, the usage of consecutive extractions with similar chemicals, and so on, can all help to avoid these issues [36]. Although time intensive, sequential extractions provide precise information regarding the source, method of occurrence, biological and physicochemical availability, mobilization, and transportation of trace metals [37].

Several SEPs are available, but some are intended to function within exact factors. In contrast, others are intended for a broader application, such as the Tessier [33], Community Bureau of Reference (BCR) [38], Short [39], Galán [40], and Geological Society of Canada (GCS) approaches [41]. Similarly, several researchers have proposed a modified version of this and applied it to the soil, sediment, and sewage. All sequential extraction procedures (SEPs) facilitate fractionation [42]. Exchangeable, carbonate bound, Fe and Mn oxide bound, organic matter bound, and residual were the names given to these fractions by [33,42]. Fractionation patterns have not been consistent, and the consequences of various methods are not consistently similar due to the deficiency of consistency in the test situations (i.e., the number of extractions, chemicals, shaking period) [43]. Even though various protocols have been described, the Tessier and BCR schemes remain the most commonly adopted [30]. Table 1 shows the operating parameters, including Tessier and BCR schemes [37]. Sequential extraction studies have proven to be useful for determining the metals linked with the main cumulative stages in sedimentary depositions [37]. XRD is also effective for determining silicate clay reactivity during the extraction process [42]. Understanding the chemical and physical features of heavy metals in soil requires identifying the chemical states (speciation) and dispersal of heavy metals released, trapped, or adsorbed on soil particles [44]. To evaluate heavy metal redistribution (Pb, Cd, Zn, and Cu), SEPs were used by the European Union Bureau of Reference Procedure (EUBCR) [45]. According to [30], sequential extraction has a bright future in the twenty-first century, but its sustained utility necessitates researchers' awareness of its limitations, particularly for environmental monitoring. The SEPs future is not as bright as initially assumed, but it is still useful [42]. It is crucial to realize that the Tessier and BCR processes will not always produce the same outcomes. For example, Mn is extracted from agricultural soils primarily by the reducible fraction of the BCR technique but mostly through the residual fraction of the Tessier procedure [30]. Tessier's approach is the most efficient when there is a high soil metal content [46]. Still, no extraction procedure is 100% effective, but we have recommended the Tessier sequential extraction method as a suitable method for estimating high metal(loid) concentrations in soils. Critical problems need critical solutions. In conclusion, the invention of a suitable extraction method is predicted to revolutionize the field of soil contamination research in the future.

Table 1. Original Tessier and BCR SESs. (Reprinted from ref [37], with permission of the publisher).

	Tessier Scheme [a]		
Stage	**Operationally-Defined Phase**	**Reagent**	**Operating Conditions**
1	Exchangeable	8 mL of $MgCl_2$ 1 mol L^{-1} (pH = 7)	1 h at 25 °C
2	Acid soluble	25 mL of NaOAc 1 mol L^{-1} (pH = 5)	5 h at 25 °C
3	Reducible	20 mL $NH_2OH\cdot HCl$ 0.04 mol L^{-1} in HOAc 25% w/w	6 h at 96 °C
4	Oxidizable	3 mL HNO_3 0.02 mol L^{-1} + 5 mL H_2O_2 30% w/v	2 h at 85 °C
		3 mL H_2O_2 30% w/v + 5 mL NH_4OAc 3.2 mol L^{-1}	3 h at 85 °C 30 min at 25 °C
	BCR Scheme [a]		
1	Acid soluble	40 mL HOAc 0.11 mol L^{-1}	16 h at 25 °C
2	Reducible	40 mL $NH_2OH\cdot HCl$ 0.1 mol L^{-1} (pH = 2)	16 h at 25 °C
3	Oxidizable	10 mL H_2O_2 30% w/v (evaporation)	1 h at 25 °C
		10 mL H_2O_2 30% w/v (evaporation)	1 h at 85 °C
		50 mL NH_4OAc 1 mol L^{-1}	16 h at 25 °C

[a] 1 g sample mass is employed for sequential extraction.

3. Pollution Levels in Various Environmental Compartments

3.1. Soil

Toxic heavy metals are deposited in soils from natural and human activities [47]. As a consequence of environmental and health issues, soil heavy metal pollution has a huge interest [35]. The A horizon is called "topsoil", in this layer, minerals are present which are generated from the parent material with the organic matter accumulating. The B horizon is called "subsoil" or "zone of accumulation", the mineral seeps down from the A or E horizons and accumulates in this layer. However, the variation of the elemental concentrations is higher in the A and B horizons. The B horizon, or the third layer of soil, contains the majority of heavy metals [48]. This layer comprises components that were dissolved in the higher layer (the A horizon) and subsequently moved down or sidelong into the inferior layer, where they were dumped, and heavy metals are drawn to the B horizon because it has a high content of iron oxyhydroxides and clay, both of which can absorb cationic aspects [48]. Microorganisms cannot degrade heavy metals in the soil; therefore, they accumulate, influence the soil's properties, and are assimilated and enhanced in biomass [49]. Cadmium (Cd) pollution is a major problem in China's agriculture [23]. Pb, Cd, polybrominated biphenyls (PBBs), polybrominated diphenyl ethers (PBDEs), and polychlorinated biphenyls (PCBs) have appeared in high quantities in rice and organic contaminants have been found in vegetables growing surrounding unmanaged e-waste recycling locations [19]. E-waste soil samples frequently contain persistent organic pollutants (POPs), including polycyclic aromatic hydrocarbons (PAHs) and PBDEs. In 2011 (0.59 mg/kg) and 2016 (0.40 mg/kg), researchers found high molecular weight polycyclic aromatic hydrocarbons (PAHs) in paddy soils close to e-waste recycling areas in Taizhou, China [50]. The principal pollutants of concern in e-waste-affected soil are Pb and Cd [50]. According to [19], the maximum Pb (629–7720 mg kg^{-1}) and Cd

(3.05–46.8 mg kg^{-1}) contents found in soils around e-waste combustion operations far surpassed Chinese farming soil requirements (Pb: 250 mg kg^{-1}; Cd: 0.3 mg kg^{-1}). Metal concentrations were highest in historic e-waste incineration locations, with an average of 17.1 mg kg^{-1} of Cd, 11,140 mg kg^{-1} of Cu, 4500 mg kg^{-1} of Pb, and 3690 mg kg^{-1} of Zn. Metals in high amounts could seep out of the locations and contaminate pond water and sediment [19].

3.2. Water

Apart from soil pollution, which can contribute to water quality degradation and various negative environmental effects, heavy metal replication across the food supply chain has serious health impacts [51]. The global demand for freshwater is steadily increasing. Because arsenic (As) pollution affects such a broad population, the toxicity resulting from As enrichment in sedimentary aquifers beyond prescribed limits, which causes drinking water contamination, is a global concern [52]. Because As species are proven carcinogens, their presence in the environment is a primary public concern and is linked to severe health hazards [53]. The possible polluting roots from agriculture, including fertilizer, urban (such as wastewater), and industrial (including spills and leaks), and groundwater contact with surface water sources including rivers and lakes, are depicted in Figure 1 [54]. This refers to the spread of new infections due to pesticides, and it is a threat to human health [55].

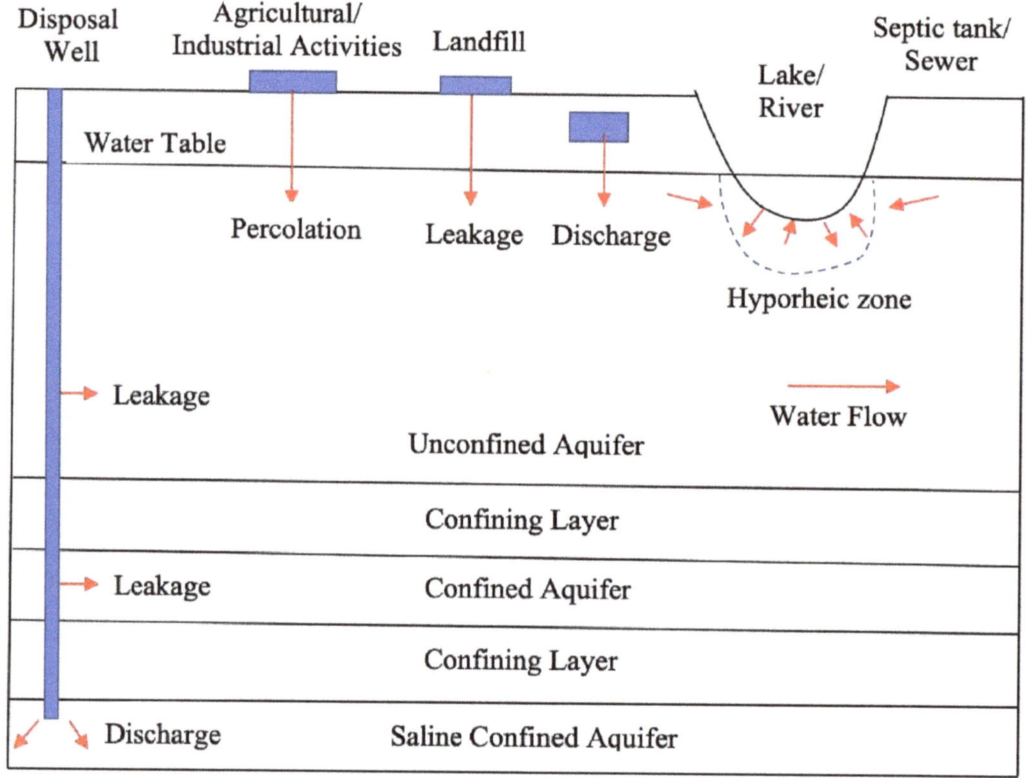

Figure 1. Groundwater interactions with contaminant pathways. (Reprinted from ref [54], with permission of the publisher).

Chronic exposure to harmful contaminants in groundwater has negative health consequences and leads to serious diseases such as cancer, neurological disorders, reproductive system damage, congenital malformations, and, more recently, diabetes mellitus [55]. Enormous amounts of wastewater are released during the liquid and solid separation [56]. Pollutants can enter the groundwater system through karstic soils [55]. Karst covers around 30% of China's land surface, and karst aquifers provide a quarter of the country's groundwater supplies (200 billion m^3 per year) [57]. Many karst locations worldwide have experienced rocky desertification, particularly in southwest China's karst area, known as the world's biggest karst area with constant carbonate rock outcrops [58]. Contamination in the atmosphere in the karst area is difficult to dissipate before precipitating again due to its unique geomorphological properties, as the environmental fragility of the karst aquifer in southwestern China is widely recognized [57]. According to [59], currently, no research has been done on the impact of karst water with various chemical properties on dissolved organic matter (DOM) leaching into karst soils. According to the various contaminants, heavy metals are numerous important and dangerous contaminants for groundwater [60]. The Lianjiang River was found to be polluted by As, Cr, molybdenum (Mo), selenium (Se), lithium (Li), and antimony (Sb), whereas the Nanyang River had higher levels of Ni, Zn, Cu, Pb, cobalt (Co), and silver (Ag) [50]. Toxic heavy metals have been found in wastewater discharged from tailing ponds: Pb, Zn, Cu, Cr, Ni, and As at average concentrations of 4.33, 269.90, 2.40, 1.69, 1.04, 11.40, and 24.62 g/L, respectively [61].

In China, 28% of groundwater examinations surpassed the WHO limit contamination level (10 mg N L^{-1}) between 2000 and 2012. Up to 36% of the river sectors and 40% of the main lakes in China did not fulfill the quality standards to be used as drinking water sources in 2010 [62]. Rainwater, surface runoff, and groundwater in karst environments frequently have high Ca^{2+} levels [59]. Surface waters (from springs and streams to rivers and lakes) can transport heavy metals across long distances, and their chemical structure varies depending on the geological characteristics through which they travel [48]. Furthermore, surplus nutrients in rivers are transferred to seas, resulting in nearly 500 instances of hazardous algae blooms in China's shore waters between 2006 and 2012, posing a threat to human health and shore ecosystems [62]. For a sustainable future, technologies that improve the efficiency of agriculture irrigation are required to grow more food or biomass with less water [63]. Because of socioeconomic and climate changes, this scenario is predicted to deteriorate in the future [62]. The hydrological cycle can forecast several climate change consequences [64]. In e-waste operations, there is still a severe lack of evidence about the origins and characteristics of heavy metal pollution. Groundwater is quickly depleting due to global climate change, and this process is threatening to overrun the entire water cycle. Because the rain cycle follows a four-step procedure in which groundwater is used in the evaporation and condensation process, the aquifer is an important aspect of the rain cycle. Unfortunately, global warming has disrupted the entire cycle.

4. COVID-19

COVID-19 has wreaked havoc on the worldwide economy and health system, resulting in >5 million lives lost [65] and enormous economic and social upheaval. Similarly, an Ebola epidemic began in middle Africa in 2013 and expanded to neighboring nations in Western Africa, resulting in 28,652 human infections and 11,325 lives lost between 2013 and 2016 [66]. COVID-19 is a disease transmitted by the recently identified acute respiratory syndrome coronavirus 2 (SARS-CoV-2) with various clinical symptoms ranging from mild flu-like symptoms to pneumonia and acute respiratory syndrome [67]. COVID-19 is the most troubling challenge humanity has ever encountered. This is owing to the fact that its consequences are both profound and worldwide. Over a couple of years after the financial crisis, the globe is dealing with the health and economic consequences of the latest crisis brought by the COVID-19 epidemic [68]. China has already faced viral outbreaks, such as the SARS outbreak in 2003. The main distinction between COVID-19 and SARS is the intricacy of the distribution networks in which China is now enmeshed [69].

If heavy metals are found to be a source of COVID-19 vulnerability, we will have a valuable tool for identifying who is at risk and a technique for proactively reducing risk [70]. According to the previous research, Cd and Pb are accountable for the COVID-19 mutation of the influenza virus, whereas As and Hg are responsible for the emergence of the COVID Beta variation [71]. Compared to other heavy metals, As has a dual role in viral infections [25]. According to [70], this revelation could save the world economy tens of trillions of dollars, in addition to saving precious human lives. For example, Zn/ZnO nanoparticles (NPs) are also employed as disinfectants and integrated into commercial items, such as food packaging. Furthermore, because the virus can stay active on plastic and stainless-steel surfaces, using nano-Zn as a disinfection agent could help to limit SARS-CoV-2 transmission by generating selfsterilizing coatings [65].

COVID-19 susceptibility and severity have been linked to various medical, lifestyle, and environmental variables (Figure 2) [25]. The novel coronavirus, SARS-CoV-2, has been recognized in wastewater [72]. The advancement of wastewater-based epidemiology (WBE) techniques is described as a harmonizing way of tracking the SARS-CoV-2/COVID-19 surveillance system [73]. According to the WHO, viruses such as SARS-CoV-2 mutate over time and will continue to alter as they circulate. There is also a lack of evidence about COVID-19 variants. Thus, continuous monitoring of SARS-CoV-2 could have a big effect on efforts to figure out how the variants change.

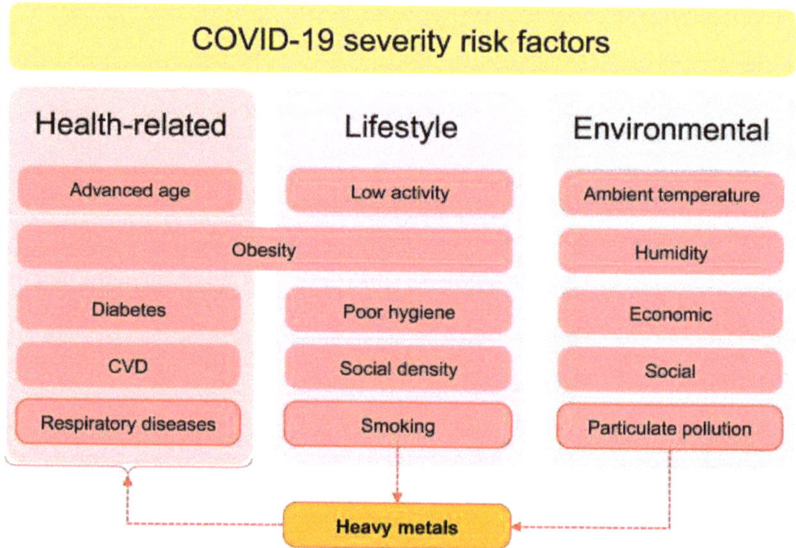

Figure 2. The proposed role of heavy metals as a link between risk factors for COVID-19 severity. Both particulate (PM2.5) pollution and smoking are associated with heavy metal exposure that at least partially mediate adverse effects of these factors on the respiratory system. In addition, heavy metal exposure was shown to be associated with higher incidence of obesity, diabetes, and cardiovascular diseases (Reprinted from ref [25], with permission of the publisher).

Our greatest concern with COVID-19 is to address the public health implications, but it is also critical to maintain our economic recovery. According to COVID-19 vaccinations, the recent findings show that the safety measurement of mRNA COVID-19 vaccinations stands out as the most thorough of any vaccine in United States record [74]. Such effectiveness suggests that the vaccines offer significant protection against infection. Further research should be focused on vaccine safety monitoring. Rapid changes in COVID-19 variants and future vaccinations will need political commitment, financial support, resource management, and multisectoral partnership. Furthermore, lack of sufficient funding is a key obstacle to success for developing countries.

CO$_2$ Emission

In the global carbon cycle, soil is both an origin and a sink [59]. The fundamental cause of global climate change is carbon dioxide emissions. Carbon dioxide or CO$_2$ originates from the burning of fossil fuels and the manufacture of cement and solid, liquid, gas, and gas flaring. Transportation, factories, refineries, and agricultural activities contribute to air pollution [60]. Globally, 6 billion tons of CO$_2$ was emitted in 1950. China is the world's leading emitter; it emits almost 10 billion tons of CO$_2$ annually [75], accounting for more than a quarter of worldwide emissions. The worldwide COVID-19 lockdown reveals a direct link between air pollution levels [60,76]. COVID-19 is expected to positively influence natural resources and cause a significant reduction in CO$_2$ emissions globally. The safety measures, which included a travel embargo and the suspension of most commercial and industrial operations, produced an economic downturn and lowered CO$_2$ and other pollution emissions while also assisting in controlling the pandemic in China [77]. Warmer winter temperatures in 2020 contributed to some of the reduction in China's power sector discharges [78]. In 337 cities across China, hazardous gas and other pollution emissions decreased by 25% at the start of 2020, and air quality improved by 11.4% compared to the start of the previous year; this modification is believed to have saved 50,000 lives in China [79]. Figure 3 depicts global and regional changes in daily CO$_2$ discharges. The numbers for other countries in the first half of 2020: global (-1550.5 Mt CO$_2$ -8.8%), the United States (-338.3 Mt CO$_2$, -13.3%), followed by the EU27 and the United Kingdom (-205.7 Mt CO$_2$, -12.7%), India (-205.2 Mt CO$_2$, -15.4%), China (-187.2 Mt CO$_2$, -3.7%), and Germany (-54.0 Mt CO$_2$, -15.1%), with significant but gradual drops in Japan (-43.1 Mt CO$_2$, -7.5%), Russia (-40.5 Mt CO$_2$, -5.3%), Brazil (-25.9 Mt CO$_2$, -12.0%), Spain (-23.1 Mt CO$_2$, -18.8%), Italy (-22.9 Mt CO$_2$, -13.7%), and France (-21.5 Mt CO$_2$, -14.2%) [78]. NASA (National Aeronautics and Space Administration) and ESA (European Space Agency) used the Ozone Monitoring Instrument (OMI) to track the sudden drop in nitrogen dioxide (NO$_2$) content during COVID-19's initial phase in China [79]. During the global shutdown, carbon monoxide (CO), NO$_2$, and "particulate matter with a diameter smaller than or equal to 10 μm" (PM10) all declined dramatically, but ozone (O$_3$) increased significantly due to the NO$_2$ decrease [76]. The reduction in NO$_2$ concentrations started in China and spread around the world [79]. The pandemic decreased pollution levels in the United States, but to a lesser extent than in China [78].

According to [80], the observed global temperature is rising faster than the simulated temperature when natural variables are taken into account alone, with human activity accounting for the entire difference; the average worldwide surface temperature has risen by 1.07 °C since 1850, with each of the last four decades being warmer than the one before it. Because soil microorganisms and the activities they facilitate are heat sensitive, global factors such as heating are instantly affecting microbial soil respiration rates. The function of increased temperature in microbial metabolism has recently received considerable attention [81]. For example, global warming has already resulted in extensive glacier and Arctic ice retreat, a 0.2 m rise in sea level, and more frequent and severe heavy precipitation occasions as well as hot extremes [80]. Since the beginning of industrialization, the world's environment has been affected, but it is the path to real development. The same scenario has occurred in south China as it also affects the middle riparian zone and the two waters are from the Yangtze River, and Yellow River. Henan Province has many rivers, the terrain is generally higher in the west and lower in the east, and the majority of the region is low plains, putting most cities in the province at high risk of flooding [82]. In central China's Henan province, the city of Zhengzhou received more than 200 mm of rain in a single hour on 20 July 2021 (denoted as "Zhengzhou flood"). More than three million people were affected by the Zhengzhou flood, which claimed the lives of 302 people and left 50 persons missing [83]. The Central Meteorological Observatory of China issued yellow rainfall signals the next morning (06:00 on 19 July), warning of severe heavy rain in areas of Henan Province, including Zhengzhou (24 h of precipitation from 08:00 on 19 July to 08:00 on 20 July, with 100–160 mm) [84]. Scientists are predicting that the rapid climatic change has had a negative impact on the environment, and these changes will affect the water resources worldwide because this can be held accountable for drought and flood in the same year.

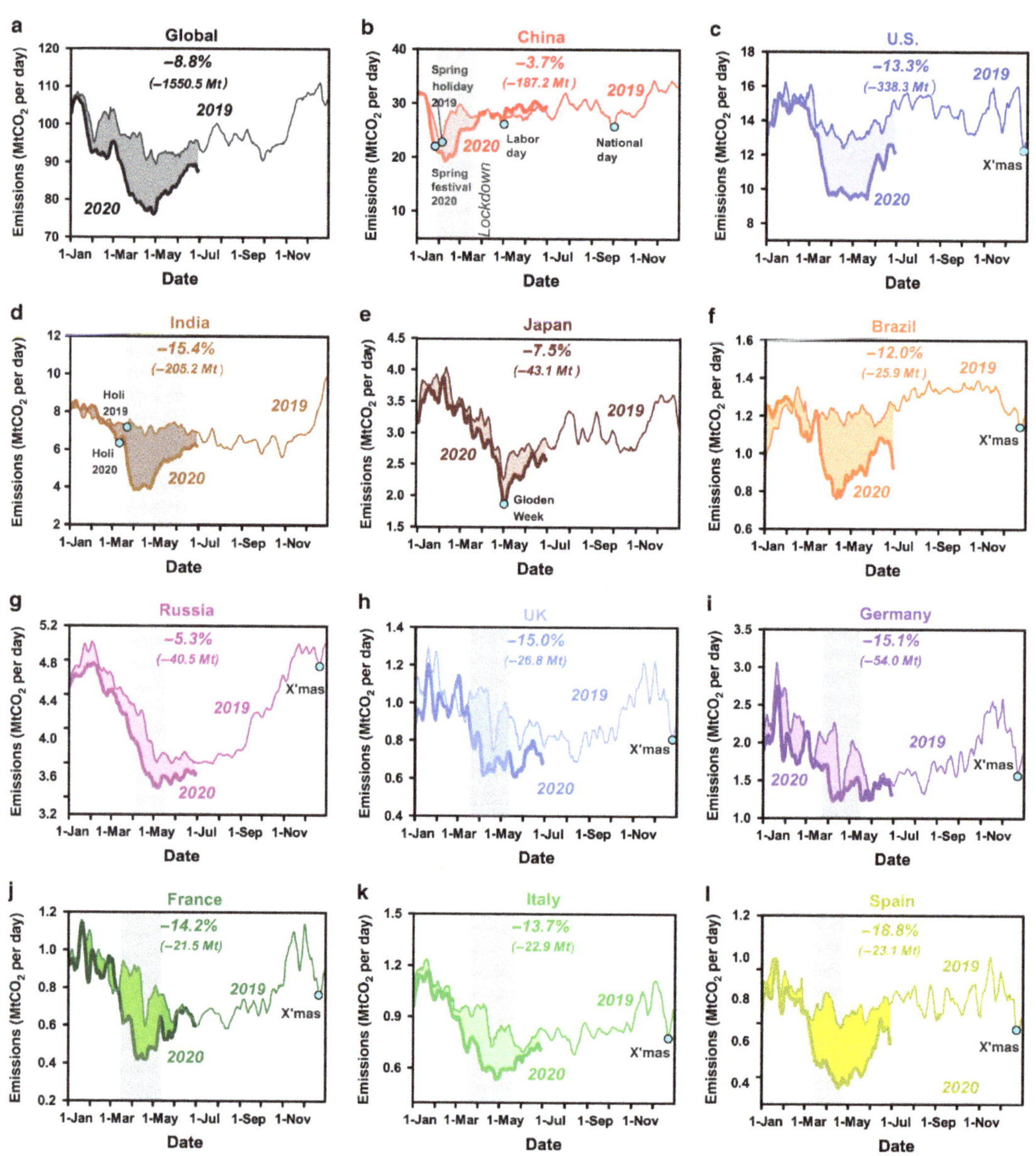

Figure 3. Daily CO_2 emissions for countries. Effects of the COVID-19 pandemic on daily CO_2 emissions globally and in each of 11 regions are reflected by the shaded differences between 1 January and 30 June of 2019 and 2020. (Reprinted from ref [78], with permission of the publisher).

Among the most controversial topics in the global research community working on waste disposal is the provision of appropriate practical resolutions to developing and growing countries [85]. Under each region's waste management standards, medical waste and plastics were usually disposed of before the COVID-19 epidemic [60]. As the epidemic advanced, changes occurred, providing challenges for healthcare facilities, with the amount of waste generated being the most difficult to control. The constant growth in the volume of medical waste following the advent of COVID-19 produced severe issues in managing plastic garbage and even immobilized the waste dumping infrastructure in several nations [60]. COVID-19 has had a terrible worldwide economic impact, with many individuals losing their employment and employers finding themselves unable to maintain their staff as their businesses decline [86]. Climate change is putting the world's food security at risk. Climate change is expected to considerably impact agriculture, affecting crops, soils, livestock, and pests directly and indirectly [87].

Furthermore, the pandemic's potential effect on food production in major food-producing nations (including China, the EU, and the United States) might have considerable impacts on global food availability and costs [88]. The COVID-19 pandemic has added tens of thousands of tons of extra medical waste to the health care waste management systems around the globe, posing a significant hazard to the environment and human health and emphasizing the urgent need to improve waste management approaches according to WHO. The current pandemic has a good effect on environmental pollution reduction, but the COVID-19 biomedical waste is a challenging issue for the world. Such waste could be infected with the virus. Medical waste disposal has become a serious global issue. Medical waste incinerators, both new and old, must adhere to increasingly stringent emission standards for a range of contaminants. There are two types of medical waste: special waste and general waste. All waste items that are not classified as toxic are included in general waste, even the potentially hazardous, and do not require special management and disposal. Special wastes require special management, treatment, and disposal, normally restricted by specialized laws and regulations. Such waste may be hazardous to one's health, safety, or the environment, or it may simply be inappropriate for dumping [89]. The climatic change affects the comparative quantity of soil community components in their physiology, temperature sensitivity, development rates, and function of the soil community [81]. The most important step in dealing with climate change is to minimize the use of fossil fuels on a daily basis.

5. Challenges Associated with Waste

5.1. Waste Management Strategy

Waste management is among the most pressing environmental issues in today's society. Waste management has become a big concern in China, as careless disposal of hazardous materials poses serious damage to the environment. The term "sustainable development" has gotten much interest in many areas, including scientific debates, daily life activities, foreign relations, local and national policies, and commercial [90]. The historic sustainable development summit in 2015 approved the 2030 Agenda, which included 17 sustainable development goals (SDGs) that would guide countries' efforts to build a sustainable world by 2030, and Goal 6 aims to guarantee that everyone has access to clean water and sanitation [91]. Heavy metal pollution and waste require a strong policy, constant monitoring, remediation, the media, and the determination to overcome with public support. As the world is working towards better wastewater treatment, it is improving; what will happen if the water sources decrease due to global climate change? Due to the closure of the manufacturing industries, the quality of the water bodies improved during the COVID-19 lockdown.

5.2. Waste Avoidance and Waste Minimization at Source

China has the world's leading electronics consumer and producer [92]. E-waste has become a severe concern in China and other Asian emerging countries since it is one of the leading causes of heavy metals and organic contaminants in municipal waste and is the most rapidly increasing waste stream [93]. The establishment of China's e-waste recycling system was founded in the early 1990s; transnational flows of e-waste from industrialized to poor nations have gotten a lot of attention because of the substantial contamination related to these recycling activities in some locations in early 2000 [92]. Electronics manufacturing, a main economic driver in China and one of the most rapidly increasing industries since the 1980s, is the third source contributing to the massive volume of e-waste [93]. E-waste is the most rapidly increasing waste source in the industrialized world, expanding at over 4% per year. The Chinese government has tightened its e-waste rules, resulting in a large amount of e-waste being held in Hong Kong's New Territories [94].

In recent years, China has created a standard e-waste recycling strategy with 109 accredited recycling manufacturers armed with the finest functional recycling machinery and supervised by highly stringent environmental safety requirements [92]. Figure 4 shows the predicted e-waste production, reported proportions of dismantled units by licensed e-waste recyclers, and the dismantling capacity of authorized recycling factories in provinces of China in 2004 [92].

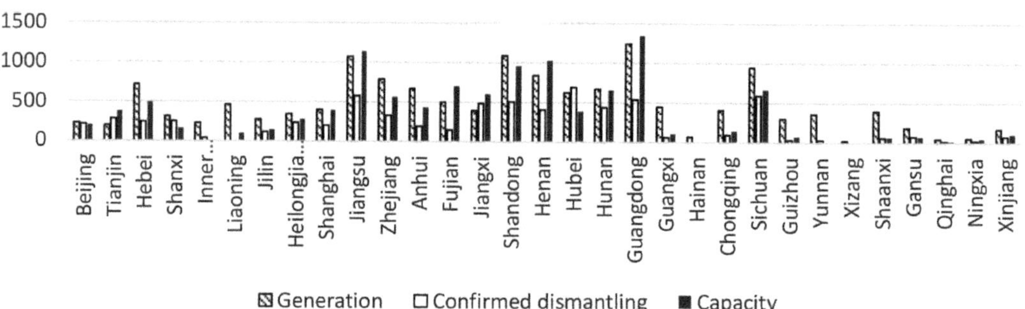

Figure 4. The generation, confirmed dismantled units, and formal recycling capacity of e-waste in each province in China in 2014. (Reprinted from ref [92], with permission of the publisher).

5.3. Reuse, Recovery, and Recycling of Hazardous Waste

According to a report by Toxics Link, 70% of the e-waste accumulated at recycling units in New Delhi, India, was shipped or dumped by industrialized nations [95]. At the same time, roughly 50–80% of the e-waste accumulated for recycling in the western United States is shipped to Asia, with approximately 90% of it going to China for recycling. For about 25 years, the Taizhou area in Zhejiang province, East China, has been e-waste recycling. It is one of China's most well-known e-waste processing facilities [96]. E-waste recycling plants utilizing rudimentary technologies are being removed quite intensively in a few sites in south China, driven by commercial motives, and in these areas, it is quickly emerging as a significant new source of pollution [19]. For example, in Tianjin, Taicang, Ningbo, Linyi, Liaozhong, Taizhou, and Zhangzhou, special resource recovery industrial gardens have been constructed to facilitate effective and ecologically friendly recycling of original and imported metal trash. Such operations are particularly common in the suburbs of major recycling cities, including Guiyu in the Guangdong region and Taizhou in the Zhejiang region, because of a lack of efficient enforcement and oversight [97]. Nonrecyclable waste is burnt or dumped directly on the ground. On a national basis, the partitioning between incineration and landfill is decided for each waste category [98]. In addition, recycling contributes to the decrease of pollutants and landfills.

Modern waste management has two basic goals: to preserve the environment and human health and to conserve resources, including materials, energy, and space [98]. E-waste recycling can provide for at least a portion of the world metal demand, particularly in areas where resources are scarce. Reuse has been encouraged to extend the life of used electrical and electronic devices [99]. Although e-waste accounts for just around 5% of global municipal waste, it is a substantial source of employment in the recycling industries of various low- and middle-income nations, including China, Pakistan, India, Malaysia, Vietnam, Thailand, the Philippines, Ghana, and Nigeria. For example, around 100,000 individuals are engaged as e-waste recyclers in Guiyu, China, likely the world's biggest e-waste recycling center [100]. During the 75th United Nations General Assembly, China declared that it would reduce carbon dioxide emissions to zero by 2030 and attain carbon neutrality by 2060. China stipulated producers' obligations in 2019, demanding them to be in charge of the logistics activities implicated in recycling and reusing lithium-ion batteries [101].

Time monitoring and applying remediation technology can also be sustainable solutions for pollution reduction. As a result, monitoring is critical for addressing current pollution most effectively. Furthermore, the world community must take a holistic strategy to address the waste management problem properly. Creating, on a national level, a pollution-control mechanism. First and foremost, establishing a new strategy based on the current COVID-19 biomedical waste is critical for environmental pollution; otherwise, it links with serious problems.

5.4. Lessons Learned from Waste Disposal

The main cause of COVID-19 is earlier waste management strategies that were ineffective. Insightful research on COVID-19 biomedical waste problems is needed for a new approach to waste management. Climate policies and action programs need to be modified from time to time, concentrating on the most contemporary situation and present necessities and demands. There must be a safe trash disposal system set in place by each government agency to preserve our natural resources and protect against potential health risks [102]. Additionally, wastewater is polluted with COVID-19, which has led to a need to establish a wastewater surveillance system to monitor COVID-19 in wastewater [103]. Such types of waste should be thrown away after being disinfected, according to the WHO recommendations [104]. Healthcare waste can contain hazardous microorganisms that can spread easily to other patients, healthcare professionals, or the general public if it is not properly handled or thrown away [105].

6. The Role of Circular Economy (CE)

Both academics and practitioners are interested in the circular economy notion because it is seen as an effective implementation for firms to apply the much-debated notion of sustainable development [106]. This energy flow paradigm of industrial operations was dubbed "extract-produce-use-dump," "take-make-waste," or "take-make-dispose" by experts [107]. Climate change, sustainable growth, nationwide legislation and policy, patron knowledge and activism, and business continuity appear to be the five guiding principles of the circular economy [108]. These three preceding domains, ecological economics, environmental economics, and industrial ecology, all played a role in developing circular economy [109]. The "Circular Economy Promotion Law," "Solid Waste Pollution Control Law," and "Clean Production Promotion Law" are three key legislations on e-waste management. These rules do not contain specific provisions, but they offer a legal foundation for handling e-waste [97]. In recent decades, recycling, production, and pollution management have received more attention because of the expanding global population and improvement in people's living standards. The 'Circular Economy Promotion Law of the People's Republic of China,' which was signed on 29 August 2008 and went into effect on 1 January 2009, is the most significant contribution to the Chinese legal system to date [110]. As a result of the COVID-19 pandemic, the world economy has been devastated, though the online market has risen significantly. Fortunately, the negative impact has diminished, and the global economy has recovered due to the efforts of developed countries around the world. Since the COVID-19 outbreak has shut down multiple mines, industries, and borders, the supply of cobalt and lithium has been interrupted.

Metal recovery from wastewater and its revaluation as precious metals brings the waste material back into the manufacturing stream, facilitating the transition from a linear to a circular economy. Invest in what we know, China provides nearly 60% of global production and works for modern, clean technologies that are the mainstay of a cleaner ecosystem. Economic development and population growth are linked. The most important aspect of this population growth is the rise of industrialization and urbanization. Then, to produce more opportunities to eliminate poverty and pollution, we must preserve our environmental sustainability. To the best of our knowledge, it is a very complex and challenging environment that requires more than a business as usual solution.

7. Conclusions

This review investigated the impacts of heavy metals, waste, and COVID-19 on the environment. Similarly, population growth is a primary source of contamination in the environment, and China is the most populous country on the planet. For the study of heavy metal pollution in soil, the Tessier sequential extraction method has a bright future. The outbreak of COVID-19 resulted in the close of industries. As a result, there was a drastic decrease in CO_2 emissions. It had a major impact on human lives as well. Millions of people have been affected, with the majority of people losing their employment, and the world continues to be plagued by this deadly virus. Because of the long-term lockdown, the global economy is currently in a perilous scenario. However, COVID-19 biomedical waste is becoming the most serious side effect. As a result of this ignorance and poor waste management practices, environmental and health disasters occur. So far, acid deposition has become a worldwide problem and has become a major issue in agriculture and forestry production, and water resource utilization. Future studies should concentrate on new policies for waste management and pollution monitoring as well as waste recycling. If the issue cannot be countered soon, then the world could face a serious crisis in the near future. At the same time, China's environmental management is still relatively weak. There is a need to conduct a specific investigation on the various factors to gain insight into the future development direction of the industry, the evolution trend of the industry competition pattern, and evaluate the degree of benefit and effect. Under the background of global climate change, the study of heavy metal pollution ecology may go beyond the category of heavy metals and pollution and conduct research in combination with other adversities. Scholars from various fields should be encouraged to collaborate on developing a new treatment approach for the COVID-19 biomedical waste.

Author Contributions: Conceptualization, M.A.; investigation, M.A.; software, M.A.; writing—original draft preparation, M.A.; writing—review and editing, S.B., P.X., P.Z. and B.X.; supervision, funding acquisition, B.X. All authors have read and agreed to the published version of the manuscript.

Funding: This study was supported by the National Natural Science Foundation of China (41773147, 41273149).

Institutional Review Board Statement: Not applicable.

Informed Consent Statement: Not applicable.

Data Availability Statement: Not applicable.

Acknowledgments: The authors wish to thank the anonymous reviewers and the editor for their comments and suggestions.

Conflicts of Interest: The authors declare no conflict of interest.

Abbreviations

WHO "World Health Organization" IQ "Intelligence Quotient" SEPs "Sequential Extraction Procedures" SESs "Sequential Extraction Schemes" BCR "Community Bureau of Reference" GCS "Geological Society of Canada" XRD "X-ray Power Diffraction" EUBCR "European Union Bureau of Reference Procedure" PBDEs "Polybrominated diphenyl ethers" PBBs "Polybrominated biphenyls" PCBs "Polychlorinated biphenyls" POPs "Persistent Organic Pollutants" PAHs "Polycyclic aromatic hydrocarbons" DOM "Dissolved Organic Matter" NPs "nanoparticles" CO_2 "Carbon dioxide" NASA "National Aeronautics and Space Administration" ESA "European Space Agency" Mt "Metric Ton" EU27 "European Union '27 Countries'" SARS "Severe Acute Respiratory Syndrome" SARS CoV-

2 "Severe Acute Respiratory Syndrome Coronavirus 2" OMI "Ozone Monitoring Instrument" O_3 "Ozone" CO "Carbon monoxide" NO_2 "Nitrogen dioxide" PM10 "Particulate Matter 10" mRNA "Messenger Ribonucleic acid" Ca^{2+} "Calcium ion" Ni "Nickel" Cr "Chromium" Cd "Cadmium" Pb "Lead" Cu "Copper" Zn "Zinc" As "Arsenic" Hg "Mercury" Mn "Manganese" Mo "Molybdenum" Se "Selenium" Li "Lithium" Sb "Antimony" Co "Cobalt" Ag "Silver".

References

1. Liu, Y.; Wang, H.; Zhang, H.; Liber, K. A comprehensive support vector machine-based classification model for soil quality assessment. *Soil Tillage Res.* **2016**, *155*, 19–26. [CrossRef]
2. Huang, Y.; Chen, Q.; Deng, M.; Japenga, J.; Li, T.; Yang, X.; He, Z. Heavy metal pollution and health risk assessment of agricultural soils in a typical peri-urban area in southeast China. *J. Environ. Manag.* **2018**, *207*, 159–168. [CrossRef] [PubMed]
3. Anaman, R.; Peng, C.; Jiang, Z.; Liu, X.; Zhou, Z.; Guo, Z.; Xiao, X. Identifying sources and transport routes of heavy metals in soil with different land uses around a smelting site by GIS based PCA and PMF. *Sci. Total Environ.* **2022**, *823*, 153759. [CrossRef] [PubMed]
4. Khalid, S.; Shahid, M.; Niazi, N.K.; Murtaza, B.; Bibi, I.; Dumat, C. A comparison of technologies for remediation of heavy metal contaminated soils. *J. Geochem. Explor.* **2017**, *182*, 247–268. [CrossRef]
5. Li, M.S. Ecological restoration of mineland with particular reference to the metalliferous mine wasteland in China: A review of research and practice. *Sci. Total Environ.* **2006**, *357*, 38–53. [CrossRef]
6. Li, G.; Sun, G.X.; Williams, P.N.; Nunes, L.; Zhu, Y.G. Inorganic arsenic in Chinese food and its cancer risk. *Environ. Int.* **2011**, *37*, 1219–1225. [CrossRef]
7. Li, Z.; Feng, X.; Li, G.; Bi, X.; Sun, G.; Zhu, J.; Wang, J. Mercury and other metal and metalloid soil contamination near a Pb/Zn smelter in east Hunan province, China. *Appl. Geochem.* **2011**, *26*, 160–166. [CrossRef]
8. Chen, R.; De Sherbinin, A.; Ye, C.; Shi, G. China's soil pollution: Farms on the frontline. *Science* **2014**, *344*, 691. [CrossRef]
9. Zhao, F.J.; Ma, Y.; Zhu, Y.G.; Tang, Z.; McGrath, S.P. Soil contamination in China: Current status and mitigation strategies. *Environ. Sci. Technol.* **2015**, *49*, 750–759. [CrossRef]
10. He, Z.; Shentu, J.; Yang, X.; Baligar, V.C.; Zhang, T.; Stoffella, P.J. Heavy metal contamination of soils: Sources, indicators and assessment. *J. Environ. Indic.* **2015**, *9*, 17–18.
11. Ernst, W.H.O. The origin and ecology ofecontaminated, stabilized and non-pristine soils. In *Metal-Contaminated Soils: In Situ Inactivation and Phytorestoration*; Landes BioScience/Springer: Heidelberg, Germany; New York, NY, USA, 1998; pp. 17–29.
12. Soubrand, M.; Joussein, E.; Courtin-Nomade, A.; Jubany, I.; Casas, S.; Bahí, N.; Martínez-Martínez, S. Investigating the relationship between speciation and oral/lung bioaccessibility of a highly contaminated tailing: Contribution in health risk assessment. *Environ. Sci. Pollut. Res.* **2020**, *27*, 40732–40748.
13. Gabarrón, M.; Faz, A.; Martínez-Martínez, S.; Zornoza, R.; Acosta, J.A. Assessment of metals behaviour in industrial soil using sequential extraction, multivariable analysis and a geostatistical approach. *J. Geochem. Explor.* **2017**, *172*, 174–183. [CrossRef]
14. Wang, J.Y.; Long, J.X.; Lu, H.W. Heavy Metal Contamination of Soil in Zhuzhou Smelting. In *Advanced Materials Research*; Trans Tech Publications Ltd.: Kapellweg, Switzerland, 2014.
15. Xing, W.; Cao, E.; Scheckel, K.G.; Bai, X.; Li, L. Influence of phosphate amend ment and zinc foliar application on heavy metal accumulation in wheat and on soil extractability impacted by a lead smelter near Jiyuan, China. *Environ. Sci. Pollut. Res.* **2018**, *25*, 31396–31406. [CrossRef]
16. Li, X.; Lv, G.; Ma, W.; Li, T.; Zhang, R.; Zhang, J.; Lei, Y. Review of resource and recycling of silicon powder from diamond-wire sawing silicon waste. *J. Hazard. Mater.* **2022**, *424*, 127389. [CrossRef] [PubMed]
17. Sharma, B.; Vaish, B.; Singh, U.K.; Singh, P.; Singh, R.P. Recycling of organic wastes in agriculture: An environmental perspective. *Int. J. Environ. Res.* **2019**, *13*, 409–429. [CrossRef]
18. Rene, E.R.; Sethurajan, M.; Ponnusamy, V.K.; Kumar, G.; Dung, T.N.B.; Brindhadevi, K.; Pugazhendhi, A. Electronic waste generation, recycling and resource recovery: Technological perspectives and trends. *J. Hazard. Mater.* **2021**, *416*, 125664. [CrossRef]
19. Luo, C.; Liu, C.; Wang, Y.; Liu, X.; Li, F.; Zhang, G.; Li, X. Heavy metal contamination in soils and vegetables near an e-waste processing site, south China. *J. Hazard. Mater.* **2011**, *186*, 481–490. [CrossRef]
20. Wu, Q.; Leung, J.Y.; Geng, X.; Chen, S.; Huang, X.; Li, H.; Lu, Y. Heavy metal contamination of soil and water in the vicinity of an abandoned e-waste recycling site: Implications for dissemination of heavy metals. *Sci. Total Environ.* **2015**, *506*, 217–225. [CrossRef]
21. Liu, Y.; Liu, Y.; Xing, P.; Liu, J. Environmental performance evaluation of different municipal solid waste management scenarios in China. *Resour. Conserv. Recycl.* **2017**, *125*, 98–106. [CrossRef]
22. Kim, M.H.; Song, H.B.; Song, Y.; Jeong, I.T.; Kim, J.W. Evaluation of food waste disposal options in terms of global warming and energy recovery: Korea. *Int. J. Energy Environ. Eng.* **2013**, *4*, 1–12. [CrossRef]
23. Tan, M.; Li, H.; Huang, Z.; Wang, Z.; Xiong, R.; Jiang, S.; Luo, L. Comparison of atmospheric and gas-pressurized oxidative torrefaction of heavy-metal-polluted rice straw. *J. Clean. Prod.* **2021**, *283*, 124636. [CrossRef]
24. Cucinotta, D.; Vanelli, M. WHO declares COVID-19 a pandemic. *Acta Bio Med. Atenei Parm.* **2020**, *91*, 157.

25. Skalny, A.V.; Lima, T.R.R.; Ke, T.; Zhou, J.-C.; Bornhorst, J.; Alekseenko, S.I.; Aaseth, J.; Anesti, O.; Sarigiannis, D.A.; Tsatsakis, A.; et al. Toxic metal exposure as a possible risk factor for COVID-19 and other respiratory infectious diseases. *Food Chem. Toxicol.* **2020**, *146*, 111809. [CrossRef] [PubMed]

26. Li, Y.; Ma, L.; Ge, Y.; Abuduwaili, J. Health risk of heavy metal exposure from dustfall and source apportionment with the PCA-MLR model: A case study in the Ebinur Lake Basin, China. *Atmos. Environ.* **2022**, *272*, 118950. [CrossRef]

27. Burges, A.; Epelde, L.; Garbisu, C. Impact of repeated single-metal and multi-metal pollution events on soil quality. *Chemosphere* **2015**, *120*, 8–15. [CrossRef]

28. Zhang, P.; Qin, C.; Hong, X.; Kang, G.; Qin, M.; Yang, D.; Pang, B.; Li, Y.; He, J.; Dick, R.P. Risk assessment and source analysis of soil heavy metal pollution from lower reaches of Yellow River irrigation in China. *Sci. Total Environ.* **2018**, *633*, 1136–1147. [CrossRef]

29. Sutherland, R.A.; Tack, F.M. Determination of Al, Cu, Fe, Mn, Pb and Zn in certified reference materials using the optimized BCR sequential extraction procedure. *Anal. Chim. Acta* **2002**, *454*, 249–257. [CrossRef]

30. Bacon, J.R.; Davidson, C.M. Is there a future for sequential chemical extraction? *Analyst* **2008**, *133*, 25–46. [CrossRef]

31. Gabarrón, M.; Faz, A.; Martínez-Martínez, S.; Acosta, J.A. Concentration and chemical distribution of metals and arsenic under different typical Mediterranean cropping systems. *Environ. Geochem. Health* **2019**, *41*, 2845–2857. [CrossRef]

32. Clevenger, T.E. Use of sequential extraction to evaluate the heavy metals in mining wastes. *Water Air Soil Pollut.* **1990**, *50*, 241–254. [CrossRef]

33. Tessier, A.; Campbell, P.G.C.; Bisson, M. Sequential extraction procedure for the speciation of particulate trace metals. *Anal. Chem.* **1979**, *51*, 844–851. [CrossRef]

34. Qayyum, S.; Khan, I.; Zhao, Y.; Maqbool, F.; Peng, C. Sequential extraction procedure for fractionation of Pb and Cr in artificial and contaminated soil. *Main Group Met. Chem.* **2016**, *39*, 49–58. [CrossRef]

35. Tang, X.-Y.; Cui, Y.-S.; Duan, J.; Tang, L. Pilot study of temporal variations in lead bioaccessibility and chemical fractionation in some Chinese soils. *J. Hazard. Mater.* **2008**, *160*, 29–36. [CrossRef] [PubMed]

36. Ariza, J.G.; Giráldez, I.; Sánchez-Rodas, D.; Morales, E. Metal sequential extraction procedure optimized for heavily polluted and iron oxide rich sediments. *Anal. Chim. Acta* **2000**, *414*, 151–164. [CrossRef]

37. Filgueiras, A.V.; Lavilla, I.; Bendicho, C. Chemical sequential extraction for metal partitioning in environmental solid samples. *J. Environ. Monit.* **2002**, *4*, 823–857. [CrossRef]

38. Ure, A.M.; Quevauviller, P.; Muntau, H.; Griepink, B. Speciation of Heavy Metals in Soils and Sediments. An Account of the Improvement and Harmonization of Extraction Techniques Undertaken Under the Auspices of the BCR of the Commission of the European Communities. *Int. J. Environ. Anal. Chem.* **1993**, *51*, 135–151. [CrossRef]

39. Maiz, I.; Arambarri, I.; Garcia, R.; Millán, E. Evaluation of heavy metal availability in polluted soils by two sequential extraction procedures using factor analysis. *Environ. Pollut.* **2000**, *110*, 3–9. [CrossRef]

40. Galán, E.; Ariza, J.G.; González, I.; Caliani, J.F.; Morales, E.; Giráldez, I. Utilidad de las tecnicas de extraccion secuencial en la mejora de la caracterización mineralogica por DRX de suelos y sedimentos con altos contenidos de oxidos de hierro. *Libro de conferencias y Resumenes de la XV Reunion Científica de la Sociedad Española de Arcillas* **1999**, *15*, 68–69.

41. Benitez, L.N.; Dubois, J.-P. Evaluation of the Selectivity of Sequential Extraction Procedures Applied to the Speciation of Cadmium in Soils. *Int. J. Environ. Anal. Chem.* **1999**, *74*, 289–303. [CrossRef]

42. Zimmerman, A.J.; Weindorf, D.C. Heavy Metal and Trace Metal Analysis in Soil by Sequential Extraction: A Review of Procedures. *Int. J. Anal. Chem.* **2010**, *2010*, 387803. [CrossRef]

43. Silveira, M.L.; Alleoni, L.R.F.; O'Connor, G.A.; Chang, A.C. Heavy metal sequential extraction methods—A modification for tropical soils. *Chemosphere* **2006**, *64*, 1929–1938. [CrossRef] [PubMed]

44. Dahlin, C.L.; Williamson, C.A.; Collins, W.K.; Dahlin, D.C. Sequential extraction versus comprehensive characterization of heavy metal species in brownfield soils. *Environ. Forensics* **2002**, *3*, 191–201. [CrossRef]

45. Awad, M.; Liu, Z.; Skalicky, M.; Dessoky, E.; Brestic, M.; Mbarki, S.; Rastogi, A.; EL Sabagh, A. Fractionation of Heavy Metals in Multi-Contaminated Soil Treated with Biochar Using the Sequential Extraction Procedure. *Biomolecules* **2021**, *11*, 448. [CrossRef] [PubMed]

46. Vilar, S.; Gutierrez, A.; Antezana, J.; Carral, P.; Alvarez, A. A comparative study of three different methods for the sequential extraction of heavy metals in soil. *Toxicol. Environ. Chem.* **2005**, *87*, 1–10. [CrossRef]

47. Elnazer, A.; Salman, S.; Seleem, E.M.; Abu El Ella, E.M. Assessment of Some Heavy Metals Pollution and Bioavailability in Roadside Soil of Alexandria-Marsa Matruh Highway, Egypt. *Int. J. Ecol.* **2015**, *2015*, 689420. [CrossRef]

48. Kobielska, P.A.; Howarth, A.J.; Farha, O.K.; Nayak, S. Metal–organic frameworks for heavy metal removal from water. *Co-ord. Chem. Rev.* **2018**, *358*, 92–107. [CrossRef]

49. Chai, Y.; Bai, M.; Chen, A.; Peng, L.; Shao, J.; Shang, C.; Peng, C.; Zhang, J.; Zhou, Y. Thermochemical conversion of heavy metal contaminated biomass: Fate of the metals and their impact on products. *Sci. Total Environ.* **2022**, *822*, 153426. [CrossRef]

50. Lin, S.; Ali, M.U.; Zheng, C.; Cai, Z.; Wong, M.H. Toxic chemicals from uncontrolled e-waste recycling: Exposure, body burden, health impact. *J. Hazard. Mater.* **2021**, *426*, 127792. [CrossRef]

51. Tóth, G.; Hermann, T.; Da Silva, M.; Montanarella, L. Heavy metals in agricultural soils of the European Union with implications for food safety. *Environ. Int.* **2016**, *88*, 299–309. [CrossRef]

52. Singh, P.; Borthakur, A.; Singh, R.; Bhadouria, R.; Singh, V.K.; Devi, P. A critical review on the research trends and emerging technologies for arsenic decontamination from water. *Groundw. Sustain. Dev.* **2021**, *14*, 100607. [CrossRef]

53. Yin, N.; Zhang, Z.; Cai, X.; Du, H.; Sun, G.; Cui, Y. In Vitro Method to Assess Soil Arsenic Metabolism by Human Gut Microbiota: Arsenic Speciation and Distribution. *Environ. Sci. Technol.* **2015**, *49*, 10675–10681. [CrossRef] [PubMed]

54. Ahmad, A.Y.; Al-Ghouti, M.A. Approaches to achieve sustainable use and management of groundwater resources in Qatar: A review. *Groundw. Sustain. Dev.* **2020**, *11*, 100367. [CrossRef]

55. Rodriguez, A.G.P.; López, M.I.R.; Casillas, D.; León, J.A.A.; Banik, S.D. Impact of pesticides in karst groundwater. Review of recent trends in Yucatan, Mexico. *Groundw. Sustain. Dev.* **2018**, *7*, 20–29. [CrossRef]

56. Kim, M.-H.; Kim, J.-W. Comparison through a LCA evaluation analysis of food waste disposal options from the perspective of global warming and resource recovery. *Sci. Total Environ.* **2010**, *408*, 3998–4006. [CrossRef] [PubMed]

57. Wang, Y.; Xu, Y.; Qi, S.; Li, X.; Kong, X.; Yuan, D.; Theodore, O.I. Distribution and potential sources of organochlorine pesticides in the karst soils of a tiankeng in southwest China. *Environ. Earth Sci.* **2013**, *70*, 2873–2881. [CrossRef]

58. Di, X.; Xiao, B.; Dong, H.; Wang, S. Implication of different humic acid fractions in soils under karst rocky desertification. *CATENA* **2018**, *174*, 308–315. [CrossRef]

59. Xiao, P.; Xiao, B.; Adnan, M. Effects of Ca 2+ on migration of dissolved organic matter in limestone soils of the southwest China karst area. *Land Degrad. Dev.* **2021**, *32*, 5069–5082. [CrossRef]

60. Yang, M.; Chen, L.; Msigwa, G.; Tang, K.H.D.; Yap, P.-S. Implications of COVID-19 on global environmental pollution and carbon emissions with strategies for sustainability in the COVID-19 era. *Sci. Total Environ.* **2021**, *809*, 151657. [CrossRef]

61. Liang, Y.; Yi, X.; Dang, Z.; Wang, Q.; Luo, H.; Tang, J. Heavy Metal Contamination and Health Risk Assessment in the Vicinity of a Tailing Pond in Guangdong, China. *Int. J. Environ. Res. Public Health* **2017**, *14*, 1557. [CrossRef]

62. Wang, M.; Janssen, A.B.G.; Bazin, J.; Strokal, M.; Ma, L.; Kroeze, C. Accounting for interactions between Sustainable Development Goals is essential for water pollution control in China. *Nat. Commun.* **2022**, *13*, 730. [CrossRef]

63. Gallo, A., Jr.; Odokonyero, K.; Mousa, M.A.; Reihmer, J.; Al-Mashharawi, S.; Marasco, R.; Mishra, H. Nature-Inspired Superhydrophobic Sand Mulches Increase Agricultural Productivity and Water-Use Efficiency in Arid Regions. *ACS Agric. Sci. Technol.* **2022**. [CrossRef]

64. Martínez-Retureta, R.; Aguayo, M.; Abreu, N.; Stehr, A.; Duran-Llacer, I.; Rodríguez-López, L.; Sauvage, S.; Sánchez-Pérez, J.-M. Estimation of the Climate Change Impact on the Hydrological Balance in Basins of South-Central Chile. *Water* **2021**, *13*, 794. [CrossRef]

65. Rodelo, C.G.; Salinas, R.A.; Jaime, E.A.; Armenta, S.; Galdámez-Martínez, A.; Castillo-Blum, S.E.; la Vega, H.A.-D.; Grace, A.N.; Aguilar-Salinas, C.A.; Rodelo, J.G.; et al. Zinc associated nanomaterials and their intervention in emerging respiratory viruses: Journey to the field of biomedicine and biomaterials. *Co-ord. Chem. Rev.* **2022**, *457*, 214402. [CrossRef]

66. Arora, N.K.; Mishra, J. COVID-19 and importance of environmental sustainability. *Environ. Sustain.* **2020**, *3*, 117–119. [CrossRef]

67. Martin, R.L.A.S.; Crochemore, T.; Savioli, F.A.; Coelho, F.O.; Passos, R.D.H. Thromboelastometry early identifies thrombotic complications related to COVID-19: A case report. *SAGE Open Med. Case Rep.* **2021**, *9*, 2050313X211033160. [CrossRef]

68. International Monetary Fund. *World Economic Outlook: The Great Lockdown*; International Monetary Fund: Washington, DC, USA, 2020.

69. Sohrabi, C.; Alsafi, Z.; O'Neill, N.; Khan, M.; Kerwan, A.; Al-Jabir, A.; Iosifidis, C.; Agha, R. World Health Organization declares global emergency: A review of the 2019 novel coronavirus (COVID-19). *Int. J. Surg.* **2020**, *76*, 71–76. [CrossRef]

70. Lee, T.X. COVID-19 Heavy Metal Hypothesis. *Qeios* **2020**. [CrossRef]

71. Hamad, M.N.M.; Al-Qahtni, A. Understand COVID-19 through Heavy Metals Pollution. *J. Pharm. Res. Int.* **2022**, *34*, 19–35. [CrossRef]

72. Larsen, D.A.; Wigginton, K.R. Tracking COVID-19 with wastewater. *Nat. Biotechnol.* **2020**, *38*, 1151–1153. [CrossRef]

73. Hart, O.E.; Halden, R.U. Computational analysis of SARS-CoV-2/COVID-19 surveillance by wastewater-based epidemiology locally and globally: Feasibility, economy, opportunities and challenges. *Sci. Total Environ.* **2020**, *730*, 138875. [CrossRef]

74. Krantz, M.S.; Phillips, E.J. COVID-19 mRNA vaccine safety during the first 6 months of roll-out in the USA. *Lancet Infect. Dis.* **2022**. [CrossRef]

75. Xu, G.; Schwarz, P.; Yang, H. Adjusting energy consumption structure to achieve China's CO2 emissions peak. *Renew. Sustain. Energy Rev.* **2020**, *122*, 109737. [CrossRef]

76. Haidar, A.; Ferdous, N.; Soma, S.K.; Hossain, M.M.; Chowdhury, N.N.; Akter, T.; Hossain, I. Attenuation of Air Pollutants: A Blessing during COVID-19 Outbreak. *Int. J. Recent Adv. Multidiscip. Top.* **2021**, *2*, 42–46.

77. Wang, R.; Xiong, Y.; Xing, X.; Yang, R.; Li, J.; Wang, Y.; Cao, J.; Balkanski, Y.; Peñuelas, J.; Ciais, P.; et al. Daily CO2 Emission Reduction Indicates the Control of Activities to Contain COVID-19 in China. *Innovation* **2020**, *1*, 100062. [CrossRef]

78. Liu, Z.; Ciais, P.; Deng, Z.; Lei, R.; Davis, S.J.; Feng, S.; Zheng, B.; Cui, D.; Dou, X.; Zhu, B.; et al. Near-real-time monitoring of global CO_2 emissions reveals the effects of the COVID-19 pandemic. *Nat. Commun.* **2020**, *11*, 5172. [CrossRef]

79. Khan, I.; Shah, D.; Shah, S.S. COVID-19 pandemic and its positive impacts on environment: An updated review. *Int. J. Environ. Sci. Technol.* **2020**, *18*, 521–530. [CrossRef]

80. Erans, M.; Sanz-Pérez, E.S.; Hanak, D.P.; Clulow, Z.; Reiner, D.M.; Mutch, G.A. Direct air capture: Process technology, techno-economic and socio-political challenges. *Energy Environ. Sci.* **2022**, *15*, 1360–1405. [CrossRef]

81. Classen, A.T.; Sundqvist, M.K.; Henning, J.A.; Newman, G.S.; Moore, J.A.; Cregger, M.A.; Patterson, C.M. Direct and indirect effects of climate change on soil microbial and soil microbial-plant interactions: What lies ahead? *Ecosphere* **2015**, *6*, 1–21. [CrossRef]

82. Deng, G.; Chen, H.; Wang, S. Risk Assessment and Prediction of Rainstorm and Flood Disaster Based on Henan Province, China. *Math. Probl. Eng.* **2022**, *2022*, 5310920. [CrossRef]

83. Wang, J.; Yu, C.W.; Cao, S.-J. Urban development in the context of extreme flooding events. *Indoor Built Environ.* **2021**, *31*, 3–6. [CrossRef]

84. Zhao, X.; Li, H.; Qi, Y. Are Chinese Cities Prepared to Manage the Risks of Extreme Weather Events? Evidence from the 2021.07. 20 Zhengzhou Flood in Henan Province. *SSRN* **2021**, *20*, 38. [CrossRef]

85. Vaccari, M.; Torretta, V.; Collivignarelli, C. Effect of Improving Environmental Sustainability in Developing Countries by Upgrading Solid Waste Management Techniques: A Case Study. *Sustainability* **2012**, *4*, 2852–2861. [CrossRef]

86. Hakovirta, M.; Denuwara, N. How COVID-19 Redefines the Concept of Sustainability. *Sustainability* **2020**, *12*, 3727. [CrossRef]

87. Pareek, N. Climate Change Impact on Soils: Adaptation and Mitigation. *MOJ Ecol. Environ. Sci.* **2017**, *2*, 26. [CrossRef]

88. Rahman, S.; Hossain, I.; Mullick, A.R.; Khan, M.H. Food security and the coronavirus disease 2019 (COVID-19): A systemic review. *J. Med. Sci. Clin. Res.* **2020**, *8*, 180–184.

89. Hasselriis, F.; Constantine, L. Characterization of today's medical waste. In *Medical Waste Incineration and Pollution Prevention*; Springer: Berlin, Germany, 1992; pp. 37–52.

90. Ziolo, M.; Fidanoski, F.; Simeonovski, K.; Filipovski, V.; Jovanovska, K. Business and sustainability: Key drivers for business success and business failure from the perspective of sustainable development. In *Value of Failure: The Spectrum of Challenges for the Economy*; Union Bridge Books: London, UK, 2017; p. 55.

91. Ferasso, M.; Bares, L.; Ogachi, D.; Blanco, M. Economic and Sustainability Inequalities and Water Consumption of European Union Countries. *Water* **2021**, *13*, 2696. [CrossRef]

92. Tong, X.; Wang, T.; Chen, Y.; Wang, Y. Towards an inclusive circular economy: Quantifying the spatial flows of e-waste through the informal sector in China. *Resour. Conserv. Recycl.* **2018**, *135*, 163–171. [CrossRef]

93. Chi, X.; Streicher-Porte, M.; Wang, M.; Reuter, M. Informal electronic waste recycling: A sector review with special focus on China. *Waste Manag.* **2011**, *31*, 731–742. [CrossRef]

94. Wong, M.H.; Wu, S.C.; Deng, W.J.; Yu, X.Z.; Luo, Q.; Leung, A.O.W.; Wong, A.S. Export of toxic chemicals–a review of the case of uncontrolled electronic-waste recycling. *Environ. Pollut.* **2007**, *149*, 131–140. [CrossRef]

95. Bhutta, M.K.S.; Omar, A.; Yang, X. Electronic waste: A growing concern in today's environment. *Econ. Res. Int.* **2011**, *8*, 474230. [CrossRef]

96. Tang, X.; Shen, C.; Shi, D.; Alam Cheema, S.; Khan, M.I.; Zhang, C.; Chen, Y. Heavy metal and persistent organic compound contamination in soil from Wenling: An emerging e-waste recycling city in Taizhou area, China. *J. Hazard. Mater.* **2010**, *173*, 653–660. [CrossRef] [PubMed]

97. Lu, C.; Zhang, L.; Zhong, Y.; Ren, W.; Tobias, M.; Mu, Z.; Xue, B. An overview of e-waste management in China. *J. Mater. Cycles Waste Manag.* **2015**, *17*, 1–12. [CrossRef]

98. Bertram, M.; Graedel, T.; Rechberger, H.; Spatari, S. The contemporary European copper cycle: Waste management subsystem. *Ecol. Econ.* **2002**, *42*, 43–57. [CrossRef]

99. Zhang, K.; Schnoor, J.L.; Zeng, E.Y. E-waste recycling: Where does it go from here? *Environ. Sci. Technol.* **2012**, *46*, 10861–10867. [CrossRef] [PubMed]

100. Heacock, M.; Kelly, C.B.; Asante, K.A.; Birnbaum, L.S.; Bergman, Å.L.; Bruné, M.-N.; Buka, I.; Carpenter, D.O.; Chen, A.; Huo, X.; et al. E-Waste and Harm to Vulnerable Populations: A Growing Global Problem. *Environ. Health Perspect.* **2016**, *124*, 550–555. [CrossRef] [PubMed]

101. Miao, Y.; Liu, L.; Zhang, Y.; Tan, Q.; Li, J. An overview of global power lithium-ion batteries and associated critical metal recycling. *J. Hazard. Mater.* **2021**, *425*, 127900. [CrossRef] [PubMed]

102. Shammi, M.; Rahman, M.; Ali, L.; Khan, A.S.M.; Siddique, A.B.; Ashadudzaman; Doza, B.; Alam, G.M.; Tareq, S.M. Application of short and rapid strategic environmental assessment (SEA) for biomedical waste management in Bangladesh. *Case Stud. Chem. Environ. Eng.* **2021**, *5*, 100177. [CrossRef]

103. Mangindaan, D.; Adib, A.; Febrianta, H.; Hutabarat, D.J.C. Systematic Literature Review and Bibliometric Study of Waste Management in Indonesia in the COVID-19 Pandemic Era. *Sustainability* **2022**, *14*, 2556. [CrossRef]

104. Kothari, R.; Sahab, S.; Singh, H.M.; Singh, R.P.; Singh, B.; Pathania, D.; Singh, A.; Yadav, S.; Allen, T.; Singh, S.; et al. COVID-19 and waste management in Indian scenario: Challenges and possible solutions. *Environ. Sci. Pollut. Res.* **2021**, *28*, 52702–52723. [CrossRef]

105. Agamuthu, P.; Barasarathi, J. Clinical waste management under COVID-19 scenario in Malaysia. *Waste Manag. Res. J. Sustain. Circ. Econ.* **2020**, *39* (Suppl. 1), 18–26. [CrossRef]

106. Kirchherr, J.; Reike, D.; Hekkert, M. Conceptualizing the circular economy: An analysis of 114 definitions. *Resour. Conserv. Recycl.* **2017**, *127*, 221–232. [CrossRef]

107. Ibn-Mohammed, T.; Mustapha, K.B.; Godsell, J.; Adamu, Z.; Babatunde, K.A.; Akintade, D.D.; Acquaye, A.; Fujii, H.; Ndiaye, M.M.; Yamoah, F.A.; et al. A critical analysis of the impacts of COVID-19 on the global economy and ecosystems and opportunities for circular economy strategies. *Resour. Conserv. Recycl.* **2021**, *164*, 105169. [CrossRef] [PubMed]

108. Rahman, S.M.; Kim, J.; Laratte, B. Disruption in Circularity? Impact analysis of COVID-19 on ship recycling using Weibull tonnage estimation and scenario analysis method. *Resour. Conserv. Recycl.* **2020**, *164*, 105139. [CrossRef] [PubMed]
109. Rahman, S.M.; Kim, J. Circular economy, proximity, and shipbreaking: A material flow and environmental impact analysis. *J. Clean. Prod.* **2020**, *259*, 120681. [CrossRef]
110. Veenstra, A.; Wang, C.; Fan, W.; Ru, Y. An analysis of E-waste flows in China. *Int. J. Adv. Manuf. Technol.* **2010**, *47*, 449–459. [CrossRef]

Review

Exploring Industry-Specific Research Themes on E-Waste: A Literature Review

Hilal Shams [1,*], Altaf Hossain Molla [1], Mohd Nizam Ab Rahman [1], Hawa Hishamuddin [1], Zambri Harun [1] and Nallapaneni Manoj Kumar [2,3,4,*]

[1] Department of Mechanical and Manufacturing Engineering, Faculty of Engineering and Built Environment, Universiti Kebangsaan Malaysia, Bangi 43600, Malaysia; altafhossain1994@gmail.com (A.H.M.); mnizam@ukm.edu.my (M.N.A.R.); hawa7@ukm.edu.my (H.H.); zambri@ukm.edu.my (Z.H.)
[2] School of Energy and Environment, City University of Hong Kong, Kowloon, Hong Kong
[3] Center for Resource Recovery, HICCER—Hariterde International Council of Circular Economy Research, Palakkad 678631, Kerala, India
[4] Swiss School of Business and Management Geneva, Avenue des Morgines 12, 1213 Genève, Switzerland
* Correspondence: hilal_shams7@yahoo.com (H.S.); mnallapan2-c@my.cityu.edu.hk (N.M.K)

Abstract: The usance of electric and electronic products has become commonplace across the globe. The growing number of customers and the demand for these products are resulting in the manufacturing of new electrical and electronic products into the market, which is ultimately generating a plethora of e-waste. The notion of a circular economy (CE) is attracting more researchers to work in the growing field of e-waste management. Considering e-waste as a prominent menace, the objective of this study was to undertake a comprehensive review of the literature by analyzing the research articles published in the MDPI *Sustainability* journal pertaining to the topic of e-waste in the context of operations and supply chain management (OSCM). This study was addressed via three research questions. A total of 87 selected papers from 2014 to 2023 were analyzed, reviewed, and categorized after data were collected from Web of Science (WOS) and Scopus academic databases with articles only published in the MDPI *Sustainability* journal. This entails identifying prominent research themes, publication trends, research evolution, research clusters, and industries related to e-waste through descriptive analysis. The field of study and methods employed were analyzed by means of content analysis by delving into the main body of the published articles. Further, four major research themes and clusters were identified: (1) closed-loop supply chains; (2) e-waste; (3) sustainable development; and (4) waste electrical and electronic equipment (WEEE). Consequently, this review can be a foundation for subsequent scholarly pursuits toward e-waste management and fresh lines of inquiry for the journal. Finally, in the conclusion section, some future research guidelines are also provided.

Keywords: e-waste; WEEE; sustainability; circular economy; electrical and electronic waste; journal review

Citation: Shams, H.; Molla, A.H.; Ab Rahman, M.N.; Hishamuddin, H.; Harun, Z.; Kumar, N.M. Exploring Industry-Specific Research Themes on E-Waste: A Literature Review. *Sustainability* **2023**, *15*, 12244. https://doi.org/10.3390/su151612244

Academic Editor: Silvia Fiore

Received: 16 June 2023
Revised: 27 July 2023
Accepted: 6 August 2023
Published: 10 August 2023

1. Introduction

Electric and electronic equipment (EEE) are a part of everyday life. Their significance is increasing in terms of providing business opportunities, but also as an emerging problem in terms of waste generation, more commonly known as e-waste or WEEE [1]. E-waste or WEEE is the generic short term used for electric and electronic devices either disposed of or discarded, or near their end of life [2]. Electrical and electronic equipment (EEE) include a wide variety of products, such as PCs, laptops, mobile phones, washing machines, televisions, and monitors [3]. The usage of such devices has proliferated as household appliances, in manufacturing, other industries, and services, and simultaneously, the waste related to such products has proportionally increased, which, as a result, is ravaging the environment. Amongst such products, mobile phones and computers comprise the most common form of waste due to their short lifespan, along with their rapid development, replacement, and widespread usage. This brings in the notion of the overconsumption of raw materials, making it a global concern. Moreover, such indulgence of overconsumption results in waste generation. Such wastes have intrinsic noxious and toxic properties, which are pernicious in nature. This is evident as e-waste has an abundance of diverse compounds, presenting opportunities in terms of materials and product reutilizations; however, concurrently, challenges and risks are associated with it in terms of possible contaminations and harmful contents. Furthermore, in terms of e-waste, there is a clear composition differentiation regarding other forms of waste, like industrial or municipal wastes, containing valuable and toxic materials, which have severe and detrimental effects on the environment and humans if not properly taken care of [4–6].

Due to the growing subtle and apparent concern about such wastes, the CE has a vital role in waste recycling and resource efficiency by encouraging circular practices and closed-loop systems to be a sustainable business model [7]. Certain CE activities, such as recycling, reuse, remanufacturing, and refurbishment, play a significant role in e-waste life extension and recovering precious materials from it [8,9]. Such roles and activities of CE are instrumental in sustainable development [10] and e-waste management [3] by maintaining utility and value for such products [11]. Therefore, the adoption of such circular practices is imperative for e-waste management due to its increasing generation of 3–5% per annum [12]. Hence, the core objective of CE is the promotion of sustainable practices to safeguard the environment within its natural domain and ensure the healthy life of living beings.

1.1. Research Motivation

Research regarding e-waste has been growing over the past decades, where researchers play a significant role in journal publications and conferences. Each journal contributes its role in minimizing the menace of e-waste and how to circulate the materials and products associated with it in terms of research through the theoretical framework, results and discussions, and practical implementations. Recent review studies about e-waste [3,13–15] present the growing interests of the research community through their studies on the different aspects of e-waste management.

The purpose of a literature review aims to map and evaluate the extant literature [16] to identify research gaps [17] and to analyze their contribution to the respective research field. Therefore, the research objective of this study is to focus on and gather the extant literature and studies based on the aforementioned *Sustainability* journal. As Govindan and Soleimani [16] as well as Nita [17] suggest, some reviews are limited to certain databases or specific journals to meet the research objectives. Therefore, we have limited this study only to *Sustainability*. The notion of focusing only on *Sustainability* to analyze its role in e-waste research is a promising one. This journal is working on and publishing research related to e-waste management. The aim of reviewing a particular publisher and its specific journal is to steer toward exposing and uncovering the fields and areas where major or minor research is needed. This is based on the proposition proposed by Webster and Watson in their article [18]. Therefore, we can draw out more insights by delving into the publications; and so, in a way, this study will categorize and analyze research themes and trends published in *Sustainability*. To our best knowledge, this is the first study to primarily focus on the said publisher while reviewing the literature and research on e-waste. This was acknowledged while a search was conducted in the month of December on the MDPI website using the exact keywords as presented in Figure 1, along with analyzing the spreadsheets obtained from WOS and Scopus databases. Furthermore, the approach is to analyze the ongoing research patterns for academics and the management team of the publisher, along with the overall research community, whilst focusing on one particular field under a journal. This will provide a thorough overview by presenting the overall research trends and themes while concentrating on the specific journal and presenting insightful information and directions for future research. This study contributes to synthesizing past and extant research findings in the e-waste category, along with presenting a knowledge gap that will pave the way for researchers whilst the journal team can focus on critical areas for revision and improvement.

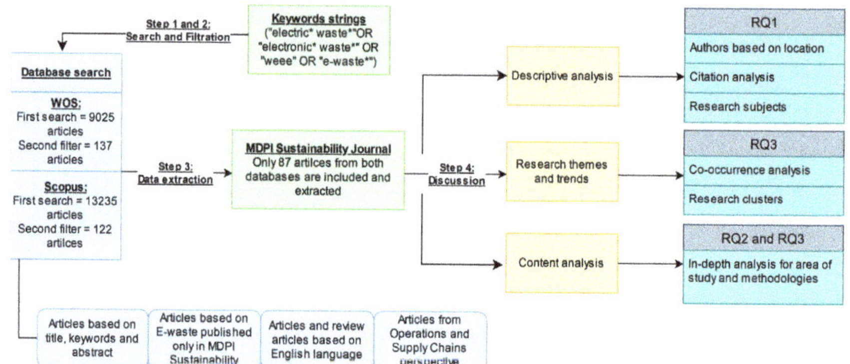

Figure 1. Flowchart of research framework and design.

1.2. Research Objectives

The main objectives of this study within the purview of the selected journal pertaining to e-waste management in the context of OSCM are as follows:

1. To examine the extent and magnitude of the extant published research.
2. To explore the key findings, publication trends, themes and topics covered, and advancements.
3. To find out the area of study and key methodologies employed by authors whilst pursuing their research.
4. To answer the research question: What are potential areas for future research?

For this purpose, three research questions were proposed to conduct this study under such pretext.

RQ1: How has the research field of e-waste evolved under *Sustainability*?

RQ2: What is the currently ongoing research on e-waste in *Sustainability*?

RQ3: What are the current underlying research themes and trends?

The obtained results illustrate that Sustainability has played a substantial role in covering the research in the field of e-waste, which is discussed in detail. For this purpose, the rest of the article is based on the following sections: a brief overview of the past literature is discussed in Section 2, while methodology is presented in Section 3; and Section 4 covers content analysis, which is discussed in several parts based on methodologies different authors have adopted in their studies.

2. Literature Review

The literature review is beneficial in providing a general overview of disparate and interconnected research fields. The existing literature provides excellent review papers covering e-waste from several aspects. Cucchiella et al. [19] perform an economic assessment of WEEE recycling based on descriptive and predictive disposed volumes in Europe. They analyze the economic assessment of 14 e-products. They mention that new products and technologies and the export of used products to developing countries from developed countries are prominent forms through which e-wastes are generated. Kiddee et al. [20] emphasize combining several e-waste management tools instead of using a single technique. For this purpose, they conducted a study by combining four e-waste management techniques, Life Cycle Assessment (LCA), Material Flow Analysis (MFA), Multicriteria Analysis (MCA), and Extended Producer Responsibility (EPR). All four tools are thoroughly discussed in their respective sections. Kumar et al. [14] conducted a correlation study concerning the United States, China, and India based on three variables: e-waste generation, gross domestic product (GDP), and population. Their results show that GDP directly correlates with the amount of e-waste generation, while the population has no significant impact. One of their crucial identification concerns the shorter life span of specific small electronic devices, like mobile phones, laptops, and tablets, indicating them as a significant source of waste generation.

Bressanelli et al. [3] systematically review the WEEE industry under the lens of CE based on four aspects, reviewing previous studies, the geographical location of focus studies, materials' life cycle phases, and 4R strategies of CE. They identify that little attention is given to the design of the practical solution; Europe is focusing on the WEEE industry mainly from a geographical perspective, and there is a lack of attention to the supply chain actors and life cycle phases. In contrast, from the CE lens, the main focus is on reducing and recycling strategies, whereas little attention has been given to the remanufacturing and reusing strategies. Bressanelli et al. [21] study 115 articles to identify CE enablers, levers, and benefits in the EEE supply chain (SC). For this purpose, they develop a framework for categorization, where enablers, levers, and benefits are discussed in detail in the findings section. In a recent review paper, Al-Saleem et al. [15] analyze the implementation process of the CE way of e-waste management based in Kuwait as a case study. Islam et al. [22] review the sustainable approaches for recovering metals from printed circuit boards of e-waste. Shittu et al.'s [23] review concentrates on the global situation regarding policies, legislation making, and WEEE generation. Europe is found to be leading in WEEE management, while the legislation trend is increasing in China, India, and Latin America. The authors identify four concern areas for further improvement; for example, WEEE management needs a formal system at the global level, and the ownership and production of EEE are rising on a global scale, which tends to generate more waste but no implementation of regulations. Furthermore, they suggest some measures for improving WEEE management on a global scale.

In a more recent paper, Ismail and Hanafiah [24] assessed four approaches for managing e-waste in Malaysia based on LCA and MFA. They proposed that energy recovery could be the ideal way to manage e-waste. Ismail and Hanafiah [25] review the present and future perspectives on sustainable e-waste generation. They perform descriptive analysis regarding worldwide e-waste generation and distribution and publication trends, research practices, and applications adopted in articles through content analysis. They study the current practices and research applications adapted for e-waste management thus far. For this purpose, the focus was on the current practices and research applications. They segment the scope and boundary and methodology of the current practices, where scope and boundary are defined through the level of analysis on the macro level and research areas. In contrast, the methodology is split into data sources, whereas the section on research application is mainly related to management practices. Also, in their previous study, Ismail and Hanafiah [26] reviewed the opportunities and challenges regarding the Malaysian recycling system related to e-waste. Their findings suggest that implementing laws on household e-waste management is still lacking. Gollakota et al. [27] review is related to studying inconsistencies in correspondence to e-waste management in developing countries. In such regard, their research identifies ten shortcomings impeding effective e-waste management pertaining to developing nations. They also analyze the micro, meso, and macro levels of e-waste management. The study analyzes the approach for recovering valuable materials from e-waste on the micro level. For addressing meso-level problems, material compatibility, its use, and reclamation are addressed along with LCA, MFA, and MCA, while the macro level is related to the role of government, consumers, and EPR. Furthermore, they have discussed the disproportion in developed and developing countries for e-waste management.

The aforementioned articles provide an overview in terms of covering various aspects of the overall research conducted on e-waste management. In addition, findings of past research are synthesized to elaborate the trends and themes. Moreover, such categorization paves the way for future research. However, this traditional approach is quite broad and comprehensive in synthesizing publications, research findings, and areas. As observed, the existing review works on e-waste in the *Sustainability* journal were lacking. Therefore, the present study was intended to fill and close this gap by gathering and reviewing current research articles published in the said journal. This will explore the existing publications on e-waste whilst providing an overview of topical research progress along with recommendations for future research to fill in gaps and limitations after identification. For this purpose, 87 articles are selected and analyzed, and the discussion is presented in detail.

3. Materials and Methods

The research objective of this study is to focus on and gather the extant literature and studies based on the *Sustainability* journal. A review is based on specific steps determined to address the diversity of knowledge for particular research inquiries [28] by formulating a research topic, data collection, and analysis [29]. This led us to adopt the methodology based on the working of [16,17,30] with some adaptations. The methodology part is divided into three sections, material collection, material filtration, and discussion; each is discussed briefly in Sections 3.1–3.3, respectively, along with a detailed discussion in Section 4. Figure 1 depicts the overall flow of this study.

3.1. Material Collection

The material was collected in three stages. First, databases were selected. In the second stage, data were filtered and exported into a suitable format for analysis. In the last step, data were cleaned to remove any discrepancies and duplication for overall analysis.

WOS and Scopus are leading database sources of information for scientific research [31,32]. Therefore, both WOS and Scopus platforms were selected to search relevant research articles based on specific keywords. For this study, the first search was conducted in December 2022 on both databases. The second and final search was conducted on 11 January 2023. As suggested by Suering and Muller [33], material delimitation and unit characterization is an important decision for conducting analysis. Under this approach, relevant publications were selected from both databases while focusing on e-waste. For this purpose, data were collected in a four-step approach:

- Step 1: A search based on Boolean operators was conducted to search relevant articles based on the following combination of strings: ("electric* waste*" OR "electronic* waste*" OR "weee" OR "e-waste*"). Keywords were limited to topics covering their titles, keywords, and abstract only.
- Step 2a: First search: Specific keywords were applied on both WOS and Scopus databases.
- Step 2b: Results were filtered and limited to: (a) *Sustainability* journal issued by MDPI, (b) articles and review articles based on the English language, and (c) articles only focusing on e-waste.
- Step 3: Relevant articles were extracted for data analysis.
- Step 4: Only articles with the above criteria were analyzed in the final list in both WOS and Scopus databases.

3.2. Material Filtration

The first search was conducted on 21 December 2022. The initial search resulted in 8977 and 13,128 articles from WOS and Scopus, respectively. A second search was made on 11 January 2023, resulting in 9025 and 13,235 articles from WOS and Scopus, respectively. The material filtration was performed in two phases. In the first phase, the WOS data were filtered and cleansed. Only articles and review papers based on the English language published in *Sustainability* were included, limiting the result to 137 publications from 2014 to 2023 for WOS. This limitation was performed as publications that were not related to the domain of this study based on the analysis of keywords and abstract; therefore, they were removed from the study, reducing the final list.

Afterward, the same method was applied to data obtained from the Scopus database, limiting the final result to 122 articles from 2016 to 2023. Also, there were no papers from

2014 and from 2015 on the Scopus database, but the time frame was set from 2014 to 2023, so both databases have the same time frame per analysis regarding time frame. The process is presented in Figure 1 for visualization.

In the second phase, duplication was removed, and overall results were analyzed. Moreover, articles were limited to 87 papers, mainly based on the authors working in the area of OSCM. The whole process was conducted and analyzed on structured datasheets in a spreadsheet. By following step 4, in Section 3.1, the final list of the selected 87 articles was analyzed in detail for this study.

3.3. Content Analysis

Out of 87 papers, only highly cited papers are discussed in Section 4.3; the top 30 papers are presented in Section 4.1.2, based on citations and average citations per year from both databases. The content of selected articles is analyzed in terms of the study's scope, methodology, subject area of research, their discussion, and results.

4. Discussion

This section is divided into two sub-sections. First, we have discussed the descriptive analysis followed by the content analysis of selected papers in detail. Descriptive analysis is regarding publications and citation trends, authors' affiliations based on location, and research subjects in Section 4.1, which will focus on RQ1. Section 4.2, on the other hand, is about research trends and themes, and Section 4.3 is regarding -content analysis, where we have described and analyzed selected publications concerning RQ2 and RQ3.

4.1. Descriptive Analysis

The overall trend suggests that research on e-waste is gaining momentum among the scientific community. For this purpose, descriptive analysis shows the publication and citation trend based on data collected from WOS and Scopus databases. The yearly publication and citation trends are presented in Table 1. The total number of articles published on WOS is 137, with a total citation of 1175. Table 1 illustrates the year 2014, which has only one article published on WOS, while no article was published in 2015. The year 2016 had only five publications, but it gradually started to increase after 2018, with the year 2022 having top publications of 41 articles. Also, it is evident that citations gradually increased after 2018 as well, with a significant portion of citations occurring from 2019 to 2022 (1093 out of 1175), indicating that citation tendency accounts for 93% of total citations alone in these four years. The h-index is 20. This indicates the inclination and growing interests of researchers.

Table 1. Publications and citations from WOS and Scopus (11 January 2023).

Year	WOS		Scopus	
	Publications	Citations	Publications	Citations
2014	1	3	0	0
2015	0	3	0	0
2016	5	3	5	1
2017	7	21	7	10
2018	13	46	8	49
2019	16	94	15	99
2020	24	208	21	242
2021	30	369	28	341
2022	41	422	38	488

Table 1. *Cont.*

Year	WOS		Scopus	
	Publications	Citations	Publications	Citations
2023	0	6	0	13
Total	137	1175	122	1263
h-index	20	-	21	-

On the other hand, data obtained from Scopus almost show the same trend. However, the years 2014 and 2015 account for no publications, where the contribution of publications started from the year 2016 with five publications, which gradually increased with time progression, and the year 2022 having 38 publications. The citation trends tend to be the same with a slight hovering above citations based on the WOS dataset. The h-index for the Scopus dataset is 21.

4.1.1. Authors Based on Locations

All authors' regional affiliations or geographical dispersion based on selected papers are presented in Table 2 in descending order. This selection is only based on authors' affiliations per research paper. It is evident from Table 2 that China is highly contributing, with 18.8% of the total 106 in such regard, followed by Italy at 8.4%, the USA at 4.71%, and Germany at 4.71%. In Asia, China is the main most significant contributor with 20 authors against a total of 41 in all of Asia, while in Europe, it is Italy with 9 affiliations against a total of 45. The publications are primarily divided globally, with countries from Asia and Europe playing a significant role in publications and research. Asia corresponds to 41, Europe to 45, North America to 9, South America to 6, Africa to 3, and Oceania to 2 articles.

Table 2. Authors' affiliations are based on countries.

Continent	Countries	Authors' Affiliations	Total Authors	Percentage of Total
Asia	China	20	20	18.8%
	Taiwan, Indonesia	4,4	8	7.5%
	Malaysia	3	3	2.8%
	UAE	2	2	1.8%
	Saudi Arabia, Singapore, Philippines, Iran, South Korea, India, Vietnam, Japan	1,1,1,1, 1,1,1,1	8	7.5%
Total			**41**	**38.6%**
Europe	Italy	9	9	8.5%
	Germany	5	5	4.7%
	Norway, UK	4,4	8	7.5%
	Romania, Poland, Ireland, Finland, Denmark, Sweden	2,2,2,2,2,2	12	11.3%
	Austria, Lithuania, Belgium, Croatia, Ukraine, Serbia, Greece, Spain, Slovakia, Malta, EU	1,1,1,1,1, 1,1,1,1,1, 1	11	10.3%
Total			**45**	**42.4%**
North America	USA	5	5	4.7%
	Mexico, Canada	2,2	4	3.8%
Total			**9**	**8.5%**
South America	Brazil	5	5	4.7%
	Colombia	1	1	0.9%
Total			**6**	**5.6%**

Table 2. *Cont.*

Continent	Countries	Authors' Affiliations	Total Authors	Percentage of Total
Africa	South Africa	2	2	1.8%
	Nigeria	1	1	0.9%
Total			**3**	**2.8%**
Oceania	Australia	1	1	0.9%
	New Zealand	1	1	0.9%
Total			**2**	**1.8%**
Grand Total			**106**	**100%**

4.1.2. Citation Analysis

Table 3 presents the details of citations and average citations per year based on the two academic databases, WOS and Scopus. This table is presented under the notion of citation reports, which is utilitarian in terms of assessing the impact and influence of publications as research papers are considered crucial research criteria.

Table 3. Author(s) and citations report on WOS and Scopus (11 January 2023).

Row	Author(s)	WOS		Scopus	
		Citations	Average Citations	Citations	Average Citations
1	Nduneseokwu et al. [34]	58	8.29	67	9.57
2	Sverko Grdic et al. [35]	54	13.5	65	16.25
3	Rocca et al. [36]	54	13.5	65	16.25
4	Thi Thu Nguyen et al. [37]	53	10.6	58	11.6
5	Isernia et al. [38]	36	7.2	45	9
6	Popa et al. [39]	34	4.86	45	6.42
7	Shevchenko et al. [40]	31	6.2	39	7.8
8	Cruz-Sotelo et al. [41]	29	4.14	32	4.57
9	Miner et al. [42]	28	7	30	7.5
10	Parajuly and Wenzel [43]	26	3.71	30	4.28
11	Vermesan et al. [44]	25	5	26	5.2
12	Cordova-Pizzaro et al. [45]	25	5	26	5.2
13	Abalansa et al. [46]	23	7.67	26	8.66
14	D'Adamo et al. [47]	23	2.88	30	3.75
15	Delcea et al. [48]	20	5	20	5
16	Yu and Solvang [49]	19	2.38	36	4.5
17	Wang et al. [50]	17	2.83	21	3.5
18	Cao et al. [51]	17	2.83	21	3.5
19	Vieira et al. [52]	16	4	17	4.25
20	Barletta et al. [53]	14	1.75	14	1.75
21	Murthy and Ramakrishna [54]	13	6.5	22	11
22	Corsini et al. [55]	13	3.25	13	3.25
23	Andersson et al. [56]	13	2.6	14	2.8
24	Magrini et al. [57]	10	2	14	2.8
25	Tu et al. [58]	12	2	12	2
26	Maheswari et al. [59]	10	2	14	2.8
27	Liu et al. [60]	10	1.67	15	2.5
28	Sari et al. [61]	9	3	11	3.6
29	Parajuly and Fitzpatrick [62]	8	2	10	2.5
30	Wang et al. [63]	8	2	12	3

The data presented in Table 3 are based on the search provided on 11 January 2023. Data are presented in descending order only of the top 30 articles; however, citation data are listed in WOS data order, which the authors initially studied for this study. The highly cited article was published by the authors of [34] with a total of 58 citations and an average citation of 8.29 on WOS, and a total of 67 citations and an average citations per year of 9.57 on the Scopus database.

4.1.3. Research Subjects

Figure 2 presents the categorizations of all 87 papers based on research subjects and methodologies adopted by authors. The survey-based methodology is the most prominent one adopted by most authors, with 21 papers, followed by mathematical modeling papers, with 9 papers. Reviews, evaluation studies, and game theory are found in seven papers. Interestingly, the authors of all seven game-theory-model papers [60,64–69] are based in China.

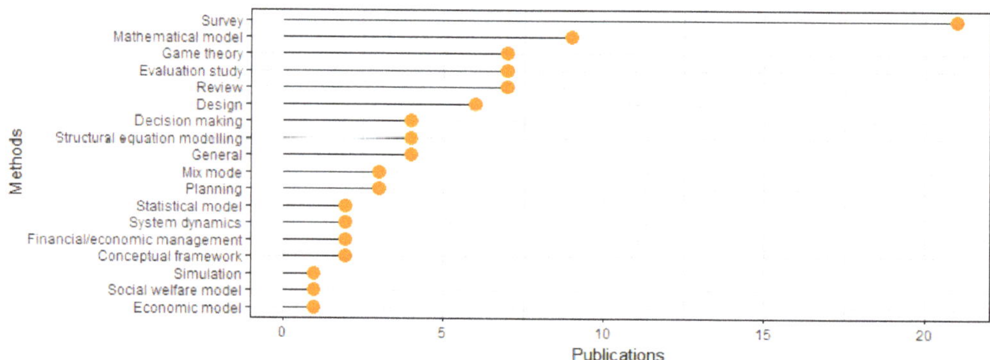

Figure 2. Methodologies and publications.

Table 4 presents the concentration of research fields pursued by the authors. E-waste management has a significant focus, along with recycling and reverse logistics (RLs). Table 5, on the other hand, presents the focus of the authors industry-wise. This table illustrates that a wide variety of studies are primarily based on e-waste, accounting for 68.96% of total publications, followed by mobile phones at 12.64% of total publications. Various other industrial items, like waste printed circuit boards and e-devices, have two publications each, while the rest comprise one publication each.

Table 4. Research fields.

Research Areas	Publications
E-waste management	[34,39,41–43,45–47,53,54,56,57,70–80]
E-waste	[81–84]
E-waste collection	[51]
Recycling	[35,37,40,44,48,60,63,66,69,69,81,85–99]
Disassembly	[36,100]
Reuse	[101,102]
Disposal	[103–105]
Repair	[106]
Consumer behavior	[55,107–112]
Life cycle assessment and material flow analysis	[113]
Supply chain	[114]
Supply chain: reverse supply chain	[68,115]
Supply chain: closed-loop supply chain	[50,64,67,116,117]
Logistics	[62]
Logistics: reverse logistics	[38,49,52,59,61,65,118]
Location: vehicle routing	[119,120]

Table 5. Focus of industries.

Industries	Publications
E-waste	[34,36–43,46,48–57,60,64–73,75,76,78–81,84,86,87,89,91,92,94–99,103,106,108–111,113,116,118,119]
Various sectors	[35]
Waste printed circuit boards	[44,47]
Mobile phones	[45,58,59,61,63,74,83,85,104,107,120]
Plastic and electronic waste	[62]
Washing machine	[101]
Notebooks	[77,88]
Computers	[102]
Car and refrigerator	[90]
Home appliances	[117]
Televisions and monitors	[114]
RFID	[82]
End-of-life vehicles	[93]
Batteries	[100]
E-devices	[105,112]

4.2. Research Themes and Trends

Based on RQ3, research themes and trends are discussed in terms of co-occurrence and research clusters to identify the topics and areas pursued by researchers.

4.2.1. Hotspot Identification Using Co-Occurrence Analysis

Figure 3 illustrates the hotspots of research using authors' keywords. Co-occurrence analysis was performed based on the authors' keywords to identify the main ideas and the border of research. Using VOSviewer 1.6.18, the minimum number of keyword occurrences was kept at 2, under which 54 hotspots met the threshold out of 509 keywords. E-waste occurrence presented 38 times, WEEE at 21, CE at 18, and recycling at 14, while sustainability showed 12 occurrences as the top five hotspots and most frequent keywords. All these five keywords are connected to each other and result in more publications and keywords as off-shoots. All 54 hotspots are mapped in Figure 3, where larger and thicker spots identify more occurrences. The purple color in the visualization represents the occurrences of the authors' keywords as being that of the year 2019, in contrast to the more recent occurrences represented in yellow found in the years 2021 and onward.

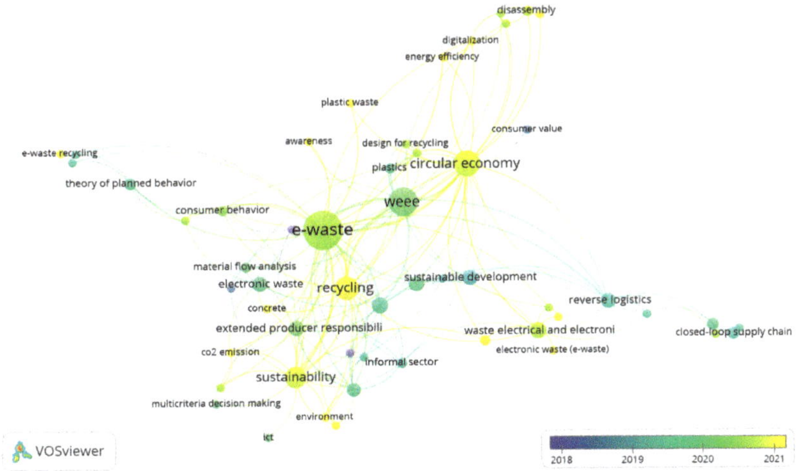

Figure 3. Authors' keywords co-occurrence network.

4.2.2. Research Themes Map by Identifying Clusters

Prominent research clusters and themes are discussed in Table 6 through content analysis. Among such, CLSC, e-waste, sustainable development, and WEEE comprise the more influential and discussed ones; therefore, these four clusters are briefly discussed here.

Table 6. Prominent research clusters.

Cluster	Cluster Category	Terms	Publications
1	Closed-loop supply chain	Closed-loop supply chain, remanufacturing, game theory, Stackelberg game, system dynamics	[50,64,65,67,74,94,117]
2	E-waste	E-waste, WEEE, CE, recycling, sustainability, waste electrical and electronic equipment, extended producer responsibility, reverse logistics, e-waste management, theory of planned behavior, consumer behavior, informal sector, plastics, circularity, design for recycling, design from recycling, end-of-life management, environment, legislation, plastic waste, waste disposal, waste management, repair, industry 4.0, material flow analysis, life cycle assessment, reverse supply chain	[34,38,40,44,46,47,49,51,52, 54,55,60–62,70– 73,76,77,81,82,85–87,89– 91,93,97,99,103,104,109, 111,114,116,118]
3	Sustainable development	Sustainable development, environmental sustainability	[35,80,95,107]
4	Waste electrical and electronic equipment (WEEE)	Waste electrical and electronic equipment (WEEE), waste management, repair	[39,56,92,101,102]

E-waste and its handling pose concerning issues to the environment. A recent report estimated that 53.6 million metric tons of e-waste was generated across the globe, and roughly less than 20% is managed under green environment policies [12]. Also, the circular transition is limited in this sector [121]. On the other hand, e-waste also has the economic potential of generating revenues, although it requires meticulous working [47]. Therefore, a systemic support and green eco-design system for restricting the menace of such wastes and acquiring economic gains has the potential for circular transition. The building blocks for such transition and sustainable development are CLSC, RL, health, policies, environmental protection, awareness, government financing, and digitalization. Such blocks work as enablers and levers for implementing CE practices.

In addition, under the CE framework, e-waste management activities can be enhanced for efficient resource usage, recovery, and recycling methods. For this purpose, CLSC activities for e-waste management operating in a CLSC system play a vital role in achieving sustainable goals from manufacturing to customer usage, recycling, and the disposal of such electrical and electronic products and components [122]. In essence, such green initiative reduces overall waste quantities, forming responsible consumption and well-designed processes linking forward and backward streams in an environmentally friendly way. E-waste appears to be one of the leading research themes covering different aspects, whereby the more prominent ones include WEEE, CE, recycling, sustainability, and RL. While the authors of [34] worked on consumers' intention, ref. [38] worked on the reverse supply chain, ref. [40] studied consumer recycling behavior, ref. [44] worked on recovery techniques from waste circuit boards, and ref. [47] identified challenges in the profitability assessment of e-waste. In CLSC research clusters, most studies are based on the game theory approach as well as on system dynamics.

4.3. Content Analysis

Figure 2 presents the various studies adopted by researchers. The 87 studies are related to OSCM. In operations management and a typical SC environment, the supplier, manufacturer, distribution channel, and customers are involved, intertwined, and related in terms of products, their consumption, collection, recycling, remanufacturing, reusing, feedback in terms of information, product delivery in the forward stream, and product flow back in terms of the upstream. Thus, in a way, all of these key stakeholders work together and are dependent on each other for their objectives, be it profit or end product for consumption and usage. Furthermore, under CE principles, CLSC, RL, and waste collection systems play their roles. Table 4 depicts the overall picture for a better understanding.

This review also identified the critical methodologies adopted by researchers and authors, as presented in Figure 2. According to the analysis, 21 out of 87 papers are based on survey methodology, followed by mathematical models with 9 papers, and finally, reviews, evaluation studies, and game theory models with 7 papers each. It is evident from Table 4 that the majority of papers are based on e-waste management (26.43%) and recycling (27.58%) subjects, while the rest are distributed on other aspects, such as: consumer behavior (10.3%); RL (8.04%); CLSC (5.74%); e-waste (4.46%); disposal (3.34%); disassembly, reuse, reverse supply chain, and vehicle routing (each with 2.29%); and e-waste collection, repair, SC, logistics, LCA, and MFA (each with 1%). It can be concluded that most studies are based on the management, SC, and R-framework of CE, like recycling.

On the other hand, Figure 2 illustrates that most of the studies are based on the qualitative type of research compared to quantitative studies. For instance, survey corresponds to 24.13% of studies; mathematical model, 10.34%; review, 8.04%; evaluation study, 8.04%; game theory, 8.04%; design, 6.89%; general, 4.59%; structural equation modeling, 4.59%; decision making, 4.59%; planning, 3.44%; mix mode, 3.44%; conceptual framework, 2.29%; financial/economic management, 2.29%; statistical model, 2.29%; economic model, 1.14%; social welfare model, 1.14%; and simulation, 1.14%. This comparative analysis shows that more research and work are required and are imperative in terms of quantitative studies.

The content analysis of 87 papers is presented below; the papers are categorized based on methodologies and publications based on Figure 1 to answer RQ2 and RQ3.

4.3.1. Survey-Based Studies

A survey gathers information and data from the sample size of people through questionnaires, phone calls, interviews, or using any web-based platform [123]. Some of the critical articles based on the survey are discussed here.

Nduneseokwu et al. [34] developed a theoretical framework based on the theory of planned behavior (TPB) to identify the influencing factors on consumers' intentions to participate in an e-waste collection system through conducting an empirical survey in Nigeria. Through studying customers' intentions for participation, the authors identified that consumer participation is vital for e-waste collection management due to the interrelation of both traits. Their research was limited to one metropolitan city, however, rather than focusing on multiple cities.

Miner et al. [42] surveyed 228 respondents about their awareness level and knowledge of e-waste management based in Nigeria. Their results indicate that cell phones and television sets are prominent devices. Also, open space dumping is the most predominant method for disposing of e-wastes, followed by storing at home and selling it, in that order. The authors only focused on household citizens in a specific city in Nigeria. Cordova-Pizarro et al. [45] analyze the economic and technical situation in Mexico by surveying the e-waste generation of cell phones using MFA. They perform fieldwork to quantify e-waste processing through personal interviews in selected twelve companies. They only address repair and recycling in terms of closing the loop for achieving circularity. Cao et al. [51] present an existing framework based on the survey to identify problems associated with the WEEE collection system in China by conducting an explorative study. They proposed four WEEE collection modes. This study, however, has focused on specific modes while ignoring the economic benefits participants can achieve form recycling systems in such different modes. Magrini et al. [57] studied the application of digital technologies, like IoT and blockchain, for the prevention of WEEE based on interviewing five companies. They performed a survey and presented a framework for enabling and using digital

e-waste management to highlight key factors, such as policies, technology requirements, and social awareness. Corsini et al. [55] explore and analyze consumer behavior related to (1) purchasing, (2) life extension, (3) recycling, and (4) the take-back participation of EoL electrical and electronic equipment by conducting a review. They adopt the TPB as a base theory.

4.3.4. Designing

Two papers primarily focused on designing a new waste collection system and consumer participation in an online recycling platform, respectively, and both of which are summarized.

Popa et al. [39] designed an IoT-based cloud platform for waste collection systems intending to develop and implement a smart system to identify and collect wastes like plastics, glass, aluminum, WEEEs, papers, cans, batteries, etc. They also propose further research in terms of virtual modeling and simulation. Wang et al. [63] perform two experiments to analyze how green information can influence the respondents participating in online recycling websites. Their study is based on the situational experiment method; however, a real-environment scenario can yield far different results, considering the seriousness level and subjects' understanding related to the experiment.

4.3.5. Frameworks

Some authors attempt to develop, adopt, and approach framework-based studies concerning an issue to pursue a solution for the said challenge. Two such approaches are discussed here.

Parajuly and Wenzel [43] propose a conceptual framework to alleviate the challenges posed by the diversity of e-waste. They investigate the quantities and management of electric and electronic products (e-products), product and material flow in the overall recovery chain, information exchange, and characteristics of e-products. They identify three aspects for improvements in the recovery chain: (1) "improved collection system", (2) "presorting and testing platform", and (3) "family-centric processing of EoL products". Andersen and Jager's [116] study is about manufacturers' ability to make their products more circular. Their findings suggest that they must explicitly bear the responsibility of all stakeholders in their product networking, from its conception up to its disposal, by creating a circular system for handling the products and their components. The authors focused on the technical parts of their model. However, encouraging actors for information sharing can be focused on through incentives and regulations.

4.3.6. Planning

Cruz-Sotelo et al. [41] suggest a planning model for the e-waste supply chain in Mexico. They analyzed actors in the WEEE recovery chain, public policies, legal regulations, existing practices for handling e-waste, opportunities, and challenges regarding waste flow management and proposed a management model. They identified one limitation and future prospective research in terms of applying uncertainty methods.

4.3.7. Decision Making

Decision-making is a fundamental process to maximize a firm's capability, resources, profits, operational capabilities, and optimality [124]. Two articles are summarized below.

Vieira et al. [52] use a multicriteria decision aid approach (MCDA) for small and medium-sized companies' (SMEs) prioritization of barriers for implementing RL for e-waste collection in Brazil. They perform the study in two steps; first, they identify the main barriers through a literature review and then apply MCDA application. The results show that the internal barriers of organizations are the main barrier to RL implementation from the government and SMEs' perspective, while consumers consider the managerial level as the main barrier. One limitation was the non-identification of e-wastes purchased by customers, as such information has the potential to shed light on the intention of reselling or repairing them. Barletta et al. [53] propose a novel methodological framework using several methods, like discrete event simulation, LCA, and stakeholder mapping, to access the sustainability dimension of e-waste management. They test their methodology using a case study.

4.3.8. Evaluation Study

Parajuly and Fitzpatrick [62] use e-waste and plastic waste as case fractions to evaluate the policy impact and assessment regarding transboundary waste movement. They recommend that policymakers should be aware of environmental and socio-economic issues. They also state that public involvement is imperative for improving the validity of any policy, as the public is ultimately one of the significant stakeholders.

4.3.9. Finance/Economic Management

Extended Producer Responsibility (EPR) is a built-in cost function for the protection of the environment as a whole from the manufacturing point of view. The background is related to producers who have to take responsibility for their post-consumer products in terms of environmental protection. For this purpose, a study is discussed here.

Cheng et al. [86] investigate the cost management functions for the recycling fee equation while considering environmental costs. Their study is based on Taiwan's EPR version for WEEE recycling. They suggest that the pricing mechanism for e-waste recycling cost has helped us by considering all perspectives, like labor, administration, and the environment itself. Their study has four players: (1) household communities generate waste, (2) the recycling industry is for processing, (3) municipalities collect e-waste, and (4) the recycling fund is for supporting incentives. They identify that the government is the most concerned authority regarding the recycling system and environmental cost compared to private authorities.

5. Conclusions

The research implications of conducting a review paper for investigating e-waste centering exclusively on a single selected journal are multifaceted and possess substantial value for involved stakeholders. This can be argued from different angles. For instance, this study primarily aimed to provide a comprehensive and confined examination of the extant literature and research activities contained precisely in the MDPI *Sustainability* journal rather than the copious literature from different sources. As a result, it offered insightful information pertaining to the progression of research on e-waste in the said journal discussed through bibliometric and content analysis in Section 4. Moreover, this review can stem from academic contribution in terms of e-waste awareness, existing techniques, approaches, and dominant research methodologies, implied along with research trends and tendencies. Hence, it could act as a reference point for researchers, scholars, and decision-makers of the said journal, sweeping for prospective research paths regarding e-waste management. Furthermore, this study could pave the way for unexplored domains concerning e-waste management within this journal or outside of its premises by stimulating multidisciplinary collaboration among researchers. In addition, such encouragement can work for both theoretical and practical contributions aiming at mitigating economic, societal, and environmental impacts posed by e-waste.

For this purpose, a thorough review is performed in this paper to analyze the selected 87 papers based on the *Sustainability* journal of MDPI in the famous research field of e-waste. Two forms of search were made based on specific keywords on WOS and Scopus databases, the first on 21 December 2022 and the second on 11 January 2023. The final list of 87 papers is examined, reviewed, and categorized based on the proposed research questions. This study aims to aid academics in the field of OSCM in understanding and analyzing the published papers under the prospect of potential future investigation. It will also help the publishers and policymakers of the said journal to assess and evaluate their publications.

Overall, this review paper has certain limitations. Firstly, we could not focus on the overall extant literature regarding e-waste, as the study was confined to MDPI's *Sustainability* journal only. This way, we could not work on the entire breadth of the available literature. This is the trade-off we had to perform in order to confine this study. The second focus was within the purview of OSCM potentially limiting the inclusivity and exhaustive examination. However, prospective researchers can work on other domains for more exploration by incorporating multiple databases and journals.

The review has provided some ideas for future research as well:

- Figure 3 presents a digitalization that was less focused. Future studies and examinations in a digital era with advanced applications can explore strategies for an effective and efficient e-waste management against the challenges and opportunities posited by e-products. E-learning platforms, big data, analytics, and subsequent digital technologies can be exploring strategies.

- Another aspect is that most underdeveloped countries have an informal market in terms of recycling, remanufacturing, and reusing. More studies on e-waste management whilst focusing on informal e-waste management processes, recycling, and remanufacturing facilities is an exploration to ponder and work on. Therefore, in such regard, the social and environmental aspect of CE practice has huge potential.

- Figure 3 illustrates that most papers are related to e-waste management and recycling, whilst other subject areas are less considered. For instance, disposal, disassembly, repair, and CLSC are crucial elements for effective e-waste management as the environmental performance of an SC affects sustainability [125].

- On the other hand, an e-waste collection center plays an essential role in the effectiveness of the SC and logistics network. We can see only one paper adequately dedicated to the collection of e-waste [51]; one paper on collection systems from an RL perspective [118]; and two papers on location problems [119,120]. Therefore, more research is needed from such a perspective, as e-waste management is very much related to collection and location centers for properly and effectively handling e-waste, as it cannot operate independently.

- Another aspect is the role of the consumer in e-waste management, as they act as a network function to supply such products. Consumer behavior and intention are the intangible aspects of sustainability. This study accounts for only 10.3% of papers; hence, more investigation is crucial and imperative.

- Only one article based on CE and I4.0 regarding e-waste, by Rocca et al. [36], and three IoT-related articles by the authors of [39,57,84] are found for this review paper during analysis. Further, this subject area can be considered by researchers for future studies based on the notion that I4.0 is the critical driving force in transforming the linear economy to a more circular method [126], which, as a result, will have profound effects on the production process and the whole SC.

- Table 5 suggests that various electronic items like computer parts, televisions parts, and end-of-life vehicle parts could be a significant future research area. Therefore, such areas need more attention from researchers for future intake.

- Another additional study is in assessing the risk management and operational capacity of such operations in terms of resource sharing and industrial symbiosis, in more generic terms, fostering a better industrial ecological system.

- The process of material and study categorization is another angle to ponder and look upon since this study is mainly focusing on the OSCM side of the research.

Author Contributions: Conceptualization, H.S. and N.M.K.; methodology, H.S. and N.M.K.; software, H.S. and A.H.M.; validation, H.S., A.H.M. and N.M.K.; formal analysis, H.S.; investigation, H.S.; resources, H.S.; data curation, H.S.; writing—original draft preparation, H.S., A.H.M. and N.M.K.; writing—review and editing, M.N.A.R., H.H., N.M.K. and Z.H.; visualization, H.S. and A.H.M.; supervision, M.N.A.R., H.H., Z.H. and N.M.K. All authors have read and agreed to the published version of the manuscript.

Funding: This research received no external funding.

Institutional Review Board Statement: Not applicable.

Informed Consent Statement: Not applicable.

Data Availability Statement: The data used in this research is enclosed within the manuscript. No external data sources are used in this research.

Acknowledgments: The authors would like to thank all anonymous reviewers and editorial team for their valuable comments to improve the manuscript.

Conflicts of Interest: The authors declare no conflict of interest.

References

1. Widmer, R.; Oswald-Krapf, H.; Sinha-Khetriwal, D.; Schnellmann, M.; Böni, H. Global Perspectives on E-Waste. *Environ. Impact Assess. Rev.* **2005**, *25*, 436–458. [CrossRef]
2. Rene, E.R.; Sethurajan, M.; Kumar Ponnusamy, V.; Kumar, G.; Bao Dung, T.N.; Brindhadevi, K.; Pugazhendhi, A. Electronic Waste Generation, Recycling and Resource Recovery: Technological Perspectives and Trends. *J. Hazard. Mater.* **2021**, *416*, 125664. [CrossRef]
3. Bressanelli, G.; Saccani, N.; Pigosso, D.C.A.; Perona, M. Circular Economy in the WEEE Industry: A Systematic Literature Review and a Research Agenda. *Sustain. Prod. Consum.* **2020**, *23*, 174–188. [CrossRef]
4. Duman, G.M.; Kongar, E.; Gupta, S.M. Estimation of Electronic Waste Using Optimized Multivariate Grey Models. *Waste Manag.* **2019**, *95*, 241–249. [CrossRef]
5. Bhattacharjee, P.; Howlader, I.; Rahman, M.A.; Taqi, H.M.M.; Hasan, M.T.; Ali, S.M.; Alghababsheh, M. Critical Success Factors for Circular Economy in the Waste Electrical and Electronic Equipment Sector in an Emerging Economy: Implications for Stakeholders. *J. Clean. Prod.* **2023**, *401*, 136767. [CrossRef]
6. Ryan-Fogarty, Y.; Baldé, C.P.; Wagner, M.; Fitzpatrick, C. Uncaptured Mercury Lost to the Environment from Waste Electrical and Electronic Equipment (WEEE) in Scrap Metal and Municipal Wastes. *Resour. Conserv. Recycl.* **2023**, *191*, 106881. [CrossRef]
7. Pollard, J.; Osmani, M.; Cole, C.; Grubnic, S.; Colwill, J. A Circular Economy Business Model Innovation Process for the Electrical and Electronic Equipment Sector. *J. Clean. Prod.* **2021**, *305*, 127211. [CrossRef]
8. Pan, X.; Wong, C.W.Y.; Li, C. Circular Economy Practices in the Waste Electrical and Electronic Equipment (WEEE) Industry: A Systematic Review and Future Research Agendas. *J. Clean. Prod.* **2022**, *365*, 132671. [CrossRef]
9. Xavier, L.H.; Ottoni, M.; Abreu, L.P.P. A Comprehensive Review of Urban Mining and the Value Recovery from E-Waste Materials. *Resour. Conserv. Recycl.* **2023**, *190*, 106840. [CrossRef]
10. Go, T.F.; Wahab, D.A.; Hishamuddin, H. Multiple Generation Life-Cycles for Product Sustainability: The Way Forward. *J. Clean. Prod.* **2015**, *95*, 16–29. [CrossRef]
11. Geissdoerfer, M.; Savaget, P.; Bocken, N.M.P.; Hultink, E.J. The Circular Economy—A New Sustainability Paradigm? *J. Clean. Prod.* **2017**, *143*, 757–768. [CrossRef]
12. Forti, V.; Balde, C.P.; Kuehr, R.; Bel, G. *The Global E-Waste Monitor 2020: Quantities, Flows and the Circular Economy Potential*; United Nations University/United Nations Institute for Training and Research: Bonn, Germany; International Telecommunication Union: Geneva, Switzerland; International Solid Waste Association: Rotterdam, The Netherlands, 2020; ISBN 978-92-808-9114-0.
13. Ding, Y.; Zhang, S.; Liu, B.; Zheng, H.; Chang, C.; Ekberg, C. Recovery of Precious Metals from Electronic Waste and Spent Catalysts: A Review. *Resour. Conserv. Recycl.* **2019**, *141*, 284–298. [CrossRef]
14. Kumar, A.; Holuszko, M.; Espinosa, D.C.R. E-Waste: An Overview on Generation, Collection, Legislation and Recycling Practices. *Resour. Conserv. Recycl.* **2017**, *122*, 32–42. [CrossRef]
15. Al-Salem, S.M.; Leeke, G.A.; El-Eskandarany, M.S.; Van Haute, M.; Constantinou, A.; Dewil, R.; Baeyens, J. On the Implementation of the Circular Economy Route for E-Waste Management: A Critical Review and an Analysis for the Case of the State of Kuwait. *J. Environ. Manag.* **2022**, *323*, 116181. [CrossRef] [PubMed]
16. Govindan, K.; Soleimani, H. A Review of Reverse Logistics and Closed-Loop Supply Chains: A Journal of Cleaner Production Focus. *J. Clean. Prod.* **2017**, *142*, 371–384. [CrossRef]
17. Nita, A. Empowering Impact Assessments Knowledge and International Research Collaboration—A Bibliometric Analysis of Environmental Impact Assessment Review Journal. *Environ. Impact Assess. Rev.* **2019**, *78*, 106283. [CrossRef]
18. Webster, J.; Watson, R.T. Analyzing the Past to Prepare for the Future: Writing a Literature Review. *MIS Q.* **2002**, *26*, xiii–xxiii.
19. Cucchiella, F.; D'Adamo, I.; Lenny Koh, S.C.; Rosa, P. Recycling of WEEEs: An Economic Assessment of Present and Future e-Waste Streams. *Renew. Sustain. Energy Rev.* **2015**, *51*, 263–272. [CrossRef]
20. Kiddee, P.; Naidu, R.; Wong, M.H. Electronic Waste Management Approaches: An Overview. *Waste Manag.* **2013**, *33*, 1237–1250. [CrossRef]
21. Bressanelli, G.; Pigosso, D.C.A.; Saccani, N.; Perona, M. Enablers, Levers and Benefits of Circular Economy in the Electrical and Electronic Equipment Supply Chain: A Literature Review. *J. Clean. Prod.* **2021**, *298*, 126819. [CrossRef]

22. Islam, A.; Ahmed, T.; Awual, M.R.; Rahman, A.; Sultana, M.; Aziz, A.A.; Monir, M.U.; Teo, S.H.; Hasan, M. Advances in Sustainable Approaches to Recover Metals from E-Waste—A Review. *J. Clean. Prod.* **2020**, *244*, 118815. [CrossRef]
23. Shittu, O.S.; Williams, I.D.; Shaw, P.J. Global E-Waste Management: Can WEEE Make a Difference? A Review of e-Waste Trends, Legislation, Contemporary Issues and Future Challenges. *Waste Manag.* **2021**, *120*, 549–563. [CrossRef]
24. Ismail, H.; Hanafiah, M.M. Evaluation of E-Waste Management Systems in Malaysia Using Life Cycle Assessment and Material Flow Analysis. *J. Clean. Prod.* **2021**, *308*, 127358. [CrossRef]
25. Ismail, H.; Hanafiah, M.M. A Review of Sustainable E-Waste Generation and Management: Present and Future Perspectives. *J. Environ. Manag.* **2020**, *264*, 110495. [CrossRef]
26. Ismail, H.; Hanafiah, M.M. Discovering Opportunities to Meet the Challenges of an Effective Waste Electrical and Electronic Equipment Recycling System in Malaysia. *J. Clean. Prod.* **2019**, *238*, 117927. [CrossRef]
27. Gollakota, A.R.K.; Gautam, S.; Shu, C.-M. Inconsistencies of E-Waste Management in Developing Nations—Facts and Plausible Solutions. *J. Environ. Manag.* **2020**, *261*, 110234. [CrossRef]
28. Tranfield, D.; Denyer, D.; Smart, P. Towards a Methodology for Developing Evidence-Informed Management Knowledge by Means of Systematic Review. *Br. J. Manag.* **2003**, *14*, 207–222. [CrossRef]
29. Seuring, S.; Gold, S. Conducting Content-analysis Based Literature Reviews in Supply Chain Management. *Supply Chain Manag. Int. J.* **2012**, *17*, 544–555. [CrossRef]
30. Barbosa-Póvoa, A.P.; da Silva, C.; Carvalho, A. Opportunities and Challenges in Sustainable Supply Chain: An Operations Research Perspective. *Eur. J. Oper. Res.* **2018**, *268*, 399–431. [CrossRef]
31. Mongeon, P.; Paul-Hus, A. The Journal Coverage of Web of Science and Scopus: A Comparative Analysis. *Scientometrics* **2016**, *106*, 213–228. [CrossRef]
32. Zhu, J.; Liu, W. A Tale of Two Databases: The Use of Web of Science and Scopus in Academic Papers. *Scientometrics* **2020**, *123*, 321–335. [CrossRef]
33. Seuring, S.; Müller, M. From a Literature Review to a Conceptual Framework for Sustainable Supply Chain Management. *J. Clean. Prod.* **2008**, *16*, 1699–1710. [CrossRef]
34. Nduneseokwu, C.K.; Qu, Y.; Appolloni, A. Factors Influencing Consumers' Intentions to Participate in a Formal E-Waste Collection System: A Case Study of Onitsha, Nigeria. *Sustainability* **2017**, *9*, 881. [CrossRef]
35. Sverko Grdic, Z.; Krstinic Nizic, M.; Rudan, E. Circular Economy Concept in the Context of Economic Development in EU Countries. *Sustainability* **2020**, *12*, 3060. [CrossRef]
36. Rocca, R.; Rosa, P.; Sassanelli, C.; Fumagalli, L.; Terzi, S. Integrating Virtual Reality and Digital Twin in Circular Economy Practices: A Laboratory Application Case. *Sustainability* **2020**, *12*, 2286. [CrossRef]
37. Thi Thu Nguyen, H.; Hung, R.-J.; Lee, C.-H.; Thi Thu Nguyen, H. Determinants of Residents' E-Waste Recycling Behavioral Intention: A Case Study from Vietnam. *Sustainability* **2019**, *11*, 164. [CrossRef]
38. Isernia, R.; Passaro, R.; Quinto, I.; Thomas, A. The Reverse Supply Chain of the E-Waste Management Processes in a Circular Economy Framework: Evidence from Italy. *Sustainability* **2019**, *11*, 2430. [CrossRef]
39. Popa, C.L.; Carutasu, G.; Cotet, C.E.; Carutasu, N.L.; Dobrescu, T. Smart City Platform Development for an Automated Waste Collection System. *Sustainability* **2017**, *9*, 2064. [CrossRef]
40. Shevchenko, T.; Laitala, K.; Danko, Y. Understanding Consumer E-Waste Recycling Behavior: Introducing a New Economic Incentive to Increase the Collection Rates. *Sustainability* **2019**, *11*, 2656. [CrossRef]
41. Cruz-Sotelo, S.E.; Ojeda-Benítez, S.; Jáuregui Sesma, J.; Velázquez-Victorica, K.I.; Santillán-Soto, N.; García-Cueto, O.R.; Alcántara Concepción, V.; Alcántara, C. E-Waste Supply Chain in Mexico: Challenges and Opportunities for Sustainable Management. *Sustainability* **2017**, *9*, 503. [CrossRef]
42. Miner, K.J.; Rampedi, I.T.; Ifegbesan, A.P.; Machete, F. Survey on Household Awareness and Willingness to Participate in E-Waste Management in Jos, Plateau State, Nigeria. *Sustainability* **2020**, *12*, 1047. [CrossRef]
43. Parajuly, K.; Wenzel, H. Product Family Approach in E-Waste Management: A Conceptual Framework for Circular Economy. *Sustainability* **2017**, *9*, 768. [CrossRef]
44. Vermeşan, H.; Tiuc, A.-E.; Purcar, M. Advanced Recovery Techniques for Waste Materials from IT and Telecommunication Equipment Printed Circuit Boards. *Sustainability* **2020**, *12*, 74. [CrossRef]
45. Cordova-Pizarro, D.; Aguilar-Barajas, I.; Romero, D.; Rodriguez, C.A. Circular Economy in the Electronic Products Sector: Material Flow Analysis and Economic Impact of Cellphone E-Waste in Mexico. *Sustainability* **2019**, *11*, 1361. [CrossRef]
46. Abalansa, S.; El Mahrad, B.; Icely, J.; Newton, A. Electronic Waste, an Environmental Problem Exported to Developing Countries: The GOOD, the BAD and the UGLY. *Sustainability* **2021**, *13*, 5302. [CrossRef]
47. D'Adamo, I.; Rosa, P.; Terzi, S. Challenges in Waste Electrical and Electronic Equipment Management: A Profitability Assessment in Three European Countries. *Sustainability* **2016**, *8*, 633. [CrossRef]
48. Delcea, C.; Crăciun, L.; Ioanăş, C.; Ferruzzi, G.; Cotfas, L.-A. Determinants of Individuals' E-Waste Recycling Decision: A Case Study from Romania. *Sustainability* **2020**, *12*, 2753. [CrossRef]
49. Yu, H.; Solvang, W.D. A Stochastic Programming Approach with Improved Multi-Criteria Scenario-Based Solution Method for Sustainable Reverse Logistics Design of Waste Electrical and Electronic Equipment (WEEE). *Sustainability* **2016**, *8*, 1331. [CrossRef]
50. Wang, W.; Zhou, S.; Zhang, M.; Sun, H.; He, L. A Closed-Loop Supply Chain with Competitive Dual Collection Channel under Asymmetric Information and Reward–Penalty Mechanism. *Sustainability* **2018**, *10*, 2131. [CrossRef]

51. Cao, J.; Xu, J.; Wang, H.; Zhang, X.; Chen, X.; Zhao, Y.; Yang, X.; Zhou, G.; Schnoor, J.L. Innovating Collection Modes for Waste Electrical and Electronic Equipment in China. *Sustainability* **2018**, *10*, 1446. [CrossRef]
52. Vieira, B.d.O.; Guarnieri, P.; Camara e Silva, L.; Alfinito, S. Prioritizing Barriers to Be Solved to the Implementation of Reverse Logistics of E-Waste in Brazil under a Multicriteria Decision Aid Approach. *Sustainability* **2020**, *12*, 4337. [CrossRef]
53. Barletta, I.; Larborn, J.; Mani, M.; Johannson, B. Towards an Assessment Methodology to Support Decision Making for Sustainable Electronic Waste Management Systems: Automatic Sorting Technology. *Sustainability* **2016**, *8*, 84. [CrossRef]
54. Murthy, V.; Ramakrishna, S. A Review on Global E-Waste Management: Urban Mining towards a Sustainable Future and Circular Economy. *Sustainability* **2022**, *14*, 647. [CrossRef]
55. Corsini, F.; Gusmerotti, N.M.; Frey, M. Consumer's Circular Behaviors in Relation to the Purchase, Extension of Life, and End of Life Management of Electrical and Electronic Products: A Review. *Sustainability* **2020**, *12*, 10443. [CrossRef]
56. Andersson, M.; Ljunggren Söderman, M.; Sandén, B.A. Adoption of Systemic and Socio-Technical Perspectives in Waste Management, WEEE and ELV Research. *Sustainability* **2019**, *11*, 1677. [CrossRef]
57. Magrini, C.; Nicolas, J.; Berg, H.; Bellini, A.; Paolini, E.; Vincenti, N.; Campadello, L.; Bonoli, A. Using Internet of Things and Distributed Ledger Technology for Digital Circular Economy Enablement: The Case of Electronic Equipment. *Sustainability* **2021**, *13*, 4982. [CrossRef]
58. Tu, J.-C.; Zhang, X.-Y.; Huang, S.-Y. Key Factors of Sustainability for Smartphones Based on Taiwanese Consumers' Perceived Values. *Sustainability* **2018**, *10*, 4446. [CrossRef]
59. Maheswari, H.; Yudoko, G.; Adhiutama, A. Government and Intermediary Business Engagement for Controlling Electronic Waste in Indonesia: A Sustainable Reverse Logistics Theory through Customer Value Chain Analysis. *Sustainability* **2019**, *11*, 732. [CrossRef]
60. Liu, H.; Wu, X.; Dou, D.; Tang, X.; Leong, G.K. Determining Recycling Fees and Subsidies in China's WEEE Disposal Fund with Formal and Informal Sectors. *Sustainability* **2018**, *10*, 2979. [CrossRef]
61. Sari, D.P.; Masruroh, N.A.; Asih, A.M.S. Consumer Intention to Participate in E-Waste Collection Programs: A Study of Smartphone Waste in Indonesia. *Sustainability* **2021**, *13*, 2759. [CrossRef]
62. Parajuly, K.; Fitzpatrick, C. Understanding the Impacts of Transboundary Waste Shipment Policies: The Case of Plastic and Electronic Waste. *Sustainability* **2020**, *12*, 2412. [CrossRef]
63. Wang, C.; Zhu, T.; Yao, H.; Sun, Q. The Impact of Green Information on the Participation Intention of Consumers in Online Recycling: An Experimental Study. *Sustainability* **2020**, *12*, 2498. [CrossRef]
64. Gong, Y.; Chen, M.; Zhuang, Y. Decision-Making and Performance Analysis of Closed-Loop Supply Chain under Different Recycling Modes and Channel Power Structures. *Sustainability* **2019**, *11*, 6413. [CrossRef]
65. Feng, D.; Yu, X.; Mao, Y.; Ding, Y.; Zhang, Y.; Pan, Z. Pricing Decision for Reverse Logistics System under Cross-Competitive Take-Back Mode Based on Game Theory. *Sustainability* **2019**, *11*, 6984. [CrossRef]
66. Guo, Q.; Li, Z.; Nie, J. Strategic Analysis of the Online Recycler's Reselling Channel Selection: Agency or Self-Run. *Sustainability* **2020**, *12*, 78. [CrossRef]
67. Lv, Y.; Bi, X.; Li, Q.; Zhang, H. Research on Closed-Loop Supply Chain Decision Making and Recycling Channel Selection under Carbon Allowance and Carbon Trading. *Sustainability* **2022**, *14*, 11473. [CrossRef]
68. Wang, B.; Wang, N. Decision Models for a Dual-Recycling Channel Reverse Supply Chain with Consumer Strategic Behavior. *Sustainability* **2022**, *14*, 10870. [CrossRef]
69. Li, S.-H.; Sun, Q. Stackelberg Game Analysis of E-Waste Recycling Stakeholders under Recovery Time Sensitivity and CRMs Life Expectancy Sensitivity. *Sustainability* **2022**, *14*, 9054. [CrossRef]
70. Blake, V.; Farrelly, T.; Hannon, J. Is Voluntary Product Stewardship for E-Waste Working in New Zealand? A Whangarei Case Study. *Sustainability* **2019**, *11*, 3063. [CrossRef]
71. Radulovic, V. Portrayals in Print: Media Depictions of the Informal Sector's Involvement in Managing E-Waste in India. *Sustainability* **2018**, *10*, 966. [CrossRef]
72. Ahen, F.; Amankwah-Amoah, J. Sustainable Waste Management Innovations in Africa: New Perspectives and Research Agenda for Improving Global Health. *Sustainability* **2021**, *13*, 6646. [CrossRef]
73. Aidonis, D.; Achillas, C.; Folinas, D.; Keramydas, C.; Tsolakis, N. Decision Support Model for Evaluating Alternative Waste Electrical and Electronic Equipment Management Schemes—A Case Study. *Sustainability* **2019**, *11*, 3364. [CrossRef]
74. Llerena-Riascos, C.; Jaén, S.; Montoya-Torres, J.R.; Villegas, J.G. An Optimization-Based System Dynamics Simulation for Sustainable Policy Design in WEEE Management Systems. *Sustainability* **2021**, *13*, 11377. [CrossRef]
75. Vaccari, M.; Zambetti, F.; Bates, M.; Tudor, T.; Ambaye, T. Application of an Integrated Assessment Scheme for Sustainable Waste Management of Electrical and Electronic Equipment: The Case of Ghana. *Sustainability* **2020**, *12*, 3191. [CrossRef]
76. Nowakowski, P.; Kuśnierz, S.; Płoszaj, J.; Sosna, P. Collecting Small-Waste Electrical and Electronic Equipment in Poland—How Can Containers Help in Disposal of E-Waste by Individuals? *Sustainability* **2021**, *13*, 12422. [CrossRef]
77. Rau, H.; Bisnar, A.R.; Velasco, J.P. Physical Responsibility Versus Financial Responsibility of Producers for E-Wastes. *Sustainability* **2020**, *12*, 4037. [CrossRef]
78. Ali, S.; Shirazi, F. A Transformer-Based Machine Learning Approach for Sustainable E-Waste Management: A Comparative Policy Analysis between the Swiss and Canadian Systems. *Sustainability* **2022**, *14*, 13220. [CrossRef]

79. Keshavarz-Ghorabaee, M.; Amiri, M.; Zavadskas, E.K.; Turskis, Z.; Antucheviciene, J. A Fuzzy Simultaneous Evaluation of Criteria and Alternatives (F-SECA) for Sustainable E-Waste Scenario Management. *Sustainability* **2022**, *14*, 10371. [CrossRef]
80. De Jager, T.; Maserumule, M.H. Innovative Community Projects to Educate Informal Settlement Inhabitants in the Sustainment of the Natural Environment. *Sustainability* **2021**, *13*, 6238. [CrossRef]
81. Abd-Mutalib, H.; Muhammad Jamil, C.Z.; Mohamed, R.; Shafai, N.A.; Nor-Ahmad, S.N.H.J.N. Firm and Board Characteristics, and E-Waste Disclosure: A Study in the Era of Digitalisation. *Sustainability* **2021**, *13*, 10417. [CrossRef]
82. Bukova, B.; Tengler, J.; Brumercikova, E. A Model of the Environmental Burden of RFID Technology in the Slovak Republic. *Sustainability* **2021**, *13*, 3684. [CrossRef]
83. Li, A.; Li, B.; Liu, X.; Zhang, Y.; Zhang, H.; Lei, X.; Hou, S.; Lu, B. Characteristics and Dynamics of University Students' Awareness of Retired Mobile Phones in China. *Sustainability* **2022**, *14*, 10587. [CrossRef]
84. Modarress Fathi, B.; Ansari, A.; Ansari, A. Threats of Internet-of-Thing on Environmental Sustainability by E-Waste. *Sustainability* **2022**, *14*, 10161. [CrossRef]
85. Martínez Leal, J.; Pompidou, S.; Charbuillet, C.; Perry, N. Design for and from Recycling: A Circular Ecodesign Approach to Improve the Circular Economy. *Sustainability* **2020**, *12*, 9861. [CrossRef]
86. Cheng, C.; Lin, C.; Wen, L.; Chang, T. Determining Environmental Costs: A Challenge in A Governmental E-Waste Recycling Scheme. *Sustainability* **2019**, *11*, 5156. [CrossRef]
87. Leader, A.; Gaustad, G.; Tomaszewski, B.; Babbitt, C.W. The Consequences of Electronic Waste Post-Disaster: A Case Study of Flooding in Bonn, Germany. *Sustainability* **2018**, *10*, 4193. [CrossRef]
88. Fang, Y.-T.; Rau, H. Optimal Consumer Electronics Product Take-Back Time with Consideration of Consumer Value. *Sustainability* **2017**, *9*, 385. [CrossRef]
89. Berwald, A.; Dimitrova, G.; Feenstra, T.; Onnekink, J.; Peters, H.; Vyncke, G.; Ragaert, K. Design for Circularity Guidelines for the EEE Sector. *Sustainability* **2021**, *13*, 3923. [CrossRef]
90. Denčić-Mihajlov, K.; Krstić, M.; Spasić, D. Sensitivity Analysis as a Tool in Environmental Policy for Sustainability: The Case of Waste Recycling Projects in the Republic of Serbia. *Sustainability* **2020**, *12*, 7995. [CrossRef]
91. Wang, R.; Deng, Y.; Li, S.; Yu, K.; Liu, Y.; Shang, M.; Wang, J.; Shu, J.; Sun, Z.; Chen, M.; et al. Waste Electrical and Electronic Equipment Reutilization in China. *Sustainability* **2021**, *13*, 11433. [CrossRef]
92. Xiao, Q.; Wang, H. Prediction of WEEE Recycling in China Based on an Improved Grey Prediction Model. *Sustainability* **2022**, *14*, 6789. [CrossRef]
93. Arnold, M.; Pohjalainen, E.; Steger, S.; Kaerger, W.; Welink, J.-H. Economic Viability of Extracting High Value Metals from End of Life Vehicles. *Sustainability* **2021**, *13*, 1902. [CrossRef]
94. Xue, R.; Zhang, F.; Tian, F. A System Dynamics Model to Evaluate Effects of Retailer-Led Recycling Based on Dual Chains Competition: A Case of e-Waste in China. *Sustainability* **2018**, *10*, 3391. [CrossRef]
95. Wang, Q.; Wang, X. An Expert Decision-Making System for Identifying Development Barriers in Chinese Waste Electrical and Electronic Equipment (WEEE) Recycling Industry. *Sustainability* **2022**, *14*, 16721. [CrossRef]
96. Kim, S.; Park, J. Network Analysis of the Disaster Response Systems in the Waste of Electrical and Electronic Equipment Recycling Center in South Korea. *Sustainability* **2022**, *14*, 10254. [CrossRef]
97. Boudewijn, A.; Peeters, J.R.; Cattrysse, D.; Dewulf, W.; Campadello, L.; Accili, A.; Duflou, J.R. Systematic Quantification of Waste Compositions: A Case Study for Waste of Electric and Electronic Equipment Plastics in the European Union. *Sustainability* **2022**, *14*, 7054. [CrossRef]
98. Ghosh, B.K.; Mekhilef, S.; Ahmad, S.; Ghosh, S.K. A Review on Global Emissions by E-Products Based Waste: Technical Management for Reduced Effects and Achieving Sustainable Development Goals. *Sustainability* **2022**, *14*, 4036. [CrossRef]
99. Pedro, F.; Giglio, E.; Velazquez, L.; Munguia, N. Constructed Governance as Solution to Conflicts in E-Waste Recycling Networks. *Sustainability* **2021**, *13*, 1701. [CrossRef]
100. Cordisco, A.; Melloni, R.; Botti, L. Sustainable Circular Economy for the Integration of Disadvantaged People: A Preliminary Study on the Reuse of Lithium-Ion Batteries. *Sustainability* **2022**, *14*, 8158. [CrossRef]
101. Johnson, M.; McMahon, K.; Fitzpatrick, C. A Preparation for Reuse Trial of Washing Machines in Ireland. *Sustainability* **2020**, *12*, 1175. [CrossRef]
102. Sánchez-Carracedo, F.; López, D. A Service-Learning Based Computers Reuse Program. *Sustainability* **2021**, *13*, 7785. [CrossRef]
103. Chen, F.; Li, X.; Yang, Y.; Hou, H.; Liu, G.-J.; Zhang, S. Storing E-Waste in Green Infrastructure to Reduce Perceived Value Loss through Landfill Siting and Landscaping: A Case Study in Nanjing, China. *Sustainability* **2019**, *11*, 1829. [CrossRef]
104. Laeequddin, M.; Kareem Abdul, W.; Sahay, V.; Tiwari, A.K. Factors That Influence the Safe Disposal Behavior of E-Waste by Electronics Consumers. *Sustainability* **2022**, *14*, 4981. [CrossRef]
105. Siddiqua, A.; El Gamal, M.; Kareem Abdul, W.; Mahmoud, L.; Howari, F.M. E-Device Purchase and Disposal Behaviours in the UAE: An Exploratory Study. *Sustainability* **2022**, *14*, 4805. [CrossRef]
106. Rudolf, S.; Blömeke, S.; Niemeyer, J.F.; Lawrenz, S.; Sharma, P.; Hemminghaus, S.; Mennenga, M.; Schmidt, K.; Rausch, A.; Spengler, T.S.; et al. Extending the Life Cycle of EEE—Findings from a Repair Study in Germany: Repair Challenges and Recommendations for Action. *Sustainability* **2022**, *14*, 2993. [CrossRef]
107. Saari, U.A.; Baumgartner, R.J.; Mäkinen, S.J. Eco-Friendly Brands to Drive Sustainable Development: Replication and Extension of the Brand Experience Scale in a Cross-National Context. *Sustainability* **2017**, *9*, 1286. [CrossRef]

108. Mohamad, N.S.; Thoo, A.C.; Huam, H.T. The Determinants of Consumers' E-Waste Recycling Behavior through the Lens of Extended Theory of Planned Behavior. *Sustainability* **2022**, *14*, 9031. [CrossRef]
109. D'Almeida, F.S.; de Carvalho, R.B.; dos Santos, F.S.; de Souza, R.F.M. On the Hibernating Electronic Waste in Rio de Janeiro Higher Education Community: An Assessment of Population Behavior Analysis and Economic Potential. *Sustainability* **2021**, *13*, 9181. [CrossRef]
110. Banaszkiewicz, K.; Pasiecznik, I.; Cieżak, W.; Boer, E. den Household E-Waste Management: A Case Study of Wroclaw, Poland. *Sustainability* **2022**, *14*, 11753. [CrossRef]
111. Guarnieri, P.; Vieira, B.d.O.; Cappellesso, G.; Alfinito, S.; e Silva, L.C. Analysis of Habits of Consumers Related to E-Waste Considering the Knowledge of Brazilian National Policy of Solid Waste: A Comparison among White, Green, Brown and Blue Lines. *Sustainability* **2022**, *14*, 11557. [CrossRef]
112. Nøjgaard, M.; Smaniotto, C.; Askegaard, S.; Cimpan, C.; Zhilyaev, D.; Wenzel, H. How the Dead Storage of Consumer Electronics Creates Consumer Value. *Sustainability* **2020**, *12*, 5552. [CrossRef]
113. Withanage, S.V.; Habib, K. Life Cycle Assessment and Material Flow Analysis: Two Under-Utilized Tools for Informing E-Waste Management. *Sustainability* **2021**, *13*, 7939. [CrossRef]
114. Wibowo, N.; Piton, J.K.; Nurcahyo, R.; Gabriel, D.S.; Farizal, F.; Madsuha, A.F. Strategies for Improving the E-Waste Management Supply Chain Sustainability in Indonesia (Jakarta). *Sustainability* **2021**, *13*, 13955. [CrossRef]
115. Andersen, T.; Jæger, B.; Mishra, A. Circularity in Waste Electrical and Electronic Equipment (WEEE) Directive. Comparison of a Manufacturer's Danish and Norwegian Operations. *Sustainability* **2020**, *12*, 5236. [CrossRef]
116. Andersen, T.; Jæger, B. Circularity for Electric and Electronic Equipment (EEE), the Edge and Distributed Ledger (Edge&DL) Model. *Sustainability* **2021**, *13*, 9924. [CrossRef]
117. Xue, R.; Zhang, F.; Tian, F.; Oloruntoba, R.; Miao, S. Dual Chains Competition under Two Recycling Modes Based on System Dynamics Method. *Sustainability* **2018**, *10*, 2382. [CrossRef]
118. Ruan Barbosa de Aquino, Í.; Ferreira da Silva Junior, J.; Guarnieri, P.; Camara e Silva, L. The Proposition of a Mathematical Model for the Location of Electrical and Electronic Waste Collection Points. *Sustainability* **2021**, *13*, 224. [CrossRef]
119. Zheng, F.; Sun, Z.; Liu, M. Location-Routing Optimization with Renting Social Vehicles in a Two-Stage E-Waste Recycling Network. *Sustainability* **2021**, *13*, 11879. [CrossRef]
120. Sari, D.P.; Masruroh, N.A.; Asih, A.M.S. Extended Maximal Covering Location and Vehicle Routing Problems in Designing Smartphone Waste Collection Channels: A Case Study of Yogyakarta Province, Indonesia. *Sustainability* **2021**, *13*, 8896. [CrossRef]
121. Rizos, V.; Bryhn, J. Implementation of Circular Economy Approaches in the Electrical and Electronic Equipment (EEE) Sector: Barriers, Enablers and Policy Insights. *J. Clean. Prod.* **2022**, *338*, 130617. [CrossRef]
122. Zhang, X.; Li, Q.; Liu, Z.; Chang, C.-T. Optimal Pricing and Remanufacturing Mode in a Closed-Loop Supply Chain of WEEE under Government Fund Policy. *Comput. Ind. Eng.* **2021**, *151*, 106951. [CrossRef]
123. Forza, C. Survey Research in Operations Management: A Process-based Perspective. *Int. J. Oper. Prod. Manag.* **2002**, *22*, 152–194. [CrossRef]
124. Dias, L.S.; Ierapetritou, M.G. From Process Control to Supply Chain Management: An Overview of Integrated Decision Making Strategies. *Comput. Chem. Eng.* **2017**, *106*, 826–835. [CrossRef]
125. Amiruddin, S.Z.; Hishamuddin, H.; Darom, N.A.; Naimin, H.H. A Case Study of Carbon Emissions from Logistic Activities during Supply Chain Disruptions. *J. Kejuruter.* **2021**, *33*, 221–228. [CrossRef] [PubMed]
126. Abdul-Hamid, A.-Q.; Ali, M.H.; Osman, L.H.; Tseng, M.-L. The Drivers of Industry 4.0 in a Circular Economy: The Palm Oil Industry in Malaysia. *J. Clean. Prod.* **2021**, *324*, 129216. [CrossRef]

*sustainability*

Review

A Bibliometric Analysis of Sustainable Product Design Methods from 1999 to 2022: Trends, Progress, and Disparities between China and the Rest of the World

Meng Gao [1,*], Ke Ma [1], Renke He [1], Carlo Vezzoli [2] and Nuo Li [3]

1   School of Design, Hunan University, Changsha 410000, China
2   Design Department, Politecnico di Milano, 20158 Milan, Italy
3   School of Engineering, University of Glasgow, Glasgow G12 8QQ, UK
*   Correspondence: meng_gao@hnu.edu.cn

Abstract: Effective product design strategies play a crucial role in promoting sustainable production, consumption, and disposal practices. In the literature, many such practices have been proposed by various researchers; however, it is challenging to understand which is more effective from the design point of view. This study employs bibliometric analysis and visualization software, CiteSpace, to comprehensively assess the literature on sustainable product design methods (SPDMs) from two major citation databases, namely, China National Knowledge Infrastructure and Web of Science, covering the period between 1999 and 2022. The objective of this review is to identify the latest research trends, progress, and disparities between China and the rest of the world in the field of SPDMs. The findings reveal that the development of SPDMs is characterized by a combination of multi-method integration and expansion, as well as qualitative and quantitative hybrids. However, research processes differ between China and other countries. Chinese studies focus on digital-driven development, rural revitalization, and system design, while research from other countries emphasizes a circular economy, distribution, additive manufacturing, and artificial intelligence. Nevertheless, both Chinese and international studies lack quantitative research methods in relation to socio-cultural sustainability. Future research should aim to deepen sustainable design methods and standards for specialized products, as well as to incorporate quantitative methods that address cultural and social sustainability dimensions. Open-source and shared SPDMs should be encouraged to promote methodological innovation that prioritizes multidimensional and systematic sustainable benefits, leveraging the strengths of new technologies.

Keywords: sustainability; design method; product design; CiteSpace; bibliometric analysis; comparative study

Citation: Gao, M.; Ma, K.; He, R.; Vezzoli, C.; Li, N. A Bibliometric Analysis of Sustainable Product Design Methods from 1999 to 2022: Trends, Progress, and Disparities between China and the Rest of the World. *Sustainability* **2023**, *15*, 12440. https://doi.org/10.3390/su151612440

Academic Editors: Nallapaneni Manoj Kumar, Subrata Hait, Anshu Priya and Varsha Bohra

Received: 14 June 2023
Revised: 9 August 2023
Accepted: 10 August 2023
Published: 16 August 2023

1. Introduction

The global population has significantly increased from three billion in 1960 to approximately eight billion in 2022, leading to an enhanced human impact on the ecosystem in pursuit of products and services necessary for production and life [1]. The unprecedented use of natural resources during this process exerts enormous pressure on the ecosystem, pushing the planet to its limits. Furthermore, the recent environmental and climate crisis, increasingly severe regional conflicts, complex and volatile international market, and the outbreak of COVID-19 have heightened the volatility and vulnerability of the socio-economic system that supports people's lives and development. Consequently, achieving sustainable socio-economic development has become a critical concern for academia. Sustainable design, first proposed during the environmental movement in the 1980s, has emerged as an effective strategic solution to this challenge [1]. Over the past four decades, scholars, designers, and engineers across the world have widely used sustainable design to guide the design of various products, services, buildings, environments, and social systems [2]. According to research, approximately 80% of sustainability is determined at the product design stage [3]. Therefore, designing and producing sustainable products is a critical strategy for achieving the sustainable development goals outlined in *Transforming our World: The 2030 Agenda for Sustainable Development* [4].

Product sustainability refers to the ability to maintain a product's sustainable service life while minimizing environmental impacts and providing socio-economic benefits for stakeholders. Sustainable product design methods (SPDMs) play a crucial role in achieving sustainability goals [5]. The widely accepted definition of sustainability is the triple-bottom-line approach, which assigns "equal importance to economic stability, ecological compatibility, and social equilibrium" [6]. The World Design Organization (WDO) updated its definition of industrial design in 2015 to emphasize its role in creating a better world through sustainable development, addressing socio-economic, environmental, and ethical issues. Sustainable design methods broadly address these issues and have become vital and increasingly focused on social innovation and cultural inheritance projects [7]. Tsinghua University, Tongji University, and Hunan University have made significant contributions in this regard [7]. In recent years, numerous research projects have been launched to explore how to guide, improve, and optimize the production and supply of sustainable products and services. Sustainable design methods are critical and involve all sectors, fields, and disciplines. To better understand these methods, we must therefore thoroughly understand past methods, which is possible only through a comprehensive literature assessment. In line with this, some researchers have already conducted literature assessments, and to move forward, we must understand how these past review assessments were done to set a path for this research.

Recent literature reviews on sustainable product design methods (SPDMs) conducted by scholars in China and beyond have identified several areas for improvement. Firstly, previous reviews have primarily focused on introducing theoretical progress in sustainable design research rather than systematically presenting the evolving trends and dynamic processes of SPDMs in recent years. Secondly, the existing research scope does not incorporate knowledge graphs or visualization regarding SPDMs from the China National Knowledge Infrastructure (CNKI) database search results, while qualitative research has been conducted on relevant literature in the Web of Science (WOS) database regarding sustainable design and relevant systems. Although some scholars have conducted bibliometric analyses and literature reviews considering WOS-based data samples on sustainable product design and production systems [8,9], the latest analysis of sustainable product design tools was performed in 2017 [10], failing to present the overall progress in SPDM research in China and around the world. Thirdly, existing review research has mainly focused on the impact of SPDMs on the environment, with little comprehensive analysis of the current status, focused topics, and dynamic evolving trends in design methods that influence environmental, socio-economic, and cultural sustainability.

In response to the above-highlighted challenges, this paper adopts a bibliometric analysis approach to visually analyze the SPDM literature from the WOS and CNKI databases over the past 24 years (1999–2022). This approach systematically explores the current status, focused topics, and research trends in China and around the world on product design methods that respond to multiple sustainability challenges. Moreover, this analysis identifies differences in research between China and other regions, providing a reference for future theoretical and practical research in sustainable design.

2. Methods and Data Sources

We used CiteSpace, a type of visualization software developed by Professor Chen Chaomei at Drexel University, to carry out a metric analysis of the scientific literature and visualize the knowledge structures [11]. The collaboration network and co-occurrence network modules in CiteSpace 5.8. and R3 were employed to conduct a visual metric analysis of global research on SPDMs. By creating the maps of collaboration networks of authors, institutes, and countries and the co-occurrence, burstness, clustering, and timeline evolution maps of keywords, the research content, movement, and focused topics in this field were comprehensively and directly presented, and relevant research progress was sorted out, which will lay a foundation of the literature and provide a methodological reference for research and practice in this field.

The Chinese literature used in this paper was sampled from CNKI, with three categories of keywords, namely, (product OR product design) and (method OR methodology OR tool OR approach OR theory OR framework OR guideline) and (sustainable OR sustainability OR green OR ecological) as the query pattern. Moreover, the Chinese literature sampled was mainly published between 1 January 1999 and 18 July 2022 in the Chinese Social Sciences Citation Index (CSSCI), Chinese Science Citation Database (CSCD), EI, and A Guided to the Core Journals of China. The literature irrelevant to this research was excluded by the following standards: (a) product design works; (b) literature with incomplete information and paper solicitation information of periodicals and magazines; (c) papers and interview articles; and (d) articles irrelevant to this research (including articles containing sustainable product design information but not design methods). Finally, a total of 531 papers was adopted. The reference in WOS was mainly sourced from the core datasets of WOS, and the time range for literature sampling was the same as that for the Chinese literature. Additionally, the language of the articles and reviews to be sampled was English, and the theme of the query was "sustainable product design method". It was ensured that the inclusion and exclusion criteria were consistent across the two databases. After the literature data were cleaned and deduplicated in CiteSpace, a total of 829 pieces of literature was sampled. This process is summarized and illustrated in an adapted PRISMA flow diagram for systematic reviews (see Figure 1).

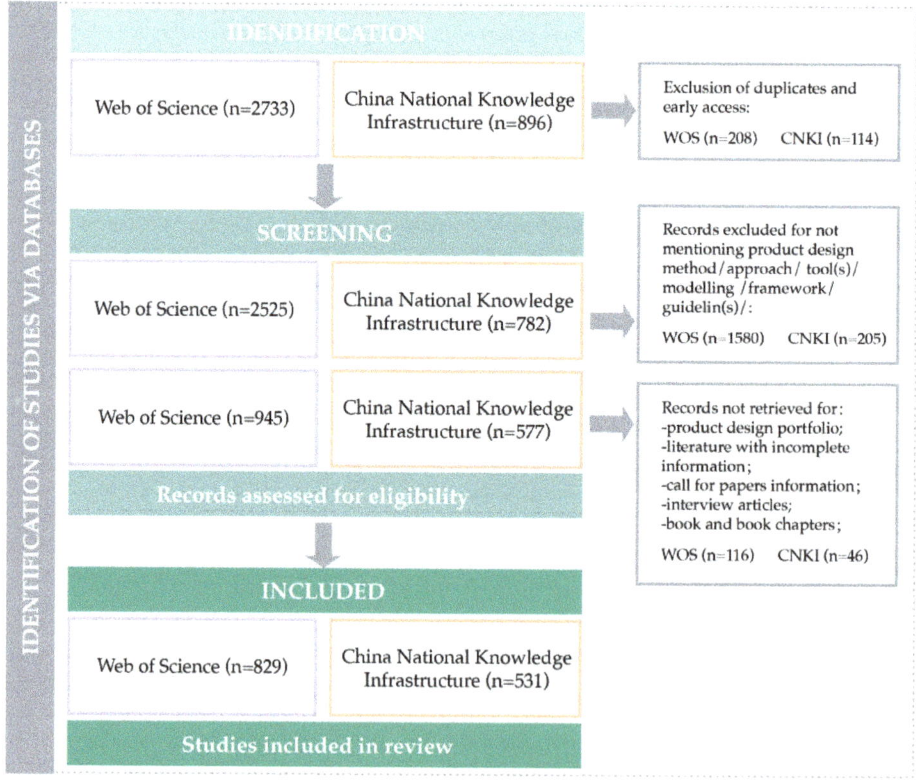

Figure 1. PRISMA flow diagram for systematic reviews.

3. Bibliometric Analysis

3.1. Analysis of Publication

3.1.1. Analysis of the Annual Number of Papers Published

A statistical graph of the number of papers published can present the dynamic evolving process of research in this field. According to the distribution and linear prediction of the publication year of the literature on SPDMs from WOS and CNKI (Figure 2), the annual number of both China and the rest of the word published in these years has leapt forward and can be divided into three phases. In the first phase, the number of annual publications in the two databases is similar between 1999 and 2007, averaging approximately eight every year, which is a preliminary exploration into this field. Then, between 2008 and 2014, this field witnessed rapid development and promotion. Accordingly, the annual number of worldwide publications increased dramatically, with the increase of the number of published papers in the WOS database being significantly higher than that of the CNKI. Statistically, the average annual number of CNKI and WOS paper s published during this period reached 21 and 32, respectively. Nevertheless, the annual number of papers published in this phase fluctuated significantly. In 2009, Ezio Manzini at the Polytechnic University of Milan established the "Design for Social Innovation and Sustainability Network (DESIS Network)" [12], which drives international cooperation and global promotion concerning research in this field. In the third phase, driven by the 17 sustainable development goals of the UN since 2015, the number of publications in these two databases has increased rapidly. Moreover, the number of WOS publications peaked in 2020, reaching 82, which surpassed the entire number of publications in these two databases in the first phase. Concurrently, the number of articles published gained constant and rapid growth in China. In December 2016, the State Council introduced the *Program for the Promotion of Extended Producer Responsibility System*, which specified that "the producer responsibility in resource consumption and environmental protection should be extended to product design [13]". Since then, the number of publications has soared in the CNKI database. In 2021, 66 references were published in CNKI, approximately eight times as many as the average annual number of papers published during the first phase. Overall, the research in this field both in China and the rest of the world is rich in content and plentiful in achievements, showing a booming trend.

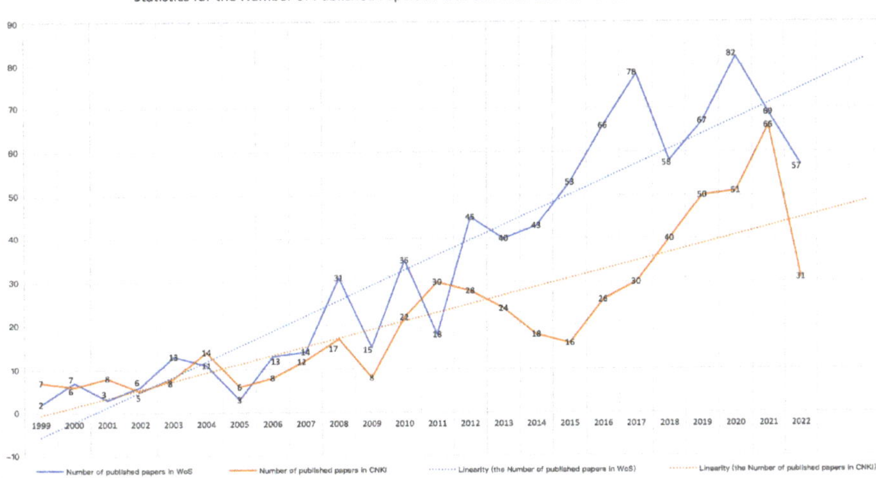

Figure 2. Distribution and linear prediction of the publication year of literature on SPDMs from WOS and CNKI.

3.1.2. Analysis of Author Collaboration Network Map

The author co-occurrence module in CiteSpace was used to create the author collaboration network maps (Figure 3) of these two databases, respectively. On the map, the nodes represent the number of published papers of authors, and the lines connecting the nodes imply the collaboration among authors. Moreover, the thickness of the lines means the degree of the collaboration, and the color indicates the time of the collaboration.

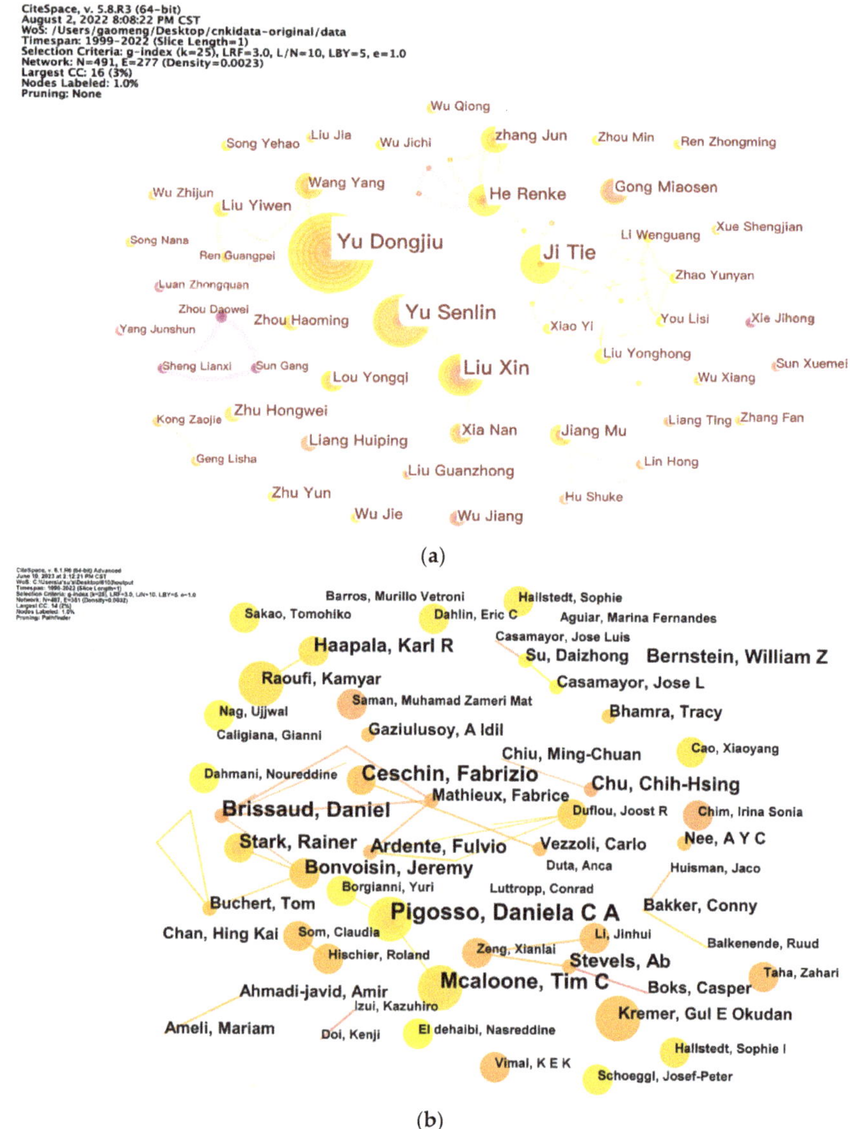

Figure 3. (**a**) Network map of collaboration among authors in CNKI. (**b**) Network map of collaboration among authors in WOS.

Among the collaboration networks of Chinese scholars, five were quite distinct (Figure 3a). The largest author base formed through the connection between the new collaboration network with Ji Tie at the core and the sub-network led by He Renke; the author base consisting of Zhou Daowei, Sun Gang, and Sheng Lianxi, who started their research earlier; the author base led by Yu Dongjiu, whose node is the biggest; and the author base consisting of Yu Senlin, Liu Xin, and Xia Nan, whose nodes are connected by only one line, indicating that though they have had quite a few published papers, the collaboration network is not formed, and the collaboration among them is not deep. There are many separate units on the map that are not connected by networks, implying that cross-unit, multidisciplinary collaboration is required for Chinese research in this sector. Scholars from the WOS database outperform CNKI scholars in terms of the number of collaboration networks. Moreover, the collaboration networks of scholars in WOS are distributed rather than centralized (Figure 3b). The author base led by Brissaud Daniel and Fabrizio Ceschin connecting nine nodes is the largest, followed by the author network with Stevels Ab at the core, and then the collaboration networks led by McAloone Tim C. and Pigosso Daniela, respectively. Except for the networks mentioned above, networks on the map connect mainly two or three nodes, indicating that despite the deep intra-base collaboration, the linkage among networks remains to be strengthened and that no large, stable author base has been formed.

Further investigation reveals that there are five scholars who have more than six publications, namely, Yu Dongjiu (15), Yu Senlin (10), Liu Xin (8), Karl Haapala (8), McAloone Tim C. (8), Pigosso Daniela (7), and Daniel Brissaud (7). Yu Dongjiu explored the application of strategies and methods for social value-oriented sustainable innovation [14] and design to the design of elderly-oriented [15] and children-oriented products [16]. Yu Senlin proposed that the focus of the methods for the innovation of sustainable product design should be extended from production to consumption, from physical products to product-service systems, and from material culture to spiritual culture [17]. Liu Xin emphasized the understanding of the sustainable design for products and services from a systematic perspective [18] and proposed creating a sustainable design assessment system based on Chinese conditions [19,20] and developing teaching tools for sustainable design of the product–service system. Karl Haapala focused on the relationship between product design and sustainable manufacturing [21] and developed a method for assessing the environmental factors behind the product design and manufacturing based on the life cycle approach [22,23] as well as a modeling method [24] that would facilitate decision-making regarding sustainable design. McAloone Tim C. and Pigosso Daniela have a tight close collaborative relationship regarding the development and implementation of multiple ecodesign tools and methods [25,26] to measure sustainability performance in product development [27], such as the ecodesign maturity model [28,29], guidelines for evaluating the environmental performance through life cycle assessment [30], and a generic process model for the early stages [31]. Daniel Brissaud, from the perspective of remanufacturing, brought forward the sustainable design method and practice [32] for redesigning products and developing new products, the product cycling strategy [33], and the multi-criteria assessment method [34] that take into consideration technological, economic, environmental, business, and social factors, and the method for the transformation [35] toward and creation [36] of a design value-driven and sustainable industrial product–service system. By Price's Law, authors with at least three papers published in this field can be defined as core authors, and when the total number of published papers of core authors represents 50% of the total number of papers in this field [37], a core author base is formed. According to the available statistics, there are 17 core authors in the CNKI database and 46 core researchers in the WOS database, and the total number of their published papers represents 17.1% (91) and 21.4% (177) of the total, respectively, neither of which has reached 50% of the total number of sampled papers published in this field. Therefore, no core author base has been formed in this field. Moreover, compared to Chinese authors, researchers contributing to WOS are more willing to carry out research through collaboration. Overall, worldwide research on this issue demonstrates the characteristics of a large number of researchers, scattered cooperative relations, and a lack of connectivity.

3.1.3. Analysis of the Academic Influence of Institutes with Papers Published

The Institution module in CiteSpace was used to illustrate the institutions that contributed to the research, and the institutions with no fewer than three published papers on this research are marked (with the threshold value being three) in Figure 4. Moreover, the top 10 institutes in China and around the world by the number of published collaborations are summarized (see Table 1). According to Table 1, the top 10 research institutes in China and around the world are dominated by universities, indicating that universities are the mainstay of global research on SPDMs. Nonetheless, the centrality values of universities are zero, which means no core team has been formed in this field worldwide.

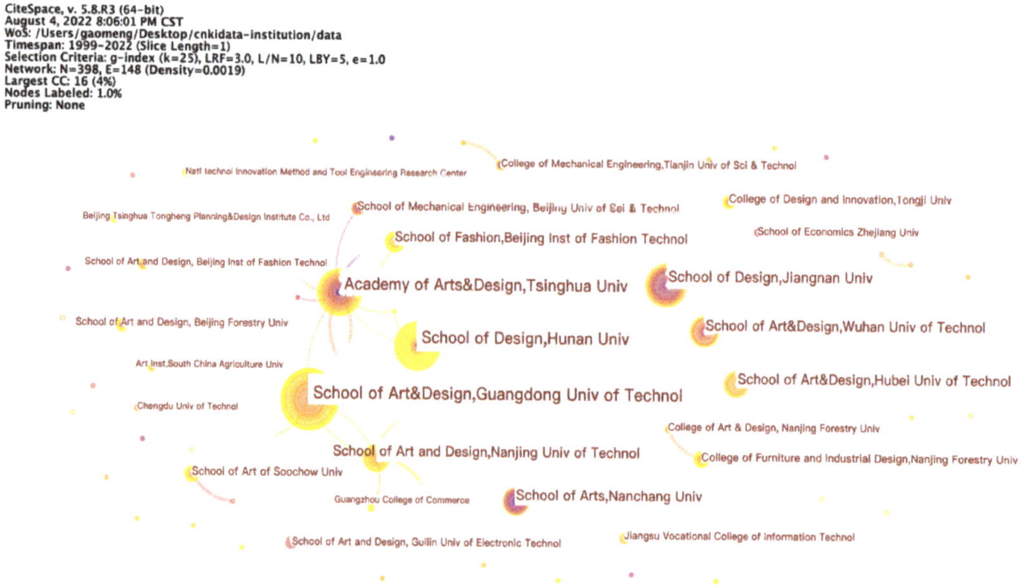

(**a**)

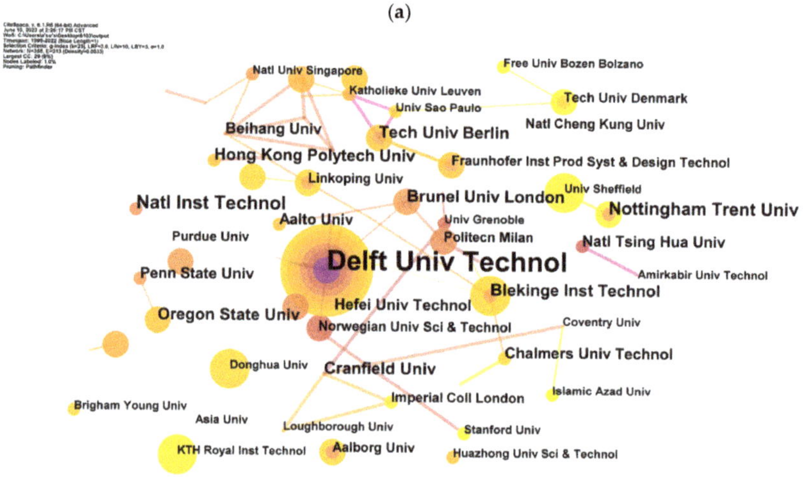

(**b**)

Figure 4. (**a**) Network map of collaboration among institutes with published papers in CNKI. (**b**) Network map of collaboration among institutes with published papers in WOS.

Table 1. Summaries of the top 10 global institutes by the number of published papers.

a. CNKI Database				
Number	Institute	Number of Published Papers	Centrality	Publication Year of the First Paper
1	School of Art and Design, Guangdong University of Technology	21	0	2014
2	Academy of Arts and Design, Tsinghua University	17	0	2003
3	School of Design, Hunan University	16	0	2004
4	School of Design, Jiangnan University	15	0	2002
5	School of Art and Design, Wuhan University of Technology	11	0	2008
6	School of Art and Design, Nanjing University of Technology	10	0	2011
7	School of Arts, Nanchang University	10	0	2001
8	School of Art and Design, Hubei University of Technology	9	0	2016
9	School of Fashion, Beijing Institute of Fashion Technology	7	0	2019
10	College of Furniture and Industrial Design, Nanjing Forestry University	6	0	2016
b. WOS Database				
Number	Institute	Number of Published Papers	Centrality	Publication Year of the First Paper
1	Delft University of Technology (the Netherlands)	28	0.01	2001
2	Oregon State University (the US)	13	0	2012
3	Polytechnic University of Milan (Italy)	10	0	2010
4	Technical University of Berlin (Germany)	9	0	2008
5	Pennsylvania State University (the US)	8	0	2012
6	Blekinge Institute of Technology (Sweden)	8	0.01	2017
7	Hefei University of Technology (China)	7	0	2004
8	Brunel University London (Britain)	7	0	2014
9	National Cheng Kung University (Taiwan, China)	6	0	2003
10	Imperial College London (Britain)	6	0	2017

In terms of the number of published papers in Chinese institutes, as shown in Table 1 (a), the School of Art and Design, Guangdong University of Technology tops the table with 21 published papers. The school has designed green design-oriented curricula for undergraduates, such as Sustainable Design Overview, Methods for Sustainable Innovation, and Upcycling-Based Low-Carbon Design. The Academy of Arts and Design, Tsinghua University; the School of Design, Hunan University; the School of Design, Jiangnan University; and School of Art and Design, Wuhan University of Technology rank second, third, fourth, and fifth in the table, respectively. They have jointly established a core team, which developed the Learning Network on Sustainability-China (LeNS-China) in 2011. So far, 19 Chinese universities and research institutes have joined LeNS-China [38] to increase local awareness, information and resource exchange, and practical engagement around sustainable design.

Figure 4a is a map of the distribution of and collaboration among Chinese research institutes with 398 nodes, 148 lines, and a network density of 0.0019. According to the map, a total of 398 Chinese organizations conducted research on SPDMs between 1999 and 2022, during which the institutions collaborated with each other 148 times. Nevertheless, the collaboration network map also indicates that no clustering phenomenon has been formed in the Chinese sustainable product design field. Although a large inter-university collaboration network led by the School of Art and Design, Guangdong University of Technology; the Academy of Arts and Design, Tsinghua University; the School of Design, Hunan University;

and the School of Art and Design, Nanjing University of Technology has been created to explore the methods for designing a sustainable product-service system, develop relevant tools, and disseminate relevant knowledge, it has only a few light-colored lines, implying that the universities have not carried out frequent, long-term, and in-depth collaboration. Moreover, the Academy of Arts and Design, Tsinghua University; the School of Fashion, Beijing Institute of Fashion Technology; and the School of Mechanical Engineering, Beijing University of Science and Technology have jointly established a small, regional network and proposed strategies for sustainable design by learning from nature [39] and sustainable design approaches for culturally innovative products [40]. Moreover, an intra-university collaboration has been formed between the College of Furniture and Industrial Design and the College of Art and Design, Nanjing Forestry University, to propose a coupled approach for furniture and product packaging design based on sustainable use [41]. In addition to the above-mentioned collaboration networks, few networks have been created among other institutes, so the institutes are not obviously linked to each other. Moreover, as research is dominated by universities, there is much room for improving academic collaboration and exchange among research institutes in this field.

In terms of the number of published works of institutes from the WOS database, as shown in Table 1 (b), the Delft University of Technology in the Netherlands leads its peers with 19 published papers. Moreover, the Faculty of Industrial Design Engineering at Delft has made sustainability its top three design research topics. As a classic in the design research field, the *Delft Design Guide: Design Strategies and Methods*, published in 2013, has systematically clarified the design methods for products and other design-related fields. Oregon State University comes second in the table with 13 papers published, most of which are contributed by its College of Engineering and focus on the development of quantitative and modeling methods for sustainable product design and manufacturing. Polytechnic University of Milan in Italy ranks third with ten published papers, all from its School of Design. The Lab of Sustainable Design and Systems Innovation (Polimi-DiS) in 2002, the Learning Network on Sustainability (LeNS) in 2007, and the international Learning Network on Sustainability (LeNSin) [42] in 2015, which were all launched by the School of Design, have led 155 universities and institutes in Europe, Asia, Africa, and South America to participate in sustainable design-related research and practice, delivering a range of cases, methods, and tools about sustainable product design. Among the top 10 institutes in WOS by the number of published papers, 2 are from the US; 2 are from Britain; 1 each are from the Netherlands, Italy, Germany, and Sweden; and the remaining 2 are from China, namely, the Hefei University of Technology and the National Cheng Kung University, with seven and six published papers, respectively. Moreover, only the research carried out by the two Chinese universities was later than that by the Delft University of Technology.

On the collaboration network map of research institutes in WOS database (Figure 4b), there is a total of 411 institutes and 202 lines, and the network density is 0.0024. Among the networks, five are quite distinct. Specifically, there is a European university collaboration group consisting of only universities, including the Delft University of Technology, Polytechnic University of Milan, Blekinge Institute of Technology in Sweden, Brunel University London, Aalto University in Finland, and the Chalmers University of Technology in Sweden. Moreover, Imperial College London, Loughborough University, Cranfield University, and Coventry University have formed a British university collaboration network. Additionally, American universities, including Pennsylvania State University and Oregon State University, have jointly built a US collaboration group. There are another two research institute–university collaboration networks. One consists of the University of California, Berkeley; the National Institute of Standards and Technology; and the Indian Institute of Science; and the other comprises the Technical University of Berlin in Germany, KU Leuven in Belgium, and Fraunhofer Institute for Production Systems and Design Technology in Germany. Except for the above-mentioned collaboration networks, there are 12 distributed small collaboration teams, displaying the "overall dispersion and small concentration" feature that characterizes the institutes contributing to WOS that have been exploring SPDMs.

3.1.4. Analysis of the Distribution of Institutes in WOS by Country and Region and of Collaboration Networks

In the country co-occurrence map (Figure 5), there are 66 countries and regions, and they have collaborated 189 times. Moreover, the network density is 0.0881. Specifically, 14 countries and regions have at least 20 published papers. Table 2 is a ranking of the top 10 countries by the number of published papers, in which China (156 published papers) ranks first, followed by the US (124), Britain (80), and then Germany (57) and Taiwan, China (51). Although China has the most published papers, it comes fifth in terms of betweenness centrality (0.13). This indicates that even though China is internationally active in SPDMs, its academic influence remains to be improved. The US (0.53) and Britain (0.24) feature high centrality, indicating that their international academic influence is significant, so they serve as a medium and propel this field forward.

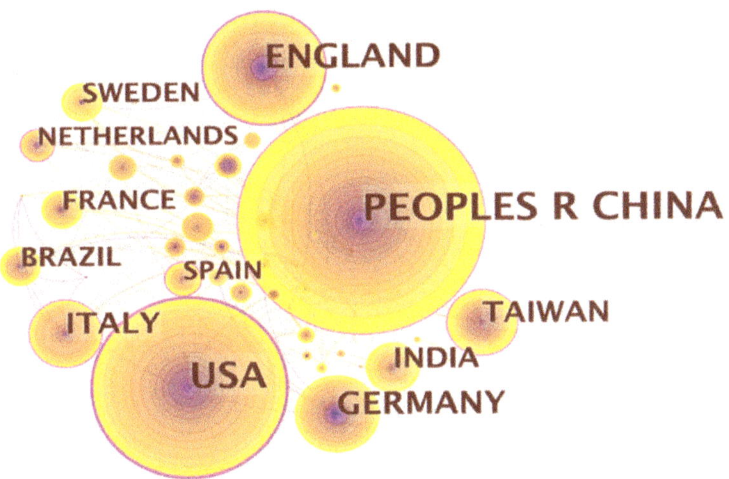

Figure 5. Network map of collaboration among countries in WOS.

Table 2. Summary of top 10 countries in WOS by the number of published papers.

Number	Country	Number of Published Papers	Centrality	Publication Year of the First Paper
1	China	156	0.13	2000
2	The US	124	0.53	1999
3	Britain	80	0.24	2000
4	Germany	57	0.09	1999
5	Italy	48	0.17	2003
6	Taiwan (China)	46	0.15	2008
7	India	36	0.09	2010
8	Brazil	30	0.01	2006
9	Sweden	30	0.06	2003
10	France	29	0.07	2007

3.2. Analysis of Research Trends and Frontiers

3.2.1. Analysis of Research Evolution Process

The temporal knowledge evolution process on SPDMs can be described as three-phase by the timeline map generated in CiteSpace's Keyword Co-occurrence Module. The horizontal axis of Figure 6 is the timeline, which is divided into multiple 1-year slices. Each slice indicates the keywords of the year, and the size of the keyword node is in direct proportion to each keyword's appearance frequency. However, the importance of the nodes in the research depends on the centrality. The higher the centrality value, the more important the nodes are. Keywords with high centrality values of no less than 0.1 represent the focused research issues in the field to some extent [43]. After the three subject terms for retrieval are excluded, in the CNKI database, "TRIZ (theory of inventive problem solving) (0.35)" is the supporting point of the keyword network and the main focused research issue concerning SPDMs, followed by "life cycle (0.34)", "innovative design (0.23)", "mechanical product (0.19)", "sharing economy (0.18)", "redesign (0.12)", "green manufacturing (0.11)", and "QFDE (0.11)", which play significant liaison and transition roles in the entire network. Specifically, keywords including "green design", "green product", and "environment" are the focused research issues appearing in the early stage, and the focus of research in recent years has shifted to "sustainable", "sharing economy", and "service design". Over the years, the focused research issues have changed significantly. The focused issues in the WOS database emphasize factors such as "environmental impact (0.19)", "barrier (0.17)", "consumption (0.16)", and "cost (0.15)" in the process of design of sustainable products, as well as the "integration (0.13)" methods employed, design "framework (0.12)" supporting sustainable strategies, the "QFD (0.11)" method, and the "LCA (0.10)" method. "Environment" is the earliest, followed by "ecodesign tool", and then "strategy", "sustainability", and "implementation".

The first phase, spanning from 1999 to 2007, is foundational for Chinese research on SPDMs. In this phase, keywords such as green design, product design, product packaging, green manufacturing, mechanical products, TRIZ, life cycle, and material selection were spawned. Research in the product design field back then mainly adhered to the green design methods and green manufacturing philosophy and focused on the R&D of green mechanical products, green mechatronic products, and green industrial products. Moreover, Chinese research between 1999 and 2007 paid attention to the environmental impacts and evaluation of the materials, structures, and functions of products, which has been a fundamental issue, namely, how to understand and cope with the influences of product design on the environment, facing the product design field since the theory of sustainable development was introduced into this field in 1987 [1].

Moreover, a review of the results of the WOS database shows that other regions of the world are also in their infancy in terms of academic research in this sector. It mainly concentrated on the analysis of environmental impacts, product performance, green product management, life cycle, and the efficiency and feasibility of product dismantling. Moreover, QFD and AHP began to be used in sustainable product design, and exploration into the tools used to support design decisions commenced.

The second phase, which lasted from 2008 to 2014, witnessed the rapid development of research interest in this field in China. In this phase, keywords such as service design, sharing economy, emotional experience, bionic design, and forms emerged. In terms of design objects, scholars paid more attention to the application of sustainable design methods in the design of shared products, furniture, children-oriented furniture, and tourism products. In terms of methodological innovation, Chinese scholars started their research from the perspectives of emotional factors, user experience, bionic appearance [44], and user-friendliness. On the other, they combined service design methods and SPDMs and explored sustainable product-service design. Nevertheless, the advancement of the rest of the world in this regard during this period was explosive. Specifically, high-frequency keywords constantly emerged and were densely connected, and focused research issues such as the LCA of products, product design, system, model, design framework, ecodesign, optimization, design decision, integration, material selection, and innovation kept emerging, deepening the research and broadening the themes.

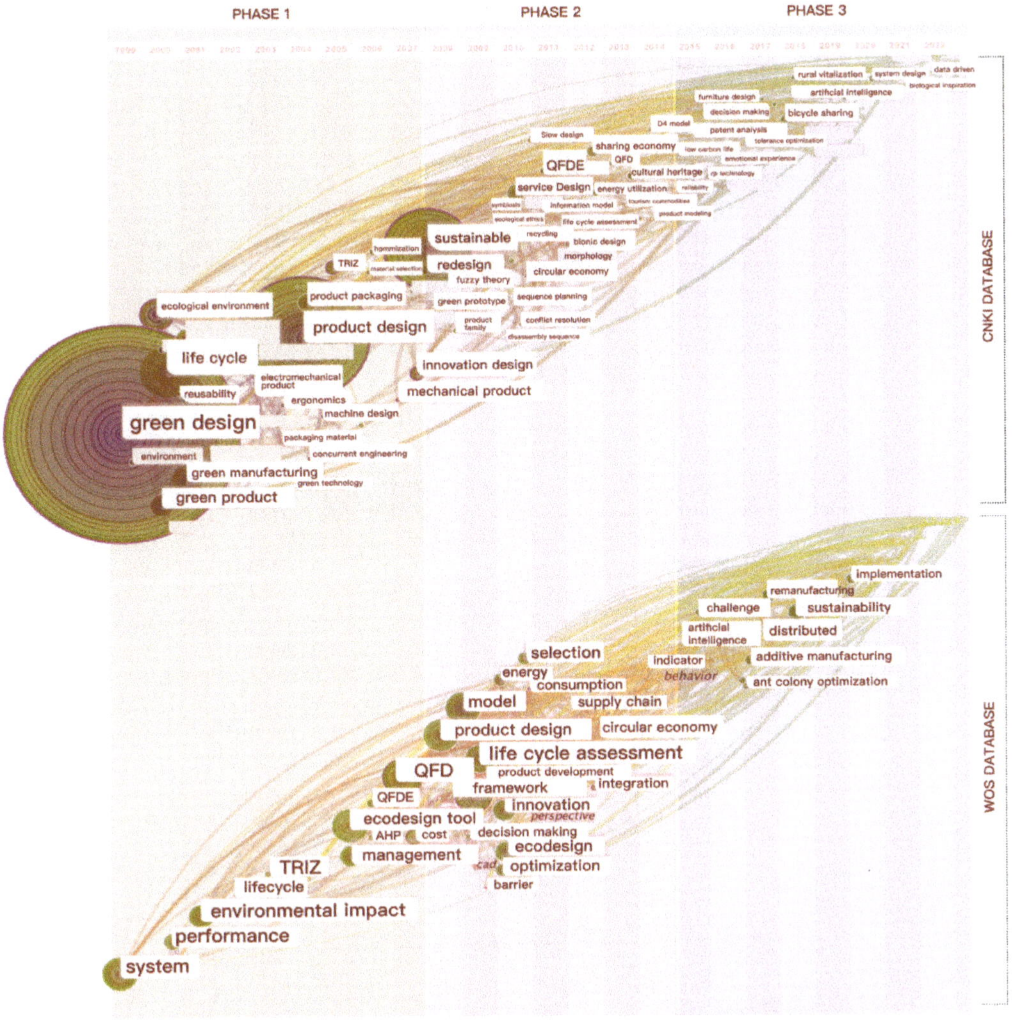

Figure 6. Keyword timeline map of research on SPDMs in CNKI and WOS databases.

Drawing on the cost–benefit analysis method after 2007, scholars developed the LCC method, improving the role of economic factors in SPDMs. Moreover, driven by the guide to the evaluation of social methods formulated by the UN Environment Program in 2009, keywords from both CNKI and WOS in the second phase demonstrated the rapid development and application of SPDMs in the economic and social dimensions. The design optimization of energy-saving products, shared products, and small household equipment, which are expected to achieve economic and social sustainability benefits in low and middle-income areas, was widely discussed.

The third phase, since 2015, is the booming period of Chinese research on SPDMs, during which the number of articles published gradually recovered. In this phase, the emerging research subjects in the second phase were explored in depth and thoroughly. China has been prioritizing the role of cultural confidence in developing a strong socialist culture since the 18th National Congress of the Communist of China. Accordingly, research on SPDMs laid great emphasis on carrying forward intangible cultural heritage and designing cultural and creative products. How to better incorporate cultural elements into sustainable product innovation became one of the principal research issues in this phase, and culture was introduced into SPDMs. Moreover, keywords including rural revitalization, system design, AI, data-driven, and bio-inspired emerged.

Research outputs available in WOS after 2015 appeared in the form of small and numerous nodes, and the number of publications rose significantly. This indicated that research in this phase was scattered and diversified and was in the phase of stable and deepening development. In this period, new, targeted hot topics such as circular economy, distributed, sustainable indicator, innovation tools, AM, AI, remanufacturing, and ant colony optimization (ACO) were spawned, in addition to the in-depth practice and research on the research themes that emerged in the second phase. The construction of sustainability standards gradually improved in more subdivided categories of products in the innovation of sustainable products, and more and more health and medical products, wearable products, personalized devices, smart products, and complex products were studied to maintain both environmental and economic and social sustainability.

3.2.2. Analysis of Method Clustering and Research Trends

To systematize the identified methods and enable their analysis, a classification framework including five main categories was iteratively developed through previous literature reviews, and especially the additional categories of purpose were derived from an inductive content analysis. Figure 7 shows how the sub-classification dimensions are related to one another; the classification has a transversal emphasis that does not typecast procedures in only one category, and it offers a spectrum of possibilities. The cultural sustainability dimension is identified separately based on the literature content analysis of the CNKI and WOS databases, on the basis of three pillars of sustainability, to further clearly understand the development of different attitudes and design methods of Chinese and foreign scholars in achieving the goal of cultural sustainability development. Analyzing the development level and nature of data allows it to be evident how the method is employed, as well as to identify the purpose of the focus when proposing the method.

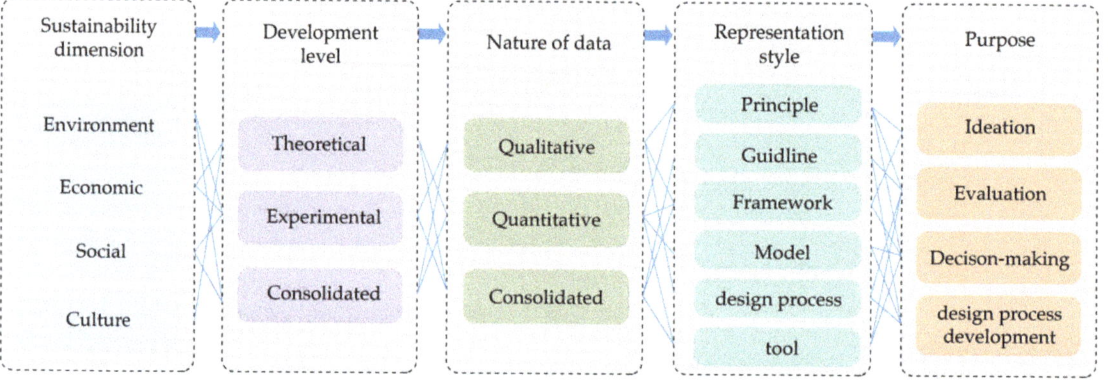

Figure 7. Framework for classifying data of methods (modified from Pigosso et al., 2011 [45], and Fernandes et al., 2020 [46]).

In particular, throughout the methodological analysis, the cases or specific product concerned are highlighted, which may enhance and optimize future design practice. The methods used in the articles were systematically reviewed according to the co-citation results of the literature and the keyword frequency and centrality. There are 79 representative methods from these two databases in total, which are listed and specified in Table 3.

Table 3. List of 79 representative sustainable product design methods based on the literature found in China National Knowledge Infrastructure and Web of Science databases between 1999 and 2022.

In the column groups below, a filled colour cell is marked ●. Development Level: Theoretical / Experimental / Consolidated. Nature of Data: Qualitative / Quantitative / Consolidated. Purpose: Ideation / Evaluation / Decision-Making / Design Process Development.

Sustainability Dimension	Representation Style	Methods	Theoretical	Experimental	Consolidated	Qualitative	Quantitative	Consolidated	Case/Example	Ideation	Evaluation	Decision-Making	Design Process Development	Time	Reference
Environment	principle	Principles and procedures for the eco-design							energy-using products					2009	[47]
		Ecological Packaging principle							energy-using products packaing					2013	[48]
		Ten Golden Rules			●				none		●			2006	[49]
	guideline	Environmentally conscious guidelines by combining reverse engineering with LCA			●			●	electric kettles				●	2010	[50]
		LCA guideline							digital products					2014	[51]
		bionic design method							sprinkling can					2014	[44]
		TRIZ-Based Guidelines for Eco-Improvement			●			●	mechanical ball		●			2020	[52]
	framework	Sustainable design guidelines for additive manufacturing applications			●			●	none					2022	[53]
		redesign method Based on Ecological Ethics			●			●	industrial product					2010	[54]
		An Integrated Approach for Eco-Design			●			●	LED Lighting		●	●	●	2020	[55]
		integration of ECQFD and LCA approaches			●			●	electronics switches			●	●	2010	[53]
		Ecodesign maturity model			●				eco-product			●	●	2013	[28]
	model	Resource Efficiency Assessment of Products method						●	liquid cristal display television		●			2014	[56]
		model of factors affecting environmental sustainability performance of PSS							ofo sharing bicycles		●			2019	[57]
		Green Product Optimization model Based on QFDE		●	●			●	mechatronic products/hobbing machine			●		2019	[58]
		F-MCDM		●	●		●		hot runner systems			●		2021	[59]
		life cycle impact assessment by openLCA		●			●		disposable face mask			●		2022	[60]
		Fuzzy AHP and modularity analysis Method						●	discrete electromechanical product			●		2000	[61]
	tool	comprehensive and simplified indexes developed based on a life cycle approach			●		●		energy-using products/washing machine		●	●		2014	[62]
		decision tables of green design knowledges by rough sets			●				mechanical product			●		2019	[63]
		PEP indicator			●		●		smartphones			●		2020	[64]
Environment-Economic	principle	TRIZ and case-based reasoning principles			●				dishwasher			●		2012	[65]
		design guidelines to meet the circular economy principles			●			●	small household electronic equipment			●		2018	[66]
		principles of 5R Based on the Fractal Structure							packaging					2018	[67]
		9R principles							none					2019	[68]
	guideline	guidelines based on Concept-Knowledge design theory supporting			●				automotive		●	●		2009	[69]
		design strategies for product life extension			●			●	electric and electronic products		●	●		2014	[70]
		ETRIZ matrix that compiled existing guidelines for remanufacturing and AM			●		●	●	remanufacturing product		●	●		2021	[71]
	framework	A framework of integration of LCA and LCC						●	diesel engine		●	●		2008	[72]
		a conceptual framework to assess energy consumption						●	mechanical part		●	●		2013	[73]
		A design framework for sustainable PSS customization							elevator				●	2017	[74]
		systematic methodology for data-driven product family green redesign							product family		●			2020	[75]
		7 Rs sustainable packaging framework			●				packaging					2022	[76]
		novel design methods of "Design for Energy Minimisation" approach			●				simple part/an elbow pipe		●	●		2010	[77]
		a mathematic energy model based systematic approach			●		●		energy-saving product/commercial blender		●	●		2010	[78]
	model	QFDE and Modular design			●			●	indoor air conditioner		●			2013	[79]
		innovative model Integration of ECQFD, TRIZ, and AHP			●			●	automotive parts			●		2014	[80]
		Integration of green quality function deployment and fuzzy theory			●			●	green mobile phone					2015	[81]
		a comprehensive evaluation model of LCC and environmental impacts based AHP			●			●	automotive door		●			2015	[82]

Table 3. *Cont.*

Sustainability Dimension	Representation Style	Methods	Development Level			Nature of Data			Case/Example	Purpose				Time	Reference
			Theoretical	Experimental	Consolidated	Qualitative	Quantitative	Consolidated		Ideation	Evaluation	Decision-Making	Design Process Development		
Environment-Economic		a hybrid optimizing method named chaos quantum group leader algorithm			●		●		drive device		●	●		2016	[83]
		an energy flow modelling approach based on Characteristics-Properties Modelling			●			●	hair dryer		●		●	2016	[84]
		QFD and Case-based Reasoning	●				●		air conditioner			●		2018	[85]
		AHP and Grey Relational Analysis Approaches			●		●		none			●		2019	[86]
		energy-aware digital twin model	●				●		energy-saving product			●		2020	[87]
		LCA and modular design							stroller					2020	[88]
	design process	a user requirements-oriented method integrated fusing Kano, QFD and FAST			●			●	baby stroller			●		2020	[89]
		process of sustainable product design based on QFDE			●				water purifier				●	2017	[90]
		design process based on TRIZ and lca analysis			●				furniture				●	2018	[91]
	tool	remanufacturable product profiles			●			●	remanufacturable product		●	●		2008	[32]
		Product design scenarios for ESFPs			●				energy-saving fashion products				●	2013	[92]
		concept circularity evaluation tool (CCET)			●			●	furniture/trampolines/ergonomic mobility aids/plastic components		●	●	●	2020	[93]
	principle	Design Principles of Resource Recycling Concept	●						sharing bicycles					2020	[94]
		Sustainable Design Strategy for Creative Products in Cultural Consumption			●				cultural and creative products					2022	[95]
		comprehensive scenario of sustainable dimensions	●						packaging					2015	[96]
	guidline	new classification system of PSS applied to Distributed Renewable Energy	●						energy product-service system		●			2016	[97]
		Evaluation Criteria of Design for Sustainability			●			●	sustainable living Lab		●			2017	[20]
		desgin theory base on Ecological Philosophical	●						furniture					2017	[98]
		guidelines for evaluating the environmental performance through life cycle assessment			●			●	bike-sharing system/lawnmower PSS/hull cleaning PSS		●	●		2018	[30]
Environment-Economic-Social		Design directions to the empowerment of end users to become co-providers	●						smart energy system					2013	[99]
		system design method based on semiotics	●						furniture					2013	[100]
	framework	A framework of a life-cycle focused sustainable new product development	●						none		●		●	2015	[101]
		a comprehensive set of process-related key performance indicators						●	none		●	●		2016	[27]
		4D system of product sustainable design	●						none					2020	[102]
		PSS model analysis		●					children's learning desk					2021	[103]
		framework Data-driven sustainable design							none				●	2021	[104]
		A trade-off navigation framework			●			●	furniture		●	●		2021	[105]
		fuzzy inference approach for evaluating sustainability			●				electrical power generation		●			2013	[106]
	model	a sustainable platform based grey relational analysis/ bayesian network/ Fuzzy			●				coffee makers		●			2017	[107]
		a Decision flow chart for bio-based products designed to be recirculated							bio-based products				●	2017	[108]
	design process	Additive manufacturing/3D printing							automotive components		●			2022	[109]
		design strategy based on LCA			●				intangible cultural heritage product				●	2020	[110]
		a generic process model							none				●	2023	[31]
	tool	Sustainable design-oriented product modularity combined with 6R concept			●			●	rotor laboratory bench		●	●		2014	[111]
		A Metrics-Based Methodology for Establishing Product Sustainability Index			●			●	electronics component		●	●		2014	[112]
		The Checklist for Sustainable Product development							automotive		●	●		2017	[113]
		a social impact checklist table						●	corrugated cardboard		●	●		2020	[114]
		an AHP-based ELECTRE I method			●			●	furniture		●	●		2021	[115]
		an Open-Source Tool for Social Impacts Assessment					●		none		●	●		2022	[116]
		product sustainability assessment tool (PSAT)			●			●	wind turbine generator		●	●		2023	[117]

In summary, sustainability in the environmental and economic dimensions shows the characteristics of easy measurement of quantitative data and mature application of methods, which have been widely implemented in a large number of practical cases. For example, a typical hair dryer is used to demonstrate the systematic approach via energy flow modelling to reduce epistemic uncertainty in the design process in the early stages of a design [84], lowering the managing complexity for designers. Duran et al., proposed a new sustainability index, namely, "potential embedded power (PEP)", determining the influences of product disposal on the product design life. By doing so, the environmental impacts of resource waste resulting from design decisions can be evaluated [64]. However, the impact of environmental sustainability is frequently associated with economic sustainability indicators, for instance, products with higher energy efficiency are better for the environment. Diverse design approaches help lessen the drawbacks of existing products' energy intensity and help achieve the objective of energy conservation and emission reduction (ESER). Ardente and Mathieux presented the Resource Efficiency Assessment of Products method for liquid crystal display televisions [56] and introduced a general index and a simplified index tested for the environmental assessment of durability of energy-using products, such as washing machines [62]. However, the contradictions and trade-offs between the impacts of environmental and economic sustainability will hinder designers from producing more sustainable solutions; certain design principles can help investigate how a product may be both economical and environmentally friendly through creative design, for example, guidelines based on Concept–Knowledge design theory [69]; enhanced TRIZ matrix [71] support solving the contradiction between environmental and economic factors; principles of sustainable energy-saving fashion products [92]; design strategies for product life extension of electric and electronic products [70]; and a conceptual framework to aid in the derivation of realistic energy consumption values from a product design perspective, allowing for an improvement of product manufacturing energy efficiency at both design and manufacturing stages [73]. Bovea and Pérez-Belis identified design guidelines for small household electrical and electronic equipment that allow for an improved product design from a circular economy perspective and convert it into a circular product design [66].

The analysis of social sustainability is mostly manifested by qualitative guidelines in the early stages of product development and design, such as eco-philosophical ideas [98], modular analysis [118], data-driven sustainable design frameworks [104], packaging design dimensions [96], principles of resource recycling design [94], and systematic frameworks for children's product [119]. The toolkit involved 15 archetypal models of PSS applied to distributed renewable energy [97] that were created to support innovative design in the energy sector while taking into account environmental, economic, and social benefits. Furthermore, methods covering quantitative and qualitative questionnaires have been gradually implemented recently, such as the checklist tool of PSAT [117], a checklist table involving eleven social impact categories [112] and tested in the automotive industry [113]. In general, social sustainability is linked to how the system benefits customers [107], with customer rights and satisfaction as essential measures that may have a beneficial impact on social sustainability. Three design directions are suggested in regard to the empowerment of end users to become co-providers as an addition to complement the ongoing development of products and services in smart grid deployment [99]. The product sustainability index (ProdSI) [114] was used to evaluate the quantitative sustainability indices of the environment, economy, and society at the product design and manufacturing stages. However, social sustainability is primarily measured by employee-related indicators such as working conditions, health, safety, training, and education [120], and applications measuring social sustainability performance in the context of various products and industrial manufacturing lack consistency [121]. For example, Kim selected the easily measurable indicator of toxic substances to represent worker health status in measuring the social sustainability of coffee machines [104], and Shin selected sedentary behavior and physical activity to analyze human-powered products [105]. Simultaneously, uncertainty of what data to utilize for social sustainability assessment and data quality [122] adds to the complexity of sustainable product design, evaluation, and decision-making. Although social sustainability indicators have been enriched and expanded in different fields and applications in recent years, such as green buildings [123] and food supply-chains [124], there is still a lack of clear definition of quantitative indicators that can be used to perform social sustainability assessment [125].

Achieving cultural sustainability is an important goal of product design. Among CNKI scholars, cultural sustainability is seen as a crucial fourth perspective for assessing product sustainability in order to support the sustainable design of intangible cultural heritage. Mou et al. analyzed the path to sustainable development of Chinese intangible cultural heritage with the life cycle method [110] and held that the incorporation of cultural factors was critical to improvements in the sustainability of intangible cultural heritage brands. Zhou proposed the 4D systematic view of sustainable product design [102], which consisted of load reduction design, persuasive design, fair design, and cultural and creative design. In recent years, China has seen a significant increase in research and practice on the cultural sustainability of product design as a response to how to shift from "going global culturally" to "cultural confidence" in the face of the contradiction between endemicity and globalization [126]. Design can serve as a vehicle and means of achieving cultural sustainability. Through design works, it can dynamically, specifically, and sustainably carry forward and internationally promote excellent traditional Chinese cultures and intangible cultural heritage, aiding in China's progress toward the establishment of a socialist culture. However, the literatures included in WOS demonstrate that the cultural sustainability dimension tends to be regarded as a minor subset of the social dimension in the field of sustainable product design [8,127]. Although Moalosi et al. proposed a culture-oriented product design method [128], they did not extend it to sustainable design methods. Ji and Lin proposed six design strategies to improve the emotional durability and lead consumer behavior toward more sustainable use of products [129], but the impact of the design practice guided by such design strategies on the ecological environment is difficult to estimate precisely. Through the analysis of sustainable design models discussed in the literature between 2000 and 2009, Rocha et al. pointed out that research on the social sustainability of sustainable design was far behind that of environmental and economic sustainability [130]. After analyzing the policies for material cultural heritage utilization, regional development of culture, and participation in art performances, Sabatini argued that culture could be fully deemed the fourth pillar of sustainable development [131], on which, however, researchers in the field of product design have not reached a consensus.

Moreover, the proposing trends can be represented visually and can be seen in Figure 8; it shows the precision and depth of the methodological development in the vertical dimension, suggesting a more accurate consolidated approach for inspiring, detailing, evaluating, and optimizing sustainable solutions. It also shows the integration and extension of the method in the horizontal dimension, signaling broader implementation and multilateral efforts between academics and industry. SPDMs are also trending toward an increasingly systematic, open-source, and sharing approach.

SPDMs are becoming more systematic, as evidenced by the expansion and deepening of sustainable design standards, such as the expansion of environmental indicators [10]; 5R [67], 6R [111], 7R [76], and 9R [68] sustainability principles; guidelines such as the top 10 golden rules [49]; and a more perfect product design process [50,132]. On the other hand, it emphasizes multi-method integration. TRIZ [65], LCA [72], LCC [82], QFD [85], QFDE [79,90], ECQFD [53], fuzzy inference [106], AHP [61], the data packet network, KANO, gray relational analysis (GRA), MCDM, F-MCDM [59], and other quantitative methods are increasingly used. They can be used to aid in design comparison, function and structural optimization, material evaluation, and decision-making. To attain more accurate environmental, economic, and social sustainable benefits, it is essential to systematically develop and employ many of the strategies mentioned above concurrently. The integration of the QFD-based method into the KANO model and function analysis system technique could bring a more sustainable stroller design [89]. Applying the combination of ECQFD,

TRIZ, and AHP in automotive parts design could lead to innovative and sustainable product design [80]. The incorporation of ECQFD into the fuzzy theory reduced the semantically subjective judgments on user demand at the product design stage [81], thereby making product design environmentally and economically sustainable. Considering environmental benefits and market value, Tan et al. used GRA and AHP [86] to optimize their decisions on the implementation of sustainable plans during the design of new products. Energy consumption is a key factor in the design of sustainable products [87], which necessitates the use of multiple strategies to calculate both environmental and economic benefits. The novel design methods of the "Design for Energy Minimization" approach [77], a systematic approach with an energy factor [78], and an energy-aware digital twin model [87] have been proposed to optimize energy-saving product design within entire product life cycles. Moreover, a hybrid optimizing method named chaos quantum group leader algorithm [83] is designed to obtain an optimal energy-consumption solution in designing products with various complexity, such as the drive device.

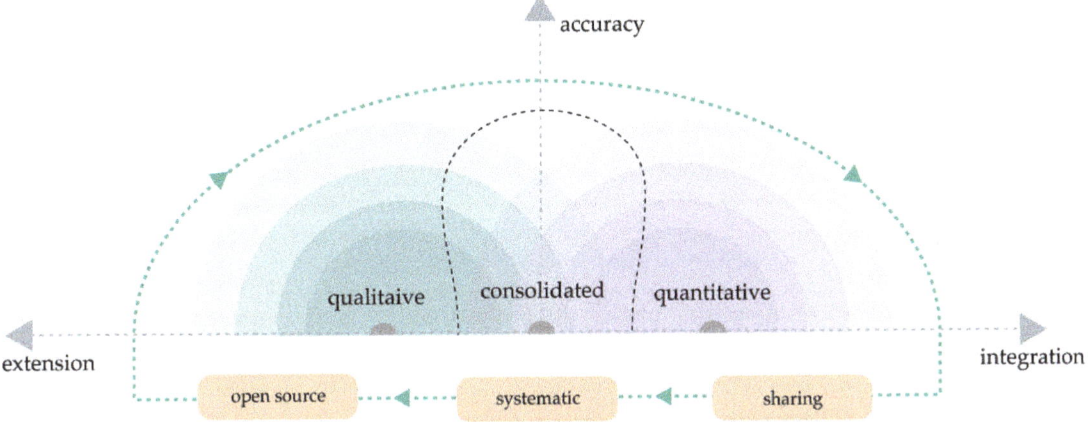

Figure 8. Graphic for proposing the trends of SPDMs.

Additionally, current research separates product sustainability evaluation and product design [5]. It is vital to develop holistic methods and tools that enable product design and sustainability evaluation simultaneously in an effort to produce more sustainable design concepts across the entire design stage. Covering a wide range and inconsistency of environmental, economic, and social sustainability assessment indicators is expected to be resolved by increasing accessibility and the sharing of methods and tools. Open data and open-source make it possible for anybody to freely duplicate research results, and the open-source philosophy holds that communities of study and practice should collaboratively construct and share tools rather than developing individual ad hoc scripts that produce incomparable indications [133]. For example, an open-source tool for social impact assessment [116] can be freely accessible to support open sharing with consistent data standards, allowing consistent measures to be produced and evaluated over time with little obstacles to participation. Simultaneously, collaboration between academics and industry is encouraged to promote the development and application of consistent indicators through multiple stakeholder participation.

In the hybrid design approach, the systematic framework that integrates multiple design tools and methods has positive significance in realizing interdisciplinary collaboration, developing holistic tools, and promoting the dialogue between designers and people from multiple industries. The systematic framework that incorporates various design tools and processes has an effective impact on realizing multidisciplinary collaboration, to generate holistic tools, and facilitate dialogue between designers and individuals from other industries. A customized design framework included ten various techniques and tools that were put to the test in elevator design [74]. Sherwood et al. presented a decision flow chart for bio-based products designed to be recirculated [108], promoting the rational circulation and utilization of biomass energy. The integrated product life cycle framework [101], comprehensive utilization of product life cycle management, LCA, social life cycle assessment (S-LCA), and LCC covering the environmental, economic, and social dimensions can realize more accurate and efficient design process management. Moreover, additive manufacturing (AM) has been increasingly leveraged to produce human-centered products in different fields to minimize material and energy spent to realize sustainability [134], such as sustainable automotive components [109], orthoses and prostheses, as well as therapeutic helmets, finger splints, and other personalized devices [135].

4. Conclusions

A bibliometric analysis of the SPDM literature indexed by CNKI and WOS from 1999 to 2022 reveals that Chinese academic interest in this field primarily focuses on TRIZ, green design, redesign, green products, sustainable philosophy, shared products, and service design. Additionally, numerous research issues and extensive themes reflect academic interest in this area. Articles indexed by WOS concentrate on environmental impacts, eco-design tools, sustainable barrier analysis, cost estimates, integration methods, QFD, and LCA. Furthermore, the centralized focused research hotspots and in-depth themes suggest a steady and advanced research stage in this field. However, SPDM research is marked by a constant expansion and enrichment of sustainable design principles, sustainability indices, and sustainable frameworks, with the integration of QFD, AHP, LCA, TRIZ, fuzzy inference, and MCDM. A single quantitative approach is no longer sufficient to adapt to increasingly complex and broadening sustainability indicators. As a result, researchers are using combined qualitative and quantitative analytic approaches and exploring how to incorporate more specific, readily quantifiable, and comprehensive sustainability indicators.

To facilitate more detailed, deep, and long-term research and provide methods and theoretical guidance for practice in the field of SPDMs, collaboration and communication between researchers and institutes in China and worldwide should be strengthened. Furthermore, researchers should widen their research subjects, construct more effective sustainability criteria, and investigate improved design approaches for various products. For example, a product design characteristic-oriented energy-accounting methodology could be developed to achieve more effective sustainability benefits, contributing to design improvement and the sustainable promotion of household and distributed energy products, such as distributed solar equipment, drip irrigation planting equipment, and water filter purifiers [136]. In the sector of health-oriented and medical care products, sustainable design guidelines and approaches have not received timely attention, and sustainable standards, methods, and design tools for diverse groups have not been successfully identified and tailored. Alfarisi et al. [60] offer a new perspective on a product's life cycle impact and highlight the nature of efforts to improve the eco-design of future facemask designs by analyzing the disposable facemask production process. Multidisciplinary and multi-stakeholder design approaches can be created and utilized to produce more systematic sustainability advantages. As the social and cultural sustainability of sustainable design methods is not well-proven, more attention should be paid to suggestions and conceptual frameworks proposed. Additionally, more quantified, practical, and standard design methods that enable sustainability across all dimensions will be needed in the future. For instance, design methods for developing culturally sustainable products based on design computation may provide designers with a new perspective. Finally, a broader and more diverse output of sustainable design solutions can be encouraged by enhancing the systematic, open-source, and sharing of design approaches in the promotion of SPDMs.

Author Contributions: M.G., conceptualization, project administration, methodology, writing—original draft; K.M., supervision, visualization, writing—review and editing; R.H., supervision, funding acquisition, resources; C.V., supervision, validation, writing—review and editing; N.L., validation, resources, visualization. All authors have read and agreed to the published version of the manuscript.

Funding: This work is supported by a grant from the National Philosophy and Social Science Foundation of China (Grant Number: 20AG011), the National Philosophy and Social Science Foundation of China (Grant Number: 13ZD03), the Hunan Provincial Postgraduate Scientific research innovation project (Grant Number: CX20200425), and the International Learning Network of networks on Sustainability project funded by the EU Erasmus + program for capacity building (www.ec.europa.eu), with 36 universities worldwide as partners.

Institutional Review Board Statement: Not applicable.

Informed Consent Statement: Not applicable.

Data availability statement: The data presented in this study are available upon request from the corresponding author. The data are not publicly available due to restrictions, e.g., privacy or ethical.

Conflicts of Interest: The authors declare no conflict of interest.

References

1. Purvis, B.; Mao, Y.; Robinson, D. Three pillars of sustainability: In search of conceptual origins. *Sustain. Sci.* **2019**, *14*, 681–695. [CrossRef]
2. Bofylatos, S. Upcycling Systems Design, Developing a Methodology through Design. *Sustainability* **2022**, *14*, 600. [CrossRef]
3. Kulatunga, A.K.; Karunatilake, N.; Weerasinghe, N.; Ihalawatta, R.K. Sustainable manufacturing based decision support model for product design and development process. *Procedia CIRP* **2015**, *26*, 87–92. [CrossRef]
4. Ameli, M.; Mansour, S.; Ahmadi-Javid, A. A multi-objective model for selecting design alternatives and end-of-life options under uncertainty: A sustainable approach. *Resour. Conserv. Recycl.* **2016**, *109*, 123–136. [CrossRef]
5. Chatty, T.; Harrison, W.; Ba-Sabaa, H.H.; Faludi, J.; Murnane, E.L. Co-Creating a Framework to Integrate Sustainable Design into Product Development Practice: Case Study at an Engineering Consultancy Firm. *Sustainability* **2022**, *14*, 9740. [CrossRef]
6. Gończ, E.; Skirke, U.; Kleizen, H.; Barber, M. Increasing the rate of sustainable change: A call for a redefinition of the concept and the model for its implementation. *J. Clean. Prod.* **2007**, *15*, 525–537. [CrossRef]
7. Huan, Q.; Chen, Y.; Huan, X. A frugal eco-innovation policy? Ecological poverty alleviation in contemporary China from a perspective of eco-civilization progress. *Sustainability* **2022**, *14*, 4570. [CrossRef]
8. Jiang, P.F.; Dieckmann, E.; Han, J.; Childs, R.N.P. A bibliometric review of sustainable product design. *Energies* **2021**, *14*, 6867. [CrossRef]
9. Jasti, N.V.K.; Jha, N.K.; Chaganti, P.K.; Kota, S. Sustainable production system: Literature review and trends. *Manag. Environ. Qual. Int. J.* **2022**, *33*, 692–717. [CrossRef]
10. Ahmad, S.; Wong, K.Y.; Tseng, M.L.; Wong, W.P. Sustainable product design and development: A review of tools, applications, and research prospects. *Resour. Conserv. Recycl.* **2018**, *132*, 49–61. [CrossRef]
11. Chen, C.M. CiteSpace II: Detecting and visualizing emerging trends and transient patterns in scientific literature. *J. Am. Soc. Inf. Sci. Technol.* **2006**, *57*, 359–377. [CrossRef]
12. Pei, X.; Gong, M.S. Status and trends of European social innovation design exploration. *Packag. Eng.* **2017**, *38*, 22–26.
13. General Office of the State Council of the People's Republic of China. *Notification of the General Office of the State Council of the People's Republic of China on the Issuance of Program for the Promotion of Extended Producer Responsibility System;* Gazette of the State Council of the People's Republic of China: Beijing, China, 2017; Volume 3, pp. 84–89.
14. Yu, D.J.; Wang, X. Social value-driven strategies for sustainable innovation and design. *J. Nanjing Arts Inst. (Fine Arts Des.)* **2016**, *2*, 171–176.
15. Yu, D.J.; Wang, X. Sustainable product design for elderly based on humanistic care. *Packag. Eng.* **2015**, *36*, 92–94, 99.
16. Yu, D.J.; Zhang, H. Sustainable design of children furniture based on usability. *Packag. Eng.* **2016**, *37*, 109–112.
17. Yu, S.L.; Yu, J. Innovative methods and case analysis of sustainable product design. *Packag. Eng.* **2018**, *39*, 15–19.
18. Liu, X.; Vrenna, M. Study on systemic design based on sustainability. *Zhuangshi* **2021**, *12*, 25–33.
19. Liu, X.; Zhu, L.; Xia, N. Study on developing a sound public health culture: Innovative design of an ecological public toilet system. *Zhuangshi* **2016**, *3*, 26–29.
20. Lv, M.Y.; Liu, X.; Xia, N. A study on evaluation criteria for sustainable design: Taking the evaluation of the "Lettuce House: Sustainable Living Lab" project as a case. *Ecol. Econ.* **2017**, *33*, 185–189.
21. Haapala, K.R.; Zhao, F.; Camelio, J.; Sutherland, J.W.; Skerlos, S.J.; Dornfeld, D.A.; Jawahir, I.S.; Clarens, A.F.; Rickli, J.L. A review of engineering research in sustainable manufacturing. *J. Manuf. Sci. Eng.* **2013**, *135*, 041013. [CrossRef]
22. Bohm, M.R.; Haapala, K.R.; Poppa, K.; Stone, B.R.; Tumer, Y.I. Integrating life cycle assessment into the conceptual phase of design using a design repository. *J. Mech. Des.* **2010**, *132*, 091005. [CrossRef]

23. Eastlick, D.D.; Haapala, K.R. Increasing the utility of sustainability assessment in product design. *Am. Soc. Mech. Eng. Digit. Collect.* **2013**, *5*, 713–722.

24. Eastwood, M.D.; Haapala, K.R. A unit process model-based methodology to assist product sustainability assessment during design for manufacturing. *J. Clean. Prod.* **2015**, *108*, 54–64. [CrossRef]

25. Pigosso, D.C.A.; McAloone, T.C.; Rozenfeld, H. Characterization of the state-of-the-art and identification of main trends for Ecodesign Tools and Methods: Classifying three decades of research and implementation. *J. Indian Inst. Sci.* **2015**, *95*, 405–428.

26. McAloone, T.C.; Pigosso, D.C.A. From ecodesign to sustainable product/service-systems: A journey through research contributions over recent decades. In *Sustainable Manufacturing: Challenges, Solutions and Implementation Perspectives*; Springer: Cham, Switzerland, 2017; pp. 99–111.

27. Rodrigues, V.P.; Pigosso, D.C.A.; McAloone, T.C. Process-related key performance indicators for measuring sustainability performance of ecodesign implementation into product development. *J. Clean. Prod.* **2016**, *139*, 416–428. [CrossRef]

28. Pigosso, D.C.A.; Rozenfeld, H.; McAloone, T.C. Ecodesign maturity model: A management framework to support ecodesign implementation into manufacturing companies. *J. Clean. Prod.* **2013**, *59*, 160–173. [CrossRef]

29. Pigosso, D.C.A.; McAloone, T.C. Maturity-based approach for the development of environmentally sustainable product/service-systems. *CIRP J. Manuf. Sci. Technol.* **2016**, *15*, 33–41. [CrossRef]

30. Kjaer, L.L.; Pigosso, D.C.A.; McAloone, T.C.; Birkved, M. Guidelines for evaluating the environmental performance of Product/Service-Systems through life cycle assessment. *J. Clean. Prod.* **2018**, *190*, 666–678. [CrossRef]

31. Sarancic, D.; Pigosso, D.C.A.; Pezzotta, G.; Pirola, F.; McAloone, T.C. Designing sustainable product-service systems: A generic process model for the early stages. *Sustain. Prod. Consum.* **2023**, *36*, 397–414. [CrossRef]

32. Zwolinski, P.; Brissaud, D. Remanufacturing strategies to support product design and redesign. *J. Eng. Des.* **2008**, *19*, 321–335. [CrossRef]

33. Alamerew, Y.A.; Brissaud, D. Circular economy assessment tool for end of life product recovery strategies. *J. Remanufacturing* **2019**, *9*, 169–185. [CrossRef]

34. Alamerew, Y.A.; Kambanou, M.L.; Sakao, T.; Brissaud, D. A multi-criteria evaluation method of product-level circularity strategies. *Sustainability* **2020**, *12*, 5129. [CrossRef]

35. Brissaud, D.; Sakao, T.; Riel, A.; Erkoyuncu, J.A. Designing value-driven solutions: The evolution of industrial product-service systems. *CIRP Ann.* **2022**, *71*, 553–575. [CrossRef]

36. Salazar, C.; Lelah, A.; Brissaud, D. Eco-designing product service systems by degrading functions while maintaining user satisfaction. *J. Clean. Prod.* **2015**, *87*, 452–462. [CrossRef]

37. Sun, D.; Li, Y.; Chen, J.J. Hot spots and frontiers of international educational technology research: Bibliometric analysis of five SSCI journals (2000–2019). *Mod. Distance Educ. Res.* **2020**, *32*, 74–85.

38. LeNS China. Available online: http://lenscn.net/ (accessed on 16 August 2022).

39. Song, J.J.; Liu, X. Learning from nature: Source, type, and case. *Ecol. Econ.* **2021**, *37*, 220–227.

40. Qin, J.Y.; Liu, X.; Zhang, Y.Y. A study on the relationship between sustainable design and cultural and creative industry development. *Mod. Commun. (J. Commun. Univ. China)* **2011**, *5*, 165–166.

41. Shi, A.Q.; Guan, H.Y. A study on the coupling design integrating furniture and packaging based on sustainable use. *Ecol. Econ.* **2016**, *32*, 220–224.

42. LeNSlab. Available online: https://www.lenslab.polimi.it/ (accessed on 16 August 2022).

43. Chen, C.M.; Chen, Y.; Horowitz, M.; Hou, H.Y.; Liu, Z.Y.; Pellegrino, D. Towards an explanatory and computational theory of scientific discovery. *J. Informetr.* **2009**, *3*, 191–209. [CrossRef]

44. Jin, H.M. A study on the bionic design of product appearance. *J. Mach. Des.* **2014**, *31*, 123–125.

45. Pigosso, D.C.A.; Rozenfeld, H.; Seliger, G. Ecodesign Maturity Model: Criteria for methods and tools classification. In *Advances in Sustainable Manufacturing: Proceedings of the 8th Global Conference on Sustainable Manufacturing, Abu Dhabi, 22–24 November 2010*; Spinger: Cham, Switzerland, 2011; pp. 241–245.

46. da Costa Fernandes, S.; Pigosso, D.C.A.; McAloone, T.C.; Rozenfeld, H. Towards product-service system oriented to circular economy: A systematic review of value proposition design approaches. *J. Clean. Prod.* **2020**, *257*, 120507. [CrossRef]

47. Gao, D.F.; Lin, L.; Huang, J.; Chen, J.H. A Brief Analysis of Environmental Conscious Design Model Standardization for Energy Using Products. *Stand. Sci.* **2009**, *12*, 32–36, 51.

48. Dai, H.M.; Dai, P.H.; Zhou, J. Design principles and methods of ecological packaging. *Packag. Eng.* **2013**, *34*, 117–120.

49. Luttropp, C.; Lagerstedt, J. Ecodesign and the ten golden rules: Generic advice for merging environmental aspects into product development. *J. Clean. Prod.* **2006**, *14*, 1396–1408. [CrossRef]

50. Telenko, C.; Seepersad, C.C. A methodology for identifying environmentally conscious guidelines for product design. *J. Mech. Des.* **2010**, *132*, 091009. [CrossRef]

51. Guo, Y.C.; Liu, M. Application of sustainable design of consumer digital products. *Packag. Eng.* **2014**, *35*, 42–45.

52. Russo, D.; Spreafico, C. TRIZ-based guidelines for eco-improvement. *Sustainability* **2020**, *12*, 3412. [CrossRef]

53. Wang, J.; Yu, S.; Qi, B.; Rao, J. ECQFD & LCA based methodology for sustainable product design. In Proceedings of the 2010 IEEE 11th International Conference on Computer-Aided Industrial Design & Conceptual Design 1, Yiwu, China, 17–19 November 2010; IEEE: Piscataway, NJ, USA, 2010; Volume 2, pp. 1563–1567.

54. Wang, R.; Gu, X.S. A study on product redesign based on ecoethics. *Packag. Eng.* **2010**, *31*, 22–24.

55. Su, D.; Casamayor, J.L.; Xu, X. An Integrated Approach for Eco-Design and Its Application in LED Lighting Product Development. *Sustainability* **2021**, *13*, 488. [CrossRef]
56. Ardente, F.; Mathieux, F. Identification and assessment of product's measures to improve resource efficiency: The case-study of an Energy using Product. *J. Clean. Prod.* **2014**, *83*, 126–141. [CrossRef]
57. Shang, H.; Chen, R.F. Factors influencing environmental sustainability performance of product-service systems: A dual case study. *J. Manag. Case Stud.* **2019**, *12*, 93–107.
58. Wang, L.M.; Li, L.; Fu, Y.; Li, F.Y.; Peng, X.; Wang, G. Green performance optimization of mechatronic products based on green features and QFD technology. *China Mech. Eng.* **2019**, *30*, 2349–2355.
59. Ayağ, Z. A comparison study of fuzzy-based multiple-criteria decision-making methods to evaluating green concept alternatives in a new product development environment. *Int. J. Intell. Comput. Cybern.* **2021**, *14*, 412–438. [CrossRef]
60. Alfarisi, S.; Sholihah, M.; Mitake, Y.; Tsutsui, Y.; Wang, H.; Shimomura, Y. A Sustainable Approach towards Disposable Face Mask Production Amidst Pandemic Outbreaks. *Sustainability* **2022**, *14*, 3849. [CrossRef]
61. Li, F.Y.; Liu, G.; Wang, J.S.; Duan, G.H.; Zhang, H.C. Application of fuzzy AHP method in the green, modular product design. *China Mech. Eng.* **2000**, *9*, 46–49. [CrossRef]
62. Ardente, F.; Mathieux, F. Environmental assessment of the durability of energy-using products: Method and application. *J. Clean. Prod.* **2014**, *74*, 62–73. [CrossRef]
63. Zhang, L.; Zheng, C.X.; Zhong, Y.J.; Qin, X. Mechanical product green design knowledge update based on rough set. *China Mech. Eng.* **2019**, *30*, 595–602.
64. Ordoñez Duran, J.F.; Chimenos, J.M.; Segarra, M.; de Antonio Boada, P.A.; Ferreira, J.C.E. Analysis of embodied energy and product lifespan: The potential embodied power sustainability indicator. *Clean Technol. Environ. Policy* **2020**, *22*, 1055–1068. [CrossRef]
65. Liu, Z.F.; Gao, Y.; Hu, D.; Zhang, J.D. Green innovation design method based on TRIZ and CBR principles. *China Mech. Eng.* **2012**, *23*, 1105–1111, 1116. [CrossRef]
66. Bovea, M.D.; Pérez-Belis, V. Identifying design guidelines to meet the circular economy principles: A case study on electric and electronic equipment. *J. Environ. Manag.* **2018**, *228*, 483–494. [CrossRef]
67. Yao, H.; Gao, Z.; Ren, Z.M. Sustainable packaging design based on the fractal structure. *Packag. Des.* **2018**, *39*, 59–64.
68. Yuan, Q.H. On the 9R principles of green design. *J. Mach. Des.* **2019**, *36*, 152–154.
69. Elmquist, M.; Segrestin, B. Sustainable development through innovative design: Lessons from the KCP method experimented with an automotive firm. *Int. J. Automot. Technol. Manag.* **2009**, *9*, 229–244. [CrossRef]
70. Bakker, C.; Wang, F.; Huisman, J.; den Hollander, M. Products that go round: Exploring product life extension through design. *J. Clean. Prod.* **2014**, *69*, 10–16. [CrossRef]
71. Kandukuri, S.; Günay, E.E.; Al-Araidah, O.; Kremer Gül, E. Inventive solutions for remanufacturing using additive manufacturing: ETRIZ. *J. Clean. Prod.* **2021**, *305*, 126992. [CrossRef]
72. Shao, X.Y.; Deng, C.; Wu, J.; Wang, L.Q. Integration and optimization of life cycle evaluation and life cycle cost in product design. *J. Mech. Eng.* **2008**, *9*, 13–20. [CrossRef]
73. Bonvoisin, J.; Thiede, S.; Brissaud, D.; Herrmann, C. An implemented framework to estimate manufacturing-related energy consumption in product design. *Int. J. Comput. Integr. Manuf.* **2013**, *26*, 866–880. [CrossRef]
74. Song, W.; Sakao, T. A customization-oriented framework for design of sustainable product/service system. *J. Clean. Prod.* **2017**, *140*, 1672–1685. [CrossRef]
75. Lai, R.S.; Lin, W.G.; Xiao, R.B. Research on the systematic methodology for data-driven product family green redesign. *J. Mach. Des.* **2020**, *37*, 85–90.
76. Jestratijevic, I.; Maystorovich, I.; Vrabič-Brodnjak, U. The 7 Rs sustainable packaging framework: Systematic review of sustainable packaging solutions in the apparel and footwear industry. *Sustain. Prod. Consum.* **2022**, *30*, 331–340. [CrossRef]
77. Rahimifard, S.; Seow, Y.; Childs, T. Minimising Embodied Product Energy to support energy efficient manufacturing. *CIRP Ann.* **2010**, *59*, 25–28. [CrossRef]
78. Zhang, H.C.; Li, H. An energy factor based systematic approach to energy-saving product design. *CIRP Ann.* **2010**, *59*, 183–186. [CrossRef]
79. Geng, K.Y.; Yu, S.R. QFDE-based product platform design methods. *Mach. Des. Res.* **2013**, *29*, 1–5.
80. Vinodh, S.; Kamala, V.; Jayakrishna, K. Integration of ECQFD, TRIZ, and AHP for innovative and sustainable product development. *Appl. Math. Model.* **2014**, *38*, 2758–2770. [CrossRef]
81. Wu, Y.H.; Ho, C.C. Integration of green quality function deployment and fuzzy theory: A case study on green mobile phone design. *J. Clean. Prod.* **2015**, *108 Pt A*, 271–280. [CrossRef]
82. Han, Q.L.; Zhang, Y. Materials Selection Evaluation Based on LCC Theory from Eco-design Perspective. *Sci. Technol. Manag. Res.* **2015**, *35*, 180–184.
83. Tao, F.; Bi, L.N.; Zuo, Y.; Nee, A. A hybrid group leader algorithm for green material selection with energy consideration in product design. *CIRP Ann.* **2016**, *65*, 9–12. [CrossRef]
84. Malmiry, R.B.; Pailhès, J.; Qureshi, A.J.; Antoine, J.-F.; Dantan, J.-Y. Management of product design complexity due to epistemic uncertainty via energy flow modelling based on CPM. *CIRP Ann.* **2016**, *65*, 169–172. [CrossRef]

85. Yang, D.; Chai, H.M. Research on green product design scheme selection based on QFD and CBR. *Sci. Technol. Manag. Res.* **2018**, *38*, 251–259.
86. Tan, Y.S.; Chen, H.; Wu, S. Evaluation and implementation of environmentally conscious product design by using AHP and grey relational analysis approaches. *Ekoloji* **2019**, *28*, 857–864.
87. Xiang, F.; Huang, Y.Y.; Zhang, Z.; Zuo, Y. Digital twin driven energy-aware green design. In *Digital Twin Driven Smart Design*; Academic Press: Cambridge, MA, USA, 2020; pp. 165–184.
88. Ding, Y.K.; Yu, S.L. Sustainable design of stroller based on product life cycle. *Packag. Eng.* **2020**, *41*, 175–179.
89. Wu, X.L.; Hong, Z.; Li, Y.J.; Zhou, F.; Niu, Y.F.; Xue, C.Q. A function combined baby stroller design method developed by fusing Kano, QFD and FAST methodologies. *Int. J. Ind. Ergon.* **2020**, *75*, 102867. [CrossRef]
90. Geng, L.S.; Kong, Z.J.; Geng, L.X. A study on the QFDE-based sustainable product design. *Math. Pract. Theory* **2017**, *47*, 40–49.
91. Kang, H.; Cao, G.Z.; Zhao, C.F. Multifunctional furniture design based on product life cycle. *Packag. Eng.* **2018**, *39*, 39–43.
92. Moon, K.K.L.; Youn, C.; Chang, J.M.T.; Yeung, A.W.-H. Product design scenarios for energy saving: A case study of fashion apparel. *Int. J. Prod. Econ.* **2013**, *146*, 392–401. [CrossRef]
93. Kamp Albæk, J.; Shahbazi, S.; McAloone, T.C.; Pigosso, D.C.A. Circularity evaluation of alternative concepts during early product design and development. *Sustainability* **2020**, *12*, 9353. [CrossRef]
94. Zhu, T.L.; Yue, H.; Xu, J.H. A study on the sustainable design based on the philosophy on sharing economy: An example of bicycle sharing system. *Ecol. Econ.* **2020**, *36*, 224–229.
95. Chen, X.; Yang, S.Y. Sustainable design strategy of cultural and creative products from the perspective of cultural consumption. *Packag. Eng.* **2022**, *43*, 320–325.
96. Ma, L. Construction of basic packaging design dimensions in the comprehensive scenario. *Zhuangshi* **2015**, *9*, 84–85.
97. Emili, S.; Ceschin, F.; Harrison, D. Product–Service System applied to Distributed Renewable Energy: A classification system, 15 archetypal models and a strategic design tool. *Energy Sustain. Dev.* **2016**, *32*, 71–98. [CrossRef]
98. Zhang, B.B. Green furniture product design based on Chinese and Western eco-philosophical ideas. *Hundred Sch. Arts* **2017**, *33*, 229–230.
99. Geelen, D.; Reinders, A.; Keyson, D. Empowering the end-user in smart grids: Recommendations for the design of products and services. *Energy Policy* **2013**, *61*, 151–161. [CrossRef]
100. Xu, X.F. A study on culturally sustainable product design methods from a semiotic perspective. *Zhuangshi* **2013**, *11*, 135–136.
101. Gmelin, H.; Seuring, S. Determinants of a sustainable new product development. *J. Clean. Prod.* **2014**, *69*, 1–9. [CrossRef]
102. Zhou, X. 4D systemic view of product sustainable design. *Packag. Eng.* **2020**, *41*, 10–15.
103. Chen, C.J.; Wang, Y.S.; Zheng, K.J.; Zhong, C. Design of shared service system for community children's learning desk based on the concept of sustainable community. *China For. Prod. Ind.* **2021**, *58*, 32–36+51.
104. Kim, S.; Moon, S.K. Sustainable platform identification for product family design. *J. Clean. Prod.* **2017**, *143*, 567–581. [CrossRef]
105. Shin, H.D.; Al-Habaibeh, A.; Casamayor, J.L. Using human-powered products for sustainability and health: Benefits, challenges, and opportunities. *J. Clean. Prod.* **2017**, *168*, 575–583. [CrossRef]
106. Hemdi, A.R.; Saman, M.Z.M.; Sharif, S. Sustainability evaluation using fuzzy inference methods. *Int. J. Sustain. Energy* **2013**, *32*, 169–185. [CrossRef]
107. Sharma, D.; Kumar, P.; Singh, R.K. Empirical Study of Integrating Social Sustainability Factors: An Organizational Perspective. *Process Integr. Optim. Sustain.* **2023**, *7*, 901–919. [CrossRef]
108. Sherwood, J.; Clark, J.H.; Farmer, T.J.; Herrero-Davila, L.; Moity, L. Recirculation: A new concept to drive innovation in sustainable product design for bio-based products. *Molecules* **2016**, *22*, 48. [CrossRef] [PubMed]
109. de Mattos Nascimento, D.L.; Mury Nepomuceno, R.; Caiado, R.G.G.; Maqueira, J.M.; Moyano-Fuentes, J.; Garza-Reyes, J.A. A sustainable circular 3D printing model for recycling metal scrap in the automotive industry. *J. Manuf. Technol. Manag.* **2022**, *33*, 876–892. [CrossRef]
110. Mou, Y.P.; Guo, M.R.; Si, X.Y.; Zhou, L.; Wang, T. Research on the sustainable growth path of Chinese intangible cultural heritage. *Chin. J. Manag.* **2020**, *17*, 20–32.
111. Yan, J.; Feng, C. Sustainable design-oriented product modularity combined with 6R concept: A case study of rotor laboratory bench. *Clean Technol. Environ. Policy* **2014**, *16*, 95–109. [CrossRef]
112. Jia, H.; Mattson, C.A.; Johnson, G. Consideration of Social Impacts During the Early Stages of Product Development for Sustainable Design. In Proceedings of the International Design Engineering Technical Conferences and Computers and Information in Engineering Conference. American Society of Mechanical Engineers, Virtual, 17–19 August 2020; Volume 83952, p. V006T06A028.
113. Schöggl, J.P.; Baumgartner, R.J.; Hofer, D. Improving sustainability performance in early phases of product design: A checklist for sustainable product development tested in the automotive industry. *J. Clean. Prod.* **2017**, *140*, 1602–1617. [CrossRef]
114. Shuaib, M.; Seevers, D.; Zhang, X.; Badurdeen, F.; Rouch, K.E.; Jawahir, I.S. Product sustainability index (ProdSI) a metrics-based framework to evaluate the total life cycle sustainability of manufactured products. *J. Ind. Ecol.* **2014**, *18*, 491–507. [CrossRef]
115. Nghiem, T.B.H.; Chu, T.C. Evaluating sustainable conceptual designs using an AHP-based ELECTRE I method. *Int. J. Inf. Technol. Decis. Mak.* **2021**, *20*, 1121–1152. [CrossRef]
116. Walters, J.; Mirkouei, A.; Makrakis, G.M. A Quantitative Approach and an Open-Source Tool for Social Impacts Assessment. In Proceedings of the International Design Engineering Technical Conferences and Computers and Information in Engineering Conference. American Society of Mechanical Engineers, St. Louis, MO, USA, 14–17 August 2022; Volume 86250, p. V005T05A016.

117. Omodara, L.; Saavalainen, P.; Pitkäaho, S.; Pongrácz, E.; Keiski, R.L. Sustainability assessment of products-Case study of wind turbine generator types. *Environ. Impact Assess. Rev.* **2023**, *98*, 106943. [CrossRef]
118. Ma, J.F.; Kremer Gül, E. A systematic literature review of modular product design (MPD) from the perspective of sustainability. *Int. J. Adv. Manuf. Technol.* **2016**, *86*, 1509–1539. [CrossRef]
119. Xue, Q.; Chen, Y. System design methods for children's products based on the philosophy of sustainability. *Dev. Innov. Mach. Electr. Prod.* **2014**, *27*, 35–37.
120. Kravchenko, M.; Pigosso, D.C.A.; McAloone, T.C. Towards the ex-ante sustainability screening of circular economy initiatives in manufacturing companies: Consolidation of leading sustainability-related performance indicators. *J. Clean. Prod.* **2019**, *241*, 118318. [CrossRef]
121. Mengistu, A.T.; Panizzolo, R. Analysis of indicators used for measuring industrial sustainability: A systematic review. *Environ. Dev. Sustain.* **2023**, *25*, 1979–2005. [CrossRef]
122. Kravchenko, M.; Pigosso DC, A.; McAloone, T.C. A Trade-Off Navigation Framework as a Decision Support for Conflicting Sustainability Indicators within Circular Economy Implementation in the Manufacturing Industry. *Sustainability* **2021**, *13*, 314. [CrossRef]
123. Tseng, M.-L.; Li, S.-X.; Lin, C.-W.R.; Chiu, A.S. Validating green building social sustainability indicators in China using the fuzzy delphi method. *J. Ind. Prod. Eng.* **2023**, *40*, 35–53. [CrossRef]
124. Desiderio, E.; García-Herrero, L.; Hall, D.; Segrè, A.; Vittuari, M. Social sustainability tools and indicators for the food supply chain: A systematic literature review. *Sustain. Prod. Consum.* **2022**, *30*, 527–540. [CrossRef]
125. Popovic, T.; Barbosa-Póvoa, A.; Kraslawski, A.; Carvalho, A. Quantitative indicators for social sustainability assessment of supply chains. *J. Clean. Prod.* **2018**, *180*, 748–768. [CrossRef]
126. Zhang, J.L. A dialectical examination of the construction of cultural confidence from the perspective of community with a shared future for mankind. *Huxiang Luntan (Huxiang Forum)* **2017**, *30*, 10–16.
127. Ceschin, F.; Gaziulusoy, I. Evolution of design for sustainability: From product design to design for system innovations and transitions. *Des. Stud.* **2016**, *47*, 118–163. [CrossRef]
128. Moalosi, R.; Popovic, V.; Hickling-Hudson, A. Culture-orientated product design. *Int. J. Technol. Des. Educ.* **2010**, *20*, 175–190. [CrossRef]
129. Ji, S.; Lin, P.S. Aesthetics of sustainability: Research on the design strategies for emotionally durable visual communication design. *Sustainability* **2022**, *14*, 4649. [CrossRef]
130. Rocha, C.S.; Antunes, P.; Partidário, P. Design for sustainability models: A multiperspective review. *J. Clean. Prod.* **2019**, *234*, 1428–1445. [CrossRef]
131. Sabatini, F. Culture as fourth pillar of sustainable development: Perspectives for integration, paradigms of action. *Eur. J. Sustain. Dev.* **2019**, *8*, 31. [CrossRef]
132. Tao, F.; Zuo, Y.; Xu, L.D.; Lv, L.; Zhang, L. Internet of things and BOM-based life cycle assessment of energy-saving and emission-reduction of products. *IEEE Trans. Ind. Inform.* **2014**, *10*, 1252–1261.
133. Boeing, G.; Higgs, C.; Liu, S.; Giles-Corti, B.; Sallis, J.F.; Cerin, E.; Lowe, M.; Adlakha, D.; Hinckson, E.; Moudon, A.V.; et al. Using open data and open-source software to develop spatial indicators of urban design and transport features for achieving healthy and sustainable cities. *Lancet Glob. Health* **2022**, *10*, e907–e918. [CrossRef] [PubMed]
134. Hegab, H.; Khanna, N.; Monib, N.; Salem, A. Design for sustainable additive manufacturing: A review. *Sustain. Mater. Technol.* **2023**, *35*, e00576. [CrossRef]
135. Liu, C.; Tian, W.; Kan, C. When AI meets additive manufacturing: Challenges and emerging opportunities for human-centered products development. *J. Manuf. Syst.* **2022**, *64*, 648–656. [CrossRef]
136. Baysan, S.; Kabadurmus, O.; Cevikcan, E.; Satoglu, S.I.; Durmusoglu, M.B. A simulation-based methodology for the analysis of the effect of lean tools on energy efficiency: An application in power distribution industry. *J. Clean. Prod.* **2019**, *211*, 895–908. [CrossRef]

 *sustainability*

Perspective

Leveraging Blockchain and Smart Contract Technologies to Overcome Circular Economy Implementation Challenges

Nallapaneni Manoj Kumar *[ID] and Shauhrat S. Chopra *[ID]

School of Energy and Environment, City University of Hong Kong, Kowloon, Hong Kong
* Correspondence: mnallapan2-c@my.cityu.edu.hk (N.M.K.); sschopra@cityu.edu.hk (S.S.C.)

Abstract: Adopting a circular economy (CE) has rapidly emerged among policymakers and business community stakeholders to promote material circularization and ensure sustainable development. While the inclination for a paradigm shift away from the linear economy is evident, many challenges have been quoted in the literature regarding its implementation. Lately, it has become common to propose Information and Communication Technologies (ICT)-based approaches to address these challenges. However, they do not question the practicality of the solutions in the context of CE. This paper aims to find an appropriate digital solution for CE implementation, which is not possible without a complete understanding of the existing challenges. A thorough literature review broadly classified the challenges under five barrier categories: Technological, Financial, Infrastructural, Institutional, and Societal, which was followed up with an investigation into the failure of ICT solutions to address CE challenges. Among the various technologies, blockchain and smart contract technologies show some promise as data-driven decision-making tools; however, they are not without their limitations when applied in the context of CE. This perspective explores the role of blockchain smart contract technology-scape in overcoming CE challenges and presents a circular economy blockchain (CEB) architecture development. The findings suggest that CEB may enable CE business models that improve trust and transparency in supply-chain networks, shared and performance economy platforms, stakeholder participation, and governance and management of organizations. Ultimately, this study highlights critical areas for research and development for the *blockchainification of CE*.

Keywords: circular economy; circular economy limitations; circular economy barriers; digital circular economy; ICT in circular economy; blockchain architecture for circular economy; circular economy blockchain; digital circular economy architecture

check for updates

Citation: Kumar, N.M.; Chopra, S.S. Leveraging Blockchain and Smart Contract Technologies to Overcome Circular Economy Implementation Challenges. *Sustainability* **2022**, *14*, 9492. https://doi.org/10.3390/su14159492

Academic Editor: Antonella Petrillo

Received: 9 May 2022
Accepted: 26 July 2022
Published: 2 August 2022

Publisher's Note: MDPI stays neutral with regard to jurisdictional claims in published maps and institutional affiliations.

1. Introduction

The circular economy (CE) paradigm is key to the life extension of products, components, and materials (PCMs). It has been viewed and understood differently by various stakeholders across the value chain [1]. For instance, while some understand it as a concept (i.e., theoretical), many others say that it is a framework (i.e., systematic and potentially a quantitative one) aimed at promoting the circularity of the PCM [2,3]. However, Ellen Macarthur defined the CE as "Looking beyond the current take-make-dispose extractive industrial model, the circular economy is restorative and regenerative by design ... " [4] and designed a framework for implementation. Though it has varied interpretations, the potential of this paradigm shift from a linear economy to CE is evident.

In early 2011, the European Commission developed a strategic roadmap for implementing a low-carbon economy in the European Union (EU) [5]. This strategic roadmap aims to minimize waste generation that characterizes the linear economy model (LEM) and maintain the continuous flux of materials [6–8]. Meanwhile, the policy initiatives by countries such as China, Japan, and Germany for CE implementation have increased attention to this concept [9–13]. Moreover, recently, the EU has explored convenient options with varying degrees of success in CE implementation, giving utmost importance to CE in meeting the sustainable development goals (SDGs) [11,12,14]. According to the Ellen Macarthur Foundation and the EU, implementing CE in the EU can create EUR 1 trillion in net economic benefits with 2 million new jobs by the end of 2030 [14–16]. Furthermore, it anticipated a 48% carbon dioxide emissions reduction. Another study from the seven European countries, called the club of Rome, has explored the options of reducing greenhouse gas emissions by up to 70% in each country by implementing CE [17]. It also concluded that the CE implementation measures could increase their workforce by 4% [3,17].

Even though traction towards CE is gaining, and many businesses and policymakers proclaim their support, its implementation seems to be in the early stages, and the reasons are many. The primary challenge that CE faces for implementation is the involvement of various stakeholders across the supply chain with different dynamics in their operations and attitudes [1,18,19]. So far, the CE implementation approach depends on various 'R' frameworks (R0-Refuse, R1-Rethink, and R2-Reduce are the strategies for improving the circularity for "smarter product use and manufacture"; R3-Reuse, R4-Repair, R5-Refurbish, R6-Remanufacture, and R7-Repurpose are the strategies in improving the circularity for "extended lifespan of the PCMs and their parts"; and R8-Recycle and R9-Recover are the strategies in improving the circularity for "useful application of materials"). The practitioners have used these 'R' frameworks in different applications based on a systems perspective [1,20,21]. Many have noted that due to the barriers and challenges associated with the CE concept, the stakeholders in the value chain could not achieve the real benefits [20–23]. This is due to the complexities seen in the technology evolution, organizational capabilities, regulations, working strategies, and market influence [24–26]. Since then, researchers in the CE field have started discussing how to overcome these barriers and challenges. Lately, some have said that digitalizing the CE implementation framework can address these barriers and challenges [27,28]. As a result, CE digitization has emerged as an active area of investigation using information and communication technologies (ICT) [29]. Although this move has facilitated the digital transition of CE stakeholders, it has failed to question the practicality of the ICT-enabled digital solutions across the CE network and for involved stakeholders when maintaining the continuous flux of resources [30]. Therefore, based on the above-highlighted issue, we need a digital solution that is more promising than conventional ICT.

Hence, this perspective aims to find an appropriate digital solution for CE implementation. The integration of ICT is challenging without a complete understanding of the existing challenges. Thus, this perspective first explored the challenges CE is currently facing; second, the role played by ICT and the reasons for its failure to enable CE transition, followed by recent trends in ICT that support CE. Considering this research scope, a critical review was conducted to find an appropriate digital solution for CE implementation.

The key contributions of this perspective include a critical review and a broad classification of CE challenges under five barrier categories: Technological, Financial, Infrastructural, Institutional, and Societal, followed by evidence on how ICT failed to focus on *ensuring traceability, transparency, durability, privacy, security, and process integrity* across the CE network and involved stakeholders. Other contributions include exploring the new ICT trends, i.e., blockchain and smart contracts. Blockchain is the trending disruptive technology based on distributed ledgers; it seems to be a potential solution [31]. With its features, blockchain technology (BCT) may enable data-driven decisions for implementing CE. Furthermore, using smart contract technology (SCT) coupled with BCT may enable the stakeholders to work as per the pre-approved rules of engagement in a CE network in an automated way. Therefore, in the quest to implement CE using these two new digital solutions, BCT and SCT, this perspective came up with a new concept called circular economy blockchain (CEB), where we thoroughly discussed the architecture development followed by CEB's potential benefits and challenges that need to be further researched.

2. Barriers and Challenges While Implementing Circular Economy

The literature on CE suggests that considerable information exists regarding the barriers and challenges to CE implementation. The first class of challenges is related to stakeholder collaboration. *Cooperation of external partners* and *partner restrictions* are critical challenges due to the lack of proper control over the circular business model (CBM). For instance, imagine the shift from a throw-away of electrical and electronic equipment, where a clear understanding of each stakeholder or partner is necessary. Accordingly, incentives can be granted to those who effectively attract electrical waste from consumers. In such cases, cooperation between the partners may not exist [32].

The second class of challenges can be described as being related to market dynamics. *Fashion change* and *fashion vulnerability* are challenges that depend on the aesthetic of the PCM, place of remanufacturing, circular PCM design, and modularity [32]. The challenge of *capital tied up* is raised as it would create risks when the financial transactions happen between the stakeholders. For example, when the infrastructure for implementing CE is considered on a rental basis, the financial inflows and outflows are shared between the investors. Another situation is that the financial transaction between the customer and producer would result in tied-up capital. However, the risk of *capital tied up* can be addressed by focusing on the customer types and their specific interests, as suggested in [32,33].

Another class of challenges can be seen as being related to demand-side issues. It is also suggested that *customer type restriction* is one of the challenges for CE where the customers may not show much interest in the remanufactured PCM [34]. While in another study, the challenge of *product category restrictions* is reported [35]. The remanufacturing *requires technological expertise* in redesigning the PCM to match the current demands as per the customer types and needs, which is another challenge for CE [36]. A challenge of *operational risk* due to the use of second or lower-grade technical solutions is highlighted in [37]. Many other studies in the literature highlighted the *return flow challenges* in CE, describing the problem of return flows to the investment and challenges of *predictability and reliability of return flow* while planning the CE solution [38–41]. The *return flow issue* challenge is assumed to have a solution when the stakeholder or customer relationships become closer [35,39]. The challenge of *"risk of cannibalization"* was reported, where the introduction of the CBM over the traditional LEM could decrease sales, as highlighted in [42,43]. Few other studies stress the challenges such as *"lack of awareness"* and *lack of support from related policy, laws, and regulations* as they could restrict the CE shift [37,44].

Various studies were built upon the above-discussed literature. In one study, six significant challenges that a CE is facing are highlighted; they include *complex international supply chains, high up-front costs, failures in company cooperation, lack of consumer enthusiasm and limited dissemination of innovation,* and *resource-intensive infrastructure lock-in* [45]. Later, a literature survey on CE challenges concerning policy options highlighted a few key factors limiting CE progress. These factors include the *economic signals that do not encourage efficient resource use, lack of awareness and information, limited sustainable public incentives, insufficient investment in technology, pollution mitigation or innovation,* and *minor consumer and business acceptance* [46]. On the other side, the managing director of Acceleratio conducted a literature survey specifically concerning the barriers and drivers towards CE implementation in a multidimensional approach and highlighted the 22 barriers [47]. Focusing on the case studies of small and medium enterprises (SMEs), the CE challenges are explored in [48]. These include *environmental culture, financial barriers, limited government support, lack of adequate legislation, information deficits, administrative burdens,* and *relatively low technical skills* [48]. The CE challenges in the view of Spanish SMEs are more or less similar to the above-discussed ones [26,49].

However, the most recent studies started exploring these challenges to better understand the CE implementation process. In line with this, a study explored the CE challenges in four categories: *cultural, regulatory, market,* and *technological* [16]. After four categories, they further listed them in 15 sub-barriers based on the extensive literature review by referring to the research articles and semi-structured interviews [16]. However, the one new challenge of CE, which is less stressed by many of the researchers, highlighted in [25,50–52], is the *inadequate information management systems (IMS)*.

Despite increasing efforts by various countries, sectors, and organizations (both industry and academic institutions) toward CE implementation, certain limitations must be considered and explored further to find a solution. For that reason, we followed the new trend of categorical classification to classify the explored challenges under five barrier categories. A summary of the various challenges CE is facing is given in Table 1.

Table 1. Challenges in circular economy categorized into five barriers.

Barrier Category	Challenges	Reference
Technological	-Limited consideration in the present product design for the EOL phase	[47]
	-Limited recycled material availability	[47,52,53]
	-No proper evidence on the quality of recycled PCMs	[47,53]
	-Lack of experience, especially the proven and demonstrated CE projects and lack of technical framework on product redesign as the linear technologies are deeply rooted	[47]
	-Existing inefficiencies to develop new business strategies and sustainable footprint (e.g., eco-design, circular design, design for reuse-repair-refurbish remanufacture- recycling, design for services instead of ownership)	[47]
	-Lacking the manufacturing ability to deliver high-quality remanufactured PCMs	[16]
	-Operational risk due to the use of cheap or lower-grade technical solutions	[37]
	-Lack of tools and data models the define the efficiency of CE projects, thermodynamic limitations	[54–58]
Financial	-High or significant upfront investment costs	[16,45,47,54]
	-Low virgin material prices and recycled materials tend to be even more costly than fresh materials	[16,54]
	-Limited funding for CE business models	[16]
	-Challenges in the predictability and reliability of return flow	[40]
	-Environmental costs	[47]
	-A shareholder with short-term agendas dominates the company governance	[47]
	- Increased management and planning costs of CE projects	[47]
	-Capital tied up between the stakeholders	[47]
Infrastructural	-Limited applications of new business models	[47]
	-Lack of secure information exchange system (IES)	[47]
	-The capacity of reverse logistics limits-Exchange of materials	[47]
	-Inadequate information management system (IMS)	[47]
	-Lack of qualified professionals in environment management	[47]
	-Limited dissemination of innovation	[45]
	-Resource-intensive infrastructure lock-in	[45]
	-Very few large-scale demonstration projects	[16,53]
	-Lack of technical resources or facilities	[36,47,48]
Institutional	-Lack of global consensus	[16]
	-Obstructing laws and regulations	[16]
	-Lack of supportive policy frameworks	[16]
	-Lack of smart regulations	[16]
	-Limited circular procurement	[16]
	-Effective integration of circularity principles into policy innovation is still lacking	[47]
	-Financial governance incentive support for the linear business models still exists	[47]
	-Recycling policies to obtain high-quality material flows are inefficient	[47]
	-Governance concerns related to duties, obligations, and ownership	[47]
	-Lack of support from the public institutions	[25,45,50]
Societal	-Hesitant culture of the organization	[16]
	-Lack of awareness of benefits and readiness to work together throughout the value chain	[16]
	-A secure attachment to a linear system	[16]
	-Lack of visionary leadership for CE transition	[7]
	-The difference in attitudes and behavior of employees	[8]
	-Changing consumer preferences	[16]
	-Lack of awareness among consumers	[16]
	-Consumers prefer new products	[10]
	-The dynamic mindset of the consumers	[16]
	-Customer type restriction	[13]
	-Cooperation of external partners	[32]
	-Challenges in collaborating with other companies due to partner restrictions	[32,54]
	-Lack of willingness to contribute to sustainability	[16]
	-Lack of confidence and trust is hampering the exchange of information	[47]

3. Role of Information and Communication Technologies

Researchers have tried to identify possible solutions to many of the listed challenges in Table 1. They mentioned that it is difficult to address each challenge. The critical issues they would encounter are the lack of support infrastructure and the data availability [59,60]. Indeed, digital technologies such as ICT may enable CE progress [27,61]. In the first European Circular Economic Summit, held in Barcelona, Ellen MacArthur introduced a report on "Intelligence Assets", highlighting the Internet of Things (IoT) role in CE. In her talk, Ellen MacArthur said, "IoT will allow the CE to develop at a much faster pace than it ever could without ICT and the IoT in telemetry" [28]. These IoT and ICT systems enable the fusion of digital innovation with CE by allowing features such as sense, store, communication, and decisions. Box 1 is provided with a brief case study showing the ICT and IoT role in electric vehicle (EV) battery charging. As said by Ellen MacArthur, the data-informed decision determines the cyclical process (e.g., repair, refurbish, and so on). By nature, CE has many interconnected cycles, which can be seen only when we visualize CE from a system perspective; these interconnected cycles are responsible for generating massive data [29]. Another study reported that using ICT to digitalize CE would help us collect data from every interconnected cycle [62]. The collected data would help the stakeholder make decisions related to the PCM life cycle. The decisions related to logistics and end-of-life (EoL) management of PCM can also be taken if the data is available [29]. The white paper by IBM highlighted that the growth of the CE is in the hands of disruptive digital innovations. According to IBM, the five key technologies, namely mobile technology (ubiquitous mobile access rules), M2M connectivity (the rise of the machines), cloud technology (dematerializing in the cloud), social media (sharing the social revolution), and big data analytics (by the numbers applying data analytics) would help in transforming the CE [63]. It is also suggested that the digitalization of CE may add intelligence to the PCM, thereby allowing the records of PCM's attributes such as location, conditions, quality, and others over the complete life cycle in line with CE [63].

Box 1. Role of ICT and IoT in EV battery charging.

Consider the concept of battery charging in EVs. Here, the EV's owners are provided with an option of renting or leasing the batteries where the manufacturer or the service provider's responsibility is to provide the fully charged battery. In this example, ICT and IoT systems help us monitor the performance of batteries very clearly; when their performance falls below a threshold, the EV owners are given an option to replace the battery on the specific condition of recovering the old battery. Once the old batteries are received, they can be put under EoL management options to recover the valuable materials. In some cases, the EV charging concept can be implemented in a community where everyone owns the EVs with a pick and drop facility for the battery. In such a business case, it is quite challenging to assess the performance, e.g., who owns the responsibility for battery failures. To address this, ICT, combined with data analytics, can be used to identify how the decrease in battery performance varied between the first and last users. Monitoring and analytics will help the service providers make appropriate decisions regarding the user and service provider compliances. On the other side, combining numerous ICT techniques, such as big data, predictive analytics, data analytics, IoT, cyber-physical systems, data mining, artificial intelligence, and others, would lead to intelligent business models. These ICT techniques bring new ways to monitor and control the CE cycles and allow optimizations, which would help improve the performance of CE practice.

The European Policy Centre (EPC), after 18 months of critical investigation on "how digitalization can boost the circular economy?" stated that, with digitalization, the CE could get benefit in terms of *better use of resources, increase the efficiency of processes, stakeholder partnership facilitation, behavior of stakeholder,* and *knowledge on materials* [64]. Although ICT and IoT facilitate the CE, few concerns regarding digitalization exist, namely *lack of resources, legal certainties, data security,* and *others* [63,64]. On the other side, the scope of ICT for sustainability beyond efficiency is suggested by focusing on the three primary digital optimizations for sustainability in CE [64]. Recently, the information technology giant 'Deloitte Finland' came up with a study connecting the CE and the current digital innovations [65]. Their study shows that the digitalization of CE will help only in certain aspects. For instance, connecting the data centers, stakeholders, or partners with devices and customers in individual/separate parts of the value chain [65]. As stated in [64,65], the CE will boost or create value for companies by selling the services rather than PCM. Even though the hype of ICT techniques adds value to the CE, there exist numerous challenges; for example, Pardo from the EU's EPC raised concerns regarding how to enable the *free flow of data across borders, fostering trust in the data economy,* and *maximizing synergies between the digital and circular economy agendas* [66]. However, in the digitalization process, many still follow the centralized databases where data-related challenges are given more importance. However, with the current digital technologies, *ensuring access to data, data ownership, data sharing, ensuring trust and transparency between the competitors, ensuring privacy,* and *property rights* are more crucial. This would become much more challenging when the marketplace is created where multiple actors will be involved in managing both the flow of information and the physical flows of PCM. However, the challenges can be addressed by shifting to decentralized systems, for example, blockchain technology. Here, blockchain adoption in the CE could improve data-driven decisions by overcoming the challenges related to *trust, traceability, privacy,* and *transparency.*

4. Blockchain in General and Scope for Circular Economy Blockchain Implementation

Blockchain, the underlying technology of bitcoin initially designed for the financial sector, has revolutionized distributed ledger technology (DLT). Regardless of its origin, blockchain has become a buzzword and started gaining attention from other sectors [67]. The popularisation and continuous attempts to explore the BCT implications in various sectors are due to "decentralized architecture, fault tolerance, and cryptographic security benefits such as pseudonymous identities, data integrity, and authentication," which are essential in CE's digital transition [31,68]. As a result, BCT has identified its technical space in various sectors beyond its original application of cryptocurrency transactions. Across the globe, many have declared that they have been engaged in BCT for various applications [69].

4.1. Blockchain Features and Implementation Steps

Blockchain is a new form of data storage formed out of distributed software architectures and computational infrastructures. It enables decentralization and a distributed nature for the data [70]. With its distributed nature, blockchain keeps track of transactions between the parties in the CE network and protects them from tampering. It also allows peer-to-peer (P2P) interactions from any device (for instance, smartphone operation in an energy-efficient manner compared to conventional desktop systems) and helps improve their trust by providing a verification facility [71,72]. In BCT-based applications, transactions signed between two or more peers or stakeholders will typically happen, and these transactions denote the type of agreements defined between the peers. Mostly, the agreements are signed using SCT [73]. Depending on the agreements, the transactions might involve a physical flow of PCMs or digital assets and sometimes the task execution alone.

In any situation, at least one of the participants or the peer in the CE network must sign this transaction to circulate within the defined network of participants in CE. In a blockchain network or functionality structure, we have a variety of terminologies, out of which a node (small or complex) is essential. Any entity or participant connected to the blockchain in a defined CE network is considered a node. Generally, it is referred to as a small node. A few nodes are needed to verify the rules whenever a transaction happens between such entities, as per the SCT-based agreements. Such nodes are referred to as complex nodes. The function of these complex nodes is to group the transactions into blocks [74]. These complex nodes manage the endorsement and decisions regarding the transaction validity and its inclusion in the blockchain. Overall, the blockchain keeps the list of completed transactions in the form of a ledger typically synchronized across the network of nodes while ensuring security. On the other side, blockchain also ensures the ledger update, which is generally based on the participant or peer approval. These approvals by the complex nodes are only implemented through the reliable consensus algorithm or protocols depending on the blockchain platform types (see list mentioned under Step 2 of Figure 1) [75].

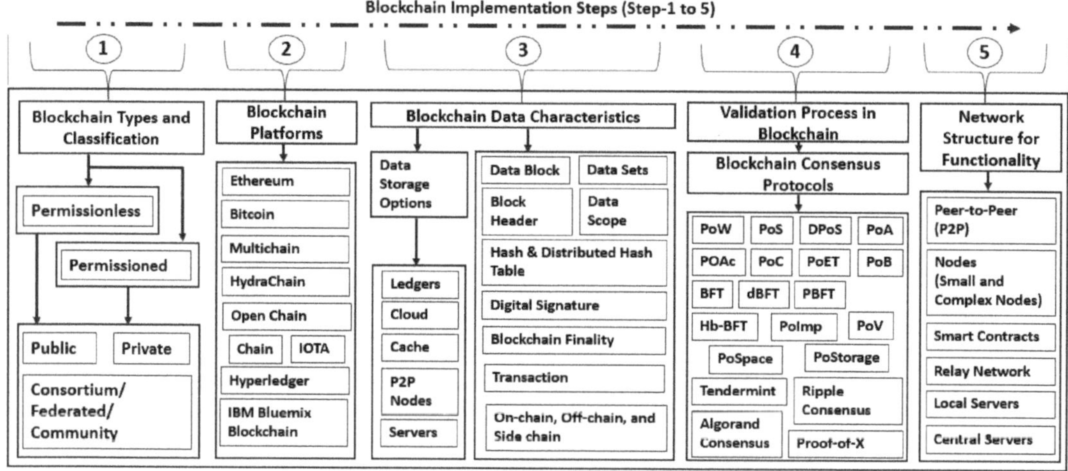

Figure 1. Implementation of blockchain in five steps considering its system based on user interest.

To understand BCT and its implementation, one can consider the blockchain system governance shown in Figure 1. We have shown the implementation process grouped into five steps. It is a set of interconnected mechanisms formed by integrating the software and hardware infrastructures with numerous cryptographic features. Details of each step are given below.

Step 1 is based on the BCT functionality and SCT-based agreements; the user can opt for one of the blockchain categories (as shown in the first step of Figure 1). As per the current literature [72,76,77], blockchain is categorized as public, private, or federated. This classification is based on the network's management and the permissions defined by the users in most cases. In general, the public blockchains are permissionless, where anyone can join as a node and can involve in performing the blockchain operations. Some of the well-known public blockchain implementations are Bitcoin, Ethereum (in Ethereum, both public and private are possible), Litecoin, and other digital currencies [78,79]. In a public blockchain, the overhead management cost and activities will be reduced for the user, mainly due to the self-maintained public node infrastructure. Other sets of blockchains are private and federated, which belong to the permission category. In the permission category of blockchains, only the specific nodes defined, or the nodes part of the consortium, can be involved in blockchain operations. Some well-known permissioned blockchain implementations are Hyperledger, Multichain, Cardano, and others. The applications of private and federated blockchain include "database management, auditing, banking, industrial sectors, and other service-oriented sectors" [70,75].

Step 2 allows the blockchain platform selection depending on the network category type, as mentioned in Step 1. Under those categories, many blockchain platforms exist. Some well know platforms are Bitcoin, Ethereum, Multichain, HydraChain, Open Chain, Chain, IOTA, Hyperledger, IBM Bluemix, and others. However, their selection should be appropriate and based on the functionality and other requirements that are bought by each involving node in the network [77].

Step 3 is the blockchain data characteristics. In blockchains, each transaction between the participants is stored with cryptographic features. For storage, numerous options exist, including the ledgers, cloud, cache, peer-to-peer (P2P) nodes, servers, and others. All such data storage options can broadly be classified under local and network storage. Depending upon the availability, users can opt for one from the wider availability. However, the most suggested one for blockchain applications is network storage. Cloud storage is an ideal network storage option mainly used for commercial applications where flexibility and easy access to data retrieval is needed. When a transaction is initiated between the nodes, as per the agreements, such a process represents the state change in the blockchain. A group of such initiated transactions which are not yet approved represents the block. The data handling capacity of a block entirely depends upon the type of blockchain platform; for example, the Bitcoin blockchain has a block size of 1 MB, which can accommodate a large number of transactions [77]. A typical block has the block header indicating the transaction time, a previous block's hash, blockchain version number, and the random nonce. These blocks are distributed over the network based on the defined agreements, which happens when the minor nodes validate the transaction. The process of transaction/block validation and its addition to the blockchain network is called mining. The mining process varies depending upon the blockchain platforms. For instance, in the case of the Bitcoin blockchain, this process is conducted by solving the complex cryptographic hash puzzle, which is generally referred to as the digital signature for the block. Here, the nodes in the network compete and solve the hash puzzle, and then the block will be added to the blockchain network. When a new block is added to the blockchain, the transaction is considered approved or completed. This approval process again varies depending on the blockchain platforms, as they are governed by different consensus protocols [80].

Step 4 is the block validation using consensus protocols. This step decides whether the blockchain is functioning according to the agreements or not, as it is responsible for maintaining the process of transactions and their approval. As shown in Figure 1, there are different consensus protocols, which mainly depend on the blockchain types and platforms. However, the most famous consensus protocol is Proof-of-Work (PoW), used in the Bitcoin blockchain. In PoW-based consensus protocol, the nodes in the network compete to solve the cryptographic hash puzzle and submit the proof to the rest of the nodes in the network and the block for validation. Attaining a solution to the cryptographic hash is difficult as it is a complex digital signature most sensitive to minor changes. Hence, it requires substantial computational power where the wealthiest person can be dominant in terms of computational power. The PoW is responsible for splitting the blocks about the available computational power needed for solving the hash (power for hash rate) [81]. Another famous consensus protocol is the Proof-of-Stake (PoS); in this, the minor nodes that help approve the transaction are selected based on splitting the available stake of blocks. The PoS consensus ensures fairness in the network of nodes rather than just being dominant based on computational power [82]. Apart from these two, there are other consensus protocols among them; a few are Byzantine Fault Tolerance (BFT)-based. In BFT-based consensus protocols, block validation is completed when an elected validator gets a two-thirds majority of votes. Then, the block is added to the blockchain. Here, the validator is elected using the round-robin approach [77,83]. In other consensuses such as Tendermint and ripple, the electing node-validator is conducted based on the elections and where two-thirds majority votes are needed. In practice, the blockchain stores the state change of something based on the consensus protocol of respective blockchain platforms, representing the final confirmation of transactions. In many blockchains, the time taken for transaction confirmation varies depending upon the type and category. In a Bitcoin-based blockchain and other public types, the transaction approval would take up to 60 min. In contrast, in the case of some private and consortium blockchains, such as Hyperledger, Tendermint, Algorand, and others of similar types, the transaction gets approved instantly [77,80].

Step 5 is the network structure functionality where the blockchain system of interest extends to cover the interactions with the physical world as per the defined agreements. For example, the smart contracts, where the system uses the nodes (either small or complex nodes) to execute a specific task based on their set agreements. Even in the P2P network, a similar strategy of smart contracts is adopted. When the blockchain is extended to human interactions, the state change of transactions would become much more complex, which essentially has to be stored and updated dynamically. These transactions in the network can only be updated using distributed computing.

Therefore, the blockchain system of interest shown in the five steps deals with selecting blockchain types, platforms, consensus, and how these come together to make up a blockchain system for specific applications.

4.2. Circular Economy Blockchain: Scope, Architecture, and Implementation Steps

Based on the brief study in Section 4.1, it can be understood that the scope for blockchain is very high in applications that need data-related security and privacy support. Currently, CE suffers numerous challenges, and many are related to data. *Blockchainification of CE* may ease certain implementation-related challenges as it ensures 'access to data', 'data ownership', 'data sharing', 'traceability, trust, and transparency between the competitors', 'privacy', and 'property rights'. Considering this, we proposed a seven-step system architecture for implementing CE, calling it a circular economy blockchain (CEB), as shown in Figure 2.

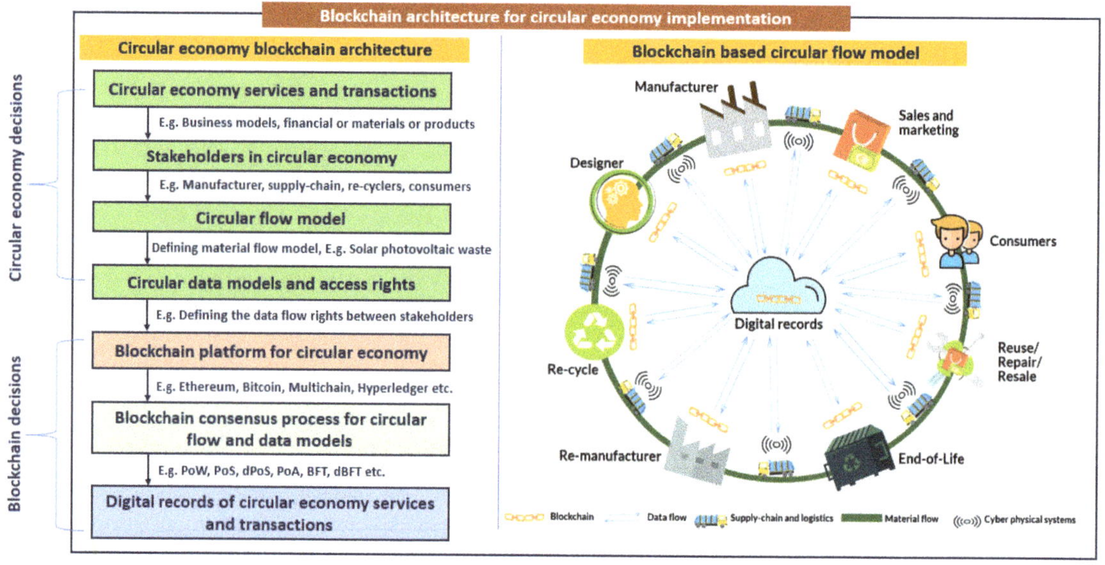

Figure 2. Circular economy blockchain architecture for implementing circular economy.

This architecture identifies the key functionalities involved in practicing CE. It then helps in categorizing the decisions based on the CE stakeholder roles. While implementing CEB, the proposed architecture is expected to be influenced by two types of decisions: CE decisions and blockchain system of interest decisions. Among the seven steps, the first four are framed based on the CE decisions, where the critical functions of CE practice and roles played by the involved stakeholder are given more importance. The next three steps are based on the blockchain system design decision for CE implementation. These three steps are essential in defining software architecture with cryptographic features.

The CEB architecture implementation steps are thoroughly explained below:

Step 1 is identifying the CE services and the type of transactions between stakeholders in the CE network. The transactions include the records of every activity of PCMs respective of the life cycle stage in a circular network. Such transactions are executed as needed by stakeholders' operation, and they form the agreements resulting in decisions for SCT implementation in CE. In most cases, the stakeholders' operations are defined as per the CBMs, for example, product as service, product life extension, and others.

Step 2 is the identification of the stakeholders concerning the CE implementation process. For any CE practice, the general stakeholders include raw material suppliers, manufacturers, sales and marketing, consumers, EoL management team, supply chain, and logistics. However, depending on the CBMs, the stakeholders mentioned above might change. Therefore, this step allows us to identify the stakeholders specific to CE practice.

Step 3 is defining the circular flow model (CFM) depending upon the CE application. For example, the CFM for solar photovoltaics application in the context of CE follows the below-mentioned life cycle stages [84].

Mining → Industrial silicon smelting → Silicon production → Ingot casting and wafer slicing → Photovoltaic cell processing → Module assembling → Marketing and use phase → EoL for material recovery → Market for recovered materials → Reuse in the manufacturing of the same product or other.

Step 4 is defining the circular data models and access rights. This step in the CEB architecture allows the data flow between the stakeholders, and the access rights were defined according to the transactions between them so that it allows privacy.

Step 5 is about the decision related to blockchain platform selection, which usually happens after considering CE decisions (as shown in Steps 1 to 4). Here, we will have various blockchain platforms, for example, public, private, consortium, or federated. Alternatively, options based on the blockchain evolution category are also possible. However, for a CE network with a defined stakeholder, private and consortium or federated would be better.

Step 6 is essential in the blockchain decisions, where it involves the selection of appropriate consensus mechanisms from the available list, PoW, PoS, BFT, and others, to name a few. However, this consensus mechanism depends again on the blockchain platform chosen in Step 5.

Step 7 generally deals with maintaining the digital records of CE transactions and services. By nature, blockchain allows cryptographic features to store the transaction and service data in the cloud. However, users will also have privileges of selecting the storage option at the node itself as per their requirements but only after a thorough discussion with the stakeholders. Sometimes, the option of on-chain and off-chain storage is also possible.

5. Benefits and Challenges of Circular Economy Blockchain

5.1. Benefits and Promises of Circular Economy Blockchain

Blockchains are, theoretically, disruptive and strongly believed to aid the CE implementation. CEB's ability of information sharing facilitates the relationship between the stakeholders involved in CE. Furthermore, the CEB enables information sharing more reliably, allowing *traceability, transparency, authenticity,* and *trusted agreements.* These CEB features help the CE prevent the fraud and falsification of compliance-related information and reduce both direct/indirect risks [85]. Finally, this addresses the CE challenge by improving the *cooperation of external partners* and *reducing the partner restrictions.* Blockchain can even support disruptions in the CE supply chain [86].

CEB records the information of PCM, mainly their origin, supply chain-related, involved stakeholders at every stage of CE and the PCM life cycle. The recorded information is made visible to all the CEB stakeholders with individual access rights. Conversely, the information in CEB will be highly secure as it is time-stamped with cryptographic features. This enables P2P with more *visibility, verifiability,* and *authenticity,* where each stakeholder in CE can track the status of imports and exports of PCM [78,87,88]. Many contracts or agreements exist among the CE stakeholders, which can be compensated with blockchain-based smart contracts. In general, the smart contract refers to the digitally signed contracts related to PCM or any other between the stakeholders (agreements might include financial transactions, monitoring, information exchange, payment transfers, payment tracing, and others) [89].

With its analytics features and predictive capabilities, blockchain can show the PCM circulation related to fashion changes and fashion vulnerability information. It keeps recording the information related to PCMs aesthetic, manufacturing place, design quality, modularity, etc. Furthermore, the challenge of *capital tied up* and other supply chain issues always affect firms and limit market services. In this situation, most firms are interlocked.

This is further extended into the CE network (for example, the suppliers, the government, employees, consumers, EoL management team, and others). Privacy is the concern here; many try to tamper with the data, but with CEB, this privacy issue may be resolved, and the same can be seen from a study focused on privacy-preserving of healthcare predictive modeling as shown in [90].

With its traceability feature, CEB can allow the firms to share and track the PCM flow in the CE network and hence can address the challenge of *capital tied up*. On the other side, the *tied-up capital* would create risks when the financial transactions happen between the stakeholders, for which a smart contract would be the solution. On the other hand, blockchain also enables us to build a strong relationship between CE participants and bring them closure on digital agreements. Mostly in CEB, the closed-looped PCM activities between stakeholders are essential [86].

Blockchain as a technology has the potential to contribute and show few benefits in creating more wealth by allowing the CE stakeholder to be a part shared economy platform. The only possibility of making that is to allow the stakeholders to monetize their transactions on a safe and secure platform [91]. On the other side, the blockchain-based value system that supports social sharing and enables the transition of the industrial economy to the information economy is also discussed in [92].

Blockchain also has the potential to contribute to the digital manufacturing of PCM. As per industry 4.0, many manufacturing industries are currently looking to automate their maintenance and control operations even with the facility of query verifications blockchain [93]. Blockchain technology, along with edge computing features, seems to have great potential. This blockchain-based distributed ledger architecture ensures the product design that is suitable for CE [94].

Blockchain can also contribute to the PCM sustainability assessment in the manufacturing stage. This is possible with the creation of a decentralized manufacturing network. In a study, 'FabRec' is proposed, where the decentralized network of machines and computing nodes are created with the capabilities of allowing transparency, verification facility, and others. These applications in PCM manufacturing will allow us to assess the PCM life cycle assessments to quantify sustainability [95].

In CE practice, due to compliance issues, many firms try to tamper with the data related to PCMs that are not sustainable (for example, the PCM deletion) [96]. Such issues can be easily traced if the manufacturing sectors are adapted to the blockchain service [86].

Even though the blockchain seems to be a very advanced technology, it could offer its promises and solutions to only a few CE challenges. Overall, it is understood that blockchain provides a unique and adequate information management system that can provide many solutions to CE digital transition and ensures *access to data, data ownership,* and *data sharing, ensuring trust* and *transparency between the competitors, ensuring privacy* and *property rights.*

5.2. Challenges with Circular Economy Blockchain

This perspective brings essential insight for CE practitioners in transiting from classic ICT-based to blockchain-based CE. Even though blockchain offers many benefits, some limits and challenges that need serious attention exist. This section explores the possible limitations of the circular economy blockchain, questioning its practicality and scalability. While identifying the limitations, the focus was made mainly on policy aspects, industry, and environmental regulators. On the other side, the technical aspects of blockchain integration with CE are also investigated to explore the challenges. Overall, its implementation will not be easy due to the issues shown in Figure 3.

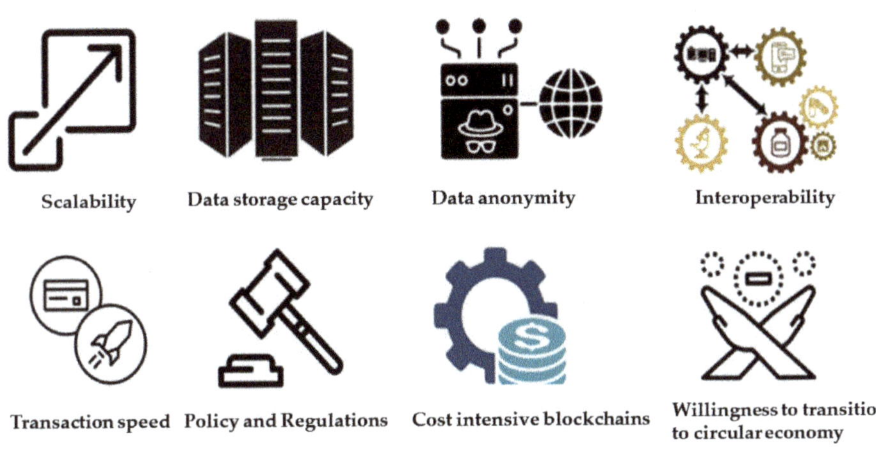

Figure 3. *Blockchainification of* CE: Challenges with circular economy blockchain.

5.2.1. Scalability, Data Storage Capacity, and Management Challenges

In the circular economy blockchain, the involved stakeholders in the network would be large in number. The transaction handling capacity of the blockchain is mostly varied depending on the chosen blockchain platform and the involved number of CE stakeholders. As the number increases, there are chances of slowing down the rate of transactions happening [97]. Considering the two popular public blockchain platforms: Bitcoin and Ethereum, where the transaction time is very slow, and in public blockchains, the transaction approval charges may go high as the involved stakeholders are high. Whereas in the private or federated blockchains, it is not the case; however, the question of scalability is still debatable [97]. Another critical concern is the data storage capacity and management; generally, the blockchain has very minimal capacity for storing transactions [87]. However, most circular economy applications are integrated with sensor systems (for example, RFID and IoT devices in the supply chain and other industrial processes) where the continuous track of information is possible. It is stated that these sensor devices can generate more than gigabytes of data on average for a real-world application [87,98]. Blockchain in current situations may not be able to handle this storage capacity and continuous transaction approval unless a minimum number of stakeholders exist on the CE network with a private blockchain platform. Some have suggested using on-chain and off-chain data storage options, but still, the question of storage capacity is debatable and needs further research [99].

5.2.2. Data Privacy, Anonymity, and Security

Data privacy and protection are concerning for most stakeholders while using blockchain [100]. However, the blockchain's main feature is data privacy and identity management. Many stakeholders in the network worry due to the anonymity that is not guaranteed in the public blockchain platforms [98]. The concerns related to data privacy, anonymity, and security could be addressed, to some extent, in a private blockchain, and it is also suggested to handle data loads [101]. However, when it comes to the CE network, where the involvement of other devices may not ensure privacy and security, we never know whether the data before feeding onto the blockchain ledgers have been tampered with or not [102–104]. Data integrity techniques, public verification, and restricted access control are suggested to ensure privacy and security [105]. It is also quite easy to hack the IoT devices that are connected to the CEB network, especially the supply chain-related ones [102].

5.2.3. Consensus, Smart Contracts, and Platforms

In the context of circular economy applications where it involves numerous sensing devices, it is quite unsuitable for some of the consensus mechanisms and some practical issues with the execution of smart contracts [98]. At present, there are many consensus mechanisms or protocols for blockchain. Among them, most of them seem to be immature, and some have not been tested in all aspects related to performance (say, for example, energy consumption, transaction approval rate, and security level). The majority of the existing consensus is on proof-of-work (PoW), which is the most energy-intensive. However, it is always advised for CE applications to opt for a private blockchain where the computational power would be less [87,98,100]. In the context of CEB, smart contracts have few challenges while modeling the logic of IoT or other sensing devices [98]. For example, the product performance assessment over its lifetime involves real-time data from multiple sources and accessing this data would overload the execution of smart contracts. On the other hand, implementing smart contracts mostly using the Oracle platform and leveraging smart contract capabilities with big data and cloud computing platforms would be difficult [98].

There are currently many blockchain platforms, and it is quite challenging to investigate their performance specific to the CE applications. Only a few platforms have smart contracts and token development features. In the context of CEB, data analytics, payment options, and integration with IoT devices are essential, and these features are not possible in all blockchain platforms. Furthermore, in CEB, stakeholders need additional functionalities related to command line control for interaction, which may not be possible in all the platforms. In the current market, many claim to have blockchain applicability, but there might be a difference in actual execution, which is the major challenge in choosing the right technology [98].

5.2.4. Cultural and Organizational Challenges

In the context of CEB, the cultural and organizational challenges hinder blockchain adoption. For example, suppose an organization wishes to transform into a blockchain-based CE. In that case, they need to embed this practice into its vision and mission [106]. Moreover, they should be ready to accept the alternation possible with blockchain adoption. However, organizations may not be ready to do so. When they are converting to a new system, organizational changes in terms of culture and individual hesitations may be seen as most common. Limited technical expertise in the new system and well-established knowledge and the market for the current system may not allow them to change to a new system [107]. Fear of market down and financial instability due to transition seems to be a challenge in adopting CEB. Regarding the cultural aspect, the CEB represents an entire digital shift and demands a considerable change in the way of handling work. This shift mostly changes the work culture and values and demands behavioral changes from both employees and customers.

In the context of CEB, some business model's execution involves the collaboration of two or more stakeholders, where the issues such as work culture, relationships, and information sharing issues as per the organization's policies and rules might limit CEB progress [108]. Lack of understanding of these cultural and organizational rules is a definite challenge, eventually affecting the entire CEB network. There is a lack of awareness related to this technology and how it works. This mainly hampers the investors and top management who plan to invest in developing CEBs.

5.2.5. Policy and Regulatory Challenges

Both of the discussed concepts, CE and Blockchain, are individually facing challenges related to policy and regulations. In the context of CEB, similar challenges that limit the practical implementation exist. Government laws and regulation are still unclear on the use of blockchain for various applications where concerns related to market and organization arises. Considering Bitcoin blockchain alone, it has faced many transaction-related issues due to adverse policies implemented by several regulatory and government groups [100,109]. Due to the lack of policies, the willingness to direct and support the CEB implementation will be limited [110]. Thus, governments need to ensure that the policies and regulations they frame promote technological growth more sustainably.

5.2.6. Financial Challenges

Blockchain technology combines software and hardware infrastructure where implementation is not relatively straightforward and involves financial decisions [100]. Considering the CEB situation, this would be further difficult due to the network size and information storage capabilities. At present, there is considerable uncertainty in the market about blockchain technology service offerings. The reasons might be the hype in technology, business competition, or the exploitation of the situation. On the other side, there is a fee collected for transaction processing in the blockchain network, and it depends upon the speed and effectiveness of the blockchain platform. Generally, the most effective and high-speed transactions would cost more transaction processing fees.

5.2.7. Lack of Skilled Workforce

As the blockchain is still in its early stages, concerns related to the skilled workforce and the financing of such training facilities may exist. In the context of CEB, both the concepts (CE and Blockchain) are in the initial stages of development. The knowledge of implementing CE networks more efficiently is itself questionable. Conversely, using blockchain technology in CE needs additional knowledge and technical expertise [87]. Although there is a growing interest in CE adoption and blockchain technology, only a limited number of skilled forces exist [100].

6. Conclusions

This perspective aimed to shed light on the current state and the future research needs for the successful adoption of blockchain and smart contract technologies to enable a digital circular economy. The literature initially called for the integration of ICT to solve the challenges of CE implementation. However, the failure of conventional ICT-based solutions has seen an emergence of literature that is now calling for the exploration of blockchain for CE. In this perspective, we thoroughly investigated the blockchain and smart contract technology and how they offer solutions to CE challenges both with and without ICT. Blockchains have credibility, ensuring the information exchange in a transparent, secure manner, and the traceability feature can significantly benefit the CE implementation and its digital transition. Based on this, we developed the CEB architecture that allows for stakeholders' interactions and data-driven decision-making across the CE network. Blockchain-enabled data-driven tools may enable CE business models that improve trust and transparency in supply-chain networks, shared and performance economy platforms, stakeholder participation, and governance and management of organizations. However, considerable challenges exist when it comes to the implementation of blockchain for CBMS, as noted in the complexities of *blockchainification* of non-digital assets. Although researchers and blockchain venture developers will continue to proclaim that digital technologies will address CE-associated challenges, this perspective may serve as a cautionary note. It presents numerous research challenges for those who think the current hype of blockchain implication in CE may automatically translate into CE implementation accomplishments. To set a path for further research and potential exploratory studies on this topic, we provided some challenges and future research propositions. We feel that this review would help build the literature around the digital circular economy and the proposed circular economy blockchain.

Author Contributions: Conceptualization, N.M.K.; methodology, N.M.K.; validation, N.M.K. and S.S.C.; formal analysis, N.M.K.; investigation, N.M.K.; resources, S.S.C.; writing—original draft preparation, N.M.K.; writing—review and editing, N.M.K. and S.S.C.; visualization, N.M.K.; supervision, S.S.C. All authors have read and agreed to the published version of the manuscript.

Funding: This research received no external funding.

Institutional Review Board Statement: Not applicable.

Informed Consent Statement: Not applicable.

Data Availability Statement: Not applicable.

Conflicts of Interest: The authors declare no conflict of interest.

References

1. Kirchherr, J.; Reike, D.; Hekkert, M. Conceptualizing the circular economy: An analysis of 114 definitions. *Resour. Conserv. Recycl.* **2017**, *127*, 221–232. [CrossRef]
2. Bocken, N.M.P.; Olivetti, E.A.; Cullen, J.M.; Potting, J.; Lifset, R. Taking the Circularity to the Next Level: A Special Issue on the Circular Economy. *J. Ind. Ecol.* **2017**, *21*, 476–482. [CrossRef]
3. Jacobi, N.; Haas, W.; Wiedenhofer, D.; Mayer, A. Providing an economy-wide monitoring framework for the circular economy in Austria: Status quo and challenges. *Resour. Conserv. Recycl.* **2018**, *137*, 156–166. [CrossRef]
4. MacArthur, E. Circular Economy Overview. 2013. Available online: https://www.ellenmacarthurfoundation.org/circular-economy/concept (accessed on 24 December 2018).
5. European Commission. (Ed.) *Communication from the Commission to the European Parliament, the Council, the European Economic and Social Committee and the Committee of the Regions A Roadmap for Moving to a Competitive Low Carbon Economy in 2050*; European Commission: Brussels, Belgium, 2011.
6. European Commission. (Ed.) *Communication from the Commission to the European Parliament, the Council, the European Economic and Social Committee and the Committee of the Regions. Accelerating Europe's Transition to a Low-Carbon Economy*; European Commission: Brussels, Belgium, 2016.
7. European Commission. (Ed.) *Communication from the Commission to the European Parliament, the Council, the European Economic and Social Committee and the Committee of the Regions. Closing the Loop—An EU Action Plan for the Circular Economy*; European Commission: Brussels, Belgium, 2015.
8. de Oliveira, F.R.; França, S.L.B.; Rangel, L.A.D. Challenges and opportunities in a circular economy for a local productive arrangement of furniture in Brazil. *Resour. Conserv. Recycl.* **2018**, *135*, 202–209. [CrossRef]
9. Mathews, J.A.; Tan, H. Circular economy: Lessons from China. *Nature* **2016**, *531*, 440–442. [CrossRef] [PubMed]
10. Ohnishi, S.; Fujii, M.; Fujita, T.; Matsumoto, T.; Dong, L.; Akiyama, H.; Dong, H. Comparative analysis of recycling industry development in Japan following the Eco-Town program for eco-industrial development. *J. Clean. Prod.* **2016**, *114*, 95–102. [CrossRef]
11. McDowall, W.; Geng, Y.-J.; Huang, B.; Barteková, E.; Bleischwitz, R.; Türkeli, S.; Kemp, R.; Doménech, T. Circular Economy Policies in China and Europe. *J. Ind. Ecol.* **2017**, *21*, 651–661. [CrossRef]
12. Gregson, N.; Crang, M.; Fuller, S.; Holmes, H. Interrogating the circular economy: The moral economy of resource recovery in the EU. *Econ. Soc.* **2015**, *44*, 218–243. [CrossRef]
13. Mathews, J.A.; Tan, H. Progress toward a circular economy in China: The drivers (and inhibitors) of eco-industrial initiative. *J. Ind. Ecol.* **2011**, *15*, 435–457. [CrossRef]
14. Schulze, G. Growth Within: A Circular Economy Vision for a Competitive Europe. Ellen MacArthur Foundation and the McKinsey Center for Business and Environment. 2016, pp. 1–22. Available online: https://www.mckinsey.com/business-functions/sustainability/our-insights/growth-within-a-circular-economy-vision-for-a-competitive-europe (accessed on 24 December 2018).
15. European Commission. *Study on Modelling of the Economic and Environmental Impacts of Raw Material Consumption*; European Commission: Brussels, Belgium, 2014.
16. Kirchherr, J.; Piscicelli, L.; Bour, R.; Kostense-Smit, E.; Muller, J.; Huibrechtse-Truijens, A.; Hekkert, M. Barriers to the Circular Economy: Evidence From the European Union (EU). *Ecol. Econ.* **2018**, *150*, 264–272. [CrossRef]
17. Wijkman, A.; Skanberg, K. *The Circular Economy and Benefits for Society—Swedish Case Study Shows Jobs and Climate as Clear Winners*; An interim report by the Club of Rome with support from the MAVA Foundation and the Swedish Association of Recycling Industries; Club of Rome: Canton Zurich, Switzerland, 2015.
18. De Vries, B.J.; Petersen, A.C. Conceptualizing sustainable development: An assessment methodology connecting values, knowledge, worldviews and scenarios. *Ecol. Econ.* **2009**, *68*, 1006–1019. [CrossRef]
19. Gladek, E. The Seven Pillars of the Circular Economy. 2017. Available online: http://www.metabolic.nl/the-seven-pillars-of-the-circular-economy/ (accessed on 24 December 2018).
20. Potting, J.; Hekkert, M.P.; Worrell, E.; Hanemaaijer, A. *Circular Economy: Measuring Innovation in the Product Chain*; PBL: The Hague, The Netherlands, 2017.
21. Blomsma, F.; Brennan, G. The Emergence of Circular Economy: A New Framing Around Prolonging Resource Productivity. *J. Ind. Ecol.* **2017**, *21*, 603–614. [CrossRef]
22. Lacy, P.; Rutqvistand, J. *Waste to Wealth: The Circular Economy Advantage*; Springer: Berlin/Heidelberg, Germany, 2016.
23. de Jesus, A.; Mendonça, S. Lost in Transition? Drivers and Barriers in the Eco-innovation Road to the Circular Economy. *Ecol. Econ.* **2018**, *145*, 75–89. [CrossRef]
24. Geng, Y.; Haight, M.; Zhu, Q. Empirical analysis of eco-industrial development in China. *Sustain. Dev.* **2007**, *15*, 121–133. [CrossRef]
25. Geng, Y.; Doberstein, B. Developing the circular economy in China: Challenges and opportunities for achieving 'leapfrog development'. *Int. J. Sustain. Dev. World Ecol.* **2008**, *15*, 231–239. [CrossRef]
26. Ormazabal, M.; Prieto-Sandoval, V.; Puga-Leal, R.; Jaca, C. Circular Economy in Spanish SMEs: Challenges and opportunities. *J. Clean. Prod.* **2018**, *185*, 157–167. [CrossRef]

27. Lacy, P. Using Digital Tech to Spin the Circular Economy. 2015, p. 105. Available online: https://www.smart-circle.org/circulareconomy/circulaire-economie/using-digital-tech-spin-circular-economy/ (accessed on 24 December 2018).

28. Ellen MacArthur Foundation. *Intelligent Assets: Unlocking the Circular Economy Potential*; Ellen MacArthur Foundation: Cowes, UK, 2016; pp. 1–39.

29. Maria Antikainen, T.U. Digitalization Accelerates the Circular Economy. 2018. Available online: https://vttblog.com/2018/06/27/digitalisation-accelerates-the-circular-economy/ (accessed on 3 January 2019).

30. Nallapaneni, M.K.; Chopra, S.S. Sustainability and Resilience of Circular Economy Business Models Based on Digital Ledger Technologies. In Proceedings of the Nature Conference on Waste Management and Valorisation for a Sustainable Future, Seoul, Korea, 26–28 October 2021.

31. Ruan, P.; Dinh, T.T.A.; Loghin, D.; Zhang, M.; Chen, G.; Lin, Q.; Ooi, B.C. Blockchains vs. Distributed Databases: Dichotomy and Fusion. In Proceedings of the 2021 International Conference on Management of Data, Virtual, China, 20–25 June 2021; pp. 1504–1517.

32. Mont, O.; Dalhammar, C.; Jacobsson, N. A new business model for baby prams based on leasing and product remanufacturing. *J. Clean. Prod.* **2006**, *14*, 1509–1518. [CrossRef]

33. Besch, K. Product-service systems for office furniture: Barriers and opportunities on the European market. *J. Clean. Prod.* **2005**, *13*, 1083–1094. [CrossRef]

34. Pearce, J.A. The profit-making allure of product reconstruction. *MIT Sloan Manag. Rev.* **2009**, *50*, 59.

35. Sundin, E.; Lindahl, M.; Ijomah, W. Product design for product/service systems: Design experiences from Swedish industry. *J. Manuf. Technol. Manag.* **2009**, *20*, 723–753. [CrossRef]

36. Berchicci, L.; Bodewes, W. Bridging environmental issues with new product development. *Bus. Strat. Environ.* **2005**, *14*, 272–285. [CrossRef]

37. Kuo, T.C.; Ma, H.-Y.; Huang, S.H.; Hu, A.H.; Huang, C.S. Barrier analysis for product service system using interpretive structural model. *Int. J. Adv. Manuf. Technol.* **2009**, *49*, 407–417. [CrossRef]

38. Seitz, M.A. A critical assessment of motives for product recovery: The case of engine remanufacturing. *J. Clean. Prod.* **2007**, *15*, 1147–1157. [CrossRef]

39. Östlin, J.; Sundin, E.; Björkman, M. Importance of closed-loop supply chain relationships for product remanufacturing. *Int. J. Prod. Econ.* **2008**, *115*, 336–348. [CrossRef]

40. Östlin, J.; Sundin, E.; Björkman, M. Product life-cycle implications for remanufacturing strategies. *J. Clean. Prod.* **2009**, *17*, 999–1009. [CrossRef]

41. King, A.M.; Burgess, S.C.; Ijomah, W.L.; McMahon, C.A. Reducing waste: Repair, recondition, remanufacture or recycle? *Sustain. Dev.* **2006**, *14*, 257–267. [CrossRef]

42. Guiltinan, J. Creative Destruction and Destructive Creations: Environmental Ethics and Planned Obsolescence. *J. Bus. Ethic.* **2009**, *89*, 19–28. [CrossRef]

43. Michaud, C.; Llerena, D. Green consumer behaviour: An experimental analysis of willingness to pay for remanufactured products. *Bus. Strat. Environ.* **2010**, *20*, 408–420. [CrossRef]

44. Linder, M.; Williander, M. Circular Business Model Innovation: Inherent Uncertainties. *Bus. Strat. Environ.* **2017**, *26*, 182–196. [CrossRef]

45. Preston, F. *A Global Redesign? Shaping the Circular Economy*; Chatham House London: London, UK, 2012.

46. Bicket, M.; Guilcher, S.; Hestin, M.; Hudson, C.; Razzini, P.; Tan, A.; ten Brink, P.; van Dijl, E.; Vanner, R.; Watkins, E. *Scoping Study to Identify Potential Circular Economy Actions, Priority Sectors, Material Flows and Value Chains*; Luxembourg Publications Office of the European Union: Luxembourg, 2014.

47. Van Eijk, F. Barriers & Drivers Towards a Circular Economy. Acceleratio BV. 2015. Available online: www.circulairondernemen.nl/uploads/e00e8643951aef8adde612123e824493.pdf (accessed on 24 December 2018).

48. Rizos, V.; Behrens, A.; Kafyeke, T.; Hirschnitz-Garbers, M.; Ioannou, A. The Circular Economy: Barriers and Opportunities for SMEs. CEPS Working Documents. 2015. Available online: https://ssrn.com/abstract=2664489 (accessed on 24 December 2018).

49. Ormazabal, M.; Prieto-Sandoval, V.; Jaca, C.; Santos, J. An overview of the circular economy among SMEs in the Basque country: A multiple case study. *J. Ind. Eng. Manag.* **2016**, *9*, 1047–1058. [CrossRef]

50. Ritzén, S.; Sandström, G. Barriers to the Circular Economy—Integration of Perspectives and Domains. *Procedia CIRP* **2017**, *64*, 7–12. [CrossRef]

51. Rizos, V.; Behrens, A.; Van Der Gaast, W.; Hofman, E.; Ioannou, A.; Kafyeke, T.; Flamos, A.; Rinaldi, R.; Papadelis, S.; Hirschnitz-Garbers, M.; et al. Implementation of Circular Economy Business Models by Small and Medium-Sized Enterprises (SMEs): Barriers and Enablers. *Sustainability* **2016**, *8*, 1212. [CrossRef]

52. Van Renswoude, K.; ten Wolde, A.; Joustra, D.J. *Circular Business Models—Part 1: An introduction to IMSA's Circular Business Model Scan*; IMSA: Amsterdam, The Netherlands, 2015.

53. Amsterdam, I. *Unleashing the Power of the Circular Economy*; IMSA: Amsterdam, The Netherlands, 2013.

54. Mont, O.; Plepys, A.; Whalen, K.; Nußholz, J.L.K. *Business Model Innovation for a Circular Economy: Drivers and Barriers for the Swedish Industry—The Voice of REES Companies*; Mistra REES: Lund, Sweden, 2017. Available online: https://lucris.lub.lu.se/ws/portalfiles/portal/33914256/MISTRA_REES_Drivers_and_Barriers_Lund.pdf (accessed on 24 December 2018).

55. Korhonen, J.; Honkasalo, A.; Seppälä, J. Circular Economy: The Concept and its Limitations. *Ecol. Econ.* **2018**, *143*, 37–46. [CrossRef]
56. Reuter, M.A.; Matusewicz, R.; van Schaik, A. Lead, Zinc and their minor elements: Enablers of a circular economy. *World Met.-Erzmetall* **2015**, *68*, 132–146.
57. Shuva, M.A.H.; Rhamdhani, M.A.; Brooks, G.A.; Masood, S.H.; Reuter, M.A. Thermodynamics data of valuable elements relevant to e-waste processing through primary and secondary copper production: A review. *J. Clean. Prod.* **2016**, *131*, 795–809. [CrossRef]
58. van Schalkwyk, R.; Reuter, M.; Gutzmer, J.; Stelter, M. Challenges of digitalizing the circular economy: Assessment of the state-of-the-art of metallurgical carrier metal platform for lead and its associated technology elements. *J. Clean. Prod.* **2018**, *186*, 585–601. [CrossRef]
59. Shi, J.; Zhou, J.; Zhu, Q. Barriers of a closed-loop cartridge remanufacturing supply chain for urban waste recovery governance in China. *J. Clean. Prod.* **2018**, *212*, 1544–1553. [CrossRef]
60. Pheifer, A.G. Barriers & Enablers to Circular Business Models. White Paper. *ValueC, Brielle.* 2017. Available online: https://www.scribd.com/document/502885354/Barriers-and-Enablers-to-Circular-Business-Models# (accessed on 9 January 2019).
61. Pagoropoulos, A.; Pigosso, D.C.; McAloone, T.C. The Emergent Role of Digital Technologies in the Circular Economy: A Review. *Procedia CIRP* **2017**, *64*, 19–24. [CrossRef]
62. Antikainen, M.; Uusitalo, T.; Kivikytö-Reponen, P. Digitalisation as an Enabler of Circular Economy. *Procedia CIRP* **2018**, *73*, 45–49. [CrossRef]
63. Townsend, J. ICT for Sustainability Beyond Efficiency: Pushing Cleantech and the Circular Economy. In ICT4S. 2018. Available online: https://ww.easychair.org/publications/download/L74j (accessed on 9 January 2019).
64. European Policy Centre (EPC). Digital Roadmap for a Circular Economy. Sustainable Prosperity for Europe. Available online: https://www.epc.eu/prog_forum.php?forum_id=77&prog_id=2 (accessed on 9 January 2019).
65. Deloitte. Circular Goes Digital. 2018, pp. 1–2. Available online: https://www.google.com/url?sa=t&rct=j&q=&esrc=s&source=web&cd=&ved=2ahUKEwjWh9695pr5AhWGEKYKHZmDCm4QFnoECBgQAQ&url=https%3A%2F%2Fwww2.deloitte.com%2Fcontent%2Fdam%2FDeloitte%2Ffi%2FDocuments%2Frisk%2FCircular%2520goes%2520digital.pdf&usg=AOvVaw3mLQI0u27FTtcP5HMkhEwF (accessed on 9 January 2019).
66. Pardo, R. How the Circular Economy can benefit from the Digital Revolution. *EPC Commentary*, 11 April 2018.
67. Jaoude, J.A.; Saade, R.G. Blockchain Applications–Usage in Different Domains. *IEEE Access* **2019**, *7*, 45360–45381. [CrossRef]
68. Makhdoom, I.; Abolhasan, M.; Abbas, H.; Ni, W. Blockchain's adoption in IoT: The challenges, and a way forward. *J. Netw. Comput. Appl.* **2018**, *125*, 251–279. [CrossRef]
69. Christidis, K.; Devetsikiotis, M. Blockchains and Smart Contracts for the Internet of Things. *IEEE Access* **2016**, *4*, 2292–2303. [CrossRef]
70. Ruan, P.; Chen, G.; Dinh, T.T.A.; Lin, Q.; Ooi, B.C.; Zhang, M. Fine-grained, secure and efficient data provenance on blockchain systems. *Proc. VLDB Endow.* **2019**, *12*, 975–988. [CrossRef]
71. Li, J.; Peng, Z.; Gao, S.; Xiao, B.; Chan, H. Smartphone-assisted energy efficient data communication for wearable devices. *Comput. Commun.* **2017**, *105*, 33–43. [CrossRef]
72. Gao, S.; Peng, Z.; Xiao, B.; Xiao, Q.; Song, Y. SCoP: Smartphone energy saving by merging push services in Fog computing. In Proceedings of the 2017 IEEE/ACM 25th International Symposium on Quality of Service (IWQoS), Vilanova i la Geltrú, Spain, 14–16 June 2017; pp. 1–10.
73. Luu, L.; Chu, D.-H.; Olickel, H.; Saxena, P.; Hobor, A. Making smart contracts smarter. In Proceedings of the 2016 ACM SIGSAC Conference on Computer and Communications Security, Vienna, Austria, 24–28 October 2016.
74. Narayanan, A.; Bonneau, J. *Bitcoin and Cryptocurrency Technologies: A Comprehensive Introduction*; Princeton University Press: Princeton, NJ, USA, 2016.
75. Ølnes, S.; Ubacht, J.; Janssen, M. Blockchain in government: Benefits and implications of distributed ledger technology for information sharing. *Gov. Inf. Q.* **2017**, *34*, 355–364. [CrossRef]
76. Buterin, V. On Public and Private Blockchains. Ethereum Blog. Posted on 7 August 2015. Available online: https://blog.ethereum.org/2015/08/07/on-public-and-private-blockchains/ (accessed on 10 January 2019).
77. Zheng, Z.; Xie, S.; Dai, S.-N.; Chen, X. Blockchain challenges and opportunities: A survey. *Int. J. Web Grid Serv.* **2018**, *14*, 352–375. [CrossRef]
78. Nakamoto, S. Bitcoin: A Peer-to-Peer Electronic Cash System. 2008. Available online: https://papers.ssrn.com/sol3/papers.cfm?abstract_id=3440802 (accessed on 10 January 2019).
79. Haferkorn, M.; Diaz, J.M.Q. Seasonality and interconnectivity within cryptocurrencies-an analysis on the basis of Bitcoin, Litecoin and Namecoin. In *International Workshop on Enterprise Applications and Services in the Finance Industry*; Springer: Berlin/Heidelberg, Germany, 2014.
80. Casino, F.; Dasaklis, T.K.; Patsakis, C. A systematic literature review of blockchain-based applications: Current status, classification and open issues. *Telematics Informatics* **2018**, *36*, 55–81. [CrossRef]
81. Vukolić, M. The quest for scalable blockchain fabric: Proof-of-work vs. BFT replication. In *International Workshop on Open Problems in Network Security*; Springer: Berlin/Heidelberg, Germany, 2015.
82. Kiayias, A.; Russell, A.; David, B.; Oliynykov, R. Ouroboros: A Provably Secure Proof-of-Stake Blockchain Protocol. IACR Cryptology ePrint Archive. 2016, pp. 1–67. Available online: https://eprint.iacr.org/2016/889.pdf (accessed on 28 September 2019).

83. Castro, M.; Liskov, B. Practical byzantine fault tolerance and proactive recovery. *ACM Trans. Comput. Syst.* **2002**, *20*, 398–461. [CrossRef]
84. Huang, B.; Zhao, J.; Chai, J.; Xue, B.; Zhao, F.; Wang, X. Environmental influence assessment of China's multi-crystalline silicon (multi-Si) photovoltaic modules considering recycling process. *Sol. Energy* **2017**, *143*, 132–141. [CrossRef]
85. Kouhizadeh, M.; Sarkis, J.; Zhu, Q. At the Nexus of Blockchain Technology, the Circular Economy, and Product Deletion. *Appl. Sci.* **2019**, *9*, 1712. [CrossRef]
86. Manupati, V.; Schoenherr, T.; Ramkumar, M.; Panigrahi, S.; Sharma, Y.; Mishra, P. Recovery strategies for a disrupted supply chain network: Leveraging blockchain technology in pre- and post-disruption scenarios. *Int. J. Prod. Econ.* **2021**, *245*, 108389. [CrossRef]
87. Swan, M. *Blockchain: Blueprint for a New Economy*; O'Reilly Media, Inc.: Sevastopol, CA, USA, 2015.
88. Cartier, L.E.; Ali, S.H.; Krzemnicki, M.S. Blockchain, Chain of Custody and Trace Elements: An Overview of Tracking and Traceability Opportunities in the Gem Industry. *J. Gemmol.* **2018**, *36*, 212–227. [CrossRef]
89. Raj, P.V.R.P.; Jauhar, S.K.; Ramkumar, M.; Pratap, S. Procurement, traceability and advance cash credit payment transactions in supply chain using blockchain smart contracts. *Comput. Ind. Eng.* **2022**, *167*, 108038. [CrossRef]
90. Kuo, T.-T.; Ohno-Machado, L. Modelchain: Decentralized privacy-preserving healthcare predictive modeling framework on private blockchain networks. *arXiv* **2018**, arXiv:1802.01746.
91. Huckle, S.; Bhattacharya, R.; White, M.; Beloff, N. Internet of Things, Blockchain and Shared Economy Applications. *Procedia Comput. Sci.* **2016**, *98*, 461–466. [CrossRef]
92. Pazaitis, A.; De Filippi, P.; Kostakis, V. Blockchain and value systems in the sharing economy: The illustrative case of Backfeed. *Technol. Forecast. Soc. Chang.* **2017**, *125*, 105–115. [CrossRef]
93. Wu, H.; Peng, Z.; Guo, S.; Yang, Y.; Xiao, B. VQL: Efficient and Verifiable Cloud Query Services for Blockchain Systems. *IEEE Trans. Parallel Distrib. Syst.* **2021**, *33*, 1393–1406. [CrossRef]
94. Isaja, M.; Soldatos, J.K. Distributed Ledger Architecture for Automation, Analytics and Simulation in Industrial Environments. *IFAC-PapersOnLine* **2018**, *51*, 370–375. [CrossRef]
95. Angrish, A.; Craver, B.; Hasan, M.; Starly, B. A Case Study for Blockchain in Manufacturing: "FabRec": A Prototype for Peer-to-Peer Network of Manufacturing Nodes. *Procedia Manuf.* **2018**, *26*, 1180–1192. [CrossRef]
96. Zhu, Q.; Kouhizadeh, M. Blockchain Technology, Supply Chain Information, and Strategic Product Deletion Management. *IEEE Eng. Manag. Rev.* **2019**, *47*, 36–44. [CrossRef]
97. Yli-Huumo, J.; Ko, D.; Choi, S.; Park, S.; Smolander, K. Where Is Current Research on Blockchain Technology?—A Systematic Review. *PLoS ONE* **2016**, *11*, e0163477. [CrossRef] [PubMed]
98. Reyna, A.; Martín, C.; Chen, J.; Soler, E.; Díaz, M. On blockchain and its integration with IoT. Challenges and opportunities. *Futur. Gener. Comput. Syst.* **2018**, *88*, 173–190. [CrossRef]
99. Eberhardt, J.; Tai, S. On or off the blockchain? Insights on off-chaining computation and data. In Proceedings of the European Conference on Service-Oriented and Cloud Computing, Oslo, Norway, 27–29 September 2017; Springer: Berlin/Heidelberg, Germany, 2017.
100. Mougayar, W. *The Business Blockchain: Promise, Practice, and Application of the Next Internet Technology*; John Wiley & Sons: Hoboken, NJ, USA, 2016.
101. Pongnumkul, S.; Siripanpornchana, C.; Thajchayapong, S. Performance analysis of private blockchain platforms in varying workloads. In Proceedings of the 2017 26th International Conference on Computer Communication and Networks (ICCCN), Vancouver, BC, Canada, 31 July–3 August 2017.
102. Lopez, J.; Rios, R.; Bao, F.; Wang, G. Evolving privacy: From sensors to the Internet of Things. *Futur. Gener. Comput. Syst.* **2017**, *75*, 46–57. [CrossRef]
103. Roman, R.; Zhou, J.; Lopez, J. On the features and challenges of security and privacy in distributed internet of things. *Comput. Networks* **2013**, *57*, 2266–2279. [CrossRef]
104. Li, Z.; Gao, S.; Peng, Z.; Guo, S.; Yang, Y.; Xiao, B. B-DNS: A Secure and Efficient DNS Based on the Blockchain Technology. *IEEE Trans. Netw. Sci. Eng.* **2021**, *8*, 1674–1686. [CrossRef]
105. Liu, C.; Yang, C.; Zhang, X.; Chen, J. External integrity verification for outsourced big data in cloud and IoT: A big picture. *Futur. Gener. Comput. Syst.* **2015**, *49*, 58–67. [CrossRef]
106. Mathiyazhagan, K.; Govindan, K.; NoorulHaq, A.; Geng, Y. An ISM approach for the barrier analysis in implementing green supply chain management. *J. Clean. Prod.* **2013**, *47*, 283–297. [CrossRef]
107. Jharkharia, S.; Shankar, R. IT-enablement of supply chains: Understanding the barriers. *J. Enterp. Inf. Manag.* **2005**, *18*, 11–27. [CrossRef]
108. Gorane, S.; Kant, R. Modelling the SCM implementation barriers: An integrated ISM-fuzzy MICMAC approach. *J. Model. Manag.* **2015**, *10*, 158–178. [CrossRef]
109. Kiviat, T.I. Beyond bitcoin: Issues in regulating blockchain transactions. *Duke Law J.* **2015**, *65*, 569–608.
110. Mangla, S.K.; Luthra, S.; Mishra, N.; Singh, A.; Rana, N.P.; Dora, M.; Dwivedi, Y. Barriers to effective circular supply chain management in a developing country context. *Prod. Plan. Control* **2018**, *29*, 551–569. [CrossRef]

MDPI
St. Alban-Anlage 66
4052 Basel
Switzerland
www.mdpi.com

Sustainability Editorial Office
E-mail: sustainability@mdpi.com
www.mdpi.com/journal/sustainability